Portland Community College Libraries
WITHDRAWN
AF868665

VARIOUS AND INGENIOUS MACHINES
VOLUME 2
MANUFACTURING AND WEAPONS TECHNOLOGY

TECHNOLOGY AND CHANGE IN HISTORY

VOLUME 6.2

THE EARLY HISTORY OF MECHANICAL ENGINEERING

Volume Two: Manufacturing and Weapons Technology

BY

BRYAN LAWTON

BRILL
LEIDEN · BOSTON
2004

This book is printed on acid-free paper.

Cover Illustration: Manufacture of weapons showing forging of a cannon and a vertical boring mill driven by a horse, Jan Breughel the Younger, *Fire*, (details), before 1620. Kingston Lacy, The Bankes Collection (The National Trust). Photograph: Photographic Survey, Courtauld Institute of Art.

Brill Academic Publishers has done its best to establish rights for the use of the illustrations printed on and in this volume. Should any other party feel that its rights have been infringed, we would be glad to hear from them.

Library of Congress Cataloging-in-Publication Data

Lawton, B. (Bryan)
The early history of mechanical engineering / by Bryan Lawton.
p. cm. — (Technology and change in history ; v. 6)
Includes bibliographical references and indexes.
Contents: v. 1. Power generation and transport.
ISBN 90-04-13609-6 (set: alk. paper) — (ISBN 90-04-13610-X (v.1) — ISBN 90-04-13611-8 (v.2)
1. Mechanical engineering—History. I. Title. II. Series.

TJ15.L39 2004
621'.09—dc22

2003065333

ISSN 1385-920X
ISBN 90 04 13609 6 (set)
ISBN 90 04 13611 8 (vol. 2)

© Copyright 2004 by Koninklijke Brill NV, Leiden, The Netherlands

All rights reserved. No part of this publication may be reproduced, translated, stored in a retrieval system, or transmitted in any form or by any means, electronic, mechanical, photocopying, recording or otherwise, without prior written permission from the publisher.

Authorization to photocopy items for internal or personal use is granted by Brill provided that the appropriate fees are paid directly to The Copyright Clearance Center, 222 Rosewood Drive, Suite 910 Danvers MA 01923, USA. Fees are subject to change.

PRINTED IN THE NETHERLANDS

CONTENTS

VOLUME ONE

List of Figures xv
List of Tables lvii
Acknowledgements lxi
Preface lxvii

Mechanics

Chapter One Machine Elements
Introduction 3
The Lever 4
The Inclined Plane 10
The Wedge 14
The Screw Thread 15
The Pulley 23
The Winch, Windlass and Capstan 29
Analysis 33
The Simple Balance 33
The Roman Balance 35
The Inclined Plane 37
A Sumerian Mathematical Exercise 38
Friction and Lubrication 39
Force Analysis on an Inclined Plane 43
Forces Acting on a Wedge 46
Efficiency of Screw Threads 48
The Simple and Compound Pulley 49
Equilibrium of a Windlass 52

Chapter Two Power Transmission
Introduction 55
Belt and Rope Drives 55
Brakes 56
Gears 59

Cams 66
Bearings 70
Analysis 74
Belt and Rope Drives 74
The Shoe or Block Brake 81
The Band Brake 83
Theory of Gearing 86
Kinematics of Cams 91
Cam Dynamics 95
Bearing Theory 98
Wear 106

POWER GENERATION

CHAPTER THREE Power, Food, and Slavery
Introduction 113
Power and Wealth 114
Agricultural Demand for Energy 119
Labour and Slavery 122
Ancient Slavery 125
Slavery in Greece 129
Slavery in Rome 140

CHAPTER FOUR Muscular Work and Power
Man Power 151
Animal Power 167

CHAPTER FIVE Muscle Technology
Introduction 181
Analysis 182
Muscle Physiology 182
Muscle Chemistry 183
Mechanics of a Single Muscle 186
Muscular Activity in Man 200
Treadmills 212
Muscular Activity in Animals 216

CHAPTER SIX Waterpower
Introduction 223
Vertical-Axis (Norse or Greek) Mills 223

Horizontal-Axis (Roman) Mills 235
Speed Governing 265
Analysis 267
Power Output of Water Wheels 267
Horizontal-Axis Undershot Wheels (Roman Mills) 269
Horizontal-Axis Overshot Wheels 271
Vertical-Axis Wheels (Greek or Norse Mills) and Pelton's Wheel 276

Chapter Seven Wind and Other Sources of Power
Introduction 283
Wind Power 283
Gravity 300
Hot Air, Gunpowder and Steam 302
Analysis 315
Wind Distribution and Location of Windmills 315
Theoretical Power of Windmills 325
Horizontal-Axis Windmills 327
Stability 334
Huygen's Gunpowder Engine 336

Transport

Chapter Eight Characteristics of Transport
Introducton 343
Invariants of Travel and Transport 345
Transport Efficiency and Energy Requirements 352
Evolution of Transport Systems 354
Development of Road Infrastructure 358
Development of Navigable Infrastructure 372
Analysis 390
Market Size, Area and Transport Cost 390
Elasticity of Demand 392
Fall-Off Analysis 393
The Logistic Growth Curve 395

Chapter Nine Land Transport
Prehistory and the Ancient Civilisations 399
Persia, Greece and Rome 417
Medieval, Renaissance and Early Modern 426

Chapter Ten Land Transport Technology
Introduction 455
Analysis 457
Energy Required in Walking and Running 457
Walking in Crowds 461
Two-Wheeled Vehicles 463
Four-Wheeled Vehicles 468
Forces on Draught Beasts: Limits of Traction 470
Soil Mechanics: The Pressure-Sinkage Relation 474
Wheels on Soft Surfaces 476
Wheels on Hard Surfaces 480
Power and Maximum Velocity 482
Lateral Stability of Vehicles 485
Steering 491
Suspension and Ride 494
Springs 502

Chapter Eleven Water Transport
Prehistory 507
Ancient Civilisations 510
Greece and Rome 526
Medieval and Renaissance 538
Early Modern 553

Chapter Twelve Ship Technology
Introduction 579
Analysis 580
Force and Power Required by Ships 580
Strength of Hulls 589
Rowing 599
Sailing 605
Stability 614
Fouling and Corrosion 622

Chapter Thirteen Undersea and Aerial Transport
Undersea Craft 625
Air Craft 631

VOLUME TWO

Manufacture

CHAPTER FOURTEEN Attitudes to Manufacture
Introduction 647
The Earliest Engineers 648
The Greek Myths 650
Greece and Rome 658
Christianity and Medieval Scholasticism 670
The Renaissance and Early Modern Periods 685

CHAPTER FIFTEEN Agriculture
Introduction 699
Ancient Agriculture 699
Greece and Rome 714
Medieval 729
Early Modern 737
Analysis 758
Ploughing 758

CHAPTER SIXTEEN Ancient and Medieval Mining
Introduction 763
Prehistoric Mines and Metallurgy 763
The Ancient Civilisations 775
Greek and Roman Mines 784
Analysis 804
Efficiency and Power of the Archimedean Screw 804
Fire Setting 806
Lift Pump 809

CHAPTER SEVENTEEN Medieval, Renaissance, and Early Modern Mining
Introduction 815
Drainage 822
Ventilation 838
Driving Tunnels and Shafts 846
Transport and Hoisting 849
Ore Dressing 853
Analysis 858

Performance of a Winch Driven Dipper Wheel 858
Analysis of a Cowl Ventilator 859
Mine Ventilation 861
Natural Convection 863
Forced Convection 868
Air Temperature Distribution in a Mine Shaft 870
Geometry of Rolls 872
Separation on an Inclined Water Plane 874

Chapter Eighteen Ancient and Medieval Metalworking
Introduction 877
Prehistory 878
Ancient Empires 882
Greece and Rome 888
Medieval 900
Analysis 918
Smelting 918
Copper Smelting 919
Tin Smelting 920
Iron Smelting by the Bloomery Process 921
Lead Smelting 924
The Blast Furnace 925
The Forge 929

Chapter Nineteen Renaissance and Early Modern Metalworking
Introduction 933
Wire Drawing, Battery, Slitting and Rolling 933
Casting 941
Wrought Iron and Steel 954
Hardening Steel by Quenching 963
Machine Tools 965
Analysis 985
Hot and Cold Working of Metals 985
Forging 991
Rolling 995
Wire Drawing 998
Carburising 1003

Chapter Twenty Textile Manufacture
Introduction 1005
Prehistory and Ancient Civilisations 1006
Greece and Rome 1011
Medieval Period 1016
The Renaissance and Early Modern Period 1027
Cotton 1030

Weapons Technology

Chapter Twenty-One Warfare
Introduction 1053
Animal and Human Aggression 1054
Warfare in Prehistory 1065
Warfare in Primitive Societies 1074
Ritual Warfare in History 1079
Major Wars 1086

Chapter Twenty-Two Ancient and Classical Warfare and Weapons
Introduction 1097
Prehistory 1100
Ancient Civilisations 1106
Greek and Roman Warfare 1125

Chapter Twenty-Three Medieval and Early Modern Weapons
Medieval Weapons and Warfare 1145
Renaissance and Early Modern Warfare 1169

Chapter Twenty-Four Weapons Technology
Introduction 1207
Analysis 1208
Analysis of a Simple Battle 1208
Flanking Geometry 1212
Wound Ballistics 1214
Penetration Mechanics 1228
Stability 1237
Temperature of Armour Plate in Sunlight 1240

External Ballistics 1244
Bow and Arrow 1249
The Constant Torque Ballista 1256
Performance of Guns 1259
Fixed Energy Propulsion and Optimum Performance 1266
Rockets 1270

Bibliography 1275

Subject Index 1289
Name Index 1307

MANUFACTURE

CHAPTER FOURTEEN

ATTITUDES TO MANUFACTURE

Introduction

"Machines are worshipped because they are beautiful, and valued because they confer power; they are hated because they are hideous, and loathed because they impose slavery." So wrote Bertrand Russell in his Sceptical Essays (1928, 83). Because of this ambivalent attitude society's disposition towards technology, and towards manufacture in particular, has changed profoundly, and more than once, since the dawn of recorded history. The word *manufacture* originally meant *made by hand*, although we now use it to mean *made by machine* and items made by hand are now usually described as *crafted*. Manufacture is closely linked to technology. The products of technology are usually admired and are in great demand but the methods used to produce them are often considered deplorable. The earliest myths record that the gods themselves might be technologists (smiths). Conversely, the first human technologists were admired hugely and came to be venerated as gods. However, during the first millennium B.C., the disadvantages of technology emerged more clearly and society, at least Greek and Roman society, turned against it. Technology was necessary, and therefore tolerated, but was given a lowly status. It became the work of slaves or the lowest class of society. Slavery stigmatised technology, which in turn stigmatised everything it touched. For this reason, technology was starved of intellectual content and suffered accordingly. It was not one of the liberal arts that were approved for study by the intellectual elite of the free and unfettered.

This state of affairs changed during the Dark Ages largely because Christianity, the new Roman religion, did not approve of slavery, with the result that industrial slavery declined and agricultural slavery was replaced by serfdom. Fewer artefacts, particularly metal artefacts, were made and they became more expensive, and greatly valued. By the Late Middle Ages and the Renaissance the association of slaves with manufacturing was forgotten. Attitudes to technology changed and work, particularly manual work, was seen as a noble undertaking that could improve the lot of humanity and glorify God. The monastic orders, particularly the Cistercians, demonstrated that mechanisation, rather than slavery, was the way to relieve the monks from excessive toil and to produce beautiful and useful products for their worship. These ideas were taken up and improved during the Renaissance. It became respectable for intelligent men to study and provide scientific support to manufacturers. To the Enlightenment, technology was one of the greatest goods and seemed to promise a golden age here on Earth. However, by the beginning of the nineteenth century the disadvantages of unrestrained technological progress were again apparent and were opposed, to some extent, by the Romantic Movement. However, industrialisation, and the Romantic reaction to it, is beyond the scope of this book. It is sufficient for us to know that industrialisation realised the worst fears of the Romantics and although these particular problems were solved, others have taken their place. We are still trying to live with technology although we surely cannot live without it.

The Earliest Engineers

In the beginning, it appears that technology was exalted and technologists, usually builders or smiths, were revered and were, as mentioned above, sometimes given the status of gods. One of the first engineers known to us by name was Imhotep who lived about 2650 B.C., see Clayton (1994, 36) for example. Imhotep was the architect and engineer to King Djoser of Egypt and was responsible for building the Step Pyramid, the immediate predecessor of the Great Pyramid. The Step Pyramid is famed as the first building to be constructed entirely of stone. It began as a normal mastaba (table) tomb but was enlarged successively by adding one mastaba onto another until it consisted of six unequal steps forming a stable, pyramidal shape, Fig. 14.1. It is 62 m high and has a base of 109 m × 125 m. There are numerous shafts and tunnels; some are original but many were made by tomb robbers. The cutting, transport, and assembly of so many heavy stones were a great feat of civil and mechanical engineering and exactly how it was done is still a matter for conjecture. Imhotep, Fig. 14.2, was a man of many parts and clearly had a high position at court as indicated by an inscription on the base of a broken statue of Djoser. Imhotep's titles are: "The treasurer

Fig. 14.1 The step pyramid, built by Imhotep about 2650 B.C., is the first building constructed entirely of stone.

Fig. 14.2 The earliest engineers. Left: a Late Period seated bronze statuette of Imhotep. Right: Amenhotep son of Hapu.

of the King of Lower Egypt, the First after the King of Upper Egypt, Administrator to the Great Palace, Hereditary Lord, the High Priest of Heliopolis, Imhotep the builder, the sculptor, the maker of stone vases—". He was a man of great wisdom and scholarship, admired as a scribe, councillor, doctor of medicine, priest, and astronomer. In the classical period, under the Ptolemys, he was worshiped as a god of architecture and medicine and was identified with Asclepius the Greek god of medicine. Bronze statuettes were cast, and his devotees made pilgrimages to Saqqara where thousands of votive offerings have been found. Presumably he was buried at Saqqara but his tomb has not been discovered. Imhotep was not the only architect-engineer to be deified in antiquity. Amenhotep son of Hapu was the master architect of Amenhotep III (1384–1349 B.C.) and presided over much of the building work of one of Egypt's finest periods, Fig. 14.2. As a great privilege, he was allowed to place statues of himself outside Pylon X in the temple at Karnak. He is shown in the traditional cross-legged position of a scribe. In the Ptolemaic period, Amenhotep son of Hapu was also worshipped as a god.

The Greek Myths

The ancient myths that survive from pre-history are another source of information on ancient attitudes to technology. Ancient myths may be thought of as the educational system of an illiterate society. They are oral stories and traditions recounted to a family or village group by an orator or bard. Eventually written down, they are generally accepted, by anthropologists and others, as an indication of the social conditions in the society generating and using the myth. Although mostly fanciful, they nevertheless contain an element of the society's history, its moral, ethical, and scientific beliefs, and much of their technology. Graves (1992) gives a good summary of the Greek myths, and Harvey (1962) is even more concise.

In the Bronze Age, the art and craft of the smith and the artisan were of supreme importance. Such craftsmen were the cornerstones on which a community's relatively more advanced civilisation depended, and which enabled them to maintain military supremacy, to fabricate the tools that made life easier, and to produce decorative wares and jewellery that made life more pleasant. The smiths were generally seen as essential to improving the quality of life for the whole community. Their knowledge was carefully guarded. In a society not accustomed to writing, this meant that the possessors of the knowledge, the smiths, were jealously retained, perhaps by giving them a high status, lands, and the finest wives. Alternatively, Graves believes that they may have been crippled to prevent their escape and he suggests that their myths provide evidence of both practices.

Prometheus, Fig. 14.3, was a smith, the son of a titan, who later acquired the status of an Olympian god. He made a model of the first man into which Mother Earth breathed life. For the Bronze Age Greeks, listening to their creation myths, it would be perfectly natural for the smith-god Prometheus to model the first man. They

Fig. 14.3 In the bottom register Prometheus carries fire, smouldering in a fennel stalk, and gives fire to mankind. Attic red-figure calyx-krater by the Dinos Painter *c.* 420 B.C.

Fig. 14.4 Atlas holds the heavens on his shoulders and a large bird attacks his brother Prometheus who is tied to a column. Laconian cup from Cervetri.

would be familiar with their own smith modelling all sorts of things in wax or clay in order to make moulds for bronze castings using the lost-wax process. Not only was the smith-god, Prometheus, considered to be man's creator, but he was also seen as his chief source of knowledge and wisdom. He stole fire from the other Gods and gave it to humankind, carrying it in a smouldering fennel stalk, Fig. 14.3. He is their protector against the Spites and the wrath of other, less kindly, gods. More importantly, he is seen as one willing to suffer great personal pain on behalf of humanity when, because of his gift of fire, he was chained to a pillar and daily had his liver pecked out by birds, Fig. 14.4. He is a precursor of the Christian idea that a god must suffer for humanity. During the Middle Ages, the anguish of Prometheus, chained to his pillar, was considered a prefiguration of Christ's suffering on the cross and helped to justify the study of classical texts. In the nineteenth century, the Romantic Movement made him a symbol of man's freedom against tyranny. All ages have

Fig. 14.5 From the Françoise Vase, *c.* 570 B.C. Hephaestus returns to Olympus on a mule (right). Aphrodite, his bride to be, stands in front of Zeus and Hera who are seated on a throne (left).

interpreted his gift of flame as the gift of divine knowledge and inspiration that sets humanity apart from the beasts. The smith was indeed an honoured member of society, but there was another aspect.

Hephaestus, also a smith-god, was a son of Zeus and Hera. Again, this is a reflection of the importance attached to the art of the smith. However, Hephaestus was crippled which is surely an indication that his work was also viewed with some revulsion, probably because continuous work at the forge disfigured and deformed. Hephaestus had his injury explained because his mother, Hera, thought him ugly and threw him from Olympus immediately after his birth. But he survived and because of his great skill and craftsmanship, was welcomed back to Olympus, married to Aphrodite, the goddess of desire (much against her will), and lived in a splendid bronze palace. Hephaestus' return to Olympus, riding on a mule, was commonly depicted on Greek pottery, Fig. 14.5. It appears to prefigure Christ's entry into Jerusalem except that Hephaestus' return to Olympus seems to have been a drunken and bawdy event. Zeus, who evidently could not stand the incessant noise from the forge, threw Hephaestus to earth a second time. He was a whole day falling to earth and his injuries were more serious. Ever afterwards, he wore golden callipers to support his crippled legs. But despite the noise and pollution from his forge, Hephaestus was again welcomed back to Olympus. It was a love-hate relationship for although the gods did not enjoy his presence amongst them, neither could they do without him. The gods required his arms and armour and the goddesses demanded his finely worked jewels. His wife, Aphrodite, was unfaithful from the beginning and preferred to sleep with more dashing, less deformed gods. Probably this is how Bronze Age peoples regarded and treated their own smiths, and perhaps they too were crippled by their arduous work or wore callipers to prevent their escape.

Both Homer and Virgil tell the story of how Hephaestus provided arms for the great hero Achilles during the siege of Troy. But it is clear that a fundamental change of attitude towards the smith gods, and towards metal work and manufacture, had occurred in the six or seven hundred years separating Homer and Virgil. In Homer, Hephaestus is a kindly, and polite god with a beautiful home and forge on Olympus. He is not ashamed to work at his forge in person or with the assistance of automatons, golden handmaidens that he made himself and respond to his requests. He produces beautiful and useful items; he is careful of his tools; he washes and changes his clothes on leaving his forge to greet visitors; he is inventive, helpful and gracious. Despite his strength, his crippled form, and his frightening occupation, he remains approachable and friendly. All this is rather surprising in a Greek god, for their behaviour was usually more disreputable. The armour he makes for Achilles is particularly fine and the famous shield depicts numerous scenes that are educationally useful to the orator's uneducated audience.

Virgil adapted this myth to glorify Rome. In Virgil, the Roman equivalent of Hephaestus is Vulcan and his wife is Venus, the Roman equivalent of Aphrodite; they are a less agreeable pair. Vulcan lives in a beautiful home but does not personally work at the anvil in a forge. Moreover, his wife must seduce him first before she can persuade him to arm Achilles. Significantly, his forge is no longer on high amongst the Gods but in the depths of the earth. When Vulcan does agree to make the arms, he must descend, by night, to his forge in an underground cavern with galleries leading from the volcano of Etna and filled with the noise of re-sounding blows on anvils. It is intended by Virgil to be an awe inspiring and fearful place. Cyclopes, one-eyed monsters who, according to Hesiod, made thunderbolts for Zeus, operate the forge. Achilles' famous shield is now made to depict numerous scenes glorifying the history of Rome.

Hephaestus's forge, by comparison with Vulcan's, is a highly civilised place where things of beauty and utility are made, and he is assisted not by one-eyed monsters but by golden handmaidens (automatons). By the first century B.C., manufacture was the province of the slave or the lowest levels of society and was not to be undertaken by the educated, leisured, upper class. This view of manufacture undoubtedly caused Virgil to adapt the story of Hephaestus to one more in keeping with his times. Where all is light and civilised in Homer, and the God may undertake the work personally, it is dark and gruesome in Virgil and Vulcan merely gives orders that the work be done; he is the manager not the smith. Hephaestus has again been ejected from Olympus and thrown back to earth, indeed beneath the earth, this time by Virgil.

In some myths, Hephaestus was also married to Aglaia, one of the three Graces, the goddesses of joy, charm, and beauty. In art, they are always represented together as a trio of nymphs dancing in a circle, as in Canova's famous statue or the wall painting found at Pompeii. This marriage is supposed to explain the long established association of joy, charm and beauty with the arts (in the broadest sense). Like the Muses, they were believed to inspire poets, musicians, and other artists and artisans. This aspect of the smith-god is strong in Homer's story of Hephaestus, but is completely lost in Virgil's Vulcan.

The principal goddess of Athens was Athena, Fig. 14.6. We do not usually think of her as an engineer or technologist but she is said to have invented the flute, the trumpet, the earthenware pot, the plough, the rake, the ox-yoke, the horse bridle, the chariot and the ship. This covers a fair proportions of mechanical engineering equipment and is a list of inventions any engineer would be proud to claim. She was supposed to be the first to teach arithmetic and women's arts such as, cooking, weaving, and spinning. Essentially, she was a goddess of war but took no pleasure from fighting and was more interested in settling disputes by peaceful means. She does not often carry arms but, once she goes to war, she usually wins, even against the god Ares, because she employed better tactics and strategy. Athena's attributes, no doubt, illustrate what Athenians of old held dear. There is a myth that Hephaestus was tricked into an attempt to make love to Athena but his advance was firmly rejected. This suggests, perhaps, that Athens may once have considered embracing technology more closely but ultimately rejected it, or kept it at arm's length.

Another myth throwing light on attitudes to manufacture is that of Daedalus and Icarus, and their escape from King Minos of Crete. Daedalus was a famous smith; a mortal who was credited with many inventions, and who fled first from Athens to Crete after he jealously murdered his more talented nephew, Talos. He found employment with King Minos. However, the scandalous work he did for

Fig. 14.6 Birth of Athena. Zeus sits on his throne holding a thunderbolt as Athena, fully armed, emerges from his head. Hephaestus moves away having split open Zeus's skull with an axe, *c.* 550 B.C.

Parsiphaë, Minos's wife, offended the King, and Daedalus, with his son Icarus, Fig. 14.7, was forced to flee again. As Minos had all the ships watched, escape from the island seemed impossible, but as everyone knows Daedalus and his son escaped by making two pairs of wings and attaching them with wax. During the escape Icarus flew too near the Sun, the wax melted, and he fell to earth. After burying his son, Daedalus made good his escape. Minos raised a fleet to search the Mediterranean and eventually found Daedalus working in Sicily for King Cocalus. Minos was murdered and Daedalus escaped again, this time to Sardinia.

This myth is best interpreted not as the flight of a single, talented, smith and his son, but perhaps as the escape of bronze workers from Crete and the spread of metal working knowledge from Crete to Italy, Sicily and Sardinia. Both Crete and Sardinia are centres of Bronze Age mining and metal work. It is conceivable that a powerful king, represented by the legendary Minos, maintained a large

Fig. 14.7 Statue of Daedalus and Icarus, Canova (Correr Museum, Venice).

body of smiths and metal workers on whom his military power and splendid lifestyle depended. Their escape from Crete to work for other masters, taking with them their knowledge and expertise, may have been seen as a real threat that justified raising a large fleet and searching the Mediterranean. Whatever the basis of the myth, it reflects the wandering lifestyle of the early metalworkers, their disputes with local kings, and the attempts of local kings to retain them, sometimes against their will. It is, perhaps, typical that we should remember Icarus, who died whilst attempting flight, rather than his father, who invented and made the wings and who flew successfully, at least in the myth. Heroic failure is much admired.

Graves (1992, 87–8) has noted that there is, in these myths, a suggestion that metal workers were routinely crippled to prevent such escapes. Hephaestus was thrown from Olympus, twice, and was seriously crippled, hobbling as he walked. Daedelus' nephew Talos was murdered by being thrown from the Acropolis and Talos' other name, Tantalus, means hobbling or lurching. There are also close connections with the partridge, a bird that hobbles in its courtship display, holding one heel ready to strike at rivals. For example, Hephaestus married Aphrodite, to whom the partridge was sacred; Daedelus' sister was Perdix, which means 'partridge'; the soul of Talos, when he was thrown from the Acropolis, flew off in the form of a partridge; and a partridge appeared at the burial of Daedelus' son Icarus. A hobbling gait was, it seems, commonly associated with Bronze Age metal workers. This persisted into historical times because Plutarch (*Pericles, 27*) tells us of Artemon, an engineer famous for his construction of siege engines, who was nicknamed Periphoretus because he was lame and had to be carried in a litter to any works that needed his attention. We are not told how he became lame.

Greece and Rome

In the time of Homer Greece did not depend very much on slavery and it was no disgrace to undertake manual work, indeed, the products of technology and the technologists themselves, if that is not too grand a title, were admired. With the growth of cities and the industry they attracted, there was a corresponding increase in the number of slaves. By the time of Solon, it was recognised that a large proportion of the citizens must be relieved of domestic occupations and manual labour, that they might devote their talents to the benefit of the state. Later, manual labour, and even those who directed manual labour, came to be despised, at least by a certain class in Greek society. This was unfortunate because it deprived technology of intellectual support and it confined intellectuals to theoretical matters that could not be tested in the real world. Neither science nor technology could progress independently.

The lowly status to which engineers had sunk is made clear by a famous passage of his Dialogues in which Plato discussed the nature of a virtuous man and criticised Kallikles for his suggestion that a virtuous man was one who, like the physician or the engine-maker, saved life. (The latter saved life because his engines defended a city when it was attacked.) Plato says, "And if he [the engineer] was to talk, Kallikles, in your grandiose style, he would bury you under a mountain of words, declaring and insisting that we ought all of us to be engine-makers, and that no other profession is worth thinking about; he would have plenty to say. Nevertheless you despise him and his art, and sneeringly call him an engine-maker, and you will not allow your daughter to marry his son, or marry your son to his daughter." (Plato, Dialogues (Georgias), trans Jowett, OUP, 1892.)

This is damning enough but Plato's abhorrence of mechanics goes much further and even prevented him from using its methods in solving geometrical problems that could not otherwise be solved. He was not interested in the solution but only in the beauty and elegance of the method of solution. He made mechanics unsuitable as a topic for philosophical study and he thus removed its intellectual support, leaving it, as he intended, in the hands of artisans and slaves. Plutarch tells us that "—the art of mechanics, now so celebrated and admired, was first originated by Eudoxus and Archytas, who embellished geometry with subtleties, and gave to problems incapable of proof by word and diagram, a support derived from mechanical illustrations that were patent to the senses.—But Plato was incensed at this, and inveighed against them as corrupters and destroyers of the pure excellence of geometry, which thus turned her back on incorporeal things of abstract thought and descended to the things of sense, making use, moreover, of objects which require much mean and manual labour. For this reason mechanics was made entirely distinct from geometry, and being for a long time ignored by philosophers, came to be regarded as one of the military arts." (Plutarch, Lives, trans. Langhorne, Chatto, 1876.) We should remember that Plato took the view that incorporeal objects generated by the mind, which he called *ideas* and Aristotle called *universals*, were more real than objects that are known only through the senses. A perfect sphere can be imagined but it is impossible to make one, therefore manufactured spheres are less 'real' than imagined spheres. Geometry, which can be imagined, is perfect, and is demeaned by the introduction of mechanical methods.

According to Aristotle, the free artisan was little better than the slave with whom he worked, see Fig. 14.8. In discussing whether an artisan needs to have virtue he says, "For the slave is a partner in his master's life, but the artisan is more remote, and only so much

Fig. 14.8 The forge of Hephaistos, from an Attic vase, sixth century B.C.

virtue falls to his share as of slavery, for the mechanic artisan is under a sort of limited slavery—" (Aristotle, Politics, 1260 a, b). Archimedes, arguably the most famous engineer and mechanic of the ancient world, also shared these opinions and would have been ashamed to be known by these titles. "And yet Archimedes possessed such a lofty spirit, so profound a soul, and such a wealth of scientific theory, that although his inventions won for him a name and fame for superhuman sagacity, he would not consent to leave behind him any treatise on this subject, but regarded the engineer, and every art that administers to the needs of life as ignoble and vulgar—". (Plutarch, Lives, trans Langhorne, Chatto, 1876.)

Despite intellectual repugnance of anything related to slavery and manufacture the Greeks well understood the advantages of the division of labour, but the advantage they saw, according to Finley (1965, 38), was not increased production and reduced costs but the improved quality of the product. He says: 'There is a passage in Xenophon's Cyropaedia (mid fourth century B.C.) which is quoted so repeatedly and with such solemnity by modern writers that I must give it in full. The context is the superiority of the meals provided in the Persian palace with its staff of kitchen specialists. "That this should be the case", Xenophon explains (8.2.5), "is not remarkable. For just as the various trades are most highly developed in large cities, in the same way the food at the palace is prepared in a far superior manner. In small towns the same man makes couches, doors, ploughs, and tables, and often he even builds houses, and still he is thankful if only he can find enough work to support himself. And it is impossible for a man of many trades to do all of them well. In large cities, however, because many make demands on each trade, one alone is enough to support a man, and often less than one: for instance, one man makes shoes for men, another for women; there are even places where one man earns a living just by mending shoes, another by cutting them out, another by sewing the uppers together, while there is another who performs none of these operations but assembles the parts. Of necessity he who pursues a very specialised task will do it best."

Finley continues, "This is an important text—But it is no cause for excitement on the subject of the division of labour. In the first place, Xenophon is clearly interested in specialisation of crafts rather than division of labour. In the second place, the virtues of both are, in his mind, improvement of quality, not increase in productivity." Finley goes on to stress that in Greek and Roman texts "Division of labour is not often discussed, but when it is, the interest is exclusively in craftsmanship, in quality. Furthermore, everything we know about ancient production argues against division of labour, or even widespread specialisation."

Xenophon, being a retired soldier and living on the profits of a great estate given him by a grateful state, would be more interested in the quality of craftsmanship than in its quantity. Much the same could be said of other Greek and Roman authors. But the artisan would see things differently. For him the increased production, reduced cost, and increased quality associated with specialisation and the division of labour would be equally important. They all helped him to make more profit and secure his old age. To take advantage of increased production, a large market was necessary. This exists in large cities, Xenophon points out, but not in small towns. Exports, in the sense of goods being transported outside their town or city of manufacture, could help to secure the market and justify the increased production.

Although the Greeks well understood the theory of the division of labour and toiled in workshops, often grouped in the same area of town, they did not develop large-scale factories in the modern sense. Numbers, except on large building sites or in mining, were low. Glotz (1965, 267–87) tells us that Timarchos employed nine or ten shoemakers, Demosthenes' workshop contained twenty cabinetmakers and thirty two or thirty three armourers, and the great factory of Cephalos employed a hundred and twenty men. On the other hand, the silver mines at Laurium employed perhaps 10,000 slaves in extraction and smelting. A concession or workshop was hired complete with personnel. The normal concession included about thirty miners but an individual investor could obtain a large number of concessions and hire the necessary slaves. Thus, Nicias hired 1,000 slaves to Sosias who paid an obel a day per slave (60 drachmas a year), and replaced any that died. Hipponicos hired out 600 and Philomenides 300 slaves on the same terms. A certain Dioclides had just one slave whom he let out at an obel per day. When the Spartans occupied Decelia in 413 B.C., Thucydides (VII, 27, 5) says 20,000 slaves from the nearby mines escaped to them, although whether this happened in a single day or over a period of years, or was only a part of the workforce, is not known. Xenophon proposed that the Athenian state should buy and let out up to 10,000 slaves to work in the mines and provide the state with an income.

The intellectual snobbishness that flourished in Greece and her colonies also infected at Rome. Seneca, a stoic philosopher and a man thought of as a Christian for much of the Middle Ages, might be expected to have a more enlightened attitude to slavery and to manual work. This was not the case. In a letter to Lucilius, he reveals that most technical inventions and improvements are the work of slaves. It is his opinion that philosophy should not be used to assist in the manufacture of arms or of tools for everyday use but should be used to train the intellect or spirit. "Some things we know to have appeared only within our own memory; the use, for example, of glass windows which let in the full brilliance of day through a transparent pane, or the sub-structures of our baths and pipes let into walls to distribute the heat and preserve an equal warmth above and below. Need I enlarge on the marbles with which our temples and houses gleam? On the columns, vast polished cylinders of stone, on which we support colonnades and halls in which whole nations could find room? Or on the shorthand, which catches even the quickest speech, the hand keeping pace with the tongue? All these are the inventions of the meanest slaves. Philosophy sits more loftily enthroned: she doesn't train the hand, but is instructress of the spirit.— She doesn't contrive arms and munitions for war: she is the friend of peace, and summons the human race to brotherhood. No, she is not, I say, an artisan producing tools for the mere everyday neces-

Fig. 14.9 Idealised industry at Pompeii. Top: perfumers. Bottom: fullers.

sities." (Seneca, Letters to Lucilius, No XC, trans Barker, Clarendon Press, 1932.) The paucity of his list of inventions discloses, perhaps, Seneca's own lack of knowledge or interest in technology.

The mention of windowpanes brings to mind a story told by Pliny, who professes not to believe the tale, but it does, perhaps, reveal Roman distrust of new inventions. The story suggests that it was better to destroy a new invention than risk its social consequences, which might destabilise society. "When Tiberius was emperor, it is said, a method of blending glass was invented to make it flexible [unbreakable?], but the craftsman's workshop was completely destroyed for fear that this might detract from the value of metals such as copper, silver, and gold." (Pliny, Natural History, XXXVI, 195.) Such an attitude, of course, was likely to discourage invention, even if the story was untrue.

An idealised view of Roman workshops and industry is illustrated in the Black Room of the House of the Vettii at Pompeii, Fig. 14.9. The owners of this fine house decorated their best rooms with industrial and commercial scenes. In deference to the sensitivities of their age, however, cupids replaced the slaves that would normally have undertaken these tasks. In the bottom picture, fullers are at work treading and washing clothes in vats, hanging them to dry, carding, inspecting and folding the finished garments. Above this is a perfumery. On the right olive oil is pressed and in the centre it is heated and mixed with essences of flowers. On the left is a customer with her slave sampling and buying perfume. The owners of this house, the Vettii, were former slaves but as freedmen they were prosperous and themselves the owners of the slaves. Probably they owned the businesses depicted on their walls.

Despite all this evidence concerning Roman and Greek attitudes to manufacture, our sample size is small and specialised, being restricted to a small group of reasonably wealthy, well-educated intellectuals. They were hardly typical of society as a whole. The Athenian Assembly in the fifth and fourth centuries B.C., for example, contained many fullers, shoe makers, masons, and metal workers, and the artisans could even form a majority. They were unlikely to share Plato's refined and tortured views even if they were aware of them. The engineers and mechanics working later in Alexandria certainly did not share these opinions. Pappus Alexandrinus, for example, considered theory and practice to be equally important in the study of mechanics. "Those who follow Hero are of the opinion that one part of Mechanics embraces mathematical demonstration and the other manual work; moreover the former part, which they term *rational* includes Geometry, Arithmetic, Astronomy, and demonstrations of Physics; while the other part which comprises manual skills, should teach the art of the metallurgist, of the iron-worker, the builder, and the wood-worker as well as the painter, and everything which concerns handwork", Klemm (1959, 41).

Carpus of Antioch attacked Archimedes for not writing more on the subject of mechanics, but Pappus defended him and at the same time maintained the rightness of introducing geometry into the crafts. "Carpus of Antioch", Pappus tells us, "wrote that Archimedes produced only the book on the construction of celestial globes because he considered the rest of the subject not worth the trouble of writing on. But this divine man, who is so famed among most people for his intellect and for his mechanical knowledge that his memory will live forever among human beings, established the foundations of Geometry and Arithmetic in the briefest and most exact fashion; and he so loved these disciplines that he could not make up his mind to introduce ought into them. Carpus himself, however, and several others, rightly introduced Geometry into certain crafts and uses in daily life. For Geometry, which can assist in many crafts and in many demands of daily life, is far from suffering injury thereby; nor on account of its contribution to these crafts is ought detracted from the honour that is due to it or from its adornment", Klemm (1959, 42). This inclusive conception of Pappus' seems so much more reasonable and civilised than Plato's instinctive reaction, which was to separate theory and exclude it from practice. Pappus' was a much healthier outlook and no doubt contributed to making Alexandria, for a time, a great centre of excellence in science, engineering, and intellectual studies. The famous Museum (Temple of the Muses) of Alexandria incorporated a huge and famous library, employed about a hundred professors, and was furnished with rooms for research, lectures, and study, and all paid for directly by the Ptolemies. It was first developed under the patronage of the first two Ptolemies (after the foundation of Alexandria in 330 B.C.). Its most important period was from Euclid, about 300 B.C., to Hipparchus, about 190 B.C., although it survived for much longer. It seems to have been largely disbanded during the reign of the ninth Ptolemy (146–117 B.C.). In the Library, most of the writings of antiquity were preserved (about 500,000 scrolls) and copies were made and disseminated throughout the civilised world. It was partially or wholly destroyed on several occasions. In 272 A.D. the Emperor Lucillius Domitius Aurelian had it destroyed, as did Theodosius I in 391, and the Muslims under Caliph Umar I in 640. The achievements of the Alexandrians, particularly in mathematics and science, were outstanding, and in mechanics they produced many ingenious toys and devices to assist in religious ritual. However, although they encouraged the use of geometry by craftsmen they did not think to devise machines, based on their clever and inventive toys and mechanisms that might also help artisans.

Writing specifically about science Farrington (1961, 301–2) declares "With the science of Alexandria and of Rome we are in very truth on the threshold of the modern world. When modern science began in the sixteenth century, it took up where the Greeks left off". In answer to the question why the Greeks and Romans did not push open the door to the modern world Farrington says: "they had already been loitering on the threshold for four hundred years. They had indeed demonstrated their inability to cross it." The fault for this state of affairs he blames, not on any lack of individual genius, but on the social circumstances, which lifted science out of the technical conditions in which it should have found application. This, he says, was the result of the universal cleavage of society into freeman and slave. The implication was that science and technology were interdependent and neither could grow and prosper until the stigma that slavery attached to all forms of manual and technical work was overcome.

However, the belief that slavery was the primary reason why ancient technology ceased to develop beyond a certain point is not the only view. Mitcham (1990, 38), for example, considers that the ancient critique of technology, which effectively prevented its growth, rested on four arguments.

i. The practice of technology often involved turning away from faith in the Gods or trust in nature.
ii. Technical affluence and processes of change tended to undermine an individual's striving for excellence and weakened political stability.
iii. Technological knowledge was not true wisdom, drew humanity into intercourse with the world, and obscured transcendence.
iv. Technical objects were less real than natural objects.

The first critique is true in that we practice technology because we are not content wholly to rely on nature or the Gods to provide life's necessities. We invented and used bellows, for example, because we are too impatient to wait for God or nature to provide a wind at exactly the right point. Most people want a better life but the ancient philosophers often advocated a simpler life, a return to nature and less demand for manufactured goods. Plato's 'Republic', Diogenes' living in a barrel, and the Stoic philosophy, all reflect the philosopher's admiration of a simpler life. The second critique is also true, for the constant change and improvements generated by technical progress lead us to expect similar advancement in politics and in social conditions and may indeed lead to political or social instability. The third critique that technological knowledge is not true wisdom, is similar to the criticism that was once made of Hesiod, Pythagoras and others by Heraclitus, namely, that 'The learning of many things teacheth not understanding'. This is true but I have always failed to see how anyone can have understanding without facts; apparently, the ancient philosophers were expected to have greater vision. The fourth critique, that technical objects are less real than natural objects, depends on how reality is defined and owes much to the Platonic conception that imaginary objects (called ideas or universals) are more real than manufactured objects. All these critiques seemed valid to ancient philosophers but we can hardly suppose that the average Athenian or Roman citizen shared them.

These arguments against engagement in technology were set aside only when faced with the extreme necessity of survival. Archimedes, as we have seen, was too high-minded to work as an engineer, or even to write on the subject, but he nevertheless invented all kinds of military devices when he was besieged at Syracuse and for which he is now greatly honoured (Plutarch, *Marcellus*, 2.252–55). Mossé (1969, 26) points out that except in a few systematic thinkers, such as Aristotle, work on the land did not incur contempt, and only the work of the artisan was despised. This, he suggests, was probably because the artisan usually worked with one of more slaves whereas the serf or peasant worked, more or less, for and by himself. The artisan was dependent on his slaves and his customers whereas the peasant was more self-sufficient and independent. Freedom and independence were much admired in ancient Greece.

The extreme position taken by Greek and Roman intellectuals coexisted with other views in which manual work was given more respect and dignity. The writer of the book of Ecclesiasticus (Jesus), in about 180 B.C., says that "The wisdom of a learned man cometh by opportunity of leisure". "How can he get wisdom that holdeth the plough, that glorieth in the goad, that driveth oxen, and is occupied in his labours, and whose whole talk is of bullocks?" He also mentions, in this respect, the carpenter, the cutters and gravers of seals, the smith "sitting by the anvil and considering the iron work", and the potter "turning the wheel about with his feet". The answer given by the writer of Ecclesiasticus is that "All these trust to their hands: and every one is wise in his work. Without these cannot a city be inhabited." He says that although such manual workers are not sought for public office, do not sit on the judges' seat, and are not to be found where parables are spoken, yet they maintain the state of the world and all their desire is in their craft (Ecclesiasticus, XXXVIII, 24–34). It is not clear that this guarded respect was extended equally to both free artisan and slave.

The so-called "barbarians" who lived outside the Roman Empire had little need of slaves. Theirs was an agricultural rather than industrial economy and was often centred on the cultivation of corn that was, as in Greece, unsuitable for economic exploitation by slaves. In their passion for gambling, the inhabitants of Germania, Tacitus tells us, would sometimes stake their freedom on a last final throw and, if losing, suffer themselves to be bound and sold into slavery. Amongst the barbarians, however, slavery was not so terrible as in Rome, "slaves are not employed after our manner with distinct domestic duties assigned to them, but each one has the management of a house and home of his own. The master requires from the slave a certain quantity of grain, of cattle, and of clothing, as he would from a tenant, and this is the limit of his subjection." We are also told that amongst the barbarians of Germania to strike, or to punish a slave with bonds, is extremely rare but they are often killed, not to enforce discipline but on the impulse of passion.

Others went further and even rejected slavery outright. The Essenes, a pre-Christian Jewish community or perhaps several communities, who seem to have influenced Christianity, endorsed this view. Josephus, Pliny the Elder and Philo have described the Essenes' way of life and their beliefs. It appears that their main settlement was on the western shores of the Dead Sea and it is assumed that the Dead Sea Scrolls, found in caves at nearby Qumran, were written by or for them. They practised extreme frugality but what little property members of the sect had was shared equally and owned in common. They rejected slavery in principle as incompatible with the equality of all men before their maker. It was an important principle.

Christianity, despite its initial appeal to the poor and the enslaved, promised much but, in fact, did very little to emancipate slaves within the Roman Empire and many continued to be used cheaply in gladiatorial contests, Fig. 14.10. However, the Judeo-Christian belief that all men are created in God's image and they must therefore be treated with respect did much to alleviate the worst conditions of slavery. Chadwick (1990, 59) points out that "What made the church conservative about slavery as an institution was not political indifference but respect for the Law as laid down in Romans XIII. That one man should possess property rights over another was recognised as an evil, and therefore held to be a consequence of the fallen state of humanity since Adam". Romans XIII made it clear that the established law and constitution of the state was to be obeyed; Christianity was not to be revolutionary in a political sense and if slavery was legal then it was to be formally upheld. However, to free individual slaves was legal, was regarded as a 'good work', and both Church and private funds, therefore, were used for the manumission of slaves. The Church, however, could not buy the freedom of all slaves within

Fig. 14.10 Gladiatorial contests against lions with the ladies of Rome shouting encouragement from their box.

the empire. The normal procedure was for a Christian master to affirm his intention to free a slave in the presence of a bishop. Constantine gave legal validity to this procedure, making it equal to the normal procedure of manumission before a magistrate. Several emancipated slaves rose to become bishops, notably Callistus of Rome in the third century, and many became saints.

With the conversion of Constantine in 338, the prohibition of pagan rites in 390, and the eventual acceptance of Christianity throughout the Roman Empire, Christianity became the established religion, and came to possess great wealth. Constantine attempted to express Christian ideals in some of his laws, such as those protecting children, slaves, peasants and prisoners. An edict of 316, for example, directs that criminals may not be branded on the face 'because man is made in God's image'. However, this was a long way from banning slavery. By the end of the fourth century, the Church had attained great status, but it became more circumspect. Chadwick (1990, 60) says, "Protests against the institution of slavery, as such, came in the fourth century when the Christians were beginning to be in a position to affect social policy. But bequests had by this time made the fourth-century Church a considerable landowner, already dependent on endowments for clergy stipends and therefore in a weak position for initiating economic changes. These protests were too little and too late to revolutionise the economy of the ancient world and only have the historical significance of outlining a programme for the future."

Christianity was eventually to remove slavery and persuade men that manual work was a noble and acceptable way to worship God; that work was morally desirable and idleness was the work of the devil. Such a change was not easily attained and it took nearly a thousand years before the technological benefits of this new attitude became clear.

Christianity and Medieval Scholasticism

In the fifth century, the invasion of Italy by Alaric, the sack of Rome, and the collapse of the Roman Empire of the West, followed rather too closely on the triumph of Christianity at the end of the fourth century. The end of the Western Empire marked the start of the Dark Ages and superficially, at least, Christianity's triumph seemed to have been a disaster for civilisation. The weakness, and ultimate collapse, of the Roman Empire was variously ascribed, at the time, to the growth of luxury, the rise of Christianity, or the continuance of paganism. Salvian, a Christian priest writing in about 440, thought that the people of the Empire were more wicked than the barbarians and this was responsible for their troubles. "Almost all the barbarians, at least those who are of one race and kin, love each other, while the Romans persecute each other.—Nay, the state has fallen upon such evil days that a man cannot be safe unless he is wicked.—So the poor are despoiled, the widows sigh, the orphans are oppressed, until many of them, born of families not obscure, and liberally educated, flee to our enemies that they may no longer suffer the oppression of public persecution. They doubtless seek Roman humanity among the barbarians, because they cannot bear barbarian inhumanity among the Romans", Robinson (1904, 28–30). Describing the symptoms of decay, Gibbon (1960, 524) held that "The Roman government appeared every day less formidable to its enemies, more odious and oppressive to its subjects. The taxes were multiplied with the public distress; economy was neglected in proportion as it became necessary; and the injustice of the rich shifted the unequal burden from themselves to the people—The severe inquisition, which confiscated their goods and tortured their persons, compelled the subjects of Valentinian to prefer the more simple tyranny of the barbarians, to fly to the woods and mountains, or to embrace the vile and abject condition of mercenary servants. They abjured and abhorred the name of Roman citizens, which had formerly excited the ambition of mankind." Gibbon goes on to say, "If all the barbarian conquerors had been annihilated at the same hour, their total destruction would not have restored the empire of the West—. The story of its ruin is simple and obvious; and, instead of inquiring why the Roman Empire was destroyed, we should rather be surprised that it had subsisted so long."

The remedies proposed at the time were to return to a simpler lifestyle, to abandon Christianity, or to forsake Paganism. The anonymous author of *De Rebus Bellicus*, Thompson (1952, 32), dates the origin of extravagance and greed in the Roman Empire to the age of Constantine when the vast quantities of gold, long stored in the temples, reached the public. Certainly this caused problems by fuelling inflation but the fourth century also saw the victory of Christianity, and the east west split of the Empire. The anonymous author, however, took an original line in solving Rome's problems. In addition to reforming the mint, rooting out the corruption in provincial governors, improving the defence of the frontiers, and removing confusion from the laws of justice, he also proposed the use of technical inventions. His own inventions included several of minor importance and five of major importance:

i. a warship driven by oxen and paddle wheels, Fig. 14.11,
ii. a chariot fitted with wheel scythes,
iii. a portable bridge,
iv. a ballista for mounting on city walls,
v. a portable ballista.

The warship driven by oxen and paddle wheels may seem rather far-fetched but horses frequently drove ferryboats and dredgers in the eighteenth and nineteenth centuries so his idea was quite feasible. Perhaps he expected to fail for he compared the Roman leisured classes, which did not contribute to technical progress, with the barbarians, who did make such a contribution. His inventions would not have saved the Roman Empire, but surely, he was right to fear the vigour and technical progress of the barbarians and the sloth

Fig. 14.11 A warship driven by oxen and paddles proposed by the anonymous author of De Rebus Bellicus. Thompson (1952).

and technical ignorance of Romans. He failed, not for any lack of realism in his inventions, but because his was a lone voice. No one, or no one that mattered in Rome, had the slightest interest in such technical matters.

With the collapse of the Roman Empire, centralised authority gave way to small pagan kingdoms, which were, eventually, Christianised. The conquerors, incidentally, took up the Christian faith almost immediately. Spiritual authority, if not political authority, returned to Rome. Town life disintegrated and the towns and cities of the Western Empire were mostly abandoned. An Early English poem, "The Ruin", written about three centuries later, records the opinions of a thoughtful and observant poet as he stumbles round a deserted and abandoned Roman city, probably Bath in Somerset. Between the withdrawal of the Romans and the arrival of the Normans, no massive work in stone was undertaken in England. The poet, unused to such things, imagined the Roman ruins to be the work of giants when he says "Snapped rooftrees, towers fallen, the work of Giants, the stone smiths, moulders". Iron rings, used to strengthen a building, amazed him: "Mood quickened mind, and a man of wit, cunning in rings, bound bravely the wall-base with iron, a wonder." Even more impressive was the hot water supply to a circular stone bathhouse: "Thence hot-springs, loosed, ran over hoar stone unto the ring tank—It is a kingly thing". The translation here is due to Alexander (1980, 28) and it admirably conveys the extent of the general decline in skill, technology, and living standards during the Dark Ages.

Slavery remained respectable so long as the enslaved were not Christians, although even they could be enslaved for debt or crime. In a Europe that was largely Christian, slaves had to be obtained from greater distances and were more difficult to obtain in large numbers. Craft and industry, practised mainly in the towns and cities, relied heavily on slave labour and suffered most from their shortage as "The Ruin" indicates. Agriculture, however, was less dependent on slavery and the manorial system, founded on serfs and peasants, slowly developed. Power driven machinery should have developed to replace slaves working in the towns and cities but those with capital had little interest in technology. Nevertheless, where techniques already existed, machinery was used. Water wheels were in use to grind corn in the first century B.C. when Antipator of Thessalonica, in 85 B.C., and Vitruvius, in 15 B.C., mentions them. They seem not to have been used very much when slaves were plentiful but by the end of the fourth century, Rome depended greatly on the mills of Janiculum for its daily supply of corn. In 536, when the Goths besieged Rome and cut the aqueducts supplying the city, the water mills of Janiculum stopped grinding corn and to avoid famine the Byzantine general Belisarius, and his engineers, devised a system of floating mills tethered to riverbanks below the bridges, see Fig. 14.12. In this way, they were able to grind sufficient corn to feed the city. Clearly, there were insufficient slaves in the city for this essential task. Belisarius' successful defence of Rome revived the Western Empire, for a time, but shortly afterwards it finally collapsed.

The Dark Ages, which I take to be from about 600 to about 1000 A.D., are usually looked on as a time of deep recession and stagnation, both of intellectual life and of commerce, but it is possible to take a slightly more generous view. If town life and manufacture declined, it did so, in part at least, because slavery declined. For example, there is only one slave mentioned in the first epic poem of English literature, 'Beowulf', namely, he who stole the cups from

Fig. 14.12 Medieval floating mills for grinding corn. The floating mills devised by Belisarius' engineers in 536 A.D. may have been similar.

the dragon's hoard. Manufactured goods came to be highly valued. In 'Beowulf', cups, weapons, and armour are clearly very precious and their owners were expected to be their guardians for the time being and eventually to pass them to others whom they considered worthy of the honour. Swords, for example, were identified by name and were handed down through the generations acquiring a heroic history that was repeated, and perhaps embellished, by the bards. We hear of 'Mimming', Waldere's sword made by Wayland the smith-god and usually said to belong to Waldere's son. When Beowulf fought with Grendel "The earls ran to the defence of their famous prince; they drew their ancestral swords—" so it seems that most swords belonging to the earls were ancestral and handed down from one generation to the next. 'Nailing' was the sword Beowulf took from Dayraven and 'Hrunting' was Unferth's sword. When Beowulf was being armed for his fight in the mere, we are told: "Not least among these mighty aids was the hilted sword that Hrothgar's spokesman, Unferth, lent him in his hour of trial. Hrunting was its name; unique and ancient, its edge was of iron, annealed in venom and tempered in blood; in battle it never failed any hero whose hand took it up.—This was not the first time that it had to do heroic work." The translations are due to Alexander (1973). Nor, amongst this list of heroic swords, must we forget 'Excaliber', the sword of the semi-mythical King Arthur.

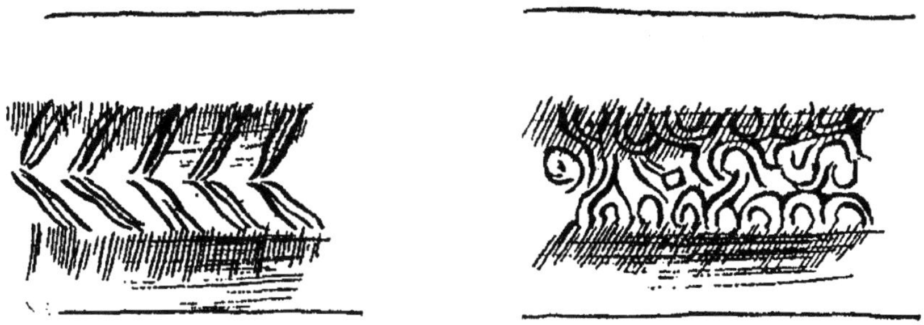

Fig. 14.13 Blade patterns in Viking swords.

Later, Hrunting is described as "this wave-patterned sword of rare hardness" which reminds us that swords at this time were made by twisting together thin bands of iron, of varying carbon content, heating and then hammering the plait to shape thus welding the separate bands into one material; a process known as pattern-welding. This procedure created a very distinctive and decorative pattern of rings and waves, Fig. 14.13.

Greek and Roman values carried no weight amongst the Anglo-Saxons and the smith, long despised in Greek and Roman society, was restored to, or rather retained, a place of honour. Tylecote (1981, 42–50) says of the smith "in Anglo-Saxon times he was treated as an officer of the highest rank and was awarded the first place in the presidency; after him ranked the maker of mead and then the physician. In Wales he sat with the king and queen next to the domestic chaplain. A similar exalted position is shown in the seating plan at the banqueting hall at Tara. Here the smith comes fourth, after the verse-maker, brewer, and teacher, on the king's right hand with the chamberlain and flute-player on his (the king's) left."

With the decline of slavery and the increased value and scarcity of manufactured goods, the stigma that slavery formerly attached to manufacture was forgotten. This was one of the great, if inadvertent, achievements of the Dark Ages, and consequently attitudes to manual work and manufacture changed, paving the way for a more rational and interested approach. Alexandrian and Roman technology might at last cross the threshold into the modern world.

This change in attitude was also promoted, before the end of the Dark Ages, by the example of the monastic orders that embraced manual work, not only for their sustenance, but also as an antidote to sin. It was largely through them and their earthly needs that technology revived. Admiration, and sometimes detestation, for monks was often intense. The first Christian hermits established themselves in Egypt in the third century, and later the Stylites (pillar saints), sought escape from their unhappy world by isolating themselves on the top of high pillars in the desert. They were esteemed and became hugely popular, often being chosen when new bishops or other dignitaries were required. Eventually, the hermits came to share common religious rites and worship but in order to ensure their seclusion each hermit had his separate cell to which he retired. St Anthony of Alexandria is considered the founder of this mode of living and thousands of followers flocked to join him and his disciples. Gibbon (1972, 511) remarks that these "—monasteries were filled by a crowd of obscure and abject plebeians, who gained in the cloister much more than they sacrificed in the world. Peasants, slaves, and mechanics, might escape from poverty and contempt to a safe and honourable profession—The brethren were supported by their manual labour; and the duty of labour was strenuously recommended as a penance, as an exercise, and as the most laudable means of securing their daily subsistence." These establishments spread to the west and Benedict of Nursia (*c.* 480–*c.* 547), who founded the famous monastery at Monte Cassino in 529, is considered as the creator of western monasticism. Benedict emphasised poverty, chastity, obedience, communal living, physical labour, a minimum of conversation, and the distribution of alms to the poor. These aims were not always attained, however, for Gibbon quotes the remarks of an unknown Benedictine Abbot: "My vow of poverty has given me an hundred crowns a year; my vow of obedience has raised me to the rank of a sovereign prince." Gibbon says he forgets the consequences of the Abbot's vow of chastity.

In the Christian faith Man was created by God in His image and was supreme over Nature. It was a duty for Man to make good use of all of God's creation. Gregory of Nyssa in the fourth century, held that technology glorified God and that the natural world was created for man's benefit. "It was the slowness and difficult motion of our body that brought the horse to supply our need, and tamed him: it was the nakedness of our body that made necessary our management of sheep,—it was the fact that we import from others the supplies for our living which subjected beasts of burden to our service: and it was the fact that we cannot eat grass like cattle which brought the ox to render service to our life;—and there has been discovered by men iron, stronger and more penetrating than prominent horns or sharp claws,—art fashions iron for this purpose too [making armour],—moreover art contrives to make our arrows feathered, and by means of the bow gives us for our needs the speed of wings—", Klemm (1959, 59).

This enthusiasm for technology paid dividends, for by the end of the Dark Ages the monasteries emerged as centres of learning and civilisation where machines, rather than slaves, freed the monks from much arduous daily work. This gave them time for their study and devotions, and time to manufacture items of great beauty for the Church. In England, St Dunstan typifies monastic enthusiasm for technology. He lived from 909–988 A.D. and was credited with great skill and dexterity as a smith, a bell-founder, a scribe, and a draughtsman. Dunstan restored monastic life at Glastonbury after the upheavals of the Danish invasions and effectively re-established monasticism in England. After banishment for a time he was recalled and in 973, as Archbishop of Canterbury, he crowned Edgar as the first king of all England. Amongst his mechanical works are two bells he cast for the church at Abingdon and crosses, censers and various vestments he made for Glastonbury. Appropriately, St Dunstan's emblem, in art, is a pair of tongs. Another English monk famous for his skill in working iron, silver, gold, jewellery, and gilding, was Anketil. He lived in the twelfth century and worked at St Alban's. A pair of gold and silver candlesticks that he made was presented to Pope Adrian IV and was consecrated at St Peter's, Rome.

Despite all this work neither Dunstan nor Anketil was adopted as the patron saint of metalworkers, blacksmiths, and farriers. That honour belongs to an earlier ecclesiastical metalworker, namely St Eloy, Fig. 14.14. Eloy was born in Chaptelat, near Limoges in 590 A.D. and was apprenticed to a goldsmith. His talent took him to Paris to work for Clotaire II who desired jewels and a golden throne. Later he became Master of the Mint under Dagobert before becoming Bishop of Noyon and Tournai, but despite such elevation he never abandoned his art and continued to use it to make fine church furnishings. He was, apparently, much tormented by the devil until finally Eloy seized him with his red-hot pincers and overthrew him. Another legend of St Eloy tells how a horse was brought into a smithy to be shod and became possessed by the devil and kicked so furiously that all fled except Eloy, who calmly cut off the horses leg,

Fig. 14.14 St Eloy in his forge. Mural painting, late fourteenth century, formerly in Highworth Parish Church, Wilts.

put it on the anvil, shod it, and then rejoined it to the body of the living horse. This is the legend depicted in Fig. 14.14 where the bishop, hammer in hand, stands in a forge. On the anvil is an amputated leg with a horseshoe and nearby is a pair of pincers. A small boy with a hoof foot seems to be running off and possibly represents the flight of the evil spirit from the horse. St Eloy died in about 660 and his relics are deposited in the cathedral at Noyon.

Klemm (1959, 67) has translated the work of Theophilus, probably an eleventh-century German Benedictine monk, who describes the many arts and crafts of which the religious houses had need. In his introduction Theophilus encourages the novice, "Act, therefore now, well intentioned man, happy before God and men in this life, happier in a future, in whose labour and study so many sacrifices are offered up to God; henceforth warm thyself with a more ample invention, hasten to complete with all the study of thy mind those things which are still wanting among the utensils of the house of the Lord.—These are the chalices, candelabra, incense burners, vials, pitchers, caskets of sacred relics, crosses, missals, and other things which useful necessity requires for the use of the ecclesiastical order." Theophilus not only encourages his young men to make things for the glory of God but he gives very detailed and specific instructions on how to proceed. For example, he describes how files are made, hardened, and tempered, how lathes are used, how moulds are made, and how bells are cast. Clearly, he is writing from personal experience.

Waterpower, already in use at the start of the Dark Ages, emerged as a major source of power by the eleventh century. In England alone, the Domesday survey of 1086 records some 5624 water mills for grinding corn, one for every fifty households. A twelfth century description of the way water was used in the Cistercian Abbey at Clairvaux, France, tells us that it drove water wheels to crush and grind wheat and to shake a sieve to separate the flour from the bran. It was drawn off to fill the vats to prepare the monk's beer, and it powered the fulling stocks to clean the wool for the monk's clothing. "How many horses would be worn out, how many men would have weary arms if this graceful river, to whom we owe our clothes and food, did not labour for us—". Before leaving the Monastery's grounds the river was also used in tanning leather, in cooking, sieving, turning, grinding, watering, and washing. "At last, in case it receives any reward for work it has not done, it carries away the waste and leaves everywhere spotless", Gimpel (1993, 5–6).

Probably the river was polluted as it left the abbey grounds but such pollution does not seem to have caused very much concern. Cistercian monasteries were all built on similar principles and so, in the twelfth century, several hundred such monasteries could, perhaps, claim a comparable degree of mechanisation. It should be noted, however, that the Cistercians were not allowed to take the profits or rents from mills built or owned by them outside their own Abbey. However, like many other monastic rules, it was frequently ignored.

The Cistercians, in fact, produced a great deal of iron in France until the middle of the fourteenth century. They were responsible for numerous technical improvements, particularly in the use of camshafts driven by water wheels and used to lift heavy hammers in the forge or to drive very large bellows so that higher temperatures could be attained in their furnaces. The forge building at Fontenay Abbey, Fig. 14.15, is a rare survival from this period. The Fontenay Forge is over 53 m long and was, in effect, a factory obtaining its ores from a nearby hill. The adjacent river was canalised and used to drive water wheels to operate drop forge hammers and bellows. Its importance to the monks may be gauged by the fact that, next to the abbey church, it is the largest building on the site and it is still considered the most beautiful. During the French Revolution Fontenay was sold and converted to a paper mill and stopped work only in 1906, almost eight centuries after it was founded.

In Italy also, the monks were at the forefront of technology. Filarete published details of a visit he made to an iron mine and smelter in 1463 at Ferriere in the Apuan Alps, see Spencer (1963, 201–6). He found that after casting pig iron they re-melted the iron in another furnace and beat it with hand-hammers to make wrought iron. This, of course, was very arduous work. He then describes how he had previously seen the monks at the abbey of Grottaferrata, about 12 miles from Rome, doing the same task using a hammer-mill driven by water. "There are plenty of woods around here and it is here that the hammer-mill is located. A little outside this circuit there is a stream that runs through the site.—It is lead through a canal in such a way that it turns wheels which blow the bellows and make the hammers beat.—They re-melt the iron in this [furnace] and then throw it on the [anvil] and beat it as they wish by means of the hammer and the water."

In England too the monasteries situated in areas where iron ore was to be found took advantage of God's gift and devoted themselves to its use. Smiles (1882, 29) notes that cinder beds are found

Fig. 14.15 The Forge at Fontenay Abbey is considered the most beautiful room in the abbey and is only marginally smaller than the church.

around the monasteries of Rievaulx and Hackness in Yorkshire and in the Forest of Dean the abbots of Flaxley, under a grant of Henry II, possessed one stationary and one itinerant forge and were allowed two oaks per week for fuel. This privilege was commuted for a wood of 872 acres in 1285 and was held until the dissolution of the monasteries. Furness Abbey, despite its remoteness from centres of population, was the second wealthiest Cistercian house in England, after Fountains Abbey, and owed much of its affluence to the profits from iron making. Iron ore deposits on their land at Dalton were exploited and the charcoal came from their land at High Furness.

Water mills were first used for fulling cloth in England in the latter half of the twelfth century, Carus-Wilson (1941, 39–60). The earliest fulling mill dates from 1185 and was built on land of the Templars at Newsham, Yorkshire. Another mill of the same date was built at Barton in the Cotswolds and Quarr Abbey on the Isle of Wight had a fulling mill in 1189, Hockey (1970, 50). Although the invention of the fulling mill cannot be ascribed to any person or place it seems likely that the Templars introduced it into England. The vast majority of the earliest mills, however, were on Abbey lands or Episcopal estates although some were built on lay holdings. By the beginning of the fourteenth century, water driven fulling mills had mostly replaced the more primitive method of treading underfoot and a profound change in English cloth manufacture was underway. The new mills were situated in the West Riding of Yorkshire, the Lake District, Cornwall, Devon, Somerset, the Cotswolds, and were almost always in hilly districts having a plentiful and rapid water supply. York, Beverley, Lincoln, Louth, Stamford and Northampton were once famous for their fine quality cloth, but they had poor water supplies and their industry declined and then vanished. Cloth manufacture moved out of towns and cities to the manors, and the restrictive Guild system of the towns lost their control of the industry. Water driven fulling mills, first used by monks for their own convenience, spread to the lay population and revolutionised the woollen industry.

The emphasis laid on manual labour in the early monasteries was relaxed somewhat in the thirteenth century, largely because such work was becoming irksome to those whose talents and skills lay elsewhere. Thomas Aquinas confirms this when he answered the query: "Must members of the Order perform manual labour?" He replied that manual work was required for four reasons: to gain a livelihood, to overcome idleness, to curb appetite by mortifying the body, and to facilitate the distribution of alms." By showing that these objectives may be attained by other methods, he justified a relaxation of this requirement. "If a man could support life without food, he would not find it necessary to work with his hands; the same applies to those who have other legitimate means of livelihood. Insofar, however, as handwork is performed to overcome idleness or to mortify the body, it does not in itself come within the scope of a command, since the body can be mortified and idleness can be overcome by many methods other than by handwork. Finally, insofar as handwork is designed for the giving of alms, it again does not fall under a command, except only in case someone is for some reason bound to give alms and knows of no other way to help the poor. If, therefore, nothing in their Rules enjoineth any special handiwork, then are members of the Orders no more bound to execute handwork than are those in the outside world", Klemm (1959, 69). This passage, alas, has all the charm of a lawyer setting out the letter of the law rather than its spirit. The rules were being bent to suit the requirements of the age.

Fig. 14.16 The seven mechanical arts of Hugo von St Victor: Weaving, Weapon Foundry, Navigation, Agriculture, the Chase, Medicine, and Drama.

Monastic scholarship was based, initially, on the seven liberal arts inherited from Greek and Roman writers, namely: Grammar, Rhetoric, Dialectic, Geometry, Arithmetic, Astronomy, and Music. To this was added some mechanics and Natural History, for Pliny's work was greatly admired and was in line with monkish taste. In the first half of the twelfth century, Hugh of St Victor, as the Head of the School of St Victor in Paris, reorganised teaching along more suitable lines so that technology might be firmly established alongside the traditional elements. To the seven liberal arts, he added the seven mechanical arts and thus he restored the intellectual support that had for so long been denied to mechanical technology. "—Mechanics", Hugh tells us, "comprises seven branches: Weaving, Weapon foundry, Navigation, Agriculture, the Chase, Medicine, and Drama [see Fig. 14.16]. Of these, three are related to the external equipment of Nature, by which Nature protecteth herself against injuries; four are related to inner equipment, that by nourishment and nurture may grow and prosper. Thus there is a definite similarity to the Trivium [Grammar, Dialectic, Rhetoric] and the Quadrivium [Arithmetic, Geometry, Astronomy, Music]", Klemm (1959, 71–2). It is not particularly extraordinary that a monk should be promoting the mechanical arts but it is, perhaps, rather surprising that a monk should be teaching weapon foundry, the chase, and drama! However, these exertions to improve the teaching of mechanical arts were not taken up generally, for the first universities (Padua, Paris, Oxford, Cambridge), founded at about this time, produced mainly clerics, physicians, and lawyers. It was not, in fact, until the middle of the nineteenth century that the first Chair in Engineering was created when Reynolds was appointed at Manchester University.

As mentioned above, medieval monks relied on mechanical technology not only in their monasteries but also they applied it in running and managing the vast estates that they accumulated, mainly from charitable donations. Every monastery had a workshop, quite close to the church and often as large, where water powered machinery was commonly used. Such a factory, as mentioned above, has survived at Fontenay. The Cistercians needed iron for their own use and they accepted any iron-ore deposits, and the necessary woodlands to provide fuel, which were offered to them. Gimpell (1993, 68) relates, "From 1250 to the seventeenth century the Cistercians were the leading iron producers of the Champagne region, and by the end of the eighteenth century they owned half the industrial iron and steel complex of the 'Plateau de Langres'. But before 1330 they already owned between eight and thirteen iron-producing factories." The map, Fig. 14.17, illustrates some of their extensive iron-ore fields and forges.

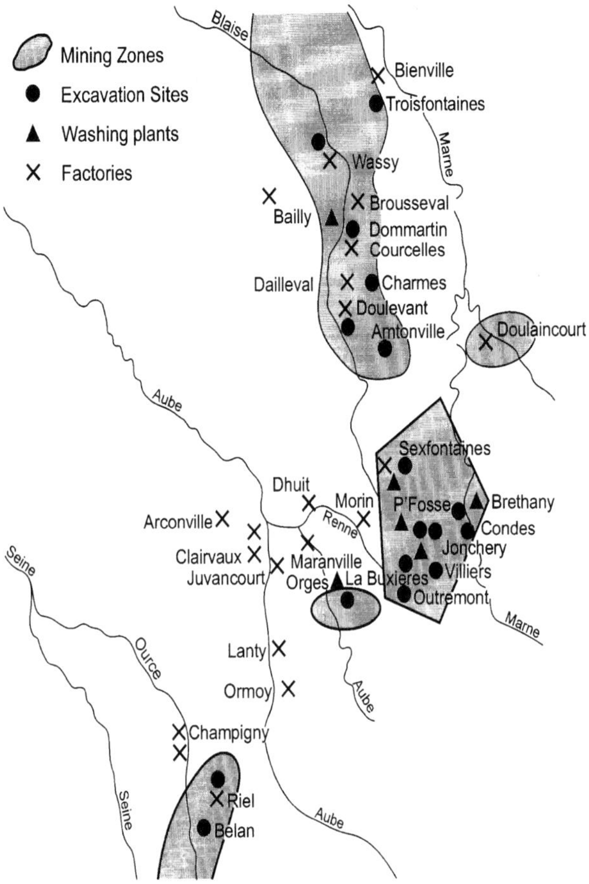

Fig. 14.17 Cistercian iron mines on the Plateau de Langres in Champagne. After Gimpell (1993, 69).

The Renaissance and Early Modern Periods

By the fifteenth century, the ancient critique of technology outlined by Mitcham (1990, 38) had been turned upside-down. Whereas in the ancient world technology was seen as undermining faith in the Gods or in nature, it was now considered that God established technology for and on behalf of humanity and that man's technical achievements greatly honoured Him. This, as we have seen, was the view of Gregory of Nyssa who, as early as the fourth century, listed the things God had created for man's benefit. Technical processes, that were once considered to threaten political stability, were now thought acceptable because they satisfied humanity's physical needs, increased sociability and thereby stabilised society. Pliny's inventor of unbreakable glass, had he lived in the fifteenth century, would have been encouraged in his endeavours, and become a wealthy man, rather than having his workshop burnt down. Technical affluence was no longer thought to undermine an individual's striving for quality and excellence. During and after the Renaissance true wisdom was not identified with abstract theory, as it was by the Greeks, but empirical and technical knowledge was recognised as truer than abstract theory. Moreover, technical products, once thought 'less real' than natural products, were given equal value with natural products for it was understood that Nature was no more real than artifice. Indeed, it was recognised that Nature operated on exactly the same principles as artifice. The common sense view was upheld that a manufactured sphere, however approximate, was more real than the ideal, imaginary sphere it sought to imitate.

This breakdown in ancient and medieval outlook began, in philosophy, with men such as William of Ockham (*c.* 1280–*c.* 1350). William was a Franciscan monk born either in Ockham, Surrey, or Ockham, Yorkshire, who led a rather controversial life, eventually being excommunicated by the Pope. He was, after St Thomas Aquinas, the most famous of the medieval schoolmen and is now most famous, perhaps, for his empirical point of view and as the inventor of 'Ockham's razor' in logic. This principle states that "It is vain to do with more what can be done with fewer", or as we should say, "the preferred explanation is that having the fewest assumptions". His empiricism was based on his opinion that knowledge was derived from experience, was made possible by reflection on experience, and depended upon experience of things through our senses. In explaining human knowledge, William of Ockham never allowed universals to be actual things. By insisting on the possibility of studying logic and human knowledge without reference to metaphysics and theology, he encouraged scientific research in a modern sense. His immediate follower, for example, was Nicholas of Oresme (d. 1382), who investigated planetary theory, and was, to a certain extent, a precursor of Copernicus. Nicholas set out both the geocentric and the heliocentric theories, and said that each would explain all the facts known in his day, so there was no way of deciding between them. Observable facts were needed to determine which theory was correct. Such an idea would have been abhorrent to Plato.

Although William of Ockham showed the way forward, the real founder of the modern scientific method, in which observation is paramount and theory is valid only inasmuch as it agrees with observation, is usually deemed to be Francis Bacon (1561–1626). In his works 'On the Advancement of Learning' and 'Novum Organum', he held that scholars should free themselves from the past and should not be overawed by the accomplishments of Greece and Rome. Neither should they waste time in vain speculations but should concentrate on practical investigations. To acquire knowledge of the world they must rely on information given by sense experience. Bacon thought there was a need for new sciences, such as: race in relation to climate, geography, natural resources, education, history of mechanics, and sciences of business and diplomacy. Scientific invention, he thought, could be encouraged by attention to negative instances and to exceptions to accepted rules. Calendars of doubts and problems should be kept, and nature studied inductively not deductively. Human learning, Bacon thought, consisted of three main parts corresponding to the three capacities of understanding: history, based on memory, poesy, based on imagination, and philosophy, based on reason. His was a more scientific and practical approach to problems and it helped in changing men's attitudes. Much more could be achieved by making and doing than by any amount of thinking. Practical science, however, can be taken too far. According to John Aubrey, Bacon himself died of a cold caught whilst experimenting with refrigeration by stuffing a chicken with snow in an attempt to preserve its flesh, Lawson-Dick (1949, 8–16). However, his published work helped to make it respectable for educated men to study practical problems and to invest in and develop mechanical technology.

Philosophy, as is often the case, lagged behind reality. Leonardo da Vinci's work in science and on machines was soundly based on observation and experiment. In science, Copernicus had already done his work and Bacon's contemporaries Kepler and Galileo, were developing theories to fit their observations rather than looking for observations that fitted their theories. Twenty years before Bacon was born Biringuccio published his very practical treatise on metals and metallurgy. A few years later Agricola published his famous work on mining and metallurgy, and Ramelli published his book of ingenious machines. All of which were based on their author's direct observation and experience and provides much of the material discussed in other chapters of this book. Philosophers merely reflected the temper of their age.

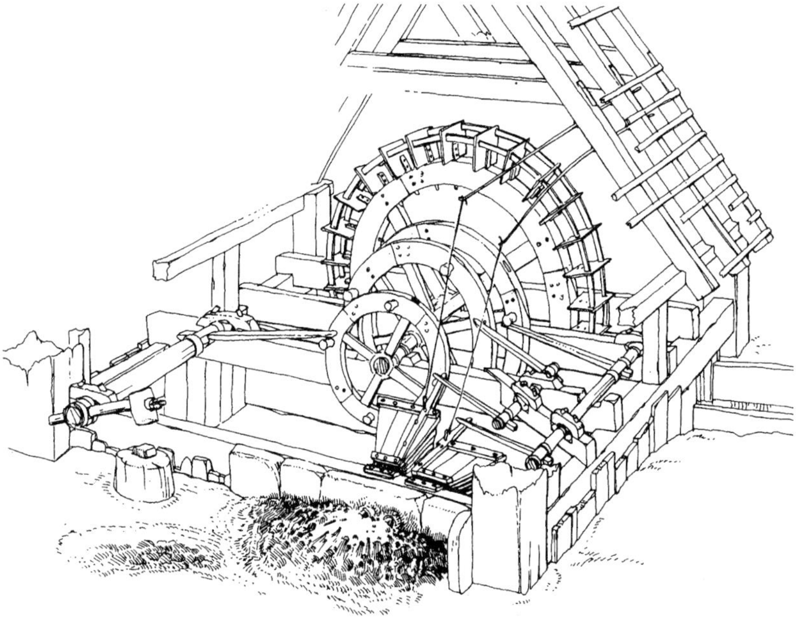

Fig. 14.18 Medieval Workshops. Top: Manual work at a waist level hearth with bellows (BM Sloane Ms 3983 f 5r). Bottom: Possible reconstruction of machinery arrangement in period 3 (Bordesly Abbey). Astill (1993). Drawn by D.R. Walsh.

Factories, in the modern sense, do not appear during the Renaissance although, as in the ancient world, mines often attained a large size and employed many hundreds, perhaps thousands, of workers. Small or medium sized workshops seemed to have been the norm, similar, no doubt, to the medieval workshops illustrated in Fig. 14.18. The reconstruction of the Bordesley abbey forge shows a degree of mechanisation that was to remain the state of the art until the nineteenth century. Some workshops, however, attained a considerable size. Biringuccio (1540, 72–3) visited a brass foundry in Milan and was astonished. "Later I saw in the same shop several apprentices and masters working at this, some of whom were hammering the said brass to make tinsel, some were laminating it to make tips for laces, some were filing it into tailor's thimbles, some were making buckles and other similar objects by casting, and there were some who were working it with the hammer into little bells, spoons, and basins, and others were tuning it into candlesticks or other vessels. In brief, some made one thing and some another, so that who ever entered that workshop and saw the activity of so many persons would, I think, believe as I did that he had entered an Inferno, nay, on the contrary, a Paradise, where there was a mirror in which sparkled all the beauty genius and power of art. Taking thought of these things with the greatest pleasure, there was not a day while I was in Milan that I did not go there and stay an hour or more. In that place I turned my eyes nowhere but I saw some ingenious novelty and beauty of craftsmanship, and thinking of the arrangement and the greatness of the new things shown to me, I was sometimes stupefied."

Biringuccio then describes the manufacture of a multi-decked mould with patterns arranged in two layers. His modern editors point out that if each built-up pattern contained 60 pieces, a single mould would contain 1200 small castings and is a good example of quantity production before the machine age. "Pondering on this", says Biringuccio, "I thought to myself that this shop alone was sufficient to furnish not only Milan but also the whole of Italy, and surely it seemed to me a splendid and fine undertaking for a single merchant. I thought that he should take great pains to keep alive all the fine undertakings that I saw in that place and continue them and surely it pleases me to see so many things being moulded continuously, and continuously being cast."

By the end of the sixteenth century, if Thomas Deloney is to be believed, England possessed at least one factory employing upwards or 850 men women and children in all aspects of the cloth trade. Deloney was, by trade, a weaver of silk but he was famous as a prolific pamphleteer and writer of ballads. "The Pleasant History of Jack of Newbury" was written in 1596 and describes a large factory belonging to John Winchcombe (Jack of Newbury) who was apprenticed to the cloth trade and by the time of his death in 1520, was one of the wealthiest men in England.

"The Pleasant History of Jack of Newbury"

Within one room being large and long,
There stood two hundred looms full strong:
Two hundred men the truth is so,
Wrought in these looms all in a row.
By every one a pretty boy,
Sat making quills with mickle joy.
And in another place hard by,
An hundred women merrily,
Were carding hard with joyful cheer,
Who singing sat with voices clear.
And in a chamber close beside,
Two hundred maidens did abide
In pettycoats of stemmell red,
And milk-white kerchers on their head:
Their smock-sleeves like to winter snow,
That on the western mountains flow,
And each sleeve with a silken band,
Was featly tied at the hand.
These pretty maids did never lin,
But in that place all day did spin:
And spinning so with voices meet,
Like nightingales they sang full sweet.
Then to another room came they,
Where children were in poor array:
And every one sat picking wool,
The finest from the course to cull:
The number was seven score and ten,
The children of poor silly [simple] men;
And these their labours to requite,
Had every one a penny at night,
Beside their meat and drink all day,
Which was to them a wondrous stay.
Within another place likewise,
Full fifty proper men he spies,
And these were shearmen every one,
Whose skill and cunning there was shown:
And hard by them there did remain,
Full fourscore rowers taking pain.
A dye-house likewise had he then,
Wherein he kept full forty men;
And likewise in his fulling mill,
Full twenty persons kept he still.
Each week ten good fat oxen he
Spent in his house for certainty:
Beside good butter, cheese, and fish,
And many another wholesome dish.
He kept a butcher all the year,
A brewer eke for ale and beer;
A baker for to bake his bread,
Which stood his household in good stead.
Five cooks within his kitchen great,
Were all the year to dress his meat.
Six scullion boys unto their hands,
To make clean dishes, pots and pans,
Beside poor children that did stay,
To turn the broachs [spits] every day.
The old man that did see this sight,
Was much amazed, as well he might;
This was a gallant clothier sure,
Whose fame forever shall endure.

It need not be doubted that such an establishment existed for Leland, some fifty years before Deloney wrote his piece, visited a very large factory at Malmesbury, Wiltshire. He says: "All the abbey buildings are now the property of a man called Stumpe, who bought them from the king [after the monasteries were dissolved]. He is an exceptionally rich clothier, whose son has married Sir Edmund Baynton's daughter. Stumpe was largely responsible for turning the abbey church into a parish church, and is its main benefactor. At the present time every corner of the enormous domestic buildings, which belonged to the abbey, is full of looms for weaving cloth, and it is Stumpe's intention to build one or two streets for clothmakers in the open ground at the back of the abbey inside the town walls. At present three thousand cloths (66,000 m) are made annually at Malmesbury." Chandler (1993, 488).

During the renaissance the ancient writings of Greece and Rome were being rediscovered by the West, usually as translations from Arabic sources. Art and architecture were advancing rapidly and learning was emerging from the monasteries to the lay population where it was seen as a pleasant and worthy occupation, as a necessity even, for the upper classes. Plato and other ancient authors were added to the writings of Aristotle, who had long been studied by medieval scholars. Greek or Roman attitudes to the mechanical arts were liable to be re-established in some quarters. The Renaissance was not wholly good, but in England, the ancient Greek and Roman disdain of experiment and the mechanical arts did not take hold. Common-sense empiricism held sway and was patronised in the seventeenth and eighteenth centuries. In 1660, the Royal Society was founded to promote experimental philosophy, physical knowledge and mathematical-experimental learning. The emphasis was on experiment. Regular weekly meetings were held at Gresham College or, during vacation, at the Temple. Robert Hooke, the most famous and ingenious mechanic of his time, the inventor of Hooke's [universal] joint, and the discoverer of Hooke's law, was their first secretary. Hooke also claims to have discovered the laws of motion usually known as Newton's laws. Apparently he informed Newton of these in a letter and was shocked when the latter published them without acknowledgement.

Typical of the new attitude to science and technology was Sir William Petty (1623–87). Petty was a founder member of the Royal Society and a promoter of the new science of political economy. His best-known work was entitled 'Political Arithmetic'. In his notebook, (Lansdown, 1927), he wrote that: "God would be much honoured:

i. By finding out the use of the fixed stars.
ii. Of the matter wherewith the Globe of the Earth is filled.

iii. The use of most animals, vegetables and minerals.
iv. The origin of man and animals.

The list continues and demonstrates the wide interests of educated men in the seventeenth century. Petty's belief that God would be honoured by these enquiries was directly contrary to the ancient view, and to the view of many clerics. Two centuries later, when Darwin published his investigations into the origins of man and animals, item four of Petty's list, the Church did not feel that God had been much honoured.

Another member of the Royal Society, although not a founder member, was John Graunt (1620–74). Together with his friend Sir William Petty, he invented the science they called political arithmetic or what we would call political economics. Graunt was a Puritan and by trade a haberdasher of small wares in the City of London. His integrity was greatly admired. According to John Aubrey, in Lawson-Dick (1949, 114–5), Graunt was "a very ingenious and studious person, and generally beloved". He made a fortune by the time he was thirty years old and was able to procure a Professorship in Music at Gresham College for his friend Dr William Petty. In 1662 he published his great work on "Natural and Political Observations made upon the Bills of Mortality—". It was a great success. Charles II himself recommended that Graunt should be elected a member of the Royal Society, saying "if they found any more such tradesmen, they should be sure to admit them all without more adoe". The Royal Society was to be open to anyone of ability, regardless of social position or wealth. After the English Civil War, the power of the monarchy was broken, although not completely destroyed, and power passed to Parliament and a newly emerging middle class with roots in business and enterprise rather than land.

In 1586, Stevin in the Netherlands had developed the important idea of force equilibrium using a "parallelogram or triangle of forces" to represent both the magnitude and direction of a force, and in the same year he published his experiment of dropping two lead balls, one ten times the weight of the other, to demonstrate that they fell through the same height in the same time, contrary to received opinion at the time. Galileo, to whom this experiment is often attributed, investigated the laws of falling bodies using inclined planes and found that velocity was proportional to time, independent of weight, and that distance fallen was proportional to the square of time. Huygens later determined the numerical value of this acceleration. Galileo investigated the strength of wooden beams and copper rods, Hooke discovered that all materials deform linearly with load up to a certain point. Newton's laws of motion set down the essential nature of force, motion, momentum and energy and the fundamental connections between them were explained. The invention of differential and integral calculus by Newton, and independently by Leibniz, and Descartes invention of co-ordinate geometry provided the essential mathematical tools of engineering. By the end of the seventeenth century, the theoretical basis of engineering was firmly established and was further enhanced in subsequent centuries.

In addition to this intellectual activity, there was a great interest in developing new forms of prime mover, based on hot air or steam. Leonardo da Vinci had already sketched a steam driven gun and proposed hot-gas or steam for propelling machines. Baptista Porta (1601), David Rivault (1608), Soloman de Caus (1615), Branca (1629), David Ramseye (1630) and Bishop Wilkins (1648) all proposed machines for raising water or experimented with steam, and the Marquis of Worcester, Huygens, Papin, and Savery, all built and tested heat engines.

The number of patents granted annually increased. This number had been steady at about six per year throughout the seventeenth century and rarely exceeded ten per year, Fig. 14.19. However, the number of patents increased steadily from about six per year in 1700 to about 100 per year in 1800 and more than 10,000 per year by 1900. 1750 seems to be the point at which a marked increase in patents occurs and on this evidence it is the period at which industrialisation begins.

The industrial revolution is the point at which I have chosen to end this history but even so it is necessary to say something on the topic. It is, after all, a high point of mechanical engineering when technological developments first became sufficiently important to displace the power politics of national governments in the popular view of history, and a few engineers became heroes that, in earlier times, may have been worshipped as gods. The industrial revolution is a term used to describe numerous factors that changed Britain into an industrial state in the period 1750–1850. French observers of the British scene first used it in the 1820's and Arnold Toynbee famously used it in 1884 in his "Lectures on the Industrial Revolution in England". It is characterised by a transition to the factory system,

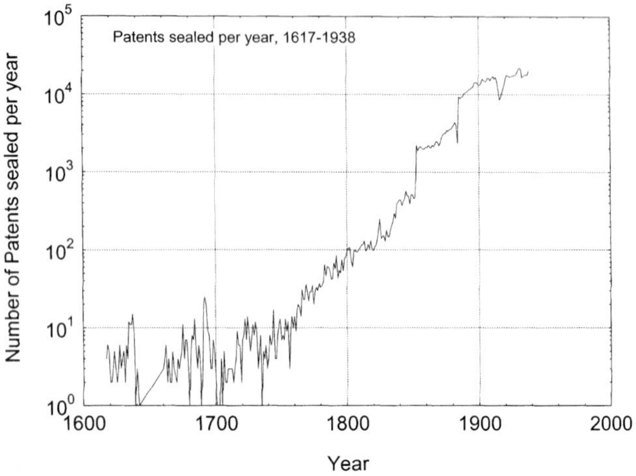

Fig. 14.19 Patents granted in England, 1617–1938.

and dependence on power driven machinery. It occurred first in the cotton and woollen industries and later in pottery manufacture, the production of iron and steel, in mining, transport, machine tools, weapons and in many other areas. The industrial revolution owed much to those engineers, inventors, and manufacturers who applied science and technology to their art to make such rapid progress that, viewed from the late nineteenth century it appeared no less than a revolution. But viewed from the twenty-first century it is clear that that the rapid progress made in technology in the period 1750–1850 has been eclipsed by that made in the period 1850–1950, which has itself been overshadowed by progress during the period 1950–2000. Technological innovation has increased continuously and it seems that no period can reasonably claim to be the most revolutionary. It would be better to refer to the process as industrial *evolution* rather than industrial *revolution*, although this is a less inspiring phrase inasmuch a book on industrial evolution is likely to sell fewer copies than the same book entitled the industrial revolution.

Associated with the industrial revolution was an "agricultural revolution", although some historians no longer think the term appropriate. The term is used to describe the transition from the medieval open-field system (with its scattered strips, two- or three-field rotation, and common pasture) to permanent consolidated enclosures. This occurred quite early in some counties but not until the nine-

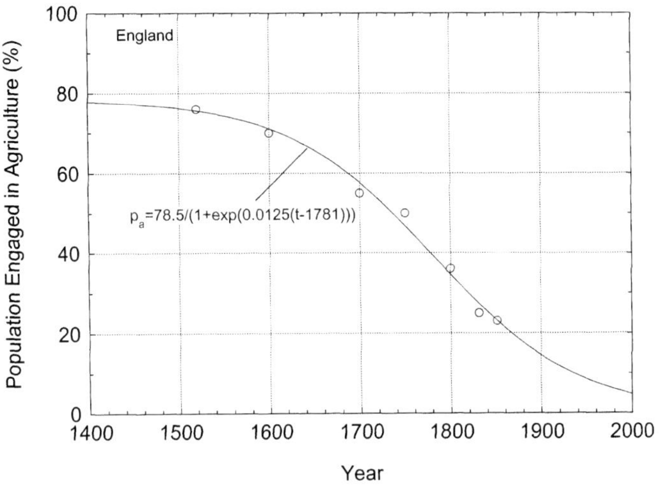

Fig. 14.20 Proportion of population engaged in agriculture in England.

teenth century in others. It also describes technical improvements in farming such as drainage and land reclamation, the use of fodder crops such as turnips and clover in place of fallow, more scientific breeding and crop rotation, the use of fertilisers and the increased use of machinery to replace human labour. The trouble is that several periods of agricultural revolution have been identified from the sixteenth century to the nineteenth century and one must accept that agriculture was in a state of permanent revolution or, more realistically, that it evolved more or less continuously over a period of several centuries. The latter view is supported by Fig. 14.20, which shows the proportion of the English population that was engaged in agriculture between 1500 and 1850 and is based on data in Overton (1996, 205–21). This shows that about 80% of the late medieval population was engaged in agriculture, the other 20% being engaged in trade, manufacture, mining, the professions, and so on. This proportion declined at an increasing rate and the decline was most marked in the period 1730–1830 and corresponds to the period usually recognised as the agricultural revolution. The mathematical function that best fits these data suggests that the most rapid decline was in 1781 and that the proportion engaged in agriculture at the present time has declined to about 4%; this is not too far from the truth. Such a smooth transition of population over many centuries

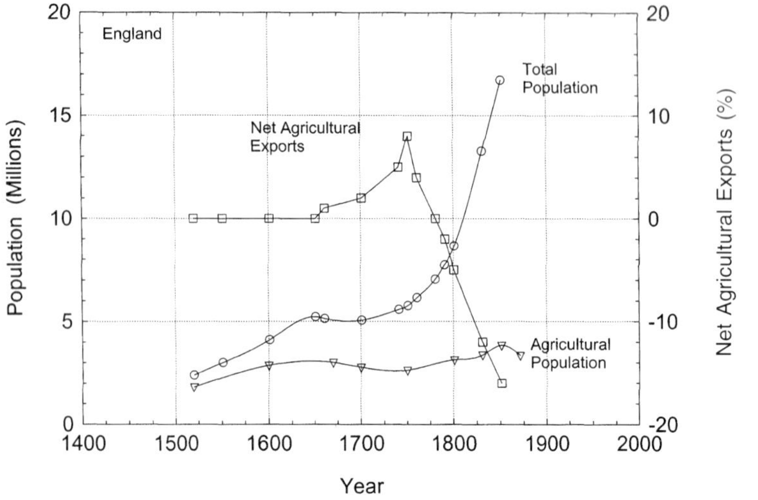

Fig. 14.21 Total population, agricultural population, and net exports/imports of agricultural produce.

suggests a steady evolution in agricultural productivity rather than a revolutionary transition.

Clearly there is a strong connection here between agricultural evolution and industrial evolution. People cannot be released from work in agriculture unless the farmers are able to grow sufficient food using a proportionately smaller workforce. Similarly, people cannot leave agriculture unless there is a demand for their labour in other sectors. Industrialisation, in general, proceeds only as fast as improvements in food production allow. Of course, if industrialisation outstrips agricultural productivity then the shortfall may be made up by agricultural imports from abroad, but this is only possible if other countries are sufficiently productive to generate a surplus. The situation is illustrated in Fig. 14.21. The population of England grew from about 2.4 million in 1500 to 5.2 million in 1650; after which it remained more or less stagnant until about 1750 before growing rapidly to about 16.7 million by 1850. The numbers engaged in agriculture did not grow anything like so fast, being about 1.8 million in 1500 and rising to only 3.8 million by 1850. The total population grew sevenfold whereas the agricultural population only doubled. It was not generally the case that people left the land to work elsewhere but that a small addition to the numbers working in agriculture were able to feed a much larger increase in the general population, who were free and able to work elsewhere in the economy. Until about 1780 English agriculture not only fed an increasing home population but also was able to export a net surplus. After this date the growth in population was too rapid and it was necessary to import agricultural produce, but even in 1850, when only about 20% of people were engaged in agriculture, food imports amounted only to about 16%.

To simplify the picture we can say that in the seventeenth century the scientific basis of engineering and manufacturing technology was first laid down and in the eighteenth-century agricultural improvement provided the food energy that enabled the ever increasing population to work in factories producing goods rather than work on the land. But already, in the seventeenth century, the damage caused by technology was apparent. Wood, needed for shipbuilding and as charcoal for metal smelting, became scarce and was replaced by coal wherever possible. Coal, brought down the coast from Newcastle and known as 'sea-coal', was commonly burnt in London in place of the meagre and expensive supplies of wood. John Evelyn, in 1661, says that the smoke from sea-coal "is so universally mixed with the otherwise wholesome and excellent air that her [London's] inhabitants breathe nothing but an impure and thick mist, accompanied by a fuliginous and filthy vapour.—While these [chimneys] are belching forth their sooty jaws, the City of London resembles the face rather of Mount Etna, the court of Vulcan, Stromboli, or the suburbs of Hell, than an assembly of rational creatures, and the imperial seat of our incomparable Monarch. Air—is here eclipsed with such a cloud of sulphur, as the Sun itself—is hardly able to penetrate—and the weary traveller, at many miles distance, sooner smells than sees the City to which he repairs. (Fumifugium, John Evelyn, 1661.)

Milton, at about the same time used the imagery of mining and the greed of the owners to depict the building of Pandaemonium in Paradise Lost. Fallen angels set to work to mine, smelt, forge, and mould metal in the soil of Hell and the devil himself is equated with Vulcan. Images of Etna, Vulcan and Hell easily spring to the critic's mind. Even enthusiasts such as Filarete say that the men working at a furnace "look like those who torment the souls of the damned". Biringuccio, as we have seen, compared the Milan brass foundry to the Inferno before correcting himself and comparing it to a Paradise, where there was a mirror in which sparkled all the beauty, genius and power of art. This ambivalent attitude has always been apparent. The smith-god Hephaestus was twice thrown from Olympus because the other Gods, while admiring and desiring the products of his workshop, apparently could not bear to have his noise and pollution. Yet he was always welcomed back. The Greeks and Romans despised technology because it was associated with slaves and undermined faith in their Gods. Medieval monks, on the other hand, embraced work and technology as a means to glorify God. In the Renaissance and Enlightenment, the intellectual support that had been denied to technology for too long was provided and theoretical science, based on observation and experiment, and the pervading empiricism of the time, enabled manufacturing technology to prosper. The scientific foundations of the industrial revolution were laid in the seventeenth century, and the necessary increase in food production to sustain an industrial workforce was provided in the eighteenth century. But already the disadvantages of these developments (consumption of scarce resources, pollution, and social discontent) were becoming clear. Technology was to remain an uneasy, if desirable, bedfellow.

CHAPTER FIFTEEN

AGRICULTURE

Introduction

Animal husbandry in the past has made few demands of the engineer, but the cultivation of crops, including animal feeds, has always demanded tools for irrigation, ploughing, harrowing, sowing, reaping, grinding, and transport, and so it is pertinent to examine the invention and spread of farming and farm implements. In this chapter we will follow the development of farm tools from the Neolithic era, when farming was first practiced, to the first half of the twentieth century, when traditional methods, represented by the horse and horse powered machinery, finally gave way to the tractor and mechanically powered machinery and when some parts of agriculture lost its traditional role as net producers of energy and became net consumers of energy.

Ancient Agriculture

The change from a hunter-gatherer economy, characteristic of the Palaeolithic period, to a food producing economy based on crop cultivation and animal husbandry, that was the defining attribute of the Neolithic period, is a fundamental change with immense implications. These implications include: a settled rather than nomadic way of life, permanent dwellings, greater wealth, increasing populations, an increased risk of warfare, and (in Europe) forest clearance. These transformations occurred neither quickly nor uniformly and therefore it is difficult to give precise dates, but the Neolithic, or New Stone Age, is usually considered to extend from about 4000 B.C. to about 2000 B.C. The rock painting shown in Fig. 15.1 illustrates remarkably well the duel economy of the Mesolithic period when these two systems coexisted. The declining Palaeolithic hunter-gatherer economy, represented on the right by the stylised hunters with their bows and by the tree with fallen fruit, was giving way to the

Fig. 15.1 Mesolithic rock picture showing the old economy (hunters and a tree with falling fruit) and the new economy (a farmer leading an animal by a halter). Albarracín, Spain.

Neolithic economy of animal husbandry, represented on the left by the man leading an animal by a halter.

It appears that farming was invented independently in the Far East, India, the Middle East, and the New World. The locations of the first farming sites in the Middle East are shown in Fig. 15.2 (top). One of the earliest sites, Zawi Chemi Shanidar, had domesticated sheep from about 9500 B.C. whereas Ali Kosh, which prospered during the eight and seventh millennia, had domesticated sheep and goats and grew wheat and barley. The remaining nine sites on the map are all of the seventh millennium and grew mainly wheat, barley and legumes. Four sites, Jericho, Beidha, Tell Ramad, and Halicar were mainly based on crop cultivation with very little evidence of animal husbandry but animals and crops are found in the remainder.

It is remarkable that farming was not invented independently in Europe but there is strong evidence that this was not the case. Carbon 14 dating of early farming sites demonstrates a chronological progression from Persia in the south-east (7500 B.C.) to Britain in the north-west (2800 B.C.). This is illustrated remarkably well in the isochronic

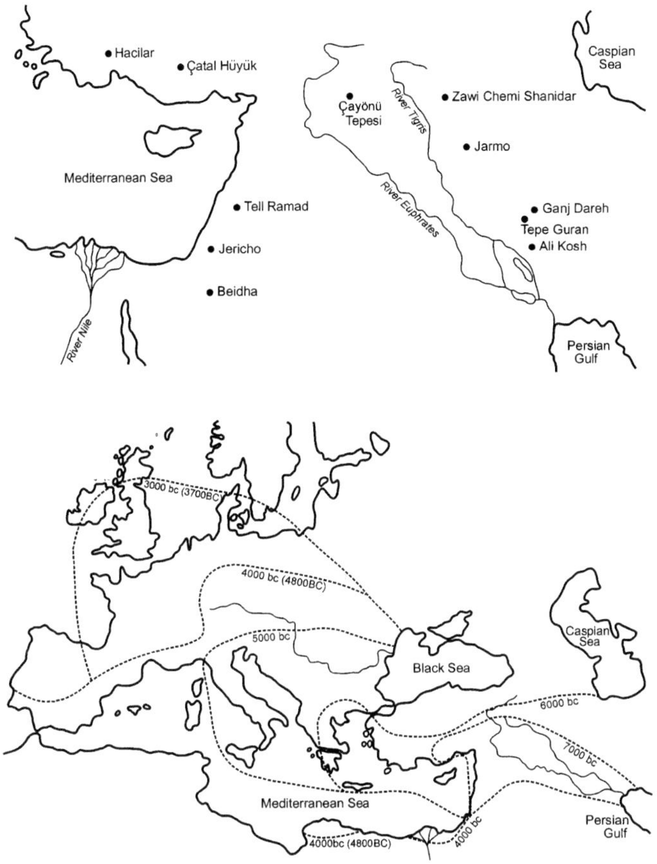

Fig. 15.2 Top: location of early farming sites. Bottom: isochronic map showing the spread of farming into Europe. After Whitehouse, in Collins (1975, 145).

map in Fig. 15.2 (bottom). Also, it is thought that the earliest crops and animals, such as emmer wheat, barley, sheep, goats, cattle, and swine, did not occur naturally together anywhere in Europe, and yet all are found in early European farming communities, suggesting that some or all were brought in from outside. Nevertheless, it is likely that some plants and animals were domesticated in Europe independently.

The mechanism for the spread of farming may be by the migration of people who brought their crops, animals and ideas with them, or it may be due to spread of farming knowledge to hunter-gatherer groups and the exchange of crops and animals. It is a characteristic of farming that it is a settled rather than a nomadic life and therefore it seems more likely that knowledge of farming spread by the conversion of hunter-gatherers than by the migration of farmers. However, there was conflict between the older Mesolithic communities and the immigrant Neolithic farmers, as discussed in chapter 21. Farming communities needed to move when the land became exhausted or if population density grew too large and this undoubtedly brought them into conflict with other groups. Of course, geography and climate were important. Mountain ranges, seas, and deserts affected the rate of spread of farming, and it may be that such influences were dominant. In Britain, there is ample evidence that the native hunter-gatherer populations learned from the Neolithic immigrants and adopted their farming methods; this process is often referred to as the Secondary Neolithic.

Some evidence for the survival of Mesolithic populations, even into modern times, was provided by the announcement, by the Oxford University Institute of Molecular Medicine, that DNA samples had been recovered from a tooth of Britain's oldest complete skeleton. This skeleton is about 9,000 years old and had been found in 1903 in the Cheddar Gorge, Somerset. Samples were also taken from the local population and it was found that the ancient DNA sample matched one taken from a history teacher, Mr Adrian Targett, and it was said to prove that he was a direct descendant of the mother of the 9,000-year-old hunter-gatherer. (Daily Telegraph, 8th March 1997.) His wife thought it might explain his culinary preference for rare-steak. Probably many such matches might be found in modern populations.

The defining characteristic of the Neolithic was farming, and in the river valleys where civilisation first blossomed it was made sufficiently productive by extensive irrigation. In Egypt the shaduf has long been used for this purpose and continues to the present time in many places in Asia, and elsewhere, Fig. 15.3. Derry and Williams (1960, 52) say that a man operating a shaduf can raise about 2.7 tonnes of water per day to a height of 2 m, although this seems rather modest. Greater lifts could be made by arranging the shadufs in series and such, apparently, was the method used to water the Hanging Gardens of Babylon. Irrigation was particularly important in the agriculture of Mesopotamia and Egypt where extensive irrigation canals and tunnels were built to carry the river water into and over the fields. Such irrigation softened the soil, so it was more easily cultivated with simple ploughs, and even carried nutrient rich alluvial deposits onto the land to keep it fertile. In other areas, where irrigation was not possible and the soil was heavier, improved ploughs were needed.

Undoubtedly farming evolved as it spread; changing to suit the varying pattern of terrain and climate. In the Near East, where farming first evolved, cereal crops were of prime importance. Hunting of wild animals continued to be a significant source of meat but this was supplemented with domesticated sheep and goats. As this spread into Europe there was little change so long as the climate and terrain remained suitable, as it mostly was around the Aegean and on some Mediterranean islands. But as farming spread northwards and westwards animal breeding became more important and sheep and goats gave way to cattle and swine. In the cooler climate of the north, barley replaced emmer wheat and einkorn as the main cereal. In the far north farming was never adopted and hunting remained the main source of food. The forms of settlements and tools also changed as eastern farming spread into Europe. The three main routes for the spread of early farming are considered to be along the northern Mediterranean coast, along the river Danube, and across the north European plain.

In the Eastern Mediterranean farming arrived about 7000 B.C. at Argissa, and about 6500 B.C. at Nikomedia. Already changes are apparent for it appears that the first domestication of cattle occurred in Greece and Anatolia although sheep and goats remained the more important. Domesticated cattle occur in almost every subsequent European culture. Broomcorn, millet and oats may also have been developed first in Greece, but otherwise the farming was similar to that of the Near East from whence it came. Another difference was in the proportion of meat provided by hunting. In the Near East this was significant but in Greece meat was supplied mainly from the domesticated herds. As in the Near East, the Neolithic Greek farmers lived in more or less permanent villages. There is no evidence for any continuation of culture from the Mesolithic to the Neolithic, which suggests that the Neolithic farms were of immigrant peoples.

On the East Coast of Italy the first farmers arrived about 6000 B.C. They were culturally different from those of Greece inasmuch as they used a different style of pottery; impressed or painted ware rather than plain ware. Farming was similar to that in Greece but

Fig. 15.3 Lifting water by shaduf (swape) in Asia, Knight (1884, 2467).

cattle and swine were more important. Single farmsteads were found but in the main farmers lived in village communities. The spread of farming again seems to be by immigration but on the West Coast of Italy it appears that the local Mesolithic population was able to adopt the new culture.

In Southeast Europe the earliest farmers were the Starcevo-Koros culture of Yugoslavia, Hungary, Rumania, and Bulgaria; they formed a link between the Neolithic cultures of the Mediterranean and those of Central Europe. This culture arrived about 6000 B.C. and spread along the Danube and its tributaries. The economy was similar to that in Greece. The most important animals were sheep and goats and the crops were emmer, einkorn, barley, perhaps millet, and some legumes. The local domestication of wild cattle was practised in Hungary. North and west of this culture, across Europe as far as Holland, there appeared the LBK (Linear Pottery) culture, formerly called Danubian I. They reached the limits of their distribution in Germany and Holland by about 4900 B.C. They were the first farming culture to be properly adapted to the temperate conditions of Europe and they were the first to concentrate on cattle breeding. Cattle accounted for some 50% to 75% of all stock. There is evidence that they practised castration of their cattle; in one area some 50% of cattle proved to be oxen. Castration enabled the size of herds to be controlled and also made them more manageable; it was an essential step towards their use as draught animals for ploughs and carts. Sheep, goats, and pigs were also bred and there was some hunting. Cereal crops were emmer wheat, einkorn, bread, club and hard wheat. Barley was relatively unimportant; peas and flax have been recorded. Although this culture spread westwards as far as Holland it did not cross the channel to reach England. This was left to the Western European cultures that spread on a more southerly route through Italy, Switzerland, and France.

The Western European cultures are Cortoillod in Switzerland, Chassey in France, Lagozza in Italy and Windmill Hill in England. The Chassey culture arrived in France about 4600 B.C. and the Windmill Hill culture reached England about 4300 B.C. These cultures are characterised by mixed farming with emphasis on cattle breeding but sheep, goat, and pig are also bred. The cereal crops are emmer, einkorn, bread and spelt wheat and barley. For the Swiss Cortoillod culture hunting and gathering of wild fruits remained important but in the Chassey and Windmill Hill cultures it was unimportant.

The Windmill Hill people are named after one of their settlements at Windmill Hill in Wiltshire, Fig. 15.4 (right), and they are likely to have come from northern France, Germany, or the Low Countries. Their site in Wiltshire is overshadowed by a causewayed camp which has been dated to 2570±150 B.C. which is already some 500 years later than their first settlement. This causewayed camp consists of three circular earth banks and ditches, broken at various points by a causeway to give easy access to the interior. Its purpose is unknown. It is not a defensive position or fort, despite its hilltop site, but is thought to be either an enclosure for the shelter of cattle or a religious centre. Around this causewayed camp were the plots of the village community. They lived in rough, timber and thatch huts, and cultivated a small patch of ground using digging sticks, Fig. 15.4 (left). Digging sticks used in primitive societies of the modern day are usually weighted and it is imagined that such sticks were also

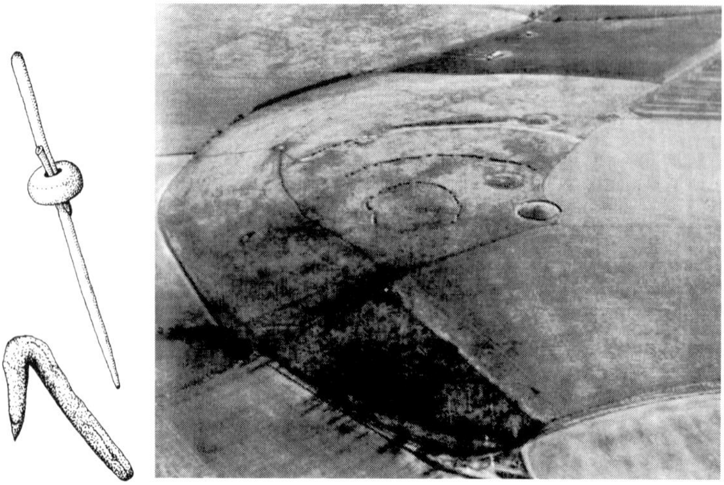

Fig. 15.4 Left: an Egyptian hoe and a weighted digging stick. No truly prehistoric digging sticks have survived. This example was used by South African Bushman. Right: aerial view of Windmill Hill, one of Britain's earliest farming communities.

used to scratch the surface of the soil in the earliest Neolithic societies. Wooden digging sticks, as such, have not survived from the Neolithic but we know that they were growing primitive species of wheat and barley and what appear to be weights for digging sticks have been found. Ploughs pulled by oxen occur in the second millennium B.C., as shown for example by a rock painting at Mont-Bégo, Italy, Fig. 15.5. This was found high in the Ligurian mountains to the north of Lake Garda, along with many other rock pictures, and at an altitude at which farming was not possible. Kühn (1956, 133–4) was much struck by the altitude at which these agricultural scenes were found and supposed that there was some realisation of the existence of a divinity living at the summits that the artist must reverence and respect.

Meanwhile, in Egypt, Mesopotamia, and Anatolia, farming practice had developed further and although their agriculture varied significantly from each other, yet from an engineering point of view they may be treated in common. They were based, mainly, on cereal production and the engineering equipment used related to irrigation, ploughing, tilling, harvesting, winnowing, grinding, and transport. Some idea of these early farms may be obtained because detailed inventories of third millennium farms and estates exist. Gurney (1990,

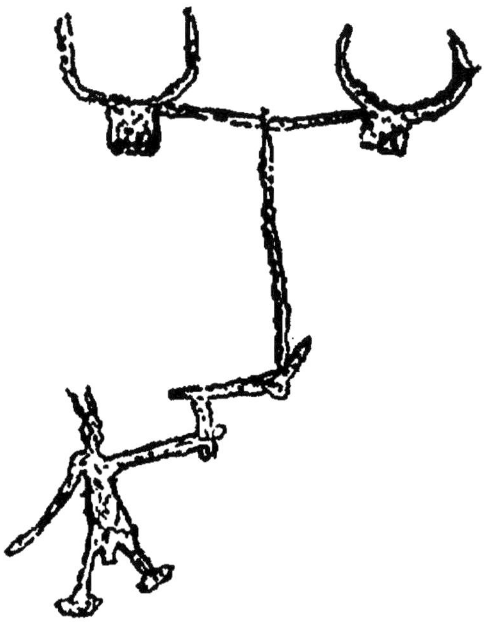

Fig. 15.5 Rock picture showing two oxen pulling a simple plough and guided by a ploughman. Mont-Bégo, Italy. Second millennium B.C.

66) quotes a typical paragraph from a Hittite land deed, which describes a small estate. "Estate of Tiwataparas: 1 man, Tiwataparas; 1 boy, Hartuwandulis; 1 woman, Azzias, 2 girls, Anittis and Hantawiyas; (total) 5 person; 2 oxen, 22 sheep, 6 draught-oxen,—; [18] ewes, and with the ewes 2 female lambs, and with the rams 2 male lambs; 18 goats, and with the goats 4 kids, and with the he-goat 1 kid; (total) 36 small cattle; 1 house. As pasture for oxen, 1 acre of meadow in the town Parkalla. 3.5 acres of vineyard, and in it 40 apple-trees, 42 pomegranate-trees, in the town of Hanzusra, belonging to the estate of Hattapis."

The farm evidently supports a family of five living in a farmhouse. A few oxen are kept, mainly as draught animals to pull the plough or the cart, but the main livestock are sheep and goats. There is a one acre [0.4 ha] pasture for the oxen in the town of Parkalla. No pasture is mentioned for the sheep or goats so probably they roamed as a mixed herd on the common land, or the fallow land, of the village. Also there is a 3.5 acre [1.4 ha] vineyard containing 40 apple trees and 42 pomegranate trees in the town of Hanzusra.

In Mesopotamia the texts demonstrate that oxen were used to plough the land before sowing and then again in the seeding process. The ox-plough was already in use in the Uruk period (4000–3200 B.C.), as the existence of the cuneiform sign proves, and it may have been used as early as the middle Ubaid levels (before 4000 B.C.) at Ras al Amiya. Cereal production dominated agriculture so every farming family needed a few cattle for this purpose, but not many, for there was no point in keeping large herds of cattle, since feeding them round the year would be difficult, and their hides are not so easily processed domestically as wool. Lists of household property often include a flock of sheep and goats, but rarely more than a few cows. Like all draught animals that work with man, they were given names. Mesopotamian texts from about 1800 B.C. also mention shrine tables, grindstones, wagons, wooden doors, ladders, and so on, see for example, Postgate (1992, Text 5.5).

Egyptian tomb paintings are prolific in agricultural scenes that tell us a great deal about their agricultural methods, as those shown in Fig. 15.6 demonstrate. The lower painting is from the tomb of Nakht at Sheikh-Abd-El-Gurna, near Thebes. A complete agricultural cycle is shown. At the top of the lower panel men are bringing land into cultivation by clearing trees, brushwood and weeds, below this agricultural workers plough Nakht's fields using a primitive plough drawn by a cow and under the watchful eye of a supervisor seated in a booth. Preparations for harvest sacrifice are before him. Like all ploughs it requires a vertical force for the share to penetrate the soil and to prevent the draught force exerted by the beast from turning the plough about its tip in the ground. This particular arrangement, however, has a lever ratio about the share tip that is close to unity so the force exerted by the beast is about equal to the force exerted by the ploughman; it seems designed to give the ploughman a backache. Later ploughs had a more generous lever ratio so a small force exerted by the ploughman applied a much larger force to the beast. In Fig. 15.6 the plough is fastened to the beast at the horns.

In the fourth panel men reap corn (right) while a girl gathers the fallen corn-ears; in the centre two men have filled a basket with corn and are about to carry it on their shoulders to the threshing floor. On the left women gather flax. The upper three panels in Fig. 15.6 are from a Theban tomb of the 18th dynasty (about 1415 B.C.) and the lower panel of which shows a group of men, rather than beasts, ploughing a field for a second time, so the soil may be presumed

Fig. 15.6 Agricultural life in Egypt of the New Kingdom.

quite soft, and sowing the seed by broadcasting. To the left workmen are hoeing a field. In the middle panel workmen are reaping corn using hand sickles, two girls are gathering the fallen corn-ears, and a women is bringing food and drink to the workers. In the top panel two men remove a full basket of corn to the threshing floor and two are returning with a carrying pole and empty basket. On the left threshing is done by means of oxen trampling over the corn spread on a threshing floor. After threshing the corn is winnowed by throwing it into the air to allow the wind to blow away the chaff. Corn may be also separated from its husk by drying, perhaps in gentle heat, until it "pops"; the Bible frequently refers to "parched corn". But a better method of consuming corn is to convert it into bread. The ample survival of saddle stones, such as that shown in use in chapter four Fig. 4.2, confirms that bread was commonly made.

Mesopotamia and Anatolia lacks the visual record of Egyptian wall paintings, but the surviving texts show that they seem usually to have been ahead of Egypt in the development of technology. Agriculture was carefully regulated as the following abstracts from the Hittite Law Code, quoted from Gurney (1990, 66), testify. "If anyone hires a plough-ox, for one month its hire is one shekel of silver. If anyone harnesses a yoke of oxen, its hire is one half 'parisu' of barley. If anyone borrows and yokes an ox, a horse, a mule, or a donkey and it dies or a wolf devours it, or it goes astray, he shall pay for its full value; but if he says 'By the hand of a god it died', then he shall take the oath. If a pig goes on the threshing-floor or a field or a garden, and the owner of the meadow, the field or the garden strikes it so that it die, he shall give it back to its owner; but if he does not give it back, he becomes a thief. If anyone [lights] a fire and the fire seizes a ripe vineyard, if vines, apples, pomegranates, or medlars (?) burn up, for one tree he shall give [six] shekels of silver and re-plant the plantation. If it be a slave he shall give three shekels of silver."

As mentioned above, cereal cultivation was fundamental to agriculture in the ancient civilisations and this is made clear in a composition called "The Farmer's Instructions", written in Sumeria in the Ur III period (2150–2000 B.C.), that sets out agricultural practice in the form of a farmer's instructions to his son. Every detail is carefully set out and it is clear that a high standard of maintenance of the farm tools and equipment is expected. In Mesopotamia, the farmer's year begins in late summer, when the field, which has lain

Fig. 15.7 Seal impression of a clay tablet from Nippur (1800–1600 B.C.) showing an old Babylonian plough in use with an attached funnel to combine the second ploughing of a field with sowing. After A T Clay, University Museum, Philadelphia.

fallow the previous year, is broken up and irrigated to soften the ground. There follows the sowing of barley, then the smaller area of emmer and wheat. The 'Farmer's Instructions' give details of the seeder-plough, see Fig. 15.7, and how it should be maintained. It appears that a funnel was attached to the plough so that at the second ploughing the seed might be dropped down the funnel and into position, one by one, at the requisite intervals. This was much more economical than the Egyptian system of broadcasting. Such traditional seeder-ploughs are still used in eastern Turkey. As we have seen, cattle and sometimes donkeys were used to haul the plough.

Early Dynastic texts from Lagash show that farmers calculated the amount of seed corn needed for a given area, Postgate (1992, 167). The distance between the seeds was known, the number of furrows in a field could be varied by increasing or reducing the space between them, and on this basis the distance to be travelled by the oxen, the time taken, and the seed corn and animal fodder needed was estimated. On Ur III estates the usual distance between rows was 60–75 cm, which is quite large. The main consideration was to achieve the best yield rate in relation to the seed corn and not the maximum yield per surface area of field. Postgate does not doubt that such calculations were carried out even at the relatively lowly level of the farm, for the widespread use of labels or tags is a clear indication of literacy. Clay lumps formed round string were attached to tablet containers, or even to cows; some with dates and names, which were presumably hung round the neck or wrist of harvesters, brick-yard workers, and craftsmen, as a record of attendance. Thus the immediate supervisors of labour, even if they could not write themselves, had a scribe with them on the job, and so the effects of writing had, by 2300 B.C., reached the most commonplace levels of society.

To minimise the time lost in turning a plough with its team of oxen the ploughman usually chose to cut long furrows and a day's ploughing thus resulted in a rather narrow strip of cultivated land. In Mesopotamia, as elsewhere, the narrow edge abutted the canal banks to ensure as many farmers as possible had direct access to water. Postgate notes that Ur III texts confirm that this was indeed the shape of their fields.

The harvest was cut using a simple sickle and Postgate (1992, 225) notes that its development illustrates the close relation between technology and agriculture. In the Natufian, 8000 B.C. or earlier, flint blades (micro-burins) were set into bone handles to create a sophisticated one-handed implement for cutting stems. Such sickles were also used in Neolithic Europe. In the Ubaid period (5000–4000 B.C.) the clay sickle was introduced and proved, by its survival into the Uruk period (4000–3200 B.C.), to be a superior product. The clay sickle's advantage lay with its easy and cheap manufacture, coupled with an inexhaustible supply of the raw material. However, in the Early Dynastic period (3000–2350 B.C.) it was replaced by a revival of the composite flint sickle, the flint blades, now serrated, being set into a wooden handle with bitumen. In their turn these were displaced by copper sickles, introduced in Early Dynastic times and very common in the Ur III period (2150–2000 B.C.). A text mentions 1083 old copper sickles and 60 copper hoes sent for refurbishment by the administration at Ur to the smiths from the storeroom. Copper sickles were much better than flint or clay sickles, and no doubt the palace and the temple authorities, as the main capitalists, were important in the substitution of metal tools and also for the introduction of refinement of the plough and its seeder funnel. The temple and palace were large employers of labour and must always have had the better tools, for they were able to invest labour and capital in the metal industry, and to initiate and search for metal supplies, whether from Oman, Cyprus, or Anatolia.

As we have seen, after harvesting, the corn was transferred to the threshing floor in baskets, Fig. 15.6 where it was threshed by trampling under the hooves of animals. Winnowing, Fig. 15.8, was undertaken when a suitable wind could be used to separate the chaff from the grain and it seems that the winnowing fan was not used although

Fig. 15.8 Agricultural life in the New Kingdom.
Top: men winnowing (top panel) and measuring grain under the supervision of Nakht. Bottom: grape harvest, treading grapes, and transfer to storage jars (top). Catching of wildfowl in a pond and preparing it for the kitchen (birds, like the wine, were stored in large jars). About 1415 B.C.

Fig. 15.9 Early forms of wringing cloth and bag press used for juice extraction.

cloths were sometimes wafted to create a suitable breeze. The grain was then collected, measured, and stored, usually under strict supervision. When required, which would be almost a daily event, corn was withdrawn from storage and ground by hand using a saddle stone, as illustrated by Fig. 4.2 of chapter 4 which shows an Egyptian servant girl grinding corn. Such saddle stones have a long history dating back to the beginning of the Neolithic period. After grinding, the flour was sieved to remove grit and other impurities before it could be used in the bake house.

Although corn was the main crop for which Egypt was justly famous many other crops were grown. Fig. 15.8 (bottom) illustrates a grape harvest with the grapes being trodden underfoot to release the juice and the juice being collected into storage jars. The bottom panel shows wildfowl being caught on a pond and their subsequent preparation and storage in jars. A certain degree of mechanisation of juice extraction was attained by the use of a wringing cloth or bag press, Fig. 15.9. Forbes[3] (1955, 110).

Greece and Rome

Cattle were all-important in Homeric Greece and their number reckoned a man's wealth. This changed, however, and cultivation of the land and consumption of bread rather than flesh came to be looked upon as more civilised. White (1967, 126) says that all ploughs are one of two basic types: the breaking plough or "ard", and the turning or mouldboard plough, each of which has numerous variations to suit local conditions. The ard is symmetrical in design and when

Fig. 15.10 Greek ploughs, men hoeing, and the transportation of oil or wine jars, from a sixth century B.C. black-figured cup. Louvre, Paris, After Rostovtzeff (1945, 198).

pulled through the ground throws up soil on both sides. The turning plough, on the other hand, is asymmetrical and a curved mouldboard on the furrow side of the plough is designed to turn the slice of earth removed by the combined action of the share and the coulter. The coulter is a vertical blade that cuts the soil ahead of the share and the share is a horizontal blade that cuts the soil at a certain depth. Complete inversion of the cut sod is necessary in wet, temperate regions where heavy soils are encountered, and here the mouldboard plough is supreme, but on light soils in semi-arid conditions, where crops are grown in soil moistened by winter rain, the ard is more useful. Usually the ard, being used on lighter soils, requires less draught and fewer animals whereas the mouldboard plough, on heavy clay soils, may require up to eight oxen. The ard was commonly used in Greece, Fig. 15.10, and the vertical force, necessary to drive the share into the soil and to prevent the ard from tipping over as it is pulled by the draught animals, is provided by the ploughman stepping on the sole of the plough. It is a marked improvement on the ancient Egyptian plough.

Hesiod (*Works and Days*, 415–600), writing in the eight century B.C., provides some information on early ploughs and ploughing. He says that in October: "—remember to hew your timber: It is the season for that work.—Hew also many bent timbers, and bring home a plough-tree when you have found it, and look out on the mountain or in the field for one of holm-oak; for this is the strongest for oxen to plough with when one of Athena's handmen has fixed the share-beam and fastened it to the pole with dowels. Get two ploughs ready and work on them at home, one of a piece, and the other jointed. It is better to do this, for if you should break one of them,

Fig. 15.11 One-piece ploughs (Roman).

you can put the oxen to the other. Poles of laurel or elm are the most free from worm, and a share-beam of oak, and a plough-tree of holm oak. Get two oxen, bulls of nine years; for their strength is unspent and they are in the prime of their age: they are best for work. They will not fight in the furrow and break the plough and then leave the work undone." The plough shown in Fig. 15.10 is a joined plough of several parts dowelled together but often a suitably shaped branch could be found to make a one piece plough, as suggested by Hesiod and as shown in Fig. 15.11.

Ploughing, Hesiod says, should begin "when you hear the voice of the crane [middle of November]", and "Pray to Zeus of the Earth and to pure Demeter to make Demeter's holy grain sound and heavy, when you first begin your ploughing, when you hold your hand to the plough-tail and bring down your stick on the backs of the oxen as they draw on the pole-bar by the yoke straps. Let a slave follow a little behind with a mattock and make trouble for the birds by hiding the seed". "But if you plough the ground at the solstice [in December], you will reap sitting, grasping a thin crop in your hand, binding the sheaves awry, dust-covered, not glad at all; so you will bring home a basket and not many will admire you.—for if you plough late, you may find this remedy—when the cuckoo first calls [in March] in the leaves of the oak and makes men glad all over the boundless earth, if Zeus should send rain on the third day and not cease until it rises above an ox's hoof nor falls short of it, then the late-plougher will vie with the early. Keep all this in mind, and not fail to mark the grey spring as it comes and the season of rain." He urges: "Pass by the smithy and its crowded lounge in winter time when the cold keeps men from the field work,—for then an industrious man can greatly prosper his house", and he advises: "Avoid the month of Lenaeon [late January early February], wretched days, all of them fit to skin an ox, and the frosts which are cruel when Boreas blows over the earth."

After its useful days before the plough were over, the ox was usually killed for meat, but a fourth century B.C. poem by Addaeus of Macedonia suggests a deeper sentiment. "His yoke-ox, growing feeble with age and years of ploughing,/Alcon its master pitied, and saved from the blood-stained knife,/Honouring its true service; mid the meadow grasses' soughing/Knee deep it lows, contented and furrow-free for life", Lucas (1951, 280).

Glotz (1965, 250) says that the great majority of Athenian citizens lived on their land or from their land. In the seventh century they almost all lived in the country and seldom went to town, except for protection in times of emergency. Agriculture had become more scientific in the fifth and fourth centuries when a three-year rotation of crops was espoused and Attica, the rural district around Athens, was usually able to provide sufficient corn, olives, and wine, for its citizens. Of 42,000 Athenian citizens in 431 B.C. 1,000 were major landowners, 22,000 had medium sized estates, and about 20,000 were poor workers having only part of a field. Attica, however, was not sufficiently large when Athens was most prosperous and crowded so corn had to be imported. Corn occupied up to 20% of the total area and the practice of fallow reduced the productive area to about 60,000 acres [24,000 hectare] in any one year. Glotz assumes a yield of about 10 bushels per acre [907 litre per hectare] and a return of five or six to one, although this was not always attained because the comedic poet Philemon has a farmer complain of his land, "For twenty medimni [sown] it gives back only thirteen. It is a thief."

In the Hellenistic period progress was made in the development of water lifting devices for irrigation. The Egyptian shaduf was simple to make but limited in its lift and in the quantity of water it provided. Vitruvius, *On Architecture 19.4–7*, describes five devices, namely: a tympanum, a noria, a bucket train, an Archimedean screw, and a force pump. The latter two devices are discussed in relation to mining and screw manufacture, and so we will illustrate here only the tympanum, the noria, and the bucket train.

The tympanum is illustrated in Fig. 15.12 (left). Vitruvius says "This device, to be sure, does not lift water to a great height but it discharges a great amount quickly. The axle is turned on a lathe or

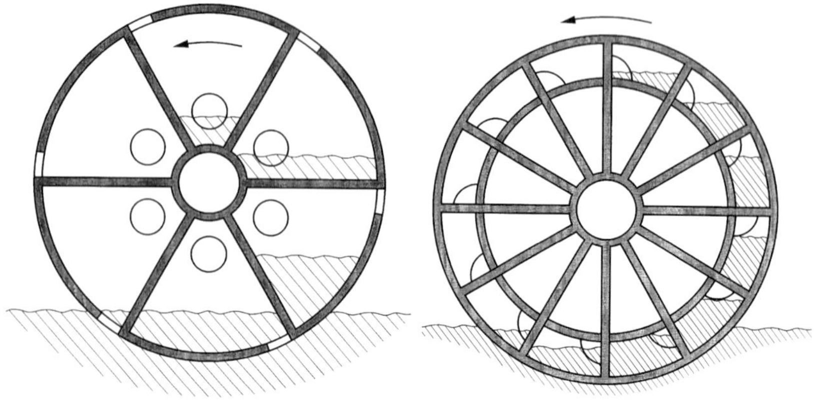

Fig. 15.12 Water lifting devices described by Vitruvius, first century B.C. Left: a Roman tympanum (drum). Right: a Roman noria.

made by means of compasses and its ends are capped with iron sheeting. Around the middle it has a drum made of planks joined together, and is mounted on beams which have iron bearings to carry the axle ends. In the interior of this drum is a set of eight radial partitions [the figure shows only six] running from the axle all the way to the circumference, dividing the interior into equal sections. Around the outer surface are fixed planks with six inch [15 cm] openings for receiving the water. Likewise, close to the axle there are small round holes in one side corresponding to each compartment. When this device has been pitched ship-fashion [sealed] it is set in motion by men treading it. It scoops up the water through the openings around the circumference and discharges it through the circular openings near the axle into a wooden trough [not shown] connected with a conduit. In this manner a quantity of water is provided to gardens for irrigation or to salt pans for dilution." Humphrey et al. [1999, 311].

The noria, Fig. 15.12 (right) is similar to the tympanum. Vitruvius says, "When, however, the water has to be raised higher, the same principle will be put to use in this manner. A wheel will be built around the axle of a large enough diameter so that it can reach the height that is required. Rectangular compartments will be fixed around the circumference of the wheel and made tight with pitch and wax. Thus when the wheel is turned by men treading it, the containers will be carried up full to the top of the wheel and on their downwards turn will pour out into a reservoir what they have themselves raised." Humphrey et al. (1999, 311).

Both the tympanum and the noria were intended to be driven by men walking and it is usually assumed that they would stand on top of the drum, opposing rotation, and walk towards the centre thus rotating the drum. Unfortunately this is an unstable position. There is a balance point at which the weight of the man (on the left in Fig. 15.12) just balances the weight of water. He must step backwards to increase his distance from the axis rotation to turn the wheel and raise the water, but the turning wheel carries him to an even greater distance from the axle and thus he accelerates away from the balance point and may be thrown off the wheel. If he steps forwards from the balance point he reduces his distance from the axis of rotation and the weight of water reverses the wheel and accelerates him in the opposite direction. Clearly, if the men treading the wheel were on the outside then they must have some independent support. A much better idea would be to have them on the inside of the treadmill, which is the normal, stable, position, but neither the tympanum not the noria lend themselves to this arrangement.

Vitruvius then describes the bucket train, Fig. 15.13. "But if a supply is required at still greater heights, a double iron chain will be set up, wound round an axle of the same sort of wheel and allowed to hang down to the lowest level, with bronze buckets the capacity of a *congius* suspended from it. Thus the turning of the wheel, by winding the chain over the axle, will carry the buckets to the top, and as they are borne over the wheel they will necessarily turn over and pour out into a reservoir what they have raised." Humphrey et al. (1999, 311). Recently archaeologists from the Museum of London have discovered such a bucket train a little to the east of St Paul's cathedral and close to the sites of London's Roman fort and amphitheatre (*Daily Telegraph*, Sept. 27th 2001, p. 18), Fig. 15.13. The machinery was found in two adjacent in-filled wells. Tree-ring dating showed that the earliest well was built about 63 A.D. and was used for about 10 years and the later one about built in about 109 A.D. Water was drawn in about 24 oak boxes each holding about two litres and were fixed in a wrought iron loop using wood and metal pins. The well was about 5 m deep. The driving mechanism has not survived but one or more slaves in a treadmill probably supplied the power. The archaeologists estimate that it was capable of lifting about 270 tonnes of water per day.

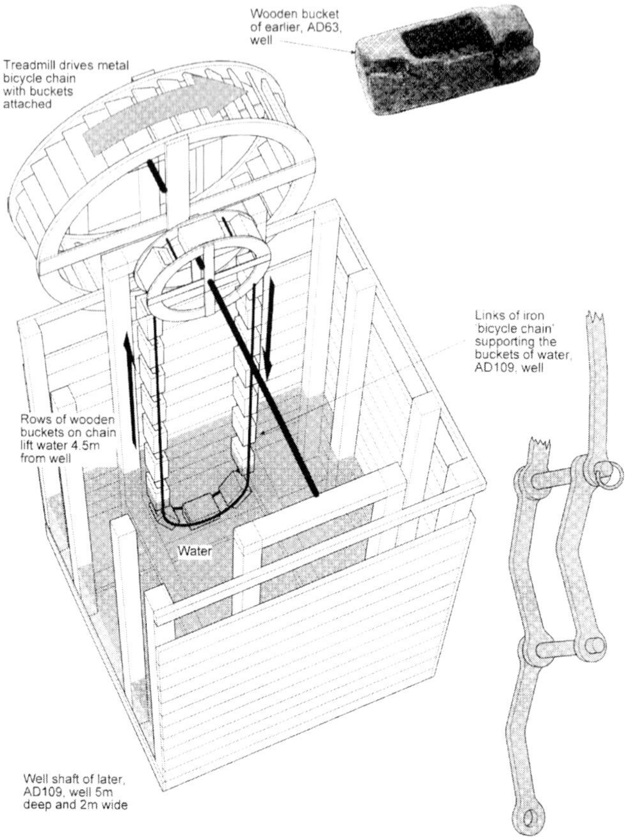

Fig. 15.13 Roman bucket train of 109 A.D. found in London.

The only description of a Roman plough is in Virgil, White (1967, 125), and by this time iron shares were commonly used to reduce wear (not to reduce friction as White, p. 130, suggests). Suitable wood was bent while still growing to give the desired shape. Virgil writes: "From the first, in the woods an elm, bent by main force, is trained for the plough beam, and receives the form of the crooked [curved] plough. To the base is fitted a pole, eight feet [2.4 m] in length, and a pair of 'ears', and a share-beam with a double back, and a stilt to turn the base of the car from the rear." It is not much but it is all we have, but although not many ancient ploughs have survived in Italy, there are numerous from Roman Britain. Moreover, Pliny (18, 171–2) has provided a very good description of the types of ploughshares used by the Romans. "There are several kinds of ploughshare. The 'knife' is the name given to the curved blade used for cutting through the dense soil before it is broken, and for marking out the tracks for the subsequent furrows with incisions which the share, sloping backwards, is to bite out in the course of ploughing. The second type is the ordinary share consisting of a bar sharpened to a point like a beak. The third type for use on light soils does not extend along the whole length of the share-beam, but has only a small spike at the extremity. In the fourth kind this spike is broader and sharper, ending in a point, and using the same blade to break up the ground and with its sharp sides to cut off the roots of the weeds." The impression given here is that the knife (coulter), which produces a vertical cut, is used as a separate operation before the main ploughing. Later ploughs, of course, combine the vertical coulter and the horizontal share on the same plough.

Pliny also mentions a newly invented wheeled-plough that was to become very common on the heavier soils of northwest Europe. "Recently a contrivance has been invented in the Raetian area of Gaul, consisting of an addition to this type of plough [the fourth type mentioned above] of two small wheels; the local name for this type is *plaumoratum*. The share is spade-shaped", White (1967, 132). It was an important invention that made it easier to plough at a fixed depth and also eased the load demanded from the ploughman and his beast. As we will see, this was not the only agricultural innovation to come from Gaul.

The most vulnerable part of the plough, apart from the coulter and share that wore out quite often unless shod with iron, seems to have been the plough-beam. Varro says "There are places where the oxen cannot break (at first ploughing) unless they are unusually powerful, and frequently they leave the plough in the field with a broken plough-beam", and Columella advises: "If you plough in a drought you will have trouble. In very dry ground the point of the plough is rejected by the hard ground; if it does enter at some point, it does not break the soil into fine particles, but rips up huge clods—(these clods affect the second ploughing) because the plough share is driven out of the furrow by the weight of the clods—so that—hard skips are left and the oxen are severely injured by the uneven strain".

The symmetrical ard is unable to form ridges like the mouldboard plough and extra mouldboards were fitted when ridges were required to cover the seed. Varro tells us: "When they plough the third time,

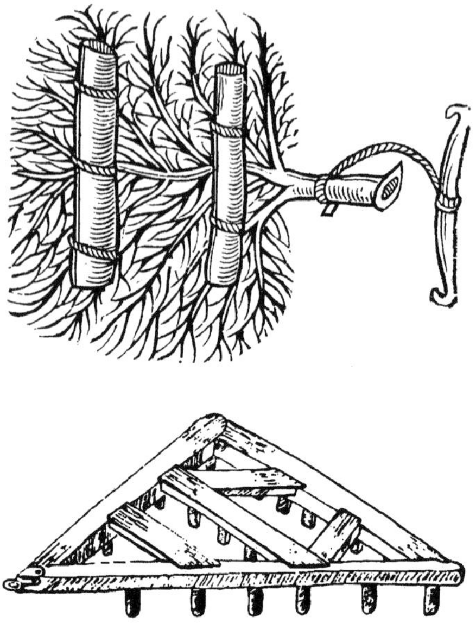

Fig. 15.14 Top: a simple bush harrow weighted with logs. Bottom: an Italian irpex (heavy harrow).

after the seed has been broadcast, the oxen are said to 'ridge', that is, with boards attached to the ploughshare they both cover the broadcast seed in ridges and at the same time cut ditches to allow the rain-water to drain off."

Pliny (18.9) says that "A parcel of land that one pair of oxen could plough in a day used to be called a *iugerum* [two-thirds of an acre or 0.27 hectare]. The distance over which oxen could draw a plough in a single haul of reasonable length was called an *actus* [40 yards or 36.6 m]." This is rather short compared to the length of furrow normally used in medieval Europe, which was 220 yards [201 m]. It appears that the ard, pulled by a pair of oxen, was relatively easy to turn at the end of a furrow whereas the heavier mouldboard plough with wheels and drawn by up to four pairs of oxen was more difficult to turn and the ploughman chose to make fewer turns per day by cutting longer furrows. It seems, therefore, that a day's ploughing using an ard and a pair of oxen amounted to about 40 × 80 yards [37 m × 73 m].

Ploughing was done twice and sometimes three times so on light soils no further work was necessary to prepare a seedbed. However, some soils required a reasonably heavy implement (a harrow) to break up the clods after ploughing and to level off the seedbed. Virgil, writing on the preparation of soils for growing wheat, says "He also much improves the fields who breaks the sluggish clods with drag-hoes and draws bush harrows over them", and Pliny writes "There are some kinds of soil which are so luxuriant that it becomes necessary to comb the crop while in leaf—the comb is another kind of harrow fitted with pointed iron teeth", White (1967, 146). The bush harrow, Fig. 15.14 (top), seems to be the earlier device and was drawn across a ploughed field by men or by animals after the clods had been broken up by hoeing. To make it more effective it was often weighted with logs. Such harrows continued into modern times. A heavier harrow, which eliminated the need for hoeing, consisted of a triangular or rectangular frame fitted with wooden or metal teeth, Fig. 15.14 (bottom), and was drawn over the ploughed field by draught animals.

When the corn was ready for cutting simple hand sickles were used, initially, but scythes eventually replace these. Sickles required about four man-days to reap an acre but the scythe reduced this to about 2 man-days per acre. However, the most famous agricultural invention of the Roman era was the "Gallic stripper", a device for reaping the ears of corn while leaving the stalks in the field, Fig. 15.15. Pliny (18.296) described this device in the first century A.D. "There are various methods of harvesting", he says. "On large estates in the provinces of Gaul huge frames with toothed edges are mounted on two wheels and driven through the corn by a team of oxen pushing from behind; the ears of corn are ripped off and fall into the frame. Elsewhere the stalks are cut through with a sickle and the ears are stripped off between two pitchforks. In some places the stalks are cut off at the root, in others the roots are pulled up as well."

Pliny does not say why the reaping machine came into use in Gaul but it survived into the fifth century A.D. when Palladius (7.2.2–4) describes this machine in more detail and says its main purpose was to save labour and time. Certainly, slave labour was scarce at this period and, in the less assured weather conditions of Gaul, it was important to gather the harvest quickly when conditions were suitable. "In the more level plains of the Gallic provinces they employ the following short cut or labour-saving device for harvesting. With the aid of a single ox the machine outstrips the effort of labourers

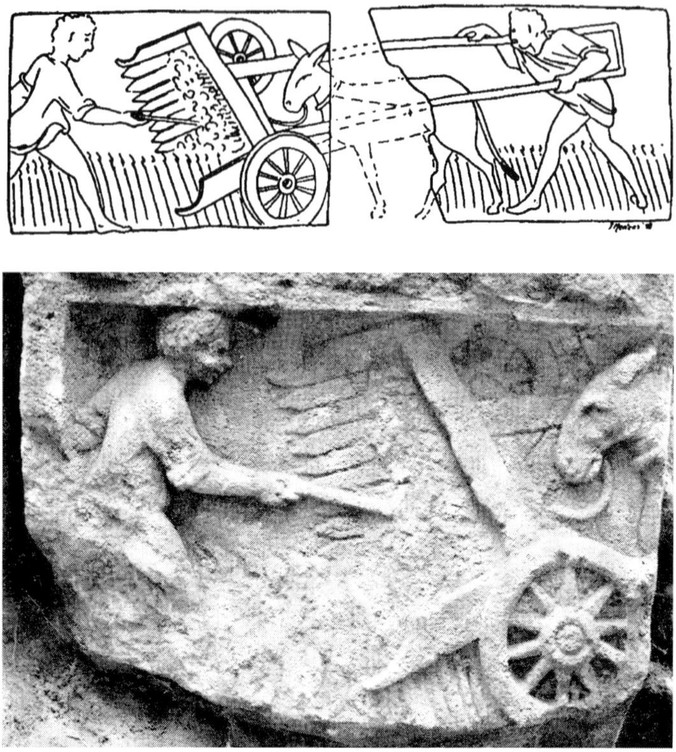

Fig. 15.15 Reconstruction of the Trier *Vallus*, which shows a Roman reaping machine in use.

and cuts down the time of the entire harvesting operation. They construct a cart carried on two small wheels. The square surface of the cart is made up of planks, which slope outwards from the bottom, and so provide a larger space at the top. The height of the planks is lower at the front of this container; at this point a large number of teeth with spaces between are set up in a line to match the height of the ears; they are bent back at the tips. At the back of the vehicle are fastened two very small yoke beams, like the poles of a litter; at this point an ox is attached by means of a yoke and chains, with his head pointing towards the cart; he must be docile, so that he will not exceed the pace set by the driver. When the latter begins to drive the vehicle through the standing corn, all the ears are seized by the teeth and piled up in the cart, leaving the straw cut off in the field, the varying height of the cut being con-

Traha (Ancient Egyptian)

Tribulum (Cypriote)

Fig. 15.16 Threshing drags of the type mentioned by various Roman authors.

trolled from time to time by the cowherd who walks behind. In this way, after a few journeys up and down the field the entire harvesting process is completed in the space of a few hours. This machine is useful on open plains or where the ground is level, and in areas where the straw has no economic value." White (1967, 158), whose translation is quoted above, says that these descriptions are strikingly confirmed by four pieces of monumental evidence, the most recent of which was discovered in 1958. These are the Tableau of the Seasons on the Porte de Mars in Rheims, a funerary relief at Arlon in Luxembourg, a relief from Montauban-Buzenol in southern Belgium, and part of a mutilated fragment from Trier in Germany. Clearly, the machine was quite common north of the Alps. White believes that two different machines are described and illustrated; a *carpentum*, and a lighter *vallus* in Fig. 15.15.

After reaping the corn needed to be threshed, to separate the ears of corn from the stalks, and winnowed, to separate the grains of corn from the chaff. Varro says "The grain should be threshed on the floor. This is done in some districts by means of a yoke of oxen and a threshing sledge. The latter is constructed either of a board roughened with stones or pieces of iron embedded in it, which separates the grain from the ear when it is dragged by the yoke of oxen with the driver or a heavy weight on it; or else of a toothed axle running on low wheels, called a punic cart; the driver sits on it and drives the oxen which pull it; this is the method used in

Fig. 15.17 Left: large Roman grain mills from Pompeii (79 A.D.). Right: second century A.D. relief on a Roman sarcophagus showing horses driving grain mills.

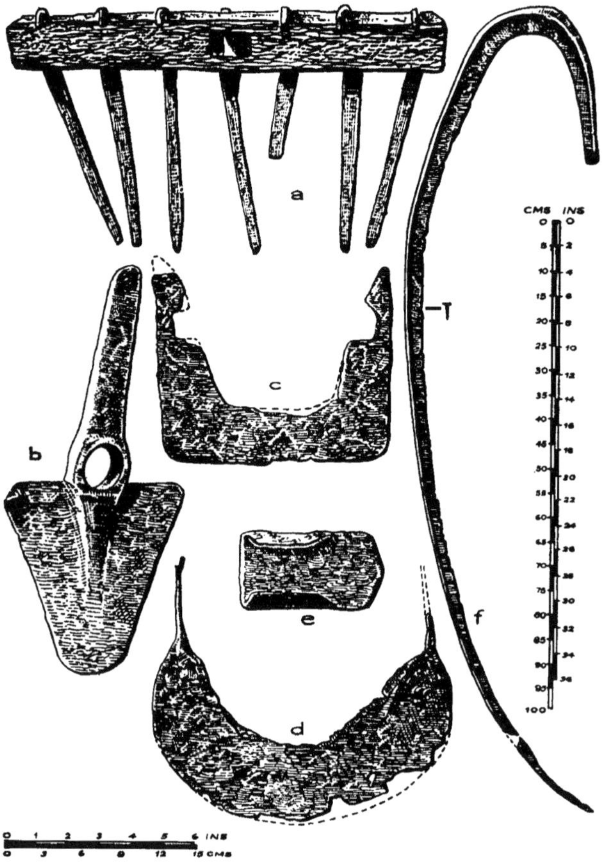

Fig. 15.18 Roman agricultural implements from Britain. All made of iron. *a*. Rake. *b*. Hoe, both from Newstead. *c*. Spade, Verulamium. *d*. Spade end, Worlington, Suffolk. *e*. Ploughshare, Silchester. *f*. Outsize scythe, Great Chesterford.

eastern Spain and elsewhere", White (1967, 153). Although no illustrations have survived from monuments numerous ancient threshing floors are known and the drags themselves are still in use in the Mediterranean area, Fig. 15.16.

After threshing the grains of corn were usually stored until needed before being ground into flour and baked to make bread. Corn grinding by hand is a difficult and arduous daily task, often the work of servants or slaves, and it was in this area that the Romans made most progress in mechanisation. Grinding by means of saddle stones was improved first by the invention of the rotary grinding wheel (quern) operated initially by hand, and later after the first century A.D. by horse power or by waterpower. A selection of three types of Roman grindstone is illustrated in chapter 4, Fig. 4.5. Hand grinding, using querns, continued in the more remote parts of Europe until modern times. For example, from the Isle of Mull, in 1772 A.D., Pennant tells us that "This method of grinding is very tedious: for it employs two pair of hands four hours to grind only a single bushel of corn [36.3 litre or about 25 kg]. Instead of a hair sieve to sift the meal the inhabitants here have an ingenious substitute, a sheep's skin stretched around a hoop and perforated with small holes made with a hot iron", Jennings (1995, 69,70). It was, of course, necessary to sift the flour to remove grit released in the grinding process. Four sets of large hourglass shaped grain mills and an oven may be seen by visitors to Pompeii, Fig. 15.17 (left) and a scene from a Roman sarcophagus of the second century A.D. shows the same type of grain mill driven by a pair of horses. A lamp on the shelf above the horses suggests that the work went on both night and day. The outer stone, usually of lava, was hollow and fitted over a conical lower stone. It had slots at the waist into which fitted beams that could turn it. Grain was poured in at the top and flour was collected at the bottom. Such stones have been found in Italy, France, and in England. Liversidge (1973, 182–3) writes that a donkey could grind 3.5 bushels [87.5 kg] of wheat in a day and that his cost amounted to 4.5 kg of wheat, about 5%. A horse could grind five or six times as much but it would be more expensive to keep. Roman and Greek water powered mills for grinding corn are discussed in chapter 6 and the only illustration of a Roman water mill is shown in Fig. 6.9 of that chapter. Although it is likely it is not definitely known that this water wheel drives a corn mill.

Miscellaneous Romano-British agricultural implements are illustrated in Fig. 15.18 and include a rake, a hoe, two spades, and iron ploughshare and a large scythe. Spades were often made of wood with the blade tipped with iron. These tips were sometimes nailed, clipped or shrunk on. Roman agriculture in Italy and the Mediterranean was based not only on cereals but also, and more importantly, on more expensive crops such as figs, olives and grapes; corn could be imported cheaply from Egypt where yields were much greater than in northern Europe. This was essentially an arboreal agriculture and the only demand it made of mechanical technology was in the development of heavy presses to extract the oil and juice. As we have seen, chapter 1 Fig. 1.6 and Fig. 1.12, they used slaves to trample grapes and stones hung on levers to compress olives, but later, chapter 1 Fig. 1.14, they developed large screws by means of which a more efficient extraction was attained.

Medieval

The heavy wheeled plough, Gallic reapers, corn mills driven by horses or by water wheels, and screw presses for extracting olive oil were the high point of mechanical development in Roman agriculture; and with the decline of their empire in the west it appears that agriculture reverted in many cases to older and simpler techniques. The large, prosperous Roman farms that abounded in England vanished and were replaced by the simpler agricultural systems of the invading Anglo-Saxons. However, Wilson (1981, 16) reminds us that our knowledge of Anglo-Saxon agriculture in Britain is slight. The most common cereals were barley, oats and wheat. Flax was grown for making cloth. Fruit and nuts were gathered from the forest and perhaps some suitable trees were planted deliberately near the forest edge. Hunting became a more important source of food. Tools were needed for all these activities but few have survived; no complete plough and only a few ploughshares (e.g. from Thetford and Westley Waterless, Cambs).

Perhaps the best description of primitive agriculture by free peasants in Britain is by Gerald of Wales, Thorpe (1978). Writing of the Welsh in the twelfth century he says: "They plough the soil once in March and April for oats, a second time in summer, and then they turn it a third time while the grain is being threshed. In this way the whole population lives almost entirely on oats and the produce of their herds, milk, cheese and butter. They eat plenty of meat, but little bread. They pay no attention to commerce, shipping or industry, and their only preoccupation is military training.—They do not have orchards or gardens, but if you give them fruit or garden produce they are only too pleased to eat it. Most of their land is used for pasture. They cultivate very little of it, growing a few flowers and sowing a plot here and there. They use oxen to pull their ploughs and carts, sometimes in pairs but more often four at a time. The ploughman walks in front, but backwards. When the bulls pull out

Fig. 15.19 A tenth century plough.

of the yoke, as often happens, he falls backwards and is in grave danger. They do not usually use sickles when they reap, for they prefer a short piece of iron in the shape of a knife, which is loosely joined to a stick at either end. If you saw it you would understand this better than you can hope to do by listening to my description."

The life of the Welsh peasant may seem hard but it was no worse, and may have been preferable, to that of the un-free ploughman described in the *Colloquies* of Aelfric, Abbot of Eynsham, writing in 1005, see Hassall (1973, 25). These are a set of questions and answers between a teacher of Latin and his pupils, who answers the part of the ploughman. We learn that the ploughman considers he has too much work. He goes out first thing to chase the oxen to the field and to yoke up for the plough. It's never so sharp a season that he dare lie in for fear of his master. But he yokes the oxen and fixes the ploughshare and coulter and every day he must plough a whole land or more. He has a helper; a child to goad up the oxen and the child is hoarse with cold and shouting. In addition the ploughman must fill the mangers with hay, and water the oxen and carry out the dung. The Master says, "Oh it is a great work!" to which the pupils reply on behalf of the ploughman, "Yes, because I am not a free man."

Fleta, a tract written in Latin about 1289, and also translated in Hassall (1973, 126) gives another view of medieval ploughmen. "The art of the ploughman consists in knowing how to draw a straight furrow with yoked oxen without striking, pricking or ill-treating them. Ploughmen ought not to be melancholy, or irritable, but gay, full of song and joyous, so that they may in a sort of way encourage the oxen at their toil with melody and song; they must take them their forage and fodder, and they ought to be attached to them, and sleep in their stable at night, and make much of them, curry and rub them down, look after them well in every way taking care that their forage and fodder is not stolen; the allowance of hay and litter should not be given out for two or three nights at once, but as much as is required for a day's feed should be given to them every day, nor should they be permitted to have a candle unless, as the saying is, it is 'held' [i.e. a lantern]. Other people's beasts found in the ploughman's pasture ought to be impounded. It is the duty of the ploughmen and husbandmen when the tillage season is over, to ditch, thresh, dig, fence, repair the watercourses in the fields, and to do other such small and useful tasks."

Fig. 15.19 illustrates a tenth century plough that is similar if not identical to the Greek or Roman ard and is suitable, therefore, for ploughing only on light soils. Certainly two Roman inventions, the wheeled plough and the watermill, survived the Dark Ages, but the Gallic stripper and the screw press disappeared. The more usual Anglo-Saxon plough was probably heavier than that shown in Fig. 15.19 and the wheeled plough illustrated on the Bayeux Tapestry, Fig. 15.20 (top) may have been more typical. A donkey draws the plough shown on the tapestry but generally they were drawn by a number of oxen. The plough is fitted with a heavy wheel, but no mouldboard, and the furrow would have been cut and turned over by a coulter that was shifted at the end of each furrow so that the next furrow would lie in the same direction. The Bayeux Tapestry also shows the ploughed field being harrowed by a horse drawn harrow, Fig. 15.20 (bottom). Curiously a donkey is doing the heavy work of ploughing and a horse the light work of harrowing, so we should be careful not to read too much into medieval illustrations, especially when their main purpose is decoration, as in this case, and not the main sequence of events.

Certain Anglo-Saxon tools and harvesting implements have survived such as four scythe-blades from Hurbuck, Durham, in a hoard of other tools, and sickles have been found in the graves of two English Vikings. Billhooks and pitchforks are illustrated in manuscripts as are spades made of wood but tipped with iron. There were, of course, 5,624 mills at the time of Doomsday, one mill for every 50 families, so there was some mechanisation in agriculture.

Fig. 15.20 Top: an eleventh century heavy, wheeled plough drawn by a donkey. Bottom: an eleventh century harrow drawn by a horse. Bayeux Tapestry.

Threshing corn by trampling under the hooves of animals or by pulling sledges over the corn was replaced by hand threshing using flails, Fig. 15.21 (top). This may seem a backward step but hand threshing was a skilled operation intended not only to separate the ears but also to preserve the stalks. Late in the twelfth century the first windmills for grinding corn appeared, Fig. 15.21 (bottom) and they proved to be a great boon to the many villages that did not have a suitable stream to power a watermill. Windmills are discussed in greater detail in chapter 7. The other major developments in medieval agriculture were organisational.

The manorial system was first developed under Charlemagne essentially to provide a surplus to support his armoured knights. After the Conquest the Normans introduced it, or promoted its spread, in England. In its simplest form the king granted land to his important earls and warriors in return for their loyalty in defending the kingdom and fighting in the king's wars. These great lords in turn granted a manor, consisting perhaps of a village and associated land and peasants, to a knight who, in return owed his loyalty and knight service to the king through his lord. Un-free peasants (villeins or serfs) worked the lord's or the knight's land (the demesne) for fixed periods, perhaps of three or four days per week plus extra duties at

Fig. 15.21 Top: winnowing corn by means of a flail.
Bottom: fifteenth century post mill.

harvest, and were able to work on their own land at other times. They were, however, tied to their manor and were not free to seek employment elsewhere. Some free tenants were attached to each manor and paid money rents to their lord or they might owe relatively light labour services. It was, of course, a protection racket. A democracy would recognise the need for defence, as the city-states of ancient Greece had, but it would never freely agree to the terms offered by the manorial system.

On most manors the land was worked by the two-field or three-field system. In the two-field system one field was ploughed while the other was left fallow and was grazed by the village animals. The following year the fallow field was ploughed and last year's ploughed field was left fallow. Allowing fields to lie fallow in alternate years, and permitting animals to graze and manure it, enabled the fields to maintain their fertility. However, manors and their customs varied greatly and there are many exceptions to the simple picture described here. This system of agriculture was not particularly productive and, although it provided the necessary surplus, it appears that it was necessary for about 80% of the population to work in agriculture leaving only 20% to work in manufacture, trade, transport, the professions, etc. Nevertheless, agricultural profits were the main source of wealth and financed the Church, the building of great cathedrals and parish churches, eight crusades to the Holy Land, and countless major and minor wars in Europe. These profits were being transferred to landlords under the emergent manorial system. Whitton (1988, 127) reports that between 940 and 980 about 80% of the donations made to Chartres cathedral were by peasant freeholders. But everywhere freeholders were under pressure to give up their freedom in return for knightly protection. This is indicated by the fall in donations of peasant freeholders to Chartres cathedral that dropped to 38% during the period 1060–90 and to only 8% in the period 1090 to 1130.

Agriculture also fed an ever-increasing population and it did this not by increasing the yield per acre (agricultural techniques remained static) but by bringing new land into cultivation and setting 80% of the increased population to work for their own sustenance. The other 20%, of course, worked in other sectors of the economy, usually in the towns. Peasants in England enclosed and cultivated the fenlands of Lincolnshire and Cambridgeshire for the first time. The denes were cut from the ancient Wealden forests and 'assarts' were cleared in the manorial woods to add to the village fields. In France, Spain and Germany, during the eleventh century, previously fallow land was ploughed for the first time. Throughout Europe forest clearance, improved drainage, and, in Flanders, land reclamation from the sea, brought new lands into cultivation to feed the rising population. In Europe the growing populace was also able to move into new lands further east, but by 1300, England's mounting population, living by a simple agriculture, had occupied most of the readily available arable land and the situation was becoming critical. To cope with the population pressure, marling and loaming the fields to increase yields was used and some manors were able to change from a two-field system to a three-field system. In the three-field system two fields were under the plough and one field lay fallow in any one year. Winter sowing (wheat or rye) in the first year was followed by spring sowing (barley, oats, peas or beans) in the second year and fallow in the third year. Thus three crops in six years from the two-field system increased to four crops in six years with the three-field system; an increase of 33.3%.

Great floods and famines in 1315–17, brought an end to population growth and the infamous bubonic plague in 1348–9, 1361, 1368, and 1375, dramatically reduced the populations throughout Europe. Many villages and manors were abandoned; some monasteries, such as the abbey of Saint-Martin at Tournai which lost 80% of its monks, were almost wiped out, and in the large diocese of Lincoln over 40% of the beneficed clergy died. Deaths were heaviest where rats, which carried the disease, were most numerous, that is, in the towns and in the corn growing areas. It is probable that an English population of about 3.5 million in 1300 had reduced to 2.5 million by 1450. Such depopulation left its mark. Today there are numerous deserted villages in most English counties; places known from the records to have existed in the Middle Ages but where only the outlines of the houses and streets can now be seen, Fig. 15.22.

Beresford (1965), in his book discusses 293 lost villages, and lists more than 600 others. They are not rare but, of course, they were not all abandoned directly because of the Black Death. There were two main periods of depopulation. Following the Black Death many villages were abandoned because many had died of the plague and the remainder fled because of fear or because the village was no longer viable, although some tenants were persuaded to stay on by reduced rents. Depopulation also caused a severe labour shortage and later, in the fifteenth century, landlords turned from arable to pasture, which was less labour intensive, and began to destroy villages

Fig. 15.22 The deserted medieval village of Gainsthorpe, Lincs, was the first abandoned village to be so recorded.

and turn the land over to sheep grazing or deer parks. Sheep not only provided meat but also wool for the thriving cloth trade. Later this type of depopulation was regarded as a scandal and legislated against, but most of the damage had been done already.

Overton (1996, 65) has shown that throughout the medieval period and until about 1800 there was a very close relation between population size and prices; they increase and decrease together. Depopulation thus reduced the prices of goods and the labour shortage doubled wages between 1340 and 1360 despite the Statute of Labourers, by which the government hoped to control wages. Nevertheless wages did rise; many lords were determined to retain their customary labour services and resentment grew among the villeins and serfs, whose labour was in great demand. The imposition of the Poll Tax led in 1381 to the Peasants Revolt, but it became clear that their main demand was for the abolition of villeinage, which was at first granted and then withdrawn. Throughout Europe similar conditions existed and there were similar peasant revolts in many countries, all of which were brutally suppressed. But villeinage, by about 1420, did die out to be replaced by copyhold tenants who held their land without labour service, and the lords divided and let their demesne lands. Despite these changes, arable farming continued on the basis of the two or three-field system.

Early Modern

After 1450 the population began to grow again, but only slowly for the plague did not die out until after London's Great Plague in 1665–6. The population of England increased to 3 million by 1550, 4.1 million by 1600 and 5.2 million by 1650. Thus by the reign of Elizabeth I the population had returned to the level of 1300 and the situation was once again becoming critical. Overton (1996, 63) says that before 1750 the English population seems to have had a natural ceiling of about 5.5 million people, given the relatively simple agriculture and the available land. The response of landowners to this was to abandon the two-field or three-field system in favour of a more efficient and specialised use of the land, and this could only be done by enclosure. Enclosure of moorland and waste were ancient practices but enclosure of the common pasture and open fields was new and caused frequent complaints after about 1500, so the practice developed of enclosure by agreement amongst the principal landowners and others with rights in the fields and commons. The drawbacks of the open field system were: time was wasted by farmers moving from one strip to another in the two large arable fields; access to the strips was difficult, especially for a plough and team of up to eight oxen; the lack of hedges gave no protection against the spread of weeds; straying animals could do immense damage before being caught; there was little chance of selective breeding of animals pastured on a common; control of farming was in the hands of the manor court not the individual farmers (this discouraged experiment); and the large number of poorly marked boundaries led to frequent disputes. Farmers wanted control of their own lands and to stand or fall because of their own efforts. The first known private enclosure agreement dates from 1607 and concerns some parishes in Herefordshire but enclosure did not become very frequent until the end of the eighteenth century when some 1,500 private enclosure acts passed through parliament.

In the sixteenth and seventeenth centuries numerous books on agriculture appeared with the objective of improvement. Sir Hugh Plat, 1552–1608, developed an active interest in natural science, mechanical invention, and in agriculture and gardening, to which he devoted the later part of his life. His investigation of several manures, especially salt and marl, proved valuable, as was his "New Founde Art of Setting of Corne", published in 1593. He was knighted for his services as an inventor in 1605. Gabriel Plattes, who flourished about 1638 was said to be of Dutch extraction and was one of the earliest advocates in England of improved husbandry. He devoted much time and money to practical experiments but his work was neglected during his lifetime, although he influenced later writers through his "Treatise on Husbandry", 1638. Apparently he died in poverty, being found dead in the streets of London during the Commonwealth period.

Another agricultural improver was Walter Blith, who wrote "The English Improver or a New Survey of Husbandry—by Walter Blith, a lover of ingenuity", London, 1649. This was re-issued in 1652 as "The English Improver Improved". He seems to have been a captain in the parliamentary army during the Civil War and the title page of his book proclaims "Vive la Republick". One of his concerns was that the profits of any improvement made by a tenant usually went to his landlord and this discouraged such experiments and disadvantaged tenant, landlord, and country alike. He thought this could be removed by legislation that would force the landlord to give reasonable recompense for improvements or would sanction the tenant to enjoy the profits for a reasonable time.

A selection of seventeenth century ploughs illustrated by Blith is shown in Fig. 15.23. The Harfordshire wheeled plough seems very heavy and has two wheels, a coulter, a share, and a mouldboard. The hitching point is adjusted by moving a peg on the beam to one of several holes to relocate the ring at the end of the draught chain. The Dutch coulter is a wheel intended both to make a vertical cut

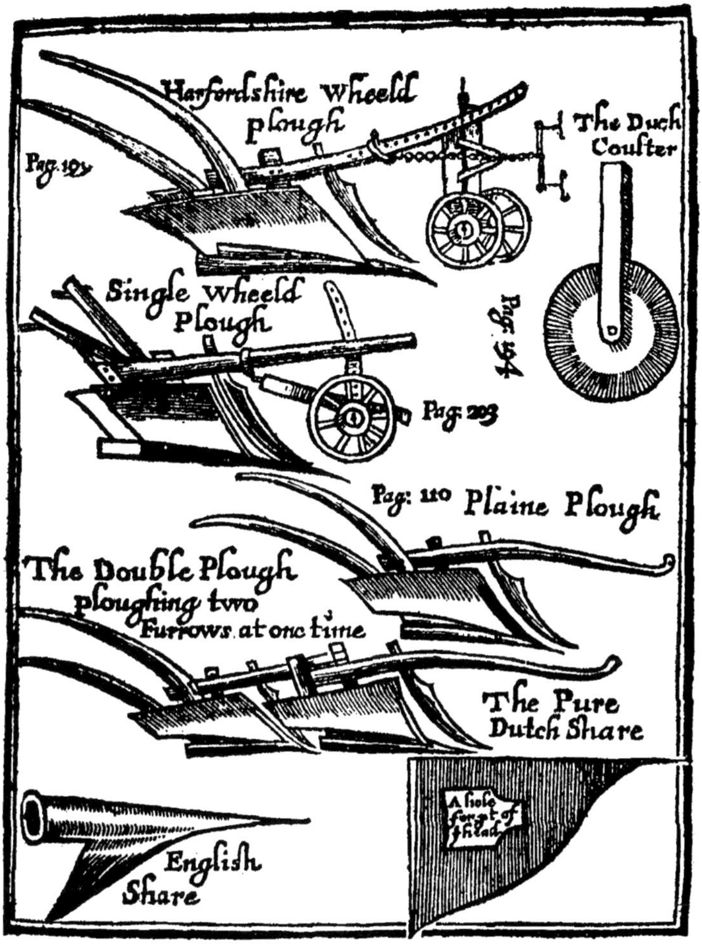

Fig. 15.23 Contemporary ploughs from Blith's *English Improver Improved*, 1653.

in the soil and to rotate to reduce frictional drag and is an admirable device much used on later ploughs. The single wheeled plough is similar to the Harfordshire plough except that the depth of the wheel, and hence the depth of the cut, may be varied. The plain plough is a simple swing plough (one having no wheels) and has no adjustable parts. It was probably the simplest, lightest and most popular plough in use at that time. The double plough, for ploughing two furrows simultaneously, was intended to save time by allowing the ploughman to double the area ploughed in a single day. Of course the

Fig. 15.24 Besson's three-shared plough, 1588.

draught load doubled and to work at the same speed it was necessary to use twice as many draught animals. On light soils this was possible without too much inconvenience but it was not possible on heavy soils that normally required a full team for a single furrow. Earlier, Besson, in about 1588, had proposed a three-shared plough that took three men and a complicated rope contrivance to operate, Fig. 15.24, but that illustrated by Blith seems simpler and much more likely to succeed. The English share and the Dutch share are also illustrated and it appears that the Dutch share may offer less resistance to movement through the soil.

Another seventeenth century agricultural writer is John Worlidge. His "Systema Agriculturae—" was first published in 1669 and such was the demand for agricultural information that it went through a number of editions (1675, 1681, 1687, 1716) before it was superseded by the numerous agricultural reference books that were to become a feature of the eighteenth century. He discusses the use of mechanical seed drills, in which respect he pre-dates Jethro Tull, and his work on the improvement of the land by parsley and his improved methods of fining cider are also original. He wrote many other books on agricultural topics, gathering together the work of previous writers such as Fitzherbert, Sir Richard Weston, Robert Child, Walter Blith, Gabriel Plattes, Sir Hugh Plat, and others.

Overton (1996, 63) quotes Gregory King's estimate, made in about 1690, that the expected population of England would grow from 5.5 million in 1700 to 6.42 million by 1800 and 7.35 million by 1900. This, of course, King based on existing agricultural capacity and the limited improvements he foresaw. It greatly underestimated the real population growth, which reached 9.2 million by 1801 and 32.5 million by 1901. This was possible because improvements in agriculture, the so-called agricultural revolution, made it feasible to feed such a large population without relying excessively on imports. It was attained largely by improving the yield of existing land, through improved technology, and partly by further recovery of difficult wastes, which was also made possible by improved technology.

One of the earliest and most successful improvers was Jethro Tull, 1674–1741. Shortly after his marriage he commenced farming on land belonging to his father. On this farm at Howbury, Wallingford, Tull invented and perfected his seed drill in about 1701, Fig. 15.25. An elementary seed funnel, as we have seen in Fig. 15.7, was used in Mesopotamia where it was desirable to maximise the return on the number of seeds sown. But where the return on sown seed was large, and the return per acre of land was more important, as it was in Europe, they were usually broadcast. Finding that his plan for sowing his farm with sainfoin in a new manner was hindered by the conservatism of his labourers Tull resolved to "contrive an engine to plant St Foin more faithfully than such hands would do. For that purpose I examined and compared all the mechanical ideas that ever had entered my imagination, and at last pitched upon a groove, tongue and spring, in the soundboard of the organ. With these a little altered and some parts from two other instruments as foreign to the field as the organ is, added to them, I composed my machine. It was named a drill because when farmers used to sow their beans and peas into channels or furrows by hand they called that action drilling." Tull appears to have been quite original in his invention although, as we have seen, others had anticipated him. His drill was a small implement, built on a frame similar to that of a wheelbarrow. Three coulters cut shallow furrows in pre-prepared soil and a notched barrel, Fig. 3 in the illustration, picks up the seed from a hopper and drops it into a guide where it falls into furrows cut by coulters. The notched barrel is driven from wheels and thus it deposits seed at regular intervals behind the coulters. This was Tull's major innovation, for earlier inventors had used little more than a seed box with a hole in the bottom. The advantage of the drill was that seed was planted in rows and at regular intervals rather than broadcast, and thus it saved seed, gave space for the crop to grow, and was easier to weed. His drill worked successfully on his own farm

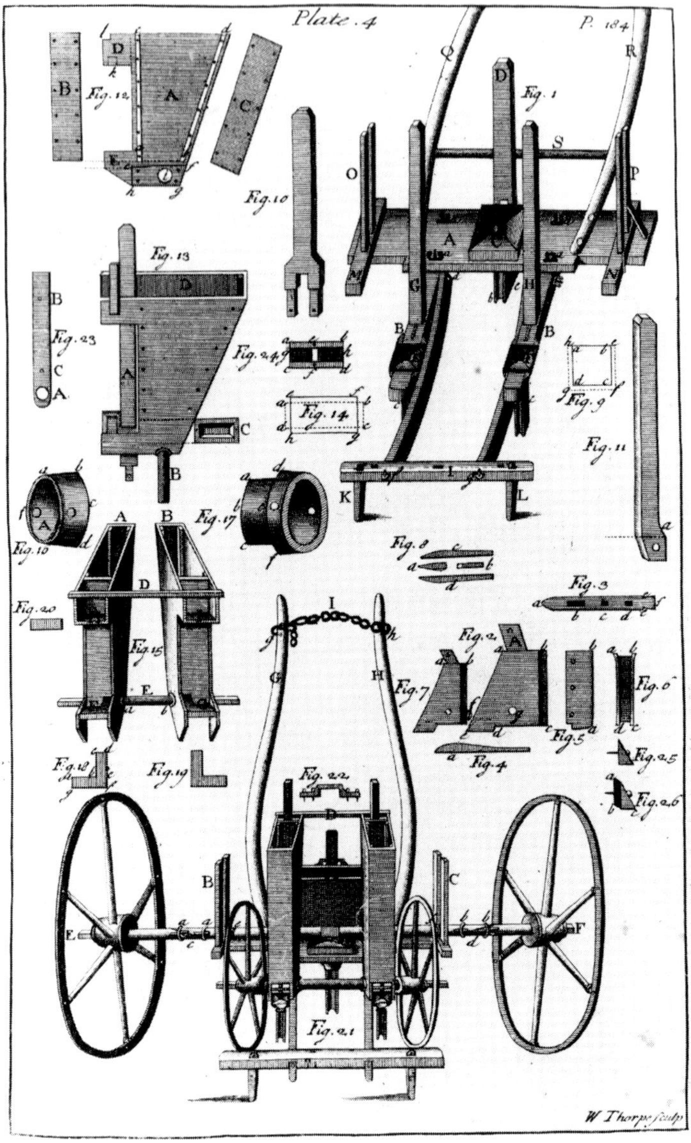

Fig. 15.25 Original plans for Jethro Tull's seed-drill of 1701.

for many years, but his labourers, whom he could not teach to use his machines properly, complicated matters.

In 1711 he was forced to travel for the sake of his health and he noticed the ploughed vineyards near Frontignan and Setts in Landguedoc, where pulverisation of the earth between the rows of vines was made to take the place of manuring the land. Because his own crops were drilled in rows rather than broadcast he was able to adopt this method himself and on his return to Prosperous Farm, near Hungerford, where he had moved in about 1709, he tried the method, first on turnips and potatoes then on wheat. Derry and Williams (1960, 71) say that he could grow above average yields of wheat from one-third the average quantity of seed on unmanured land for thirteen years in succession. At the suggestions of many noblemen and gentlemen who visited his farm he published a specimen of his "Horse Hoeing Husbandry" in 1731, which was at once pirated to Dublin. Tull determined to print no more but was dissuaded, especially by a letter from an unnamed peer, and consequently the complete "Horse Hoeing Husbandry" was published in 1733. Stephen Switzer, the editor of a monthly publication, the "Practical Husbandman and Planter", at once attacked it and Tull was accused of having plagiarised from Fitzherbert, Sir Hugh Plat, Gabriel Plattes, and John Worlidge, and his theories concerning the value of manure and the practice of pulverising the earth were contested. Tull was hurt but defended himself in subsequent smaller publications and some hard words were exchanged, but later editions of the "Horse Hoeing Husbandry" appeared in 1743 and 1751 after his death.

In 1753 a French translation was made and his methods were taken up in France; Voltaire, for example, cultivated his land at Ferney according to Tull's system. In 1822 William Cobbett edited the book. However, Tull's ideas, and his drilling machine, were not taken up immediately in England. In 1784 Forbes said "Many who had neglected to practise the new husbandry, from Mr Tull's own success were prevailed upon to engage in it upon the recommendation of these foreign gentlemen and it is now making considerable progress among farmers in the culture of beans, peas, cabbages, and in some measure of wheat."

The reluctance of farmers to use Tull's seed drill and hoe was similar to that he had found amongst his own labourers. It was more complicated than a plough and in its relatively undeveloped state it

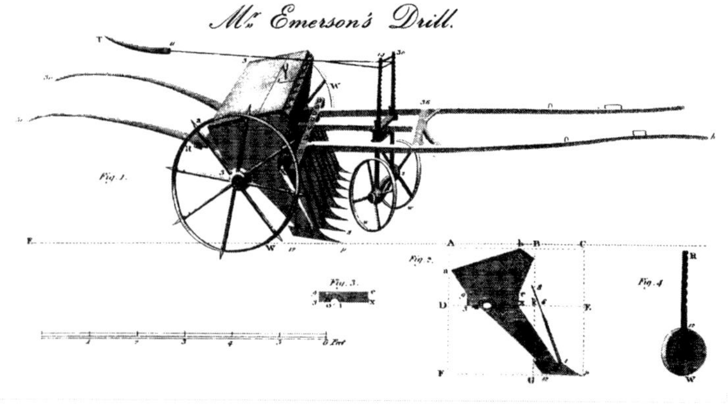

Fig. 15.26 Mr Emerson's seed drill, 1813.

was, no doubt, unreliable and in need of constant attention. But throughout the eighteenth century seed drills continued to be developed and many were illustrated in books on agriculture, even if they were not always taken up on the farm. One such development, from the 1750 edition of William Ellis's "The Farmer's Instructor", is a four-wheeled drill with separate hoppers for seed or manure. Young (1813, 354) illustrates Mr Emerson's seed drill, Fig. 15.26, but he notes, p. 135, that there was: "Scarcely any drilling in the northern part of the division [Oxfordshire]. But a little in the southern part of that district".

A better view of the working parts of a seed drill is illustrated in Fig. 15.27. This machine is based on the Reverend James Cooke's patent of 1782. The seed is lifted from the box by cups attached to a barrel driven by gears from the axle of the large wheels. The cups, emptying themselves into small hoppers and communicating with the coulters by tubes, seem to be most suitable for an even flow of seed. The coulters, which formed a bed for the seed, were sometimes jointed to move up and down to permit a uniform pressure and deposit the seed at a consistent depth. A test of this seed drill was described in *The Manchester Mercury*, 26th August 1783, see Morsley (1979, 106). "On Monday last were reaped two equal but separate crops of wheat adjoining each other, belonging to Mr Benjamin Johnson, of Top-of-Bank, near Manchester, and measured, while growing, by Mr Thomas Wright, of Manchester, one of which said crops of wheat was planted in November by the Rev. Mr James

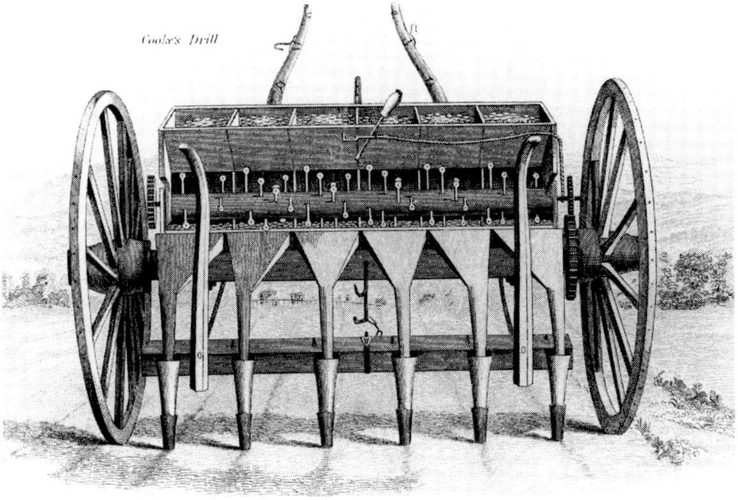

Fig. 15.27 Six furrow seed drill based on the Rev. James Cooke's patent of 1782.

Cooke's patent machine; the other was sown broadcast the same day, when, to the astonishment of all the reapers and a number of spectators,—the crop planted by the above machine (straw and grain together) was one fourth heavier than the crop sown broadcast, and on Wednesday the twentieth instant, the above crops were separately thrashed, when the grain of the planted crop weighed upwards of one third more than the grain of the sown crop." But the agricultural writers and innovators Arthur Young and William Marshall found very few farmers using drills when they examined the state of farming in the 1790's, some 60 years after Tull's original invention and a decade after Benjamin Johnson's demonstration of their utility. Mechanical seed drills, apparently, were unable to deal with uneven ground, and liable to become clogged and sow unevenly. They were expensive to buy and maintain so crop returns had to be high to make them worthwhile. Consequently they were not used outside a few intensively cultivated districts in Suffolk, Isle of Thanet, and in Northumberland.

One of the most important advantages of drilling crops in straight lines was the ease by which they could be kept free of weeds; this was particularly important in the days before weed-killers were generally available. A hoe towed by a horse more cheaply exterminated weeds than hand weeding, regardless of any advantage such hoeing

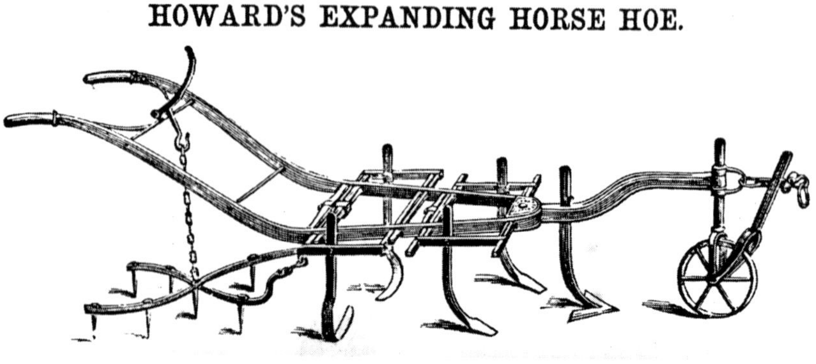

Fig. 15.28 Howards expanding horse-hoe, 1840–50, was of the type that became very common in the mid-nineteenth century for weeding between three rows of corn or one row of root crops.

might have had on increasing the fertility of the soil. The 1783 report on Cooke's drill in the *Manchester Mercury*, mentioned above, also mentions the "singular advantage to be derived from the use of a new constructed hand hoe invented by the said James Cooke, whereby one man will do the work of six or seven men, in the same time, but more effectively". The great improvements that were made in the nineteenth century in drills and horse-hoes eventually rendered these implements of great service in cleaning drilled crops. Fig. 15.28 shows a mid-nineteenth century horse hoe. The wheel allowed the cutting depth to be set accurately and there is ample adjustment to suit different row widths.

Table 15.1 Horses and Labourers in English and Welsh Agriculture, Overton (1996, 126).

	Horses (millions)	*Arable (million acres)*	*Horses per 100 arable acres*	*Adult males in Agriculture (millions)*	*Man-hours of horse-power per man**
1700	0.50	9.0	5.58	0.9	2.79
1800	0.70	11.5	6.09	1.0	3.75
1850	1.00	15.3	6.54	1.1	4.55

* One horse is taken equivalent to five men. 1 acre = 0.4 hectare.

The use of machinery on farms, usually driven by horses, increased rapidly after about 1800 and can be illustrated by the growth in the use of horses in English and Welsh agriculture, Table 15.1. Between 1700 and 1850 the use of horses in agriculture doubled and the area of arable land increased by 70% whereas the number of males employed increased by only 20%. Consequently the effective man-hours per man, allowing one horse to be equivalent to five men, increased from about 2.8 man-hours per man in 1700 to about 4.6 man-hours per man in 1850. The extra horses were not used continuously in ploughing but were also needed to drive the increasing number of machines that were being used to raise farm output and reduce farm labour. The use of horses on farms did not die out until the middle of the twentieth century although steam power, for driving threshers, was introduced as early as the 1840's.

One of the most labour intensive activities in farming is harvesting and it is surprising that mechanisation of reaping and winnowing did not occur earlier than it did, especially considering that the Romans had found it necessary to invent and use a reaper, see Fig. 15.15. Certainly the scythe replaced the sickle for reaping wheat, which increased productivity, and from 1780 onwards, mechanical reapers of one sort or another were investigated. However, it was not until the 1820's that a Scottish Presbyterian, Patrick Bell, developed a successful machine, Fig. 15.29, which was, in fact, similar to the Gallic reaper of the Romans. It was pushed through the corn by a pair of horses but the chief innovation was that rotating blades guided the corn onto the cutter bar and laid it out to one side for the stackers to gather into stoops. A man, walking alongside the reaper controlled the height of the blades by means of a control line. The liability of its cutting knives to break was its chief defect. But this was overcome by adopting an arrangement of oscillating knives driven from the wheels to cut the corn, and in this form it was quite popular until the 1850's.

However, another reaper had been invented and developed in America by Obed Hussey and an even better machine by Cyrus McCormick, Fig. 15.29 (bottom). The original patent dates from 1834 and manufacture, in Chicago, started in 1848, but they were not known in Britain until they were exhibited at the Great Exhibition of 1851, where they generated great interest and admiration. McCormick's reaper had the cutting bar to one side so horses could pull the machine without trampling the corn before it was cut. The machine exhibited in 1851 deposited the corn behind it but that shown in Fig. 15.29 (bottom), which is a later design of 1862, followed

Fig. 15.29 Top: Bell's reaping machine of 1826.
Bottom: McCormick's reaper, 1862.

Bell's reaper and is arranged to throw the corn out to the side. Derry and Williams (1960, 672) say that by 1860 some 4,000 machines a year were being produced by McCormick. Although not initially suitable for British conditions they were very soon adapted and were made under licence. In 1842 the section on "Agriculture" in Chambers "Information for the People" assumed that reaping would be done by sickle or scythe, but by 1874 it was able to claim that "Reaping-machines have now fairly displaced hand labour either by hook or

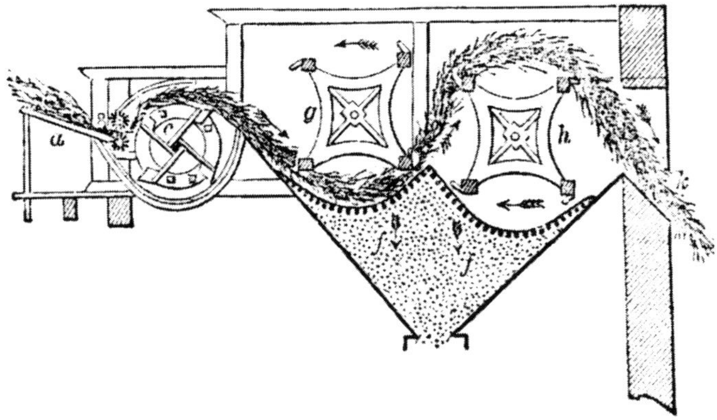

Fig. 15.30 Meikle's threshing machine, 1786.

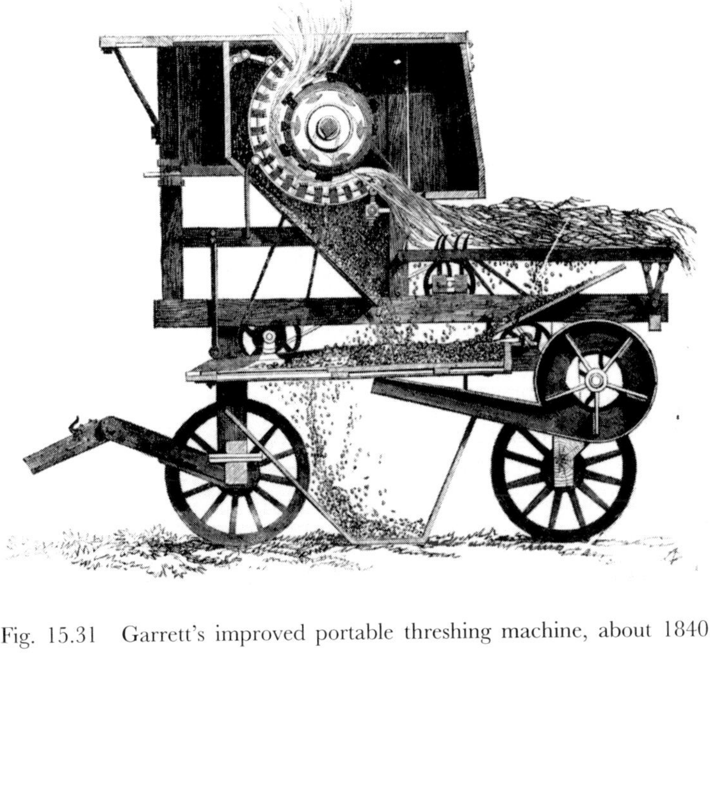

Fig. 15.31 Garrett's improved portable threshing machine, about 1840.

by scythe, except so far as cutting round fields to enable the machines to begin working.—On an ordinary farm it has been found that the number of hands required for harvest work has been reduced betwixt a third and a half."

After reaping, the corn was gathered into stoops in the field for a few days to allow the weeds to die off before being transferred to the stack-yard where it could be threshed, as required, either by flail or by threshing mill. Menzies in Scotland invented a threshing machine in 1732 and Stirling of Dumblane invented another in 1758 but neither seems to have been very successful. Meikle of Tyningham, East Lothian, attained greater success when he invented the modern type in 1786, Fig. 15.30. The grain and straw is fed in at *a* between two fluted rollers to a beater-cylinder *c*; thence it passes to a beating cylinder *g*. A third cylinder *h* raises and loosens the straw and delivers it at *k*. These cylinders operate above a curved grating through which the separated grain falls. Chillingham added circular rakes *g* and *h* in 1789 and in 1800 a fan was added to assist in separating the straw.

Threshing machines proved to be particularly unpopular with the farm labourers who rose in revolt, in December 1830, against their low wages and poor living conditions, which threshing machines exacerbated by removing a useful source of employment during the winter months. *The Coventry Mercury* on 5th December 1830 reported on the state of the riots in several southern counties, leaving the impression that southern England was everywhere ablaze. "In the country around [Banbury], the demolition of machines continues, but in every case I believe the work is being accomplished in the open fields—to which indeed the farmers have many of them conveyed their machines, leaving the labourers to destroy them at their pleasure, and out of reach of other property." In other places too the farmers simply handed over their machines to prevent more serious damage and the magistrates issued printed notices to Norfolk farmers urging them "to discontinue the use of Thrashing machines and to take them to pieces", Morsley (1979, 249). But although farm labourers sometimes won the battles in 1830, they paid dearly for it, and they lost the war. A Hampshire curate tried to calm the men of his village but they ignored him and a letter he wrote was reproduced in several newspapers. He writes that two days after his remonstrations the men's condition was: "John Symonds (wife and four children) skull fractured in five places, quite insensible and life in imminent danger—Francis Dorking, a widower (five children) committed to Devizes jail, for machine breaking, will be tried for his life—William Farmer, left cheek cut down to the bone with a yeoman's sword—three cuts in the arm, so bad that his arm was taken off that night—John Freeling, absconded with three warrants out against him for capital offences, and seven children and wife chargeable to the parish—Abraham Tucker, trampled upon by cavalry, cannot possibly recover, was just married. Daniel Jones and John Vigor, two excellent young men, became intoxicated, plundered houses—are committed to take trial, and will, in all probability, perish upon the gallows, leaving nine children to deplore the madness and folly of that fatal day", Morsley (1979, 250).

Threshing machines were recommended for farms greater than 100 acres [40 hectares] and from the 1840's horse, water, or steam-power could be used to drive them. Fig. 15.33 illustrates Garrett's improved portable threshing machine of the 1840's. The un-threshed corn was made to pass around a large cylinder or drum, which separates the grain from the straw and the grain falls through a sieve into a hopper. Winnowing is done by simple vane blower, rotating at high speed, which draws in air axially and blows it across the falling grains to remove lighter material such as small straws and dust. Threshing was often done in a barn and the most convenient source of power on most farms, even as late as the middle of the twentieth century, was the horse. When it was not used for ploughing, harrowing, drilling, hoeing, or reaping, it was used to drive the threshing and winnowing machine, Fig. 15.32 (top). Brown (1989, 38) says the horses walked about 2.5 circuits per minute while the gearing increased this to an output speed of about 400 revolutions per minute. This type of drive was very popular and could be adapted to power a large range of barn machinery that had previously been operated by hand. Much of this barn machinery was used in the preparation of animal feeds and litter and included chaff cutters, invented in the 1770's and turnip cutters, invented in the 1820's. In Fig. 15.32 (bottom) a geared whim is shown driving a combined crushing and grinding mill.

Remarkable progress had occurred since William Marshall and Arthur Young investigated the state of British agriculture in the 1790's. At that time virtually the only machine they found on most farms was the plough, and most of those they considered of inferior design although some did meet with their approval. Fig. 15.33, taken

Fig. 15.32 Top: a two-horse threshing machine at work in a barn driven by a geared horse whim, 1851. Bottom: details of a geared horse-whim of about 1910 driving a combined crushing and grinding mill.

from the *Universal Magazine*, 1748, demonstrates the wide variety of ploughs then in use. Every area of the country had its own traditional design that local farmers deviated from at their peril. There was the Lincolnshire plough *(a)* with a large wheel coulter that was thought suitable for ploughing fenlands; the Hertfordshire plough *(h)*; the two-wheeled plough of Berkshire, Oxfordshire and Wiltshire *(l)*; ploughs with four coulters *(d and k)*; and the only other machine likely to be found on a few farms, namely, William Ellis's seed drill *(c)*.

The principles of good design were known and were used by James Arbuthnot, a Norfolk farmer, who used mathematics and applied mechanics in the design of his plough and to find the best shape of a mouldboard. He designed a swing plough intended for only two

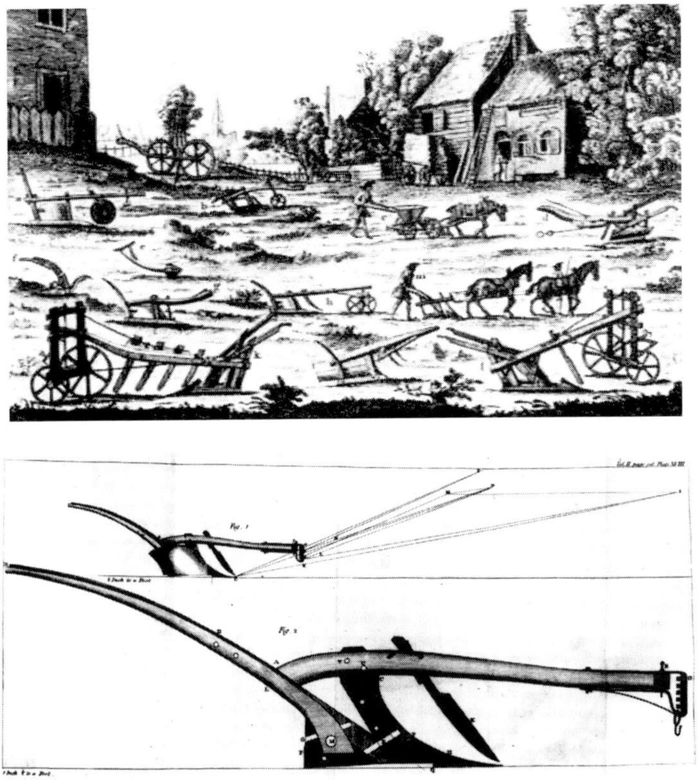

Fig. 15.33 Top: the great variety of ploughs in use in 1748 and illustrated in the *Universal Magazine*. Bottom: Arbuthnot's plough in Arthur Young's *Farmer's Tour Through the East of England* (1771).

horses that was illustrated and approved of by Arthur Young in his "Farmer's Tour Through the East of England", 1771. Arbuthnot's plough, Fig. 15.33 (bottom) is simple, light and strong and has some means of adjusting the line of the draught force so that it forms a straight line from the horse collar to the point of the coulter/share. In this way the plough may be pulled through the soil without rotating around the tip of the coulter/share, and without the ploughman having to press down to resist such rotation. In fact, it is preferable to have the line of draught intercept the coulter a little above the tip so the ploughman may generate a small downward force with which he can control the draught on the horses and control the depth of the furrow. The most important consideration is to cut the

Fig. 15.34 The famous Rotherham plough of Joseph Foljambe, patented 1730.

furrow with as little draught as possible. This means that: the share and coulter should be sharp, smooth and streamlined to cut through the earth with the minimum of effort; the friction generated as the earth moves over the share, coulter and mouldboard should be small; and the plough itself should be light and easy to handle.

These principles had been used already in the design of the famous Rotherham plough, first patented by Joseph Foljambe in 1730. Marshall had approved of what he called the Yorkshire plough when he first saw it at the end of the eighteenth century in the Midland counties. These were lighter and easier for the ploughman to handle than most other ploughs and were probably the Rotherham plough, Fig. 15.34. It was a simple swing plough and Foljambe's main innovation was a rigid triangular frame to which the share and mouldboard were attached. This replaced the use of a separate share beam, which contributed to the unwieldiness of the old ploughs. Foljambe not only designed an impressive new plough but he also established a factory in Rotherham where he was able to produce 300 ploughs a year using standard wooden parts derived from master patterns. His factory however remained unique in the eighteenth century for large-scale production of agricultural implements did not occur until there was sufficient demand.

Chambers (1842, 324) says "The object of ploughing is to delve and turn over the soil in the ridges, to destroy the surface vegetation by burying it underground, where it rots to become a kind of manure; to bury the dung spread on the land; to form furrows for different purposes [drainage, for example]; and generally speaking to prepare the land for cropping. In old times ploughs were exceedingly clumsy in construction, and dragged with much difficulty." In Scotland, James Small first introduced new and lighter swing ploughs in the 1770's and these proved very popular; he also published a *Treatise of Ploughs and Wheel Carriages* in 1784. Small's plough is described as "an elegantly shaped instrument formed on scientific principles, was originally composed of wood and iron and did not weigh altogether above 76 lbs [35 kg]; it was afterwards made of malleable iron, and of light appearance; but latterly the practice of making it of wood and iron has again become pretty general. The chief merit consists in the fore part being formed in such a slender and tapering wedge like manner, as to cut the land with the least possible resistance. The mouldboard for turning over the furrow is beautifully curved from the point of the stock to the heel of the wrest so it turns over the mould with a small degree of friction and in the best manner. In its own proper form it is particularly well suitable for light soils, and proceeds actively through the ground, cutting to a depth of from 7 to 9 inches [18 to 23 cm]; but it may be made to go much deeper if additional power be attached to it. In Scotland and other countries in which it has come into use it is almost invariably drawn by two horses, yoked abreast, and is guided and tended by only one ploughman, the reins coming to each handle of the plough. Considerable skill is required to steady and guide this sharp instrument as it advances through the ground."

In Scotland the use of wheeled ploughs was scorned, at least by Finlayson in Chambers (1842, 325), and were thought suitable only for unskilled ploughmen in backward districts. They had, of course, been used from a remote period in England but by 1874 we read, "Mr Finlayson, Bridge of Allan, has recently attached a couple of wheels to the common plough, by which the regularity and equality of the English wheel-plough are attained", Chambers (1874, 519). Finlayson laid down the following general rules for ploughing:

1. The horses should be yoked as near the plough as possible, without too much confining, or preventing them from taking a free step.
2. When at work they should be kept going at a good pace.
3. The chains or theets should, from where they are suspended over the backs of the horses, point in a direction leading through the muzzle to the centre of the cutting surfaces of the coulter and share.

Fig. 15.35 Kentish turnwrest plough, from John Boys, *General View of the Agriculture of Kent (1805).*

4. The implement, when taking the form and dimensions required, should stand upright, and glide onwards in the line of progression, without swerving in any particular way.
5. The ploughman should walk with his body upright, and without using his force to one point, or showing appearance of inclination.

It was generally recognised that two horses yoked abreast did more work than when yoked in-line, one behind the other. Where four horses were required it was usually necessary and more convenient to yoke them "two and two", as shown in Fig. 15.35, rather than four abreast or four in-line.

Most mouldboard ploughs turn the earth to the right, which means that on the return passage the earth was moved in the opposite direction and in the course of many years this can markedly change the surface of the field by creating the characteristic ridge and furrow of arable fields. The turnwrest plough was commonly used in Kent, Surrey and the Sussex Weald, Fig. 15.35, and avoided this. Its mouldboards and coulter could be moved from one side of the beam to the other by means of a simple lever and thus all the furrows were turned in the same direction across the field. The same idea is used on modern ploughs. Marshall, however, thought the turnwrest plough was particularly heavy and unwieldy although well suited to the flinty soils and steep hillsides of the downs. This is confirmed in Fig. 15.35 where it is shown being drawn by four horses.

As we have seen in Fig. 15.23, double furrow ploughs had been in use in the seventeenth century and, although never very popular,

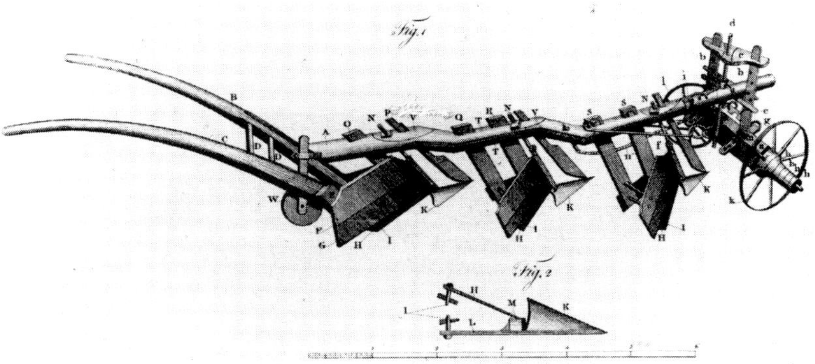

Fig. 15.36 Duckett's three furrow plough, 1770's.

improved designs appeared from time to time. Duckett, in the 1770's, according to Brown (1989), achieved some repute with his three-furrow plough, Fig. 15.36. It had a long crooked beam supported by wheels similar to a gallows plough. However, a three-furrow plough was probably too ambitious for horses except on light soils. Double-furrow ploughs were introduced into Scotland about a century later, in 1869, and were adopted rapidly. By this time all the well-known makers offered them and it was said that about half the tillage of Britain was suitable for their use. However, most farmers used three horses on a double-furrow plough and even then found it severe work on the majority of soils; they preferred to restrict double-furrow ploughs to seed furrows.

For centuries agricultural implements were made of wood with the critical parts, which needed to be sharp or free from wear, made from wrought iron by the local blacksmith, as and when required. The introduction of cast iron in the nineteenth century was a great step forward and allowed, interchangeable parts to be made from standard patterns. Implements could now be produced in quantity in factories and sold at modest process. All-metal implements appeared. Chill casting enabled the wearing surface of cast-iron implements to be hard whilst retaining a strong core, and by chill casting one side of a cutting tool it became self-sharpening because the softer unchilled side wore more quickly and helped to retain a sharp edge.

Another advantage of cast iron was realised in Crosskill's clod-crusher, Fig. 15.37. It consists of a series of metal rollers placed loosely on an axle and revolving independently of each other to

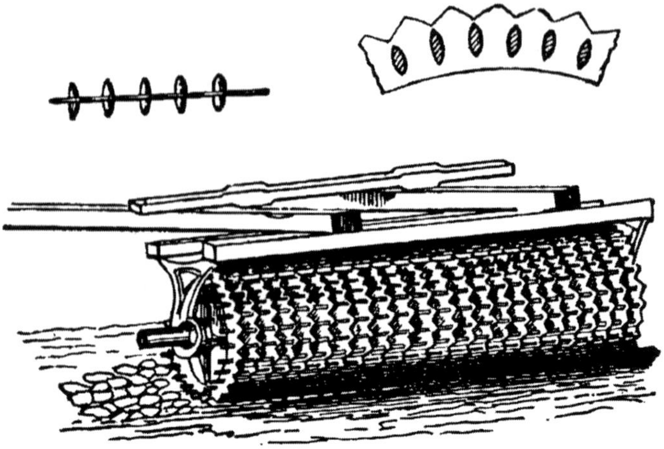

Fig. 15.37 Crosskill's clod-crusher, 1841.

produce a self-cleaning action and to make turning easy. The outer surfaces have serrated edges to break up the soil. Two larger wheels are fitted to transport the machine between fields or over the roads.

Analysis

Ploughing

A plough is a heavy implement requiring considerable force to pull it through the soil. This force is made up of two main components, namely, the force required to cut the soil with the share and the force required to turn the soil as it moves over the mould boards, assuming there are any. It is customary to refer to the draught in terms of the nominal pressure, that is, the draught force per unit cross section of furrow. Furrows are usually cut between 15 cm and 30 cm deep and the width may be 25 cm to 35 cm. The wider and deeper the furrow the greater is the draught.

Fig. 15.38 shows the required draught pressure when ploughing a furrow 18 cm × 25 cm at various speeds in light sandy soils, medium soils, and heavy clay soils. Superimposed on this is the draught pressure that can be generated by 2, 4, 6 or 8 oxen when ploughing the same furrow at various speeds. It is assumed that each ox could generate 430 W indefinitely. The steady ploughing speed

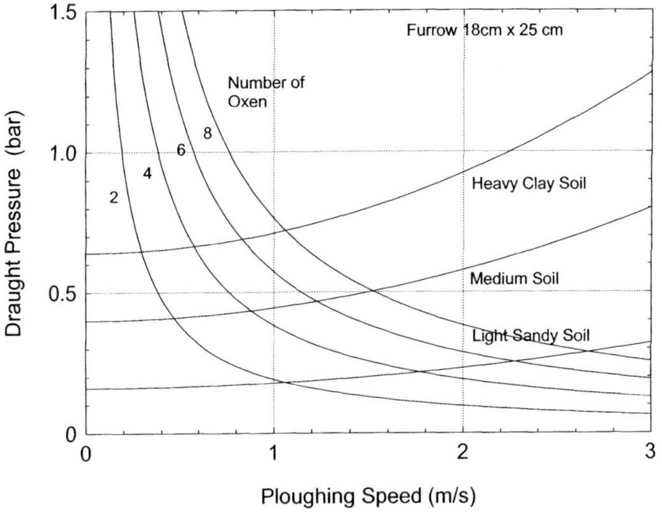

Fig. 15.38 Draught pressure and ploughing speed for 2, 4, 6, and 8 oxen, ploughing on various grades of soil.

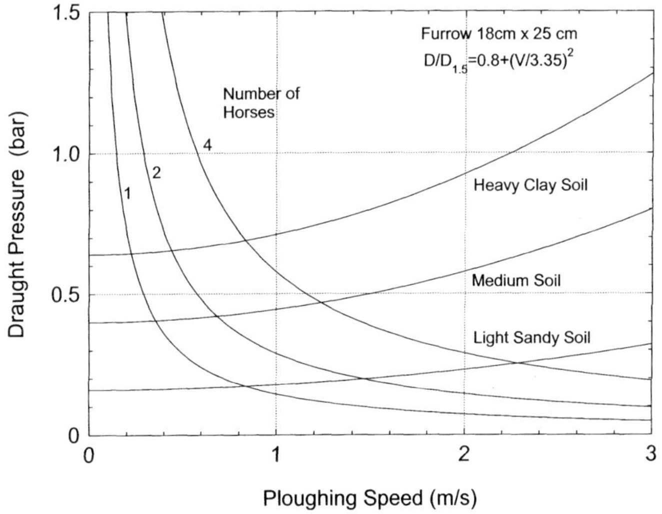

Fig. 15.39 Draught pressure and ploughing speed for 1, 2, and 4 horses, ploughing on various grades of soil.

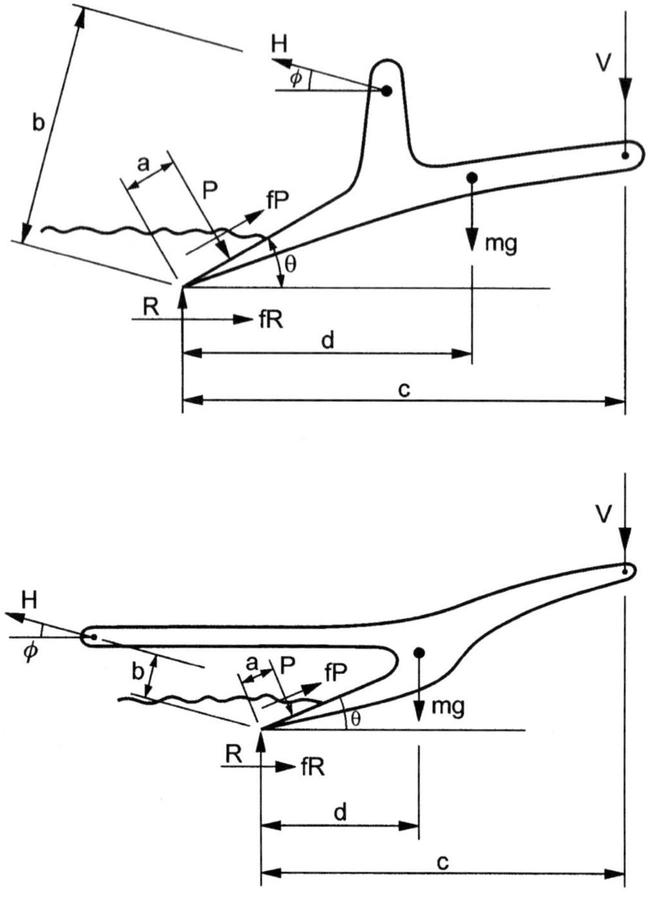

Fig. 15.40 Forces and geometry of a simple plough (left) and a later swing plough. On the swing plough the hitching point is forward and low so the dimension *b* is reduced and the load exerted by the ploughman *V* is reduced for a given draught force *H*.

and draught pressure is defined at the point where the two sets of curves intersect. The ideal ploughing speed is about 1 m/s. At 2 m/s, the ploughman would be nearly running and in addition to exhausting himself, he would find the team unmanageable, whereas 0.5 m/s he would find too slow. Thus on light sandy soils a pair of oxen could plough at about 1.1 m/s and on a medium soil two pair of oxen could plough at about 0.9 m/s. A heavy, clay soil, requires a full team of oxen. Three pair of oxen could manage 0.9 m/s and four pair could manage 1.1 m/s. Of course, the size of furrow would have considerable bearing on these speeds.

A similar calculation for ploughing by horse is shown in Fig. 15.39. In this case, it is assumed that each horse is capable of generating 650 W indefinitely. On light sandy soils, a single horse could pull the plough at about 0.9 m/s. On medium soils, two horses could attain about 0.8 m/s and four horses could attain about 1.3 m/s. On heavy, clay soils four horses are required to plough at about 0.9 m/s.

The draught pressure depends on the type of soil, the ploughing speed, the sharpness of the shares, and the shape of the mouldboard. Soil type is probably the most important factor. The draught pressure varies from 0.2 bar in sandy soils and up to 0.8 bar or more in heavy clay. The moisture content of the soil is also important because a moist soil reduces the draught and reduces wear. Draught increases with ploughing speed. Generally, the draught force increases by about 50% as the ploughing speed increases from 1.5 to 3 m/s. About 40–50% of the total draught force is required to cut the furrow and the remainder is required to turn the soil. A sharp and well-maintained share, and a smooth and well-polished mould board, is important if the cutting force is not to be excessive. The shape of the mould board also influences the draught. Shorter, deeper mould boards having a steeper and more concave curve are more effective in breaking up the soil but require a larger draught.

A useful insight into the mechanics of ploughing may be gained from an analysis of the forces acting on a simple plough during steady operation when they are in equilibrium. A simple plough consists of three main elements: a share that cuts and turns the soil, a connection point from where the plough may be pulled, and a handle with which the ploughman controls the process. In Fig. 15.40 the share enters the soils at an angle θ and depth d from the centre of gravity. The force required to move the soil is P and acts at a distance a from the tip, and the frictional force caused as the soil moves over the share is fP. There is also a vertical reaction R at the tip and its frictional component fR. The horizontal force to pull the plough is H and acts at a distance b above the tip. This force tends to rotate the plough anti-clockwise about the tip and must be balanced by the mass of the plough acting a distance d from the tip and a downward, vertical force V acting at distance c from the tip and generated by the ploughman.

For vertical equilibrium the vertical reaction at the tip is

$$R = V + mg + P(\cos\theta - f\sin\theta) - H\sin\varphi \qquad 15.1$$

The draught force H is obtained from the condition of horizontal equilibrium, and eliminating R by use of Equ. 15.1, gives

$$H(\cos\varphi - f\sin\varphi) - P[(1 - f^2)\sin\theta + 2f\cos\theta)] + f(V + mg) \qquad 15.2$$

Taking moments about the share tip gives

$$Hb = Pa + Vc + mgd \qquad 15.3$$

If a is small compared to the other dimension (b, c, d), and the angles φ and θ are small then Equ. 15.1 and 15.2 may be written as

$$H = V\frac{c}{b} + mg\frac{d}{b} = \frac{P[1 - f^2)\theta + 2f] + fmg(1 - \mathrm{d/c})}{1 + f(\varphi - b/c)} \qquad 15.4$$

The force required to move the plough through the soil is proportional to the cross-sectional area of the furrow, and the quantity $f(\varphi - b/c)$ is usually close to zero, thus we may simplify the equations further to give

$$H = V\frac{c}{b} + mg\frac{d}{b} = pA + fmg(1 - d/c) \qquad 15.5$$

p is the nominal pressure, which depends on the type of soil and the angle of the share, and A is the cross-sectional area of the furrow. Thus the draught load H on the beasts is determined by two equations (Equ. 15.5). In the first equation H depends on the weight of the plough, mg, the downward force exerted by the ploughman V, and the geometrical ratios c/b and d/b. For a given plough the only variable is the ploughman's force V and thus he controls the load on the beasts by varying V. If the ratio c/b is made larger then a small force exerted by the ploughman generates a much larger force on the beasts and the ploughman's fatigue is reduced. In the second equation, the draught load depends on the force required to move the plough through the soil pA, the coefficient of friction f, the weight of the plough mg, and the geometric ratio d/c. The nominal pressure p depends on the type of soil being ploughed and varies from about 0.2 bar on light soils to 0.8 bar on heavy soils. Thus as the ploughman increases the vertical load V he increases the draught load H and the plough sinks into the soil increasing the area A of the furrow so both equations are satisfied.

CHAPTER SIXTEEN

ANCIENT AND MEDIEVAL MINING

Introduction

Mining has a long history and has been so fundamental in human development that we characterize earlier ages by the chief metal that was worked, as copper, bronze or iron. Hesiod (*Works and Days*, 110–180) imagined previous ages that he depicted by gold, then silver, then bronze and finally his own age of iron. Although this may seem a descent from an idealized golden age to the grim reality of Hesiod's time it is, in fact, the order in which metals were discovered and used in ancient Egypt and perhaps elsewhere. But mining is older even than this, for flints were mined during the pre-historic Neolithic and Mesolithic Eras. The products of the mine (gold, silver, gemstones, copper, tin, iron, and coal) have always been greatly prized and sometimes highly valued, but the conditions under which they were won have been mostly appalling. Driven by the desire for such materials, mining has been a major influence on the development of mechanical engineering, largely because mines require ventilation and drainage, transportation for heavy ores, surface treatments to separate and concentrate the product, and transportation of the product to the smelting site. In this and the following chapter we trace something of the history of mining technology and the demands it made on mechanical engineering.

Prehistoric Mines and Metallurgy

The clearance of forest for planting during the Mesolithic and Neolithic periods required a better tool than rough Palaeolithic hand axes and led to the development of the fully polished stone axe, Fig. 16.1. Evidently, the time spent polishing the axe was made up by the time saved in its use. Until man learned how to bore holes through hard materials such polished axes had to be fastened onto their shafts using thongs and adhesives but the resulting axe was very effective.

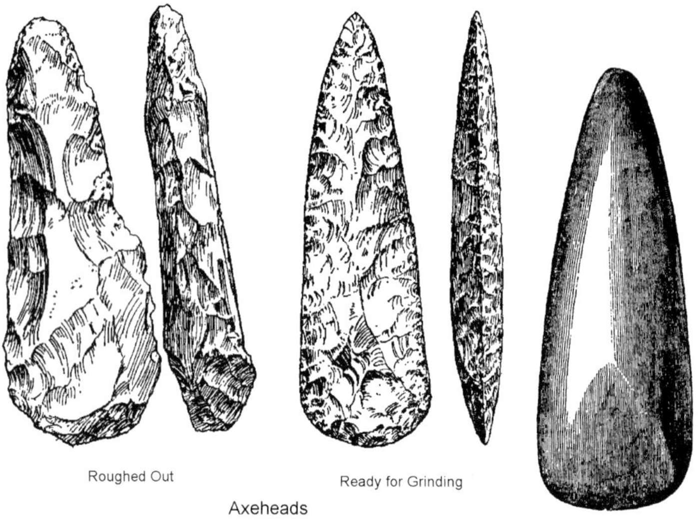

Fig. 16.1 Left and Centre: roughed-out and partly finished flint axe-heads, after Green (1977, 4). Right: a fully polished stone axe.

Modern experiments with Neolithic stone axes have shown that three men can clear 500 m^2 of birch forest in four hours (about 40 m^2 per man hour) and a 180 mm diameter pine tree can be felled in five minutes.

The demand for such axes seems to have been very great and suitable flints were mined at many sites in England, Belgium, France, Denmark, Portugal, and Sicily. Grimes Graves in Norfolk is one such site, see Green (1977). It is not the oldest English mine but it is maintained and is usually open to the public. The older English mines are to be found on the South Downs near Worthing (Cissbury, Blackpatch, Findon, and Harrow Hill) and cover several hectares. These flint mines were worked over a long period. The Windmill Hill people, who arrived in England about 3500 B.C., started the mines and they continued into the Bronze Age when they were taken over by Beaker people. Such numerous and large production facilities, coupled with their use over a very long period, suggests that the mine satisfied more than purely local needs for flint tools and there must have been a lively trade. The production of partly finished axes, requiring only final working, developed to industrial proportions at Grimes Graves and typical products are illustrated in Fig. 16.1.

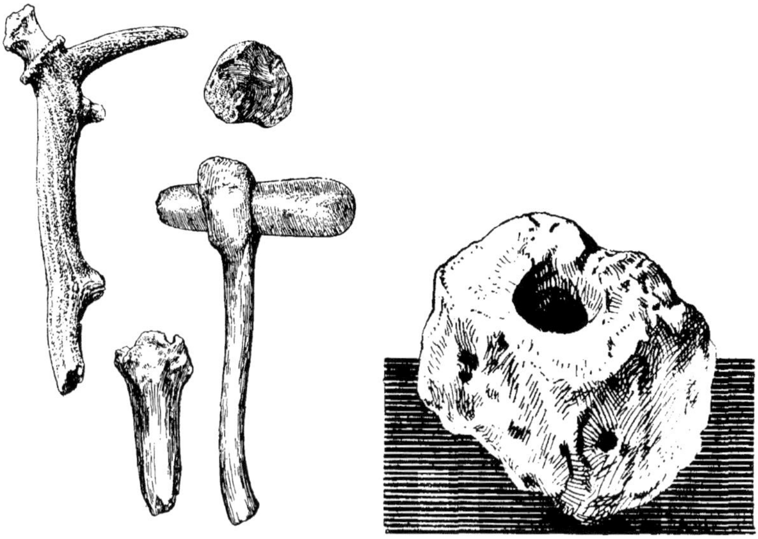

Fig. 16.2 Miners' tools from Grimes Graves. Left: a red deer antler pick, a flint hammer-stone, a bone pick, and a hafted polished flint axe-head. Right: a chalk cup used as a lamp. After Green (1977, 5, 6).

The best quality flint, known as floor-stone, is hard and black in colour. It occurs in the chalk in distinct layers at some distance below the surface and Neolithic man had to dig shafts to reach it. These shafts are of two sizes. The shallow shafts are about 2 m diameter, have vertical sides, and must be dug 4 m to reach the floor-stone. The deep shafts are up to 10 m diameter at the top and are up to 13 m deep; they are usually bell shaped to give them a larger working area at the level of the floor-stone. They were sunk through chalk covered with sand and were, at other sites, dug out with shovels made of ox shoulder blades. However, at Grimes Graves, such shovels have not been found and it is supposed that wooden shovels were used but have since perished. An *L* shaped pick made of the antler of red deer, Fig. 16.2, could remove the harder chalk. In use, the antler's tine was knocked into a crack in the chalk using a flint hammer-stone. Pressure on the pick handle then loosened the chalk. Such picks have been found on the site and pick marks are detected on the chalk walls of the shafts. A hafted flint axe was used to cut the more stubborn blocks of chalk. Much of this work was done in the darkness of tunnels and simple chalk cups, Fig. 16.2 (right), which could be used as lamps, have been found.

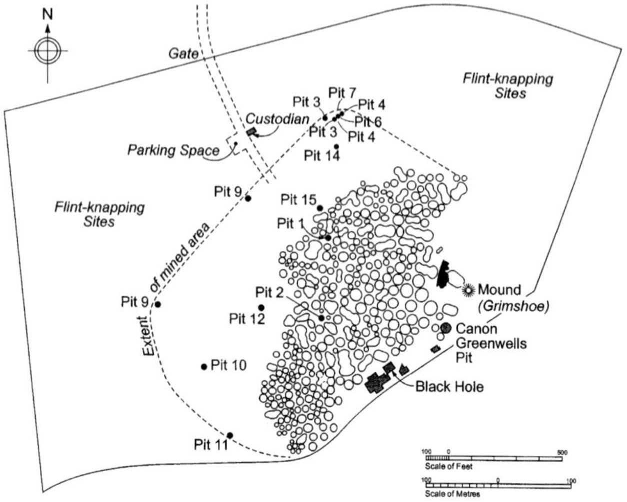

Fig. 16.3 Flint pits and knapping sites at Grimes Graves. After Green (1977).

The removal of chalk from the pits required some ingenuity and a number of ways have been suggested. However, it seems likely that skin bags were used and were drawn up the shaft sides using rope of sinew or some other material. Entrance and exit to the shafts is assumed to be by rope ladder. Removal of material from the bell shaped sides of the deeper pits is thought to be by a tree trunk laid across the top of the shaft from which rope ladders and skin bags could be suspended.

Some idea of the extent of this industry may be had from Fig. 16.3, which shows a plan of the site. It is likely that 700 to 800 shafts and shallow pits were dug over an area of some 14 hectares (34 acres). Many of the pits were connected at the floor-stone level by galleries like those shown in Fig. 16.4, and in consequence much of the work was done in comparative darkness. In one pit, which proved not to have satisfactory flints, a chalk goddess was discovered on a ledge cut in the chalk and seven red deer antler picks were laid on a heap of flints. It is thought that seven miners cut this pit but, on finding poor quality flints, they abandoned it leaving a small offering to the goddess in the hope that they would have better luck elsewhere. Neolithic miners could, perhaps, have dug such a pit and removed all the flints in about six months. As might be expected the pits occasionally collapsed, sometimes trapping or crush-

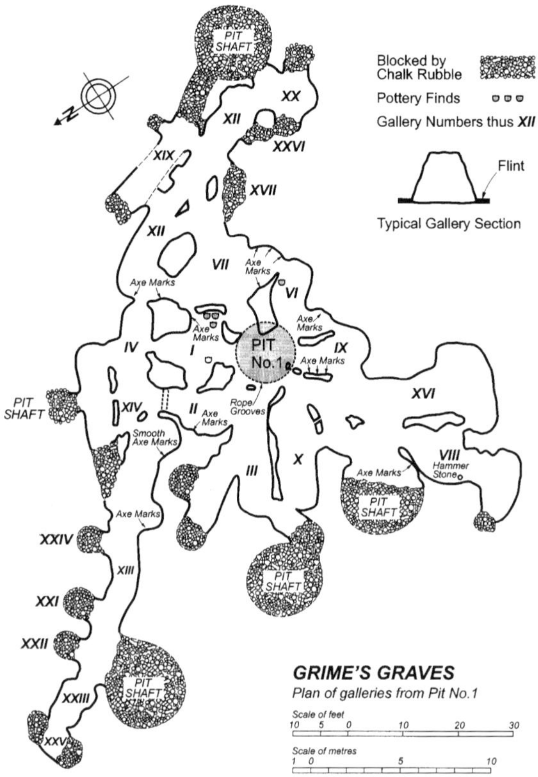

Fig. 16.4 Top: plan of galleries from Pit No. 1 at Grime's Graves. Bottom: bottom of Pit showing entrance to galleries of Pit No. 1. After Green (1977, 10).

ing the miners working at the bottom. At Obourg in Belgium the skeleton of a Neolithic flint miner was found under a rock fall. Near him was the antler pick he had been using.

The areas around the pits are flint-knapping sites where preliminary roughing and shaping of the mined flints occurred and the partly finished product was made. A mined flint was knapped into a roughly shaped axe by striking in a particular direction with a flint hammer-stone until it was reduced to about the right size and shape. Further trimming was done with small flakes or with wooden sticks and hammer stones. The waste material from these operations built up to depths of about 1.5 m on some of the knapping sites.

The invention and spread of metal axes and blades eventually put an end to flint mining but for a long time metal and flint axes co-existed. Presumably metal axes were relatively more expensive, in the beginning, but as they became more plentiful they eventually captured the market. The earliest metal mines are prehistoric although in most cases, the evidence of early mining has been destroyed by subsequent activity. However, ancient mines from the second millennium B.C. have survived, for example, at Cwmystwyth and Nant-yr-Eira in Mid-Wales, at Parys Mountain on Anglesey, on the Great Orme's Head in North Wales, and at Mount Gabriel in south-west Ireland. Radiocarbon dates have been obtained from samples of charcoal because in these mines fire setting was commonly used to excavate the rocks; Ambers (1990, 59). It appears that these mines were worked from about 2100 B.C. (Mount Gabriel), 2000 B.C. (Parys Mountain), 1700 B.C. (Nant-yr-Eira and Great Orme's Head) and 1600 B.C. (Cwmystwyth). The carbon samples give dates over a long period and show that the mines were in use for many hundreds of years.

The most extensive of the surviving early copper mines is at Pyllau on the Great Orme at Llandudno, see James, (1990, 1) and Lewis (1990, 5). It was contemporary accounts of nineteenth century miners breaking into older workings that first generated interest. The Liverpool Mercury in 1849 reported that "On Wednesday the miners at Llandudno broke in the course of their labours into what appears to be an extensive cavern, the roof of which being one mass of stalactite reflected back their lights with dazzling splendour. On examination the cavern turned out to be an old work, probably Roman, the benches, stone hammers etc used by that ancient people having been found entire, together with many bones of mutton,

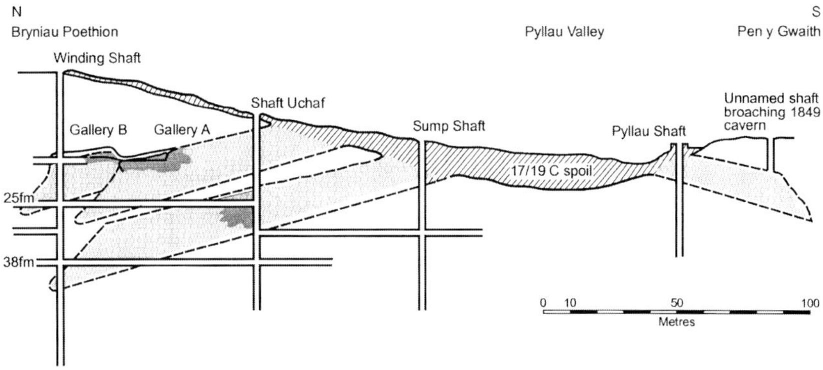

Fig. 16.5 Sketch section across the Pyllau valley, showing the principal nineteenth century shafts and levels and showing the putative extent of prehistoric stoping (broken lines), with back filled pack (light stipple) and the relative positions of surveyed early stopes (slid lines) and pack (dense stipple). After James (1990, 1).

which had been consumed by those primitive miners. The bones are to all appearance fresh, though impregnated with copper, as they were when denuded of their fleshy covering, after remaining nearly 2,000 years in the bowels of the earth. The cavern is 40 yards long and must be a subject of great interest to those investigating the remains of bygone ages."

Davies (1935) was also of the opinion that these mines were Roman but the mines were sealed when he visited them in 1948 and he based his opinion on material found in the waste dumps that included stone hammers and Roman coin and pottery. In 1976, an amateur archaeologist, Duncan James, discovered two ancient galleries containing stone hammers, bone tools, and charcoal that had been used for fire setting to crack and fracture the rock. Subsequent excavations, from 1987, have revealed a complex network of Bronze Age galleries, many of which have survived undisturbed and indicate that many thousands of tonnes of ore had been mined. James (1990, 1) prepared a simplified cross section of the mine, Fig. 16.5. He says that the working styles and techniques of the recent and ancient mining are quite distinct and wherever the two converge the intrusive element is always the modern deep-level mining.

The total length of underground tunnels that have now been surveyed is greater than 6 km and Fig. 16.6 gives some idea of their complexity and the locations of various underground-finds. Some 900

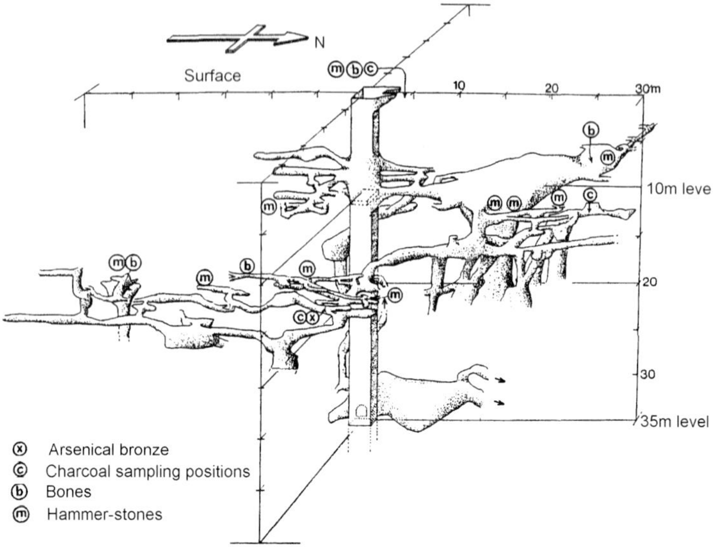

Fig. 16.6 The Bronze Age Copper Mine on the Great Orme, Llandudno showing a simplified view of Vivian's Shaft. After Lewis (1990, 8).

stone hammers, weighing between 0.25 kg and 26 kg, have been found and exhibit the characteristic percussion marks where they have been used. Over 8,000 pieces of bone have also been found, and ten samples of charcoal have yielded radiocarbon dates between 1800 B.C. and 600 B.C. There is no evidence of Roman activity at the mines and it seems they were not worked after about 500 B.C. until the seventeenth and eighteenth centuries A.D.

Like early farming, metallurgy was assumed to have spread into Europe from the Near East where it was discovered. This, certainly, was Childe's view and to some extent the isochronic map showing its early development and spread, Fig. 16.7, supports it. Comparing Fig. 16.7 and Fig. 15.2 of Chapter Fifteen, and allowing for the difference between calendar dates (B.C.) and carbon 14 dates (b.c.), we see that metallurgy developed in most places about 2,000 years after the arrival of agriculture. Smelted copper was in use in the seventh millennium B.C. in Anatolia, in the fifth millennium it was employed in the Mesopotamia, Palestine, and Iran, and by the fourth millennium it was practised throughout the Near and Middle East and in Egypt. The earliest European metallurgy is found in the Balkans in the early fifth millennium and so it does look as if it

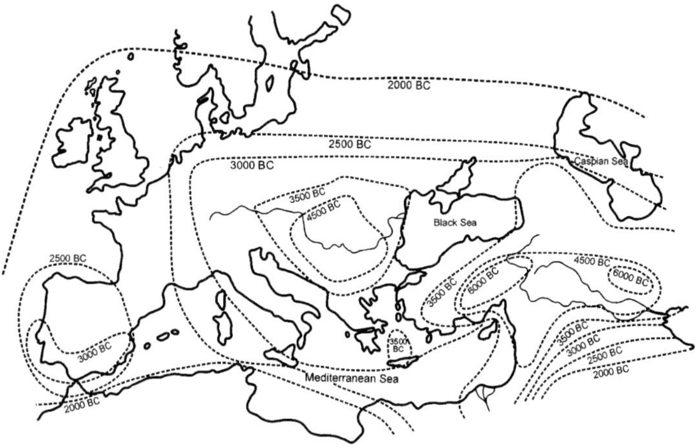

Fig. 16.7 Isochronic map showing the development and spread of early metallurgy. After Whitehouse in Collins (1975, 184).

spread to Europe from the Near East. However, it seems that Balkan metallurgy was more developed than that in the Near East. For example, Balkan metallurgy in the fifth millennium used bivalve moulds and produced tools with shaft-holes by using cores. These are quite sophisticated techniques and are considerably in advance of the contemporary methods used in the Near East. For this reason, Renfrew (1969, 12–47) has suggested that European metallurgy was discovered and developed independently. However, if this is the case then we need to explain how it developed so rapidly from simple open cast moulds to bivalve moulds and to moulds with cores.

Early farming spread into the Balkans from the Near East before 5000 B.C. (6000 b.c.), see Fig. 15.2 of Chapter Fifteen, and by 4500 B.C. copper metallurgy had been developed to a remarkable degree. Fifteen hundred years is, of course, a long time, and it is quite possible that Balkan metallurgy progressed in this remarkable fashion whereas in the Near East it remained relatively stagnant. In the fertile valleys of the Tigris, Euphrates, and the Nile, which supported a relatively dense population and where civilisation and urban life first developed, copper ore does not occur naturally and had to be imported, whereas in the Balkans the ore is plentiful. It may be that, despite a late start, the local availability of relatively inexpensive ore enabled Balkan metallurgy to develop very quickly.

In Europe there were no great concentrations of population, as there were in the Near East, and the early development of metallurgy occurred in the regions where the ores were found, namely, in the Balkans, the Aegean, Italy, and Iberia. As mentioned above, Balkan metallurgy developed in the fifth millennium B.C. and was based on Transylvanian ores. The Balkan smiths had mastered the art of casting in closed, rather than open, moulds and they produced shaft-hole axes and axe-adzes. Probably they were the first to produce these goods.

Metallurgy was practised along the north Aegean coast by about 3500 B.C., and this may be considered a southward extension from the established industry in the Balkans, and is confirmed by cultural and other similarities. Smelted copper was also in use in Thessaly, Crete and the Cyclades but it did not become common until the first half of the third millennium B.C., when it also occurs on mainland Greece, Cyprus, and western Anatolia. Tools were made of copper alloyed with arsenic, and include daggers, which were sometimes flat and sometime strengthened with a midrib, spearheads, and axes. Good sources of copper are known in eastern Crete, in Cyprus, and in more than one of the Cyclades. A great variety of ornaments were also made in copper, silver, or gold.

Smelted copper metallurgy was known in Italy, as in the Aegean, in the fourth millennium B.C.; the earliest copper slag comes from the island of Lipari before 3500 B.C. On the Italian mainland there were two principal groups practising metallurgy, namely, the Remedello group in the Po valley, who exploited Alpine ores, and the Reginaldone group in central Italy using Tuscan ores. Italian smiths made flat axes, triangular and tanged daggers (flat and with midrib), halberds, awls, and a few simple ornaments, some of silver or antimony. The halberd appears to have been an Italian invention arising from the attempt to mount a dagger at right angles to its shaft in the way stone battle-axes had been made. In Iberia, metallurgy is thought to have originated before 3500 B.C., although there are very few carbon 14 dates available. By 2500 B.C. smelted copper metallurgy, using local ores, was flourishing in southern Spain and Portugal. The usual flat axes, flat daggers, daggers with midribs, and ornaments were produced.

The dissimilarity between the polished stone axe, which continued in use during the Bronze Age, and the bronze axe, is illustrated in Fig. 16.8 and 16.9. The stone axe is a brutal instrument, and is

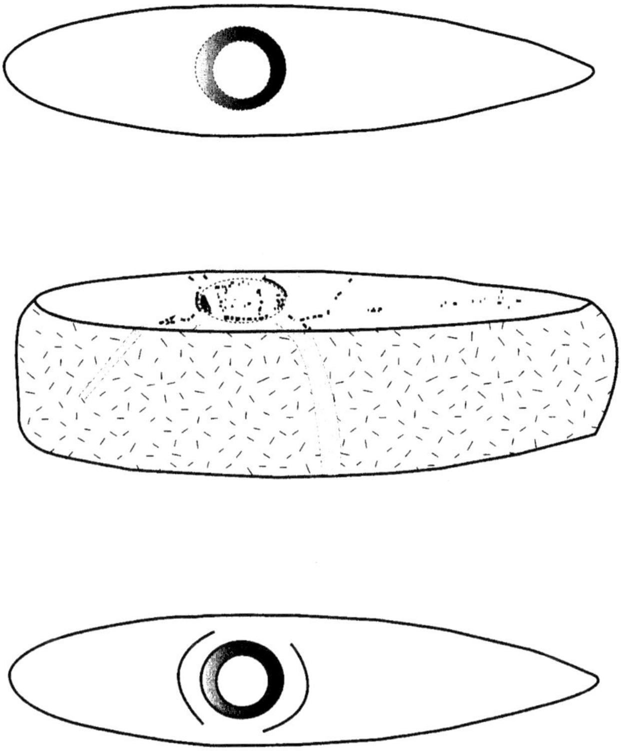

Fig. 16.8 Battle-axe. Hard bluish stone with white veins 170 mm long, 45 mm broad. Late Beaker date. After Thurman, 1872.

referred to as a battle-axe although its use is not really known. This example is made of a hard, bluish stone with white veins and has an estimated weight of about 500 grams. There is a neat hole, tapering towards centre and one side has a roughly scratched circular groove made no doubt, when the hole was cut. Probably the hole was bored from opposite sides using a bow-drill.

In contrast to the stone axe, Fig. 16.9 (left) shows a cast, flanged, bronze axe, weighing 125 grams, and 95 mm long. It has gently concave sides. A stop bevel defines lower limit of haft and bounds a panel of horizontal fluted ornaments on the blade face. Such axes, of the early Bronze Age, have a widespread distribution in England. The flat axe shown in Fig. 16.9 (right) is 183 mm long and weighs 813 grams. It has a thin, narrow, arched butt and the blade curves gently from the butt to an expanded curved cutting edge which

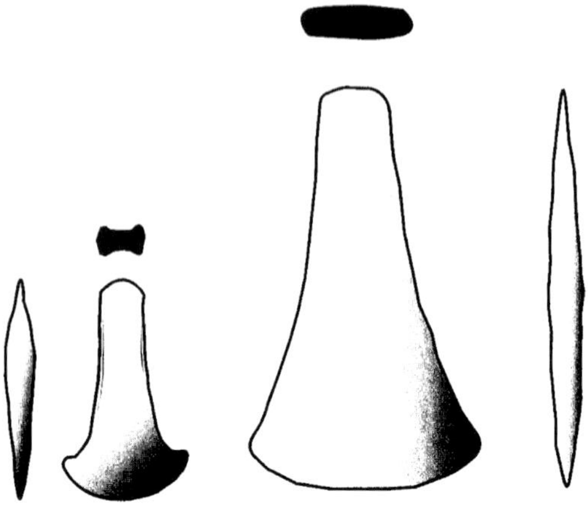

Fig. 16.9 Left: cast, flanged, bronze axe. 95 mm long, 53 mm wide, 125 grams. Right: flat, bronze axe. 183 mm long, 17 mm thick, 813 grams. After Saunders (1975–6, 125).

appears to have been work-hardened and provided with a steeper edge by hammering, perhaps when the axe was re-sharpened. Normally such axes were cast in tin-bronze in open moulds, and subsequently heated and forged to their final shape. Flat axes are also widely distributed in Britain.

The stone axe is much the same length as the flat axe illustrated in Fig. 16.9 (right) but is lighter despite its greater volume. The angle of the cutting edge is also much larger (blunter) than either of the metal axes and it would, therefore, have been considerably less effective than the metal axes. A mould, used for casting bronze axes is illustrated in Chapter Eighteen, Fig. 18.2.

The transportation of ores was a problem to all the early metal working societies. The earliest mines would have been open cast and the best ore was separated from rock by crushing, washing, and handpicking. All this work would undoubtedly be done near the mines and it is very likely that smelting and casting, to form either ingots or finished products, was also done locally. These products would then be carried for trade. Early metalworkers may have been nomadic, travelling from place to place in search of suitable ores and customers. For this purpose two-wheeled carts were constructed, the earliest from 3500 B.C. in Mesopotamia. Where possible, the smith's heavy loads were moved by boats along the rivers and must have stimulated boat building.

The period following the introduction of metallurgy in Europe appears to have been one of great social and economic change. The millennium 3500–2500 B.C. is characterised in Europe by a series of population movements, usually from east to west. These migrations are different from the earlier migrations that first brought agriculture to Europe, for the earlier groups expanded into a relatively unsettled continent whereas the later migrants moved into settled, agricultural communities. Such intrusions may have been peaceable, if there was adequate land, but conflict and warfare are also possible. The situation that developed was complex but has been simplified, by Whitehouse in Collins (1975, 195), into two main groups: the Eastern Group and the Western Group.

The Eastern group is often described by archaeologists as Globular-Amphora, Corded-Ware, Battle-axe, or Single-Grave cultures; names that lack charm but have considerable practicality. Such people are found all over northern and Eastern Europe and may have originated in the Kurgan (Pit Grave) culture of the Russian Steppes. They are characterised by single graves, often under a barrow, wheeled vehicles, domesticated horses, battle-axes, and copper metallurgy. The Western Group is more usually called the Bell Beaker culture after the characteristic shape of their pots. They had a quite dense but discontinuous distribution over eastern and central Europe, the Low Countries, Britain, Brittany, southern France, and coastal Iberia. They are also found in Italy, Sardinia, and Sicily. Like the Eastern Group, the Bell Beaker culture is characterised by single graves under barrows, by copper metallurgy, and by survivable archery equipment such as flint arrowheads and polished stone wrist-guards. The westward movement of the Bell Beaker people is also thought to have resulted from pressure by the Kurgan (Pit Grave) culture. They were in existence shortly after 3000 B.C., and in the Low Countries they seem to have met and intermixed with the Corded-Ware people. The British Isles was characterised by mixed Bell Beaker and Corded Ware cultures.

The Ancient Civilisations

The ancient civilisations were essentially of the Copper and Bronze Age, and the metalworker was a man of some importance whose expertise and products were transforming the nature of society through their effect on agriculture, warfare, and transport. Copper technology, as we have seen, was first exploited in quantity during the fourth millennium B.C., throughout the Near East. In pre-dynastic Egypt, about 3200 B.C., flint knives and tools of delicate workmanship were fitted with sheet gold handles and coexisted with copper weapons and implements. It was an age of transition from flint and stone to metals, mainly copper. There were no ores to be had in the Nile valley so Egypt had to import metals through trade or send out mining expeditions to more mountainous regions. The first dynasty king, Semerkhet, in about 3010 B.C., carried on mining operations for copper at the Wadi Maghâra in the Sinai peninsular, Breasted (1912, 48). Local tribes attacked Semerkhet's expedition but they were suppressed and the victory was recorded by a surviving relief cut into the rock walls of the Wadi.

In the Old Kingdom, 2686–2181 B.C., gold was obtained by mining the quartz veins of the granite mountains along the Red Sea coast. A road ran east from Coptos, which was on the Nile in Upper Egypt, to the Red Sea. This road passed through the Wadi Foakhir where the Egyptians mined their gold. Further south, in Nubia, gold was also mined by the independent Nubians and traded into Egypt. Without doubt much gold was mined and traded in the Old Kingdom, but little of their workmanship has survived. In Cairo Museum there is an exquisite gold container, of about 2660 B.C., made in the form of a shell with a delicate hinge and clasp, and of about 2350 B.C. there is a famous gold and obsidian falcon's head. There is also wooden furniture covered in gold and some very elegant gold vessels and personal jewellery from the tomb of Queen Hetep-heres, of about 2585 B.C. Mostly the artefacts of the Old Kingdom have been robbed or re-worked, but reliefs in the tomb-chapels often depict the goldsmith at his work, see Fig. 18.8, Chapter Eighteen. Silver was imported from further afield and was probably obtained from Cilicia in Asia Minor. It was, in consequence, exceptional and more valuable than gold. Large works in copper were undertaken, such as the life size, copper statue of Pepi I, 2332–2283 B.C. The statue is hollow and found inside it was a smaller statue of his son. It appears to have been made by hammering the metal into shape over a wooden core, to which it was then nailed.

The gold-rich country of Nubia was subjugated, at least as far as the second cataract, by Senusret I, 1971–1926 B.C., and the local

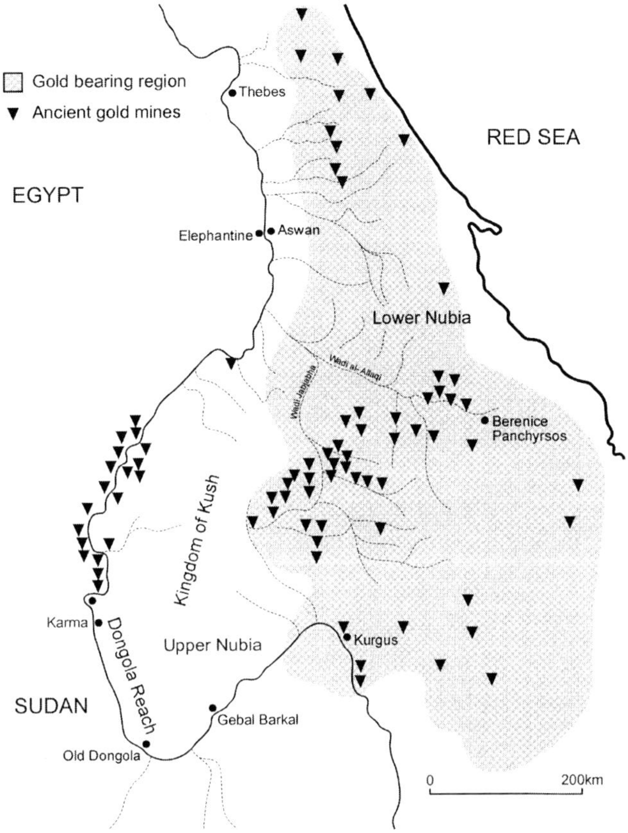

Fig. 16.10 Location of gold mines in the Eastern Desert between the Nile and the Red Sea. After Davies and Friedman (1998, 130).

chiefs were forced to work the mines around the Wadi Alâki to provide tribute to Egypt. Ameni, nomarch of the Oryynx-nome, was dispatched to Nubia at the head of four hundred of his own troops to bring back the output of gold. Similarly the gold-country to the east of Coptos was again exploited and the faithful Ameni, with six hundred troops, guarded the vizier, who had been sent there to convey the precious metal safely to the Nile valley.

Under Senusret I's son, Amenemhet II, and grandson, Senusret II, Egypt enjoyed great prosperity and the copper mines in the southwest of the Sinai Peninsula were reopened. The route from these mines was from the west coast of Sinai, southwards down the Red Sea to the Egyptian desert port of Leucos Limen, and then east along the desert road through the Wadi Foakhir to Coptos on the Nile, a little above Thebes. The road across the desert to Coptos took five days but it had already been supplied with wells and stations by the Theban kings of the eleventh dynasty, 2134–1991 B.C. It was an important route.

Egypt, however, did not acquire large amounts of gold until the famous eighteenth Dynasty and the start of the New Kingdom. Once the foreign invaders, the Hyksos, were defeated in the Delta, using the very weapons that the invaders had introduced into Egypt (horse drawn chariots and the composite bow) the Pharaoh's turned their attention once again to Nubia in the south. *Nubia* means *land of gold* and their mines, see Fig. 16.10, were the main attraction. Gold mines occur everywhere between the Nile and the Red Sea and there is no doubt that they were ruthlessly exploited, probably using slave labour, prisoners of war, or convicts. It is recorded that in a period of three years some 794 kg of gold were received into the treasury of Tuthmosis III, 1504–1450 B.C.

Seti I, 1291–1278 B.C., also sought to increase his revenue and that of the Abydos temple by increasing his supply of gold. He turned his attention to the resources from the mountains of the Red Sea region around Gabel Zebâra. Here serious difficulties beset the route, especially in relation to its water supply. The road to the mines left the river a few miles above Edfu. Seti visited the place himself to discover what might be done to improve matters. Some two days into the desert he found there was an old and probably disused caravan station of the eighteenth dynasty (1570–1293 B.C.). Here he searched for water and a well was dug, yielding a plentiful supply. Seti erected a small temple by the well and established a small settlement. Breasted (1912, 416) tells that the thirsty caravaners sang his praise: "You gods dwelling in the well give to him your duration; for he has opened for us the way to march in, when it was closed before us. We proceed and are saved; we arrive and are preserved alive. The difficult way which is in our memories has become a good way". It was now possible to reach and work the mines, and with their income, Seti made a permanent endowment for his temple at Abydos. Seti attempted to repeat his success at Gabel Zebâra with the gold mines further south in the Wadi Alâki. Here he dug a well 70 m deep on the road leading southeast from Kubbân in

Nubia, but he failed to find water, and the attempt to open the mines was abandoned.

Seti's son Ramesses II (1279–1212 B.C.) also searched for new sources of income. In his third year he visited Memphis to consult with his officials and the viceroy of Kush (Nubia) about opening up the gold mines at the Wadi Alâki in Nubia, the very mines his father had failed to exploit. He was told that when caravans attempted the journey "it was only half of them that arrived there; for they died of thirst on the road, together with the asses they drove before them". Travellers were obliged to take enough water for the round trip, as they could obtain none at the mines, "Hence no gold was brought from this country for lack of water." The viceroy's advice that another attempt be made to supply the route with water was accepted. A subsequent letter from the viceroy announced the complete success of the enterprise and the discovery of a copious spring of water at a depth of only 7 m. At Kubbân, where the road leading to the mines left the Nile valley, Ramesses had the viceroy erect a stela commemorating the achievement and bearing a record of these events.

Expeditions in search of copper were also launched under Ramesses III, 1182–1151 B.C. Copper had long been imported from Cyprus where the ancient world's richest deposits were found. Ramesses III found a closer source at Atika, a now unknown place in the Sinai Peninsula. He sent a special expedition of galleys up the Red Sea, no doubt using the old route from Coptos. They returned with great quantities of the metal and the Pharaoh was so proud of this achievement that he had it displayed under the palace balcony that all the people might see it.

Mesopotamia, like Egypt, also used metals despite having no local ores, so they too obtained supplies by trading or by sending out mining expeditions. Postgate (1992, 227) tells us that the cuneiform sign for a 'smith', was also used for the verb 'to pour', implying the practice of casting in moulds, and was found in tablets of the Uruk IV period (about 3100 B.C.). This technology is confirmed by archaeological finds of contemporary artefacts.

Pure copper is rather soft, but ores often contained small quantities of arsenic that produced a stronger, harder alloy. The addition of tin, a relatively rare metal, to form a true bronze that was much harder than copper, was already in use by the middle of the third millennium, but not universally. Recent work has shown the extensive use of arsenical copper and pure copper for casting elaborate artefacts for which bronze had been thought essential. The quality of some metal work from the Akkad Dynasty is superb, and implies mastery of large-scale and sophisticated techniques. A translated text from the Ur administration, in Postgate (1992, 226), mentions 1,083 old copper sickles and 60 copper hoes, sent from the storeroom to the smiths for refurbishment. Metal artefacts, even when their useful life was done, remained valuable and were easily re-cycled into other products, consequently much of the early metal work was lost.

From these texts, it is certain that the palace and temple in Mesopotamia maintained large metalworking establishments, although none have yet been identified by excavation. However, two early second millennium workshops are known, one in the kārum at Kaneš, and the other in the small town of Tell adh-Dhibai, now inside modern Baghdad. The excavators found the equipment of a coppersmith, complete with clay moulds, crucible, tuyeres and bellows.

Standard measures were in use and were made in copper or bronze. From the prologue to the Ur III law code the king proclaims "—I fashioned a copper bushel, and fixed it at 60 quarts, I fashioned a copper measuring-vat and fixed it at 10 quarts, I fashioned the regular vat of the king and fixed it at 5 quarts. I fixed the weight-stone of 1 shekel of silver at 1-sixtieth of a mina; I fashioned a bronze (vessel) of 1 quart and fixed it at 1 mina." Postgate (1992).

The metal and metal ores used in Mesopotamia were not mined locally but were imported. The chief sources of ore appear to have been Anatolia and Cyprus. The mountains of Anatolia are rich in minerals and were worked in ancient times. Copper was the main export of Assyrian traders in Cappadocia just before the foundation of the Hittite kingdom, and silver seems to have been plentiful from that time onwards, since it served as a currency. Hittite and Assyrian texts give the names of the mountains and cities from which these metals were obtained but the places have not yet been identified. Gurney (1990, 67) tells us that there are Neo-Hittite monuments near the present day mining area of Bolkar Maden in the Cilician Taurus, so perhaps it has always been a principal mining centre for silver and lead.

We are told in a religious text that copper was brought from Alasiya (modern Cyprus) where the richest copper mines in the ancient world were to be found. The Hittite king Amuwandas I claimed Alasiya as a province of his empire. Copper was relatively plentiful. Its price, compared with silver was 1:240, and there was a surplus for export. There is, however, very little direct evidence, either in ancient texts or from excavations, of the commercial dealings of the Hittites either with each other or with the world at large but the evidence relating to Cyprus is very considerable. Cyprus and copper were synonymous in the ancient world and bars of Cypriot copper bearing the characteristic ox-hide shaped stamp, have been found all over the Mediterranean area.

Constantinou (1992, 69–71) gives us some idea of the size of this industry, and his estimate of the environmental impact on the island is revealing. The present day remains of ancient slag in Cyprus amount to 4×10^6 tons from which Constantinou estimates that 200,000 tons of copper metal were produced over a period of 3,500 years (an average of 57 tons per year). This seems credible. He estimates 300 tons of charcoal to produce a ton of copper metal and 14 tons of pine to produce a ton of charcoal using relatively inefficient ancient methods. Thus 57 ton of copper per year consumes 240,000 tons of pine per year. Constantinou allows an average production rate from Cypriot woodland of 80 m^3 per hectare (56 tons per year per hectare) from 80–100 year old trees. Thus the annual consumption of woodland was 4,286 hectare and as it takes about 90 years to re-grow, the required area of woodland was 386,000 hectare. This is 42% of the land area of Cyprus, and over 3,500 years the area of woodland destroyed amounted to about 16 times the size of the island. The consumption of woodland to provide fuel for smelting and other metalworking processes was to remain a major problem in all countries and at all times.

The Late Bronze Age (1600–1050 B.C.) was the most prosperous age for copper mining for they declined somewhat after the discovery of iron. Copper, even arsenical copper, is relatively soft but it can be hardened by the addition of about 12% tin to form bronze, and this was how copper was normally used. Tin, however is a quite rare metal and supplies were difficult to find. Anatolian copper was exported in exchange for Mesopotamian woven materials and tin. The latter was probably obtained from somewhere to the east of the Hittite kingdom and imported via Assur, Gurney (1990, 71). It is possible that the tin deposits of the Caucasus region were becoming available to the Hittites and a source of tin has even been identified near Nighde on the Anatolian plateau. Constantinou (1992, 17) says "The Cypriot miners and metallurgists ventured far in search for tin, a metal found in the Iberian Peninsular within the Mediterranean realm.—The superiority of bronze over copper led to voyages beyond the confines of the Mediterranean—Cornwall in the west, Afghanistan to the east, the Caucasus to the northeast-in search of both copper and tin." But copper and bronze lost their predominance as iron technology spread.

Silver seems to have been available in quantity for it was the common medium of exchange throughout the ancient Near East. For small denominations, lead was used. Silver, in bars or rings, was measured by weight, though the more primitive method of reckoning in terms of measures of barley was also found. The units of weight were the shekel and the mina, in the proportion of 1:40 in Anatolia, as compared to the Babylonian and Hebrew proportions of 1:60 and 1:50 respectively. Thus if the weight of Hittite shekel was the same as in Babylon (8.4 grams), the Hittite mina would have been much lighter. That money, in the form of weighed silver, was in common use is demonstrated, for example, by the Hittite Law Code that contains a table of prices, and of fines for wrongdoing of which an extract is given by Gurney (1990, 78). "If anyone breaks a freeman's arm or leg, he pays him twenty shekels of silver and he (the plaintiff) claims for it against his house. If a freeman kills a serpent and speaks the name of another [a form of sorcery], he shall give one pound of silver; if a slave does it, he shall die. If a man fouls a pot or a cistern, formerly they paid six shekels of silver; he who fouled it paid three shekels of silver (to the owner), and into the palace, they used to take three shekels of silver. But now the king has remitted the share of the palace; the one who fouls it pays three shekels of silver only and he (the plaintiff) claims for it against his house."

The relatively high value of metal artefacts, even as scrap when they had completed their useful lives, meant that many articles were re-melted and re-used and thus much of the evidence of early metal working is lost. Nevertheless, artefacts have survived in sufficient numbers so there is adequate evidence of what was made and of how it was made. Iron was a very valuable commodity in the early second millennium. Although the techniques required for smelting were not known, iron goods certainly occur and are mentioned in Sumerian texts. Presumably, at this early date, the iron was derived from meteoric sources. Iron was a valuable commodity and was smuggled, as the following Sumerian text shows. "Please do not smuggle anything. If you pass through Timilkia, leave your iron, which you are bringing through, in a friendly house in Timilkia, and leave one of your lads whom you trust, and come through yourself and we can discuss it here", Postgate (1992, text 11.4).

Iron was not used in the third millennium and throughout the second millennium, it was treated as a precious metal. This was not from any shortage in the supply of ores but rather it was because the technique of producing the high temperature required for smelting and working it was not generally understood. The Hittites are usually credited with the discovery of iron smelting and working. Iron ores are abundant in the Anatolian Mountains. Iron swords, iron writing-tablets, and iron statues of gods and animals are mentioned in Hittite texts, but only a few iron objects of the Hittite period have been found. These articles were all special and expensive products for use in the temples or intended as royal gifts. The earliest reference of this kind appears to be an inscription of Anittas (early second millennium B.C.), who records that he received as tribute from the city of Puruskhands an iron sceptre and an iron throne. However, some part of the document is known to be a later addition so the passage is not firm evidence for the manufacture of these large objects at such an early date. Small iron objects, such as pins and small ornaments, have been found in earlier contexts.

A few skilled artisans mastered the technique of iron working during the Hittite period. Consequently, they were able to charge a high price for their products. The prominence of the Hittites in iron technology in the thirteenth century is substantiated by a well-known passage from a letter. The letter was from the Hittite king Hattusilis III (1264–1239 B.C.) to a contemporary, probably the king of Assyria. "As for the good iron which you wrote about to me, good iron is not available in my seal-house in Kizzuwatna. That it is a bad time for producing iron I have written. They will produce good iron, but as yet, they will not have finished. When they have finished, I shall send it to you. Today now I am dispatching an iron dagger-blade to you." This passage suggests that the Hittite king was placing an embargo on the export of iron, probably for military reasons. Gurney (1990, 67–8) is not convinced of this. He thinks that the reference to the 'bad time for producing iron' may be explained by the assumption, observed in other societies, that iron smelting was carried out by peasants in their homes during the winter season when there was no work to be done in the fields. So it might well occur that in the late summer or autumn the stocks of smelted iron were very low. This, however, seems to be a very unlikely explanation. If iron were a valuable metal, those who knew how to work it would surely not waste that talent by working in the fields. They would smelt iron, were it available, all year round and pay others to work in the fields. It is known that Kizzuwatna, mentioned in the text, was to the south and it is probable that the iron was mined somewhere in the Taurus Mountains, where it is known to exist.

Greek and Roman Mines

Finley (1965, 30) notes that in mining: "—the ancient world reached its maximum achievement very early, in the Athenian silver mines of the fourth and fifth centuries B.C. The tunnelling, ventilating, and lighting in those mines, the washing, crushing, and smelting in the near-by mills and furnaces, and the utilisation of the by-products were all as competent and efficient as anything to be found in the next thousand years, and better than most." This is certainly true, and it is from this period that more extensive archaeological sites and written references to mines and mining techniques occur.

The number of Roman mines is astonishing. Davies (1935, Map IV) has recorded 120 in Wales and Cornwall and a further 66 in the remainder of England and Scotland; see Fig. 16.11. Table 16.1 lists the number of mines in various regions of Rome's European Empire. It appears that the number of mines in Europe amounts to 1034 and there were, of course, many more in Asia and North Africa. In Britain, surface outcrops of coal were exploited and the coal was sometimes carried a considerable distance. It was used for drying unripe grain, for heating hot baths, for domestic heating, and for iron-smelting, Liversidge (1973, 200). Numerous iron mines were worked, particularly in the Forest of Dean and on the Weald but the Cornish tin mines were not developed much by the Romans until the third century A.D. Probably, the Cornish mines could not compete, until the third century, with those of Northern Spain. Lead, copper, and gold were also mined in Britain.

The only known British gold mine worked by the Romans is at Dolaucothi in Wales. This was opened up by Mitchell in 1909 and revealed something of the Roman use of fire setting. The remains of old fires seem to have been lit several feet away from the rock

Table 16.1 Number of Roman Mines in Europe (Davies, 1935, Map IV).

Location	*Number of Roman Mines*	*Proportion (%)*
Italy	76	7.4
Greece	113	10.9
France and the Rhineland	198	19.1
North West Spain	44	14.6
Southern Spain	82	
Rest of Spain	25	
Wales and Cornwall	120	18.0
Rest of England and Scotland	66	
Germany and Middle Danube	60	5.8
Illyricum, Moesia, and Macedonia	251	24.2
Total	1034	100

Fig. 16.11 Roman mines in England, Scotland and Wales. After Davies (1935, Map IV).

face and the hard quartz veins had been reduced to uniform, granular sand, without crushing. Experiments by Holman (1927, 239–245) showed that the quartz undergoes a phase transformation at about 575°C, which causes it to crumble. Better results were obtained if the rock was heated quickly to a temperature not too far above 575°C and then quickly cooled or quenched from this temperature. Repeated firing to temperatures of 300 to 400°C was also found effective in crumbling the quartz. At Dolaucothi the Romans worked two large outcrops of gold bearing pyrites to a depth of about 10 m, and underground galleries, 2 m square, were dug to a depth of about 25 m. Nash-Williams (1950). Two crosscuts, which may have been intended for drainage, extended about 55 m into sterile rock. Rock pillars and heavy timbers were used to support the overburden. The lower galleries were drained by large horizontal-axis waterwheels operated by men sitting above them and pushing against a treadmill on the hub of the wheel. Buckets were fastened to the outer rim of the wheel and as they rotated about the horizontal axis they scooped water from the floor level and released it into drainage channels at a higher level. A piece of the rim of such a waterwheel was found at Dolaucothi, and such wheels remained in use, in the Cornish tin mines for example, until the twentieth century, although treadmills were not used.

The ore was washed and cleaned by water from the River Cothi and brought to the mine by a rock-cut aqueduct 11 km long. Wooden pan cradles, millstones, a wooden haulage tray, part of a stone mortar, and several crucibles were also found and show that the gold was processed at the mine. Unusually, pithead baths for the miners were found in the nearby settlement and the discovery of a hoard of jewellery suggests that goldsmiths lived and worked on the site.

The Romans also mined iron in Britain as illustrated by that at Lydney, south of the Forest of Dean, Gloucestershire, Fig. 16.12. This mine was cut in the third century A.D. and a hut was later erected over the entrance. Characteristic tool marks were found on the walls and suggest that they were made by short pick-hammers similar to those commonly used until the introduction of gunpowder in early modern times. Such an iron pick was found in the floor of a late third century A.D. hut that adjoined the mine. Numerous other mining shafts in the Forest of Dean are ascribed to the Romans.

Pliny (Natural History, XXXIII, 66) in the first century A.D. describes the Roman mining technique. Typically, he scorns far-

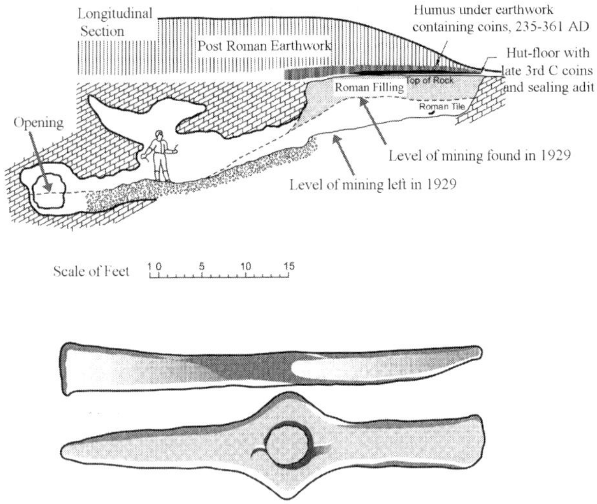

Fig. 16.12 Top: section of a Roman mine at Lydney Park, Gloucestershire. Bottom: miner's pick found in the floor of an adjoining hut. After Schubert (1957, 42).

fetched and fanciful ideas and gets straight to the point. "Leaving aside tales of Indian gold obtained by ants or the gold dug up by griffins in Scythia, gold is obtained in our part of the world in three ways. First, it can be obtained from placers in rivers—this is gold in its most perfect state, thoroughly polished by the friction induced by the current. A second method is to sink shafts. And thirdly it is sought in the debris of collapsed mountains." He goes on to describe these methods in some detail and in particular, the rather spectacular way in which parts of mountains are made to collapse. Pliny notes that Spain has long been the main gold working area of the world but by an old senatorial decree Italy is protected from over-exploitation; a decree prohibits contractors in the region of Vercellae from employing more than 5,000 men.

Placers are deposits of sand, gravel or other alluvium, containing particles of valuable minerals in workable amounts. Native gold is the most important placer mineral but platinum and cassiterite (stream tin) also occur in gravel. Occasionally native silver, copper and other minerals or gems are found. Such placers form naturally. The movement of water through rock picks up finely divided ore stuff and concentrates it because the ore is heavier and more difficult to move than the lighter waste. The specific gravity of cassiterite, for example, is 6.8–7.1 whereas quartz is only about 2.65. The heavier ore tends to settle into pockets of the streambed, or at irregularities such as boulders and bends where eddy currents form. Placers are most readily seen on the surface and were, as Pliny explained, the ancient source of gold. However, placers also occur underground and drift mining is simply the exploitation of such placers by underground methods.

Strabo (Geography 11.2.19) mentions mining for gold dust in placers when he describes the country of the Soanes. Jason and the Argonauts had visited this country and Strabo thought that the technique used there might explain the myth of the Golden Fleece. "It is said that in their country gold is carried down the mountain by torrents, and that the barbarians obtain it by means of perforated troughs and fleecy skins, and this is the origin of the myth of the Golden Fleece". Had Strabo actually seen such gold washing he may have written a more detailed and informative account.

Ore stuff also occurs in lodes or veins in the hard rock, Fig. 16.13. Often this could be seen and worked from the surface but it also extended deep underground. Mining the surface lode by trenching could be done only to a certain depth, after which it was easier to drive horizontal tunnels or levels into the hillside to expose the lode. These tunnels or adits were used both to drain the mine and to transport ore from the workface. Where the lode was deeper, and could not be reached by horizontal shafts one or more inclined or vertical shafts were required. Pliny says that "Gold mined by means of shafts is called 'channelled' or 'trenched'; it [gold] is found adhering to quartz—. Traces of veins appear here and there along the walls of the underground galleries and the overburden is supported by wooden props."

In the pre-Roman and post-Roman times, ores were often mined by pitting, that is, by shafts at frequent intervals. Sometimes this was for the legal reason that a separate shaft had to be sunk into each concession, as in the Athenian silver mines at Laurium. The Romans, however, avoided pitting and their mines are usually orderly systems of underground workings with frequent communication between galleries and to the surface. Their mines sometimes had two, three or four levels although some were a disorderly maze. They made great use of drainage adits, which need to be level or slightly inclined.

The ancient galleries were normally cramped; even at Laurium they were only 60–90 cm high and 60–75 cm wide. However, here the section was regular and the corners neatly cut. Trapezoidal and

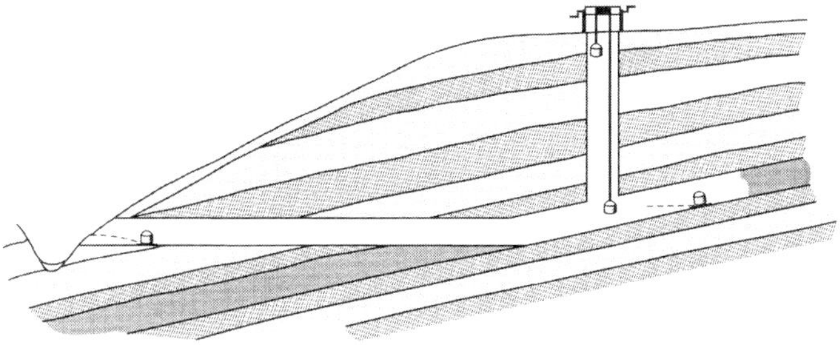

Fig. 16.13 Mining a lode by surface trenching and horizontal shaft (adit). After Taylor (1859, pl. 1).

arched sections also occur, for example, at Rio Tinto. Usually the galleries were cut using iron tools. In pre-historic times, fire setting was commonly used in driving the galleries through rock because chipping with inferior tools was too laborious. Traces of fire setting have been found in Etruria, at Aramo, and at Mitterburg. Pliny writes, "In deep-vein mining masses of flint are encountered. These can be split by fire setting which involves the use of vinegar. Fire setting in galleries usually makes them suffocatingly hot and smoke-filled. Instead, therefore, the rocks are split by means of crushers, which carry 150 pounds [68 kg] of iron." Rapid cooling with water enhanced the effects of the fire by creating thermal stresses in the rock. At Mitterburg, wooden pipes were used to carry water to the work face, and in other places, a closed trough was cut alongside a stepped adit. Vinegar, mentioned by Pliny, was believed to be more effective than water because it was thought to be colder and perhaps had a chemical action on the rock. The fumes created by fire setting were objectionable and wherever possible it was avoided. Fire setting was rarely used at Laurium.

Shafts were cut with picks and were either rectangular or circular depending on how they were lined. Wood linings require a rectangular shaft but stone linings are stronger if they are circular. The entrance shaft to Roman mines is usually quite steep so ladders were fitted, usually to one side of the shaft to leave room for buckets of ore to be hauled up. Shafts were often joined by an adit at a lower level.

The third method mentioned by Pliny (the collapse of mountains) he compares to the achievements of giants. "By the light of lamps, long galleries are cut into the mountain. Men work in long shifts [the time of which is] measured by lamps, and may not see the daylight for months on end. Local people call these mines 'deep vein'. The roofs of these are liable to give way and crush the miners—. Arches are left at frequent intervals to support the mountains above."

Pliny continues: "There is a kind of earth consisting of clay and gravel called 'conglomerate' which is almost unworkable. The method used in this case is to attack it with iron wedges and crushers—. When the work is complete the miners cut through the tops of the arches, beginning with the last. The opening of a fissure gives warning of the impending collapse, but this is only seen by a watchman perched on top of the mountain. With a shout and a wave the lookout gives the order for the miners to be called off and at the same time rushes down from his vantage point. The ruptured mountain falls asunder with an unimaginable crash, and is accompanied by an equally incredible blast of air. Like conquering heroes, the miners contemplate their triumph over Nature. Yet even at this juncture there is no gold visible, nor did they have any positive indication that there was any when they began to dig. The mere hope of obtaining their coveted prize was sufficient reason for embarking on such danger and expense." However, the rewards were very considerable because "Gold mined from deep veins does not require heat treatment but is pure gold. Nuggets are found in such mines, and similarly in pits; some weigh more than ten pounds [5 kg]."

Miners, as opposed to 'conquering heroes' are illustrated on a fragment of a bas-relief, found at Linares, Spain, Fig. 16.14. Eight miners in two files march into a gallery. One holds a pick and another a lamp. The taller figure is the foreman and he carries a large pair of tongs and a lantern. They are naked except for a short tunic and leather belt. The Roman mines at Linares were rich in silver and lead and the town was prosperous.

A technique used by the Romans was hushing, in which the softer beds were broken down by a strong current of water, Davies (1935, 18). They built canals several miles long to fill storage cisterns and discharged water from these cisterns onto soft gold bearing clays. The gold was deposited in settling tanks. Pliny, (Natural History, XXXIII, 74–75), mentions "the feat of bringing streams along mountain ridges, often a distance of 100 miles [160 km], to wash away the debris from mining operations.—On the ridge above the head of the mine reservoirs are excavated, measuring more than 200 feet

Fig. 16.14 Miners on a fragment of bas-relief from Linares, Spain.

[61 m] each way and 10 feet [3 m] deep. Five sluices each about 3 feet [1 m] square are constructed in the walls. When the reservoir is full, the sluices are knocked open so that a violent downward surge of water is sufficient to sweep away the rock and debris. The miners then carry the ore out of the workings on their shoulders, each man forming part of a human chain in the dark; only those at the ends of the line see daylight." Some indication of these conditions is illustrated in Fig. 16.15, which shows Greek miners digging at the work face, filling bags, and hauling the bags from the mine. The illustration is from a Corinthian terracotta plaque. Even when fire setting was not used, ventilation was an obstacle, particularly in shaft mines. Some mines were especially deadly and were closed, not for humanitarian reasons but because of the great cost of replacing the slave workforce when they died. Strabo (Geography 12.3.40), writing at the end of the first century B.C., tells us that "Mt Sandaracurgium is hollowed out in consequence of the mining done there, since the workmen have excavated great cavities beneath it. The mine used to be worked by publicans, who used as miners the slaves sold in the market because of their crimes; for, in addition to the painfulness of the work, they say that the air in the mines is both deadly and hard to endure on account of the grievous odour,

Fig. 16.15 Greek miners, digging the work face, filling bags, and lifting them from the mine. Corinthian terracotta plaque.

so that the workmen are doomed to a quick death. What is more, the mine is often left idle because of the unprofitableness of it, since the workmen are not only more than two hundred in number, but are continually spent by disease and death." The most deadly gases encountered in mining are: methane, carbon monoxide and carbon dioxide. Methane is dangerous because it forms an explosive mixture with oxygen. Carbon monoxide is very poisonous, even in small amounts less than a fraction of one percent, and carbon dioxide is less lethal but can cause death if miners are exposed to levels greater than 11% for periods of several hours. Methane does not occur in mines other than collieries unless the workings are in contact with carbonaceous strata. Carbon monoxide forms when combustion occurs in the presence of insufficient oxygen and was particularly dangerous in ill ventilated mines when fire setting was used to fracture the rock and naked flames were used for lighting. Carbon dioxide always forms as one of the main products of hydrocarbon combustion. It seems likely that the slaves who mined Mt Sandaracurgium died from carbon monoxide poisoning.

Methods of ventilating mines were known and practiced. Theophrastus (*Concerning Fire* 24) says that miners make ventilating shafts to thin the air and Pliny (*Natural History* 31.49) tells us that: "When well shafts have been sunk deep, fumes of sulphur or alum rush up to meet the diggers and kill them. A test for this hazard is to see if a burning lamp lowered down the shaft is extinguished. If so, air-shafts are dug next to the well shaft on both sides to drain off the oppressive fumes. Even in the absence of these pernicious substances the air itself becomes oppressive as a result of the depth. Continuous ventilation by waving linen cloths remedies this", Humphrey et al. (1999, 190).

The Romans experienced great difficulties in obtaining adequate ventilation. Sometimes the configuration of the ground allowed the miners to use adits and shafts at different levels. If the shafts and adits were connected then they generally produced a draught. This natural convection occurs because the air temperature in a mine is usually higher than that outside the mine and the denser air from the outside tends to sink to the bottom of the mine displacing the lighter mine air. The effect may be enhanced by building a chimney to increase the height of the up-shaft or by lighting a fire to heat the air in the up-shaft. Even if a single shaft is used then the temperature difference between the mine and the outside causes dense air to descend near the wall while lighter air ascends in the centre of the shaft. Fluctuations in ambient temperature may decrease, or even reverse, the flow of air. Mass transfer by diffusion, which is caused by differences in the concentrations of mine gases such as oxygen, also refreshes the air in the mine but this is a relatively slow process and is much less important than natural convection. Fans, to blow air into a mine, were not developed until the Renaissance.

The only ancient account of gold mining in Egypt is by Diodorus Siculus, in the first century B.C., although he appears to have relied on Agartharcides' description from the second century B.C. The first stage of mining that Diodorus describes is fire setting. "The gold-bearing earth which is hardest they first burn with a hot fire, and when they have crumbled it—they continue the working of it by hand; and the soft rock which can yield to moderate effort is crushed with a sledge by myriads of unfortunate wretches. And the entire operations are in the charge of a skilled worker who distinguishes the stone and points it out to the labourers,—the physically strongest break the quartz-rock with iron hammers, applying no skill to the task but only force."

"The boys who have not yet come to maturity, entering through the tunnels into the galleries formed by the removal of the rock, laboriously gather up the rock—piece by piece and carry it out into the open to the place outside the entrance. Then those who are above thirty years of age take this quarried stone from them and with iron pestles pound a specified amount of it in stone mortars, until they have worked it down to the size of a vetch. Thereupon the women and older men receive from them the rock of this size and cast it into mills of which a number stand there in a row, and taking their places in groups of two or three at the spoke or handle of each mill they grind until they have worked [it] down—to the consistency of the finest flour."

"In the last steps the skilled workmen receive the stone which has been ground to powder;—they rub [it] upon a broad board which is slightly inclined, pouring water over it all the while; whereupon the earthy matter in it, melted away by the action of the water, runs down the board, while that which contains the gold remains on the wood because of its weight. And repeating this a number of times, they first rub gently with their hands, and then lightly pressing it with sponges of loose texture they remove—whatever is porous and earthy, until there remains only gold-dust."

Shaft mining in the Egyptian desert or some other places in the Near East was unlikely to encounter water, but elsewhere in the Roman Empire the shaft was liable to penetrate below the water table, so flooding frequently occurred. Often this was attributed, superstitiously, to the will of the gods and the mine was abandoned. For example, the island of the Syphnians "contained gold mines, and the gods ordered them to pay a tithe of the revenue to Delphi. So they built a treasury, and continued to pay the tithe until greed made them omit the tribute, when a sea flooded their mines and hid them from sight", (Pausanias, Description of Greece, 10.11.2). Elsewhere, superstition gave way to practicality and active efforts were made to drain a mine if it flooded, usually by the simple expedient of hand bailing. Pliny says, "—It is a remarkable fact that shafts begun on Hannibal's initiative all over the Spanish provinces are still in existence; they are named after their discoverers. One such mine, known today as Baebelo, provided Hannibal with 300 pounds [136 kg] of silver a day. The galleries ran for between one and two miles into the mountain, and along the whole of this distance water-men stood day and night in shifts measured by lamps,

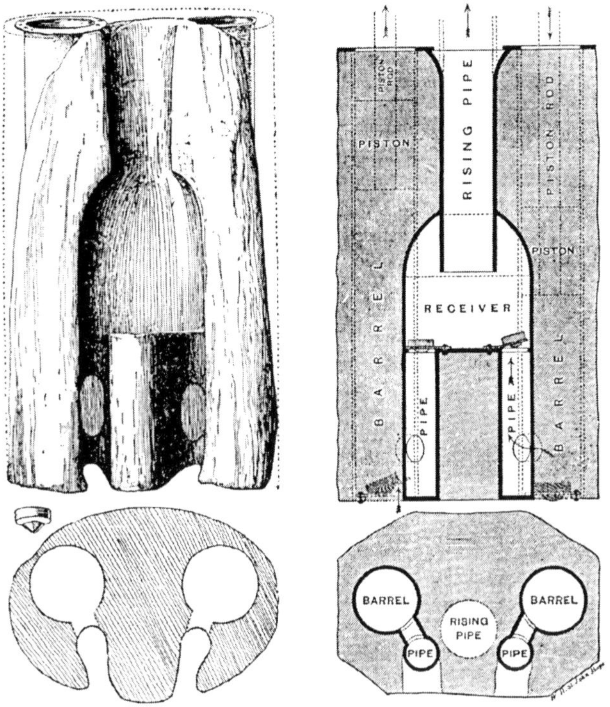

Fig. 16.16 Roman force pump, and its reconstruction, found at Silchester.

bailing out water and making a stream." (Natural History, XXXIII, 96). Esparto buckets of 150 litre capacity, made watertight by soaking in tar, were used for bailing at Cartagena, Davies (1935, 25). Similar pails have been found in the Sierra Morena and Davies (1935, 25) reports bronze or copper buckets from Cartagena, Sotiel Coronada, Rio Tinto, Centenillo, and other mines.

In addition to bailing the Roman engineers developed and used two devices to drain their mines when this could not be attained by natural discharge along sloping adits. These were the water wheel (noria) and the Archimedean screw, both of which continued in use through the medieval period and into modern times. Force pumps were certainly known and used in the Greek and Roman worlds to pump water from deep wells and there seems no reason why they should not also be used to pump out mines. A double acting force pump was found at the bottom of a well in Silchester, Fig. 16.16, and consists of two lead cylinders, 560 mm long and 75 mm diameter mounted each side of a wooden block. Two pistons operated by handles from the surface reciprocated alternately. On the up-stroke, water was sucked in through a leather flap valve at the lower end. On the down-stroke the suction valve closed under the internal pressure and the water was forced through a connecting hole to a pipe containing a leather flap valve that opened under the pressure and the water was delivered to a receiver and through a rising pipe to the surface. It is estimated that this pump could raise about 10 gallons per minute (45 litre/min, or 2.7 tonnes/hour), which would make it suitable for mining use. However, they do not appear to have been used for mine drainage.

The water wheel (noria) at the Roman gold mine at Dolaucothi has already been mentioned. Davies (1935, 26) describes finds of several waterwheels of the first century A.D. varying in diameter from 3.65 m to 4.87 m. Typical wheels are illustrated in Fig. 16.17 and 20. On the rim next to each spoke are bracket shaped treads, much worn at the top, and used by a man driving the wheel with his heel. Vessels (dippers) were attached to the rim and the wheels rotated in a vertical plane causing the vessels to take up water from a lower trough and discharge it into an upper trough. At Rio Tinto, the wheels were fitted with twenty-four such vessels. Sometimes the wheels operated singly but often they operated in pairs, rotating in opposite directions. Typically, they raised water between 2 m and 3.7 m, depending on their diameter. At Sto. Domingos they were set along a gallery having a 40 degree slope. As mentioned above, one or more men sitting above the wheels and pushing the treads with their feet provided the power to rotate them. Near one of the wheels a ladder was found, by which the operator climbed to his place. Palmer estimates that one man driving one wheel could lift 3.66 tonnes of water per hour through a height of 3.66 m. This is a power of 84 W at 61% efficiency.

The Archimedean screw was mentioned in chapter 1 where Vitruvius' description of its manufacture, proportions, and use, is quoted in full. Vitruvius was writing in the first century B.C. at much the same time as Diodorus Siculus, who also mentions the use of the Archimedean screw in Egyptian irrigation. Some three hundred years later Athenaeus of Naucratis (Deipnosophistae, V) says that such a screw was used to pump bilge water out of the ship *Syracusia*. Its use in mining was not specifically mentioned in surviving texts but archaeological finds in the Spanish mines leave no room for

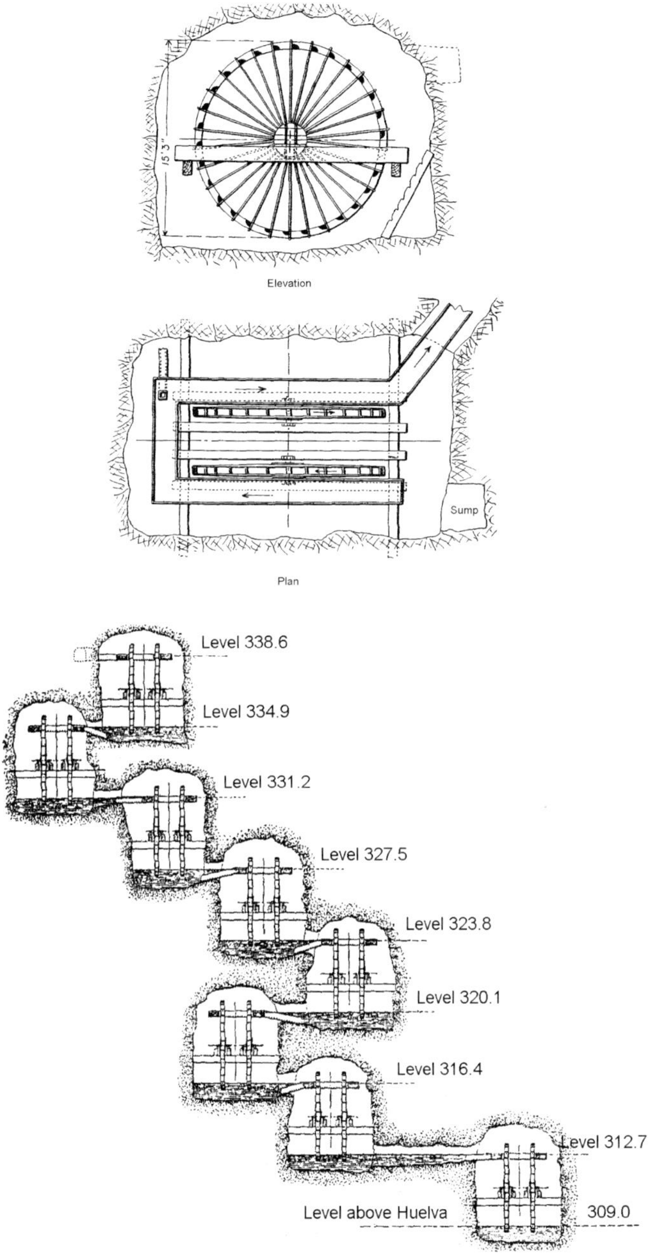

Fig. 16.17 Top: elevation and plan of a Roman water wheel. Bottom: general arrangement of Roman water wheels for lifting water from a low level to a higher level at the Rio Tinto Mine. Palmer (1926–7, 300, 303).

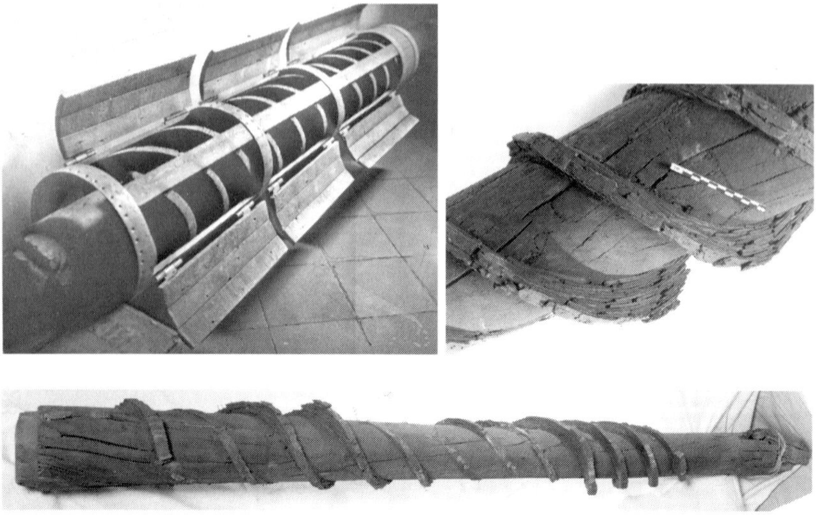

Fig. 16.18 Top left: Roman Archimedian screw as originally donated to the Liverpool Institute of Archaeology. Top right: detail of screw with vanes attached (1969). Bottom: screw with vanes attached (1969).

doubt. A fine example is in the collection of the Liverpool Museum (accession number 50.34) and shown in Fig. 16.18. Bienkowski (1987, 135–40) has described its present condition. The screw originally came from the Roman mine in Sotiel Coronada in the Rio Tinto district of Spain where it was found with two or three other screws in 1883–4. It is now thought that only the shaft and parts of the screw vanes are original and that the rest of the vanes and the outer casing are later additions of the nineteenth or twentieth century. Nevertheless, it is a particularly interesting example of a Roman screw and its original proportions are similar to those recommended by Vitruvius. The diameter of the shaft should be 1/16th of its length. As the length of the Liverpool shaft is 3226 mm long its diameter should be 202 mm whereas its actual diameter is 197 mm, an error of only 5 mm. Similarly, Vitruvius states that the diameter of the vanes should be twice that of the shaft. Thus, the diameter of the vanes should be 403 mm. The highest surviving point of the Roman vanes indicates that its diameter is at least 355 mm and it is, therefore, 48 mm too small. This may be attributed to damage of the vanes but even so the Liverpool screw confirms that the proportions recommended by Vitruvius were applied in practice.

The method of constructing the screw vanes, using short lengths of pinewood about 3 mm thick, 10–17 mm wide and 30–100 mm long, is different from the method described by Vitruvius and the number of threads is different. Vitruvius states that willow withes smeared with pitch should be used and he specifies an eight-start thread whereas the Liverpool screw is a two-start thread. However, the vanes appear to have been coated in bitumen to make them waterproof. The remains of other Roman water screws and their mountings have been found in the Spanish mines and some are constructed differently, having copper rather than pine vanes. In addition, the Spanish water screws have metal mounting rods protruding from either end of the axle. These are missing on the Liverpool water screw. A terracotta plaque in the British Museum shows a man rotating his water screw by pushing the treads round the middle of the barrel. One of the water screws from Alcaracejos was fitted with an iron crank.

The Roman water screws were used like the waterwheels, each raising water from a lower tank to a higher one. They were set at an angle of 35–40 degrees although on an Egyptian terracotta plaque and at Sotel Coronada they were set at 15–20 degrees. Vitruvius states that the inclination should correspond to the Pythagorean right-angled triangle having sides of 3, 4, and 5 units, which gives an inclination of about 37 degrees. The screw was only partially filled with water that was trapped in a pocket formed between the screw and the casing. Its three-dimensional shape is often hard to imagine but Fig. 16.19 illustrates the situation quite well. In Spain, the Romans obtained their ore from depths of more than 200 m and as each screw was no more than 5 m long and raised water only 1.5 m we must assume that large numbers were made and used.

Singer et al. (1956, 689), reports that the earliest reference to an Archimedean screw driven by a windmill occurs in Holland in 1404. In 1634, a patent was granted to Symon Hulsbos of Leiden for an Archimedean screw without a casing. A brick trough reaching into the water was constructed in which the screw revolved and a sluice-gate was placed at the top to prevent the water flowing back. The screw fitted the sides of the trough closely. The screws had a diameter of 1.5 to 1.8 m, rotated at 40 to 50 revolutions/minute, raised water 4 to 5 m, and pumped at a rate of 40 m^3/minute.

For winning and moving ore the Greeks and Romans normally used iron tools but often there is little difference between pre-Roman,

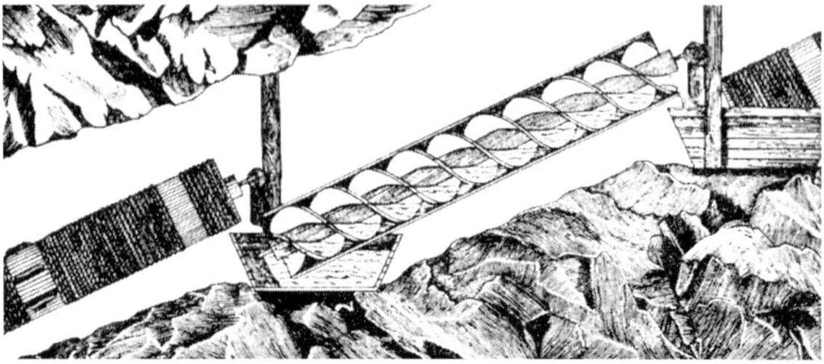

Fig. 16.19 Archimedean screw for raising water. After Klemm (1959, 49).

Roman, and medieval tools for they were all developed for the same tasks. Stone picks or wedges are sometimes found and spades are sometimes of wood. Picks were used to cut earth or soft rock but for harder rock a gad and hammer was used. The iron gad was fitted with a wooden handle into a socket near its head so it could be held steady by one miner while another miner struck it with a hammer. Perhaps a dozen or more were used in each shift and were re-sharpened between shifts. Rock surfaces cut in this way have a characteristic pattern. Suitable pieces of rock were levered off using crowbars and the ore was scraped into baskets using rakes or spades.

Haulage of ore from the mine was always difficult and heavy work. Davies (1935, 29–30) tells us that ore was transported in leather bags at several Roman mines but these, of course, would quickly wear out on rough surfaces. Certainly, bags were in use at the Greek mine shown in Fig. 16.15. Bronze bowls, as used at Laurium, were preferred, at least as far as the shaft. At other places, wooden trays, sometimes bound with iron bands, were dragged along the gallery floor. Wheeled trolleys, which require horizontal galleries in all parts of a mine, were not known until the Renaissance. Haulage up the shafts was sometimes by means of rope attached to baskets. In Etruria and at Aramo the rope marks can be seen on the lips at the top of the shaft but at Laurium and most Roman mines these marks are absent. It seems that a windlass was used to raise the baskets for pulley-axles and blocks have been found; see Fig. 16.20.

As we have seen already, the ore was usually concentrated and cleaned at the surface near the mine entrance. Transport costs were

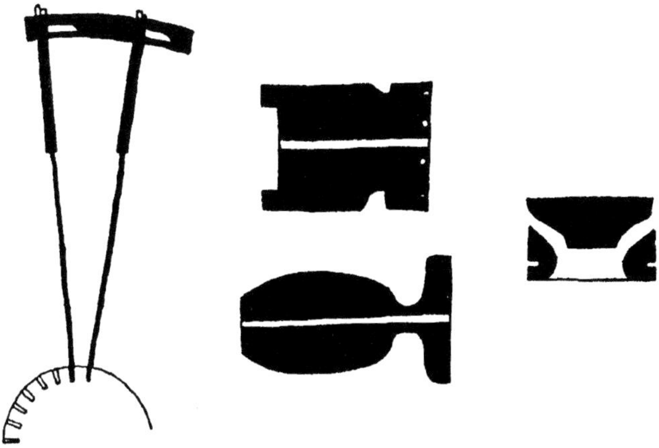

Fig. 16.20 Left: part of water wheel (Tharsis). Centre: pulleys (Ruda). Right: mill stones (Laurium). After Davis (1935, figs 25, 26, 28).

high and there was little point in transporting the unwanted gangue. In pre-historic times, ore was pounded with rilled hammers or spherical or pear-shaped mullers on cup-marked querns. This was continued by the Romans and continued even into the Middle Ages before being replaced by water driven stamps. In Egypt, Spain, and at Laurium, the ore was ground to a fine powder in hand mills of hard stone. Although the Romans invented water driven millstones for grinding corn, they do not seem to have applied them to the onerous task of grinding ore.

When first extracted from the mine and at various stages in crushing and grinding, the ore was washed by water running down inclined tables to remove the lighter gangue and leave behind the heavier ore-bearing material. Such a process tends to separate small, dense particles, along with larger, less dense particles, so a preliminary process of sieving to give particles of approximately equal size was desirable. Such sieves, made of hazel twigs, have been found. Washing is particularly effective for ores having a much higher density than the gangue. The washing tables are usually rectangular having a cemented, slightly sloping surface. Usually the water was stored in cisterns at the head of the table. The heavier material containing the ore was left on the table while the lighter material was carried away and caught in settling tanks. Tables of different sizes, slopes and plans were used to deal with the variety of ores. Ores were washed many times before it was considered sufficiently concentrated to be passed to the smelters; at least three washes were used at Laurium and five washes at Cartagena. Sometimes the concentrated ore was given a preliminary roasting before smelting. This removed the moisture and made some ores porous and lighter for transport to the smelting plant.

These processes, to clean and concentrate the ore, were not always done too thoroughly and often the refuse from older workings still contained valuable material. For example, Strabo (Geography 9.1.23) tells us that "The silver mines of Attica were originally valuable, but now they have failed. Moreover, those who worked them, when the mining yielded only meagre returns, melted again the old refuse, or dross, and were still able to extract from it pure silver, since the workmen of earlier times had been unskilful in heating the ore in the furnaces."

Security at mines was always an important consideration as the following quotation from Pliny (Natural History, XXXIII.118) confirms, "—The most famous cinnabar mine providing revenue for the Roman people is that of Sisapo in Baetica. The security precautions are second to none. Smelting or refining of the ore are not allowed locally, but as much as 2000 pounds of crude ore per year are sent to Rome under seal and there purified. To prevent the price going sky-high, it is fixed by law at 70 sesterces for about a pound. The mineral is adulterated in many ways—a source of illegal profit for the mining company."

Cinnabar is sulphide of mercury and is a heavy, bright red substance from which mercury could be obtained. It is identical in composition with vermilion and was used to colour lettering in books and for colourful lettering of inscriptions on marble walls and even on tombs. However it was a dangerous material for Pliny goes on to tell us that "Those who polish cinnabar in workshops tie loose masks made of bladder skin over their faces to prevent inhalation of dust as they breathe; the dust is a very serious health hazard. The masks, however, allow them to see over the top."

Mining seems to have reduced with the decline of Roman power. Derry and Williams (1960, 128) state the situation bluntly when they say that "Practically nothing is known about mining in the Dark Ages, not even whether coins issued by Anglo-Saxon and Carolingian monarchs were made from old metal or from the products of new mining." Nevertheless, there was a clear demand for metal to make tools, weapons and jewellery. In the manufacture of weapons, swords in particular, and in jewellery, the Anglo-Saxons excelled and produced some very fine work that will be considered in the proper place. There were, as we have seen, well over a thousand mines in Europe used by the Romans and it is very likely, although not inevitable, that the most profitable remained in production after the withdrawal of the Roman legions.

Meyer (1997, 64–68) has recently investigated a fifth and sixth century A.D. Egyptian gold mine. This is at Bir Umm Fawakhir in the Eastern Desert of Egypt and the remains there are of a gold-mining town of some 200 buildings and four ancient mines. The gold occurs in quartz veins that run through the mountain and Meyer was able to confirm some of the techniques described by Diodorus in the first century B.C. The ancient miners used two techniques to obtain this quartz, namely:

1. open cast trenches following the quartz veins from the surface, and
2. shafts sunk horizontally or diagonally into the mountainsides.

A number of shafts had stone walls reinforcing the entrances or platforms at the edge, presumably to aid in lowering or raising men, boys, baskets, tools and ore. The largest mine investigated by Meyer ran about 100 m horizontally into the mountain and is roughly 2 m high. It has two short side galleries, an airshaft, and oblong holes pounded in the rock at the work faces. It shows no sign of fire setting; there is neither charcoal nor ash, and none of the characteristic splintered out niches. There seems to have been no need of fire setting for although the quartz is tough, the surrounding rock can be splintered away.

Iron tools, or metal of any sort, have yet to be found at Bir Umm Fawakhir. Mortars are uncommon but those that have been found are of limestone and unsuitable for crushing quartz. However, hundreds of crushing stones occur on the site. They are made of a rough basalt, granite, or porphyritic granite with a smooth upper surface that measures about 20 cm × 20 cm square, with a depression pecked in the middle. It seems that ore was immediately reduced at the mouth of the mine and the richer pieces selected.

Both upper and lower stones of numerous rotary millstones or querns have been found on the site. Rotary mills are probably a late Ptolemaic or Roman introduction; earlier millstones were flat or concave. Saddle querns, which are grinding stones of granite or porphyritic granite that measure up to 80 cm long and have a shallow dished-out depression, remain unexplained. These were presumably worked with oblong hand-stones. It is not known whether the stones were used at an intermediate stage of grinding, used with something else, such as wheat flour, or were of an earlier date. Gold washing tables have not been found but Meyer (1997, 64–68) says the accounts of nineteenth century travellers do mention them at Bir Umm Fawakhir. Probably they have been destroyed by modern mining activity. It seems unlikely that final refining was carried out on the site.

Bir Umm Fawakhir is a relatively rare example of a fifth and sixth century mine but the fact that there are so few surviving records relating to mines or mining in this period does not prove that they did not exist. The same situation applies to watermills for they also have very scanty records during the Anglo-Saxon period. Despite this, the Domesday Book records some 5624 water mills in 1086, of which almost nothing was known from earlier documents. A handful of these mills paid their rents in weight of iron rather than money so presumably they, at least, were used in metal processing of some kind.

Analysis

Efficiency and Power of the Archimedean Screw

The theoretical volume of water pumped per unit time by an Archimedean screw, built to the proportions laid down by Vitruvius, and assuming it runs completely full, is

$$V_t = \frac{\pi(D^2 - d^2)}{4} PN \qquad 16.1$$

D, d are the outer and inner diameters of the screw, P is the pitch of the screw and N is the rotational speed. For a Vitruvian screw $D/d = 2$ and $P = \pi d$ so the theoretical volume flow rate is

$$V_t = \frac{3\pi^2 D^3 N}{32} \qquad 16.2$$

The volumetric efficiency may be defined as the ratio of the actual flow rate through the screw to the theoretical flow rate and it is a measure of how effectively the screw is filled. The volumetric efficiency is thus

$$V = \eta_V V_t \qquad 16.3$$

V is the actual flow rate through the screw. From an examination of Fig. 16.19, it is clearly impossible to completely fill the screw and one would expect the volumetric efficiency to be quite low. Using the data of Hulsbos, mentioned by Singer et al. (1956, 689), for an Archimedean screw driven by a windmill we may take $V = 40$ m^3/min, $D = 1.65$ m, and $N = 45$ rev/min, thus the volumetric efficiency appears to be about 21% and Hulsbos' data seems reasonable.

The power required to drive an Archimedean screw may be written as

$$W = \rho_w Vgz / \eta_m \qquad 16.4$$

η_m is the mechanical efficiency, ρ_w is the density of water, g is local gravity, and z is the height through which the water is raised. The mechanical efficiency allows for power loss between the screw and the casing, losses in the bearings, and leakage between the screw and the casing. Taking the mechanical efficiency as 75%, the density of water as 1000 kg/m^3, the volume flow rate as 40 m^3/min, local gravity as 9.81 m/s^2, and the lift to be 4.5 m, we find that that the power required to drive Hulsbos' screw is about 39.2 kW. This is, perhaps rather high for a windmill of the usual size. The power produced by a windmill is given by

$$W = \eta \rho_a A V^3 / 2 \qquad 16.5$$

η is the efficiency of the windmill, typically about 20%, ρ_a is the density of air, about 1.2 kg/m^3, A is the area swept by the sails, which may be taken as 200 m^2 (8 m radius) and v is the wind velocity. Thus to develop 39.2 kW we would need a wind velocity of about 11.7 m/s. This is a very high wind speed and is available for only a fraction of the year so we must assume that Hulsbos' claim is for the maximum that can be pumped, given a very high wind speed, rather than the average.

Fire Setting

Fire setting, as we have seen, was commonly used in pre-history and in ancient civilisations to drive tunnels through hard rock. The main indications that a tunnel has been excavated by fire setting and not simply dug out is that fire setting imparts a concave surface to the rock face and adits driven by fire setting typically form an arched profile; Timberlake (1990, 49). The recognition of firesetting waste is less easy although rock may appear reddened, burnt, and cracked.

The transient temperature distribution in a solid when heat enters at a surface is described by the heat diffusion equation. If the heat flow is in one direction and the properties of the solid are constant then the temperature distribution is given by

$$\frac{\partial T}{\partial t} = K \frac{\partial^2 T}{\partial x^2} \qquad 16.6$$

T is the temperature rise at any distance x from the surface and at any time t, and κ is the thermal diffusivity of the material, namely the thermal conductivity divided by the density and the specific heat. The thermal diffusivity of most metals, which are good thermal conductors, is about 10^{-5} m^2/s and for insulators is less than 10^{-6} m^2/s. Rock is a reasonably good insulator and the thermal diffusivity is about 8×10^{-7} m^2/s.

The solutions of Equ. 16.6 are known for many simple problems and Carslaw and Jaeger (1990) discuss them in detail. Perhaps the most appropriate solution is that which assumes the surface of the rock is held at a constant temperature T_0 by the fire during the firing operation. This, of course, is not strictly accurate but it is a reasonable approximation and gives some insight into the process. The solution of Equ. 16.6 for a semi-infinite solid when the temperature at the surface $(x = 0)$ is suddenly and permanently increased from some initial uniform temperature T_i to temperature T_0 is

$$\frac{T - T_i}{T_0 - T_i} = 1 - erf(x / \sqrt{4\kappa t} \qquad 16.7$$

The values of the error function, *erf*, must be determined from appropriate tables or calculated from an infinite series. The temperature distribution in granite at various times after the start of the fire setting process is displayed in Fig. 16.21. The temperature rise at the

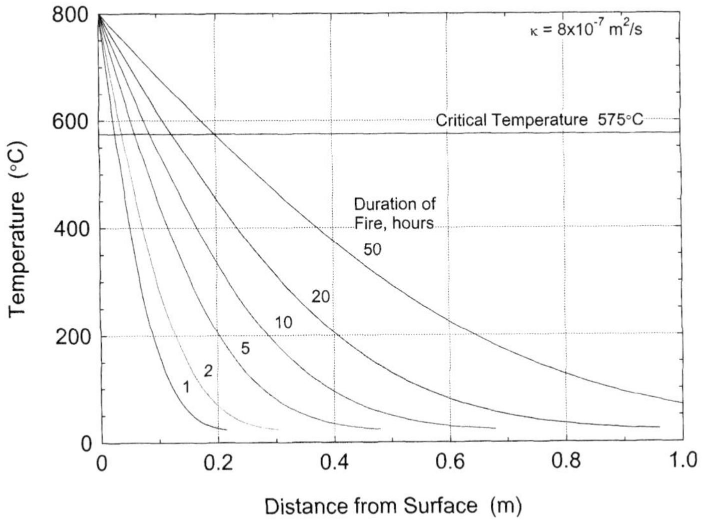

Fig. 16.21 Temperature distribution below a rock face at various times. The surface temperature caused by the fire is assumed to be 800°C. The critical temperature rise at which rock crumbles is assumed to be 575°C, the initial rock temperature is 20°C and the thermal diffusivity is, $\kappa = 8 \times 10^{-7}$ m²/s .

surface, x = 0, remains constant but the temperature beneath the surface slowly increases with time as the heat from the fire diffuses into the rock. If it is assumed that at some critical temperature, T_c, the rock crumbles then Fig. 16.21 defines the depth of rock that may be removed after various periods of heating. The temperature rise is insignificant for values of $x/(4\kappa t)^{0.5} > 1$, and this defines the upper limit of the thickness of rock that can be heated in a specified time. It is, in fact, the thickness of the thermal boundary layer.

As an approximation, the depth of rock, x_c, raised to a temperature greater than the critical temperature may be determined from

$$x_c = \frac{T_0 - T_c}{T_0 - T_i}\sqrt{\pi \kappa t} \qquad 16.8$$

The upper limit to the thickness of rock that can be heated in a given time is given by

$$x_{\max} = \sqrt{4\kappa t} \qquad 16.9$$

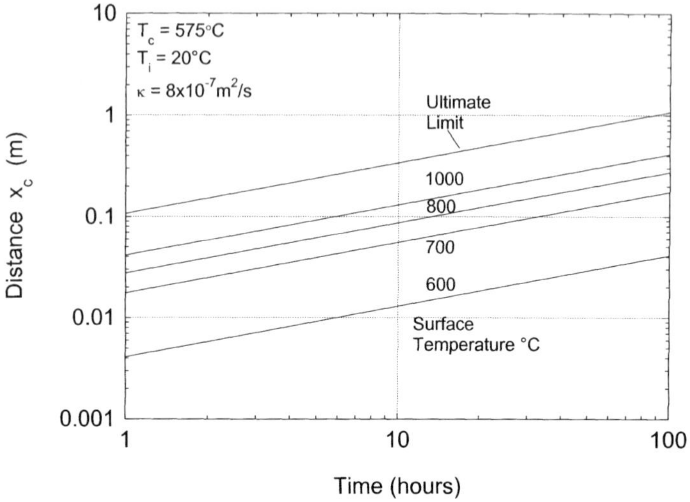

Fig. 16.22 Influence of time and temperature on depth of rock crumbled during fire setting.

Equ. 16.8 and 9 are plotted in Fig. 16.22 assuming the same thermal properties as the rock in Fig. 16.21. If the rock face is held at a temperature of 800°C for a period of 24 hours then about 0.14 m thickness of rock is above the critical temperature and can be expected to crumble from the wall. If we allow a day for stacking the firewood, a day for the fire, and a day for clearing the ashes and the crumbled rock, then the average rate of driving is about 4.7 cm per day. This may not seem very large but it is, in fact, very close to the rate realised in practice. Data quoted by Timberlake (1990, 51) and obtained originally by Collins in 1893 is shown in Table 16.2. The first trials comparing gunpowder with fire setting took place in 1860–64 whereas the second trials, comparing dynamite with firesetting, took place in 1881–85. Fire setting was faster and cheaper than gunpowder packed into hand-drilled holes, but it was both slower and more expensive than dynamite packed into hand-drilled holes. In these trials, the rate of tunnelling by fire setting was 4.9 and 4.8 cm per day, which compares very well with the theoretical rate of 4.7 cm per day.

Table 16.2 Comparison of tunnelling rates by fire setting or blasting. Tunnel face is 198 cm × 152 cm.

	Number of days work per fathom (183 cm)	*Total Cost (Wages +materials)*	*Rate of Tunnelling cm/day*
Gunpowder	47.5	£9.55	3.9
Fire setting	37.5	£7.52	4.9
Dynamite	16.5	£5.79	11.1
Fire setting	38.5	£8.13	4.8

Lift Pump

Fig. 16.23 shows the arrangement of a simple reciprocating lift pump. The water surface is distance z_s vertically below the pump (suffix s refers to the suction pipe), and the water is pumped to a reservoir

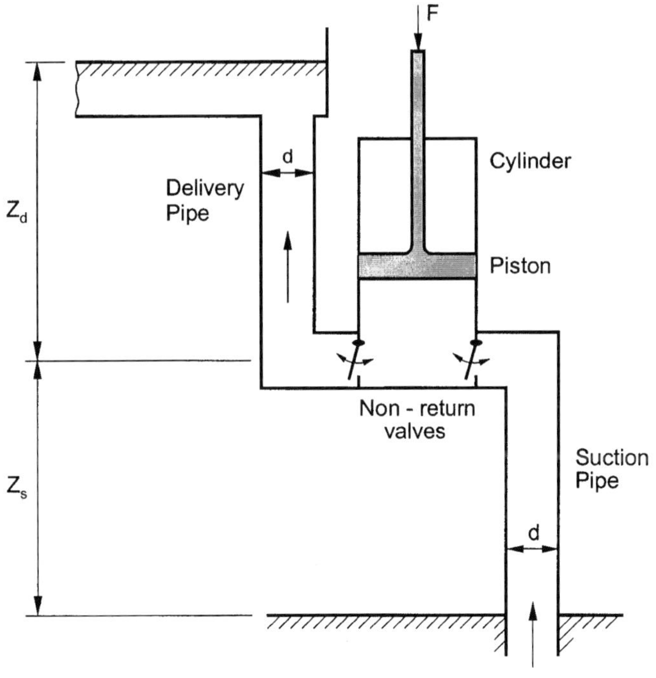

Fig. 16.23 A simple reciprocating lift pump.

distance z_d vertically above the pump (suffix d refers to the delivery pipe). When the piston moves from minimum volume to maximum volume a low pressure is created in the cylinder, which sucks the delivery valve to the closed position and sucks open the inlet valve. The water in the suction pipe is then drawn into the cylinder. When the piston reverses and moves from the maximum volume to the minimum volume the pressure in the cylinder increases and this closes the suction valve and opens the delivery valve. Continued motion of the piston forces the water out of the cylinder into the delivery pipe, and because water is incompressible the same volume of water is delivered to the upper reservoir. The theoretical power to lift the water from the sump to the delivery reservoir, ignoring friction, is given by the simple relation

$$W = mg(z_s + z_d) \qquad 16.10$$

W is the theoretical power, m is the mass flow rate of water (kg/s), and g is the gravitational acceleration (9.81 m^2/s). The quantity z_s+z_d is the difference between the water levels of the two reservoirs and is called the total head.

In practice the suction head, z_s, is limited to about 8 m because at about this value the low pressure (2 m absolute pressure head) causes the dissolved gases in the water to form bubbles that seriously reduces the volume of water that is delivered per stroke, and the subsequent collapse of these bubbles can cause cavitation erosion. However, the only restriction on the delivery head, z_d, is due to the strength of the delivery pipe, which must be sufficiently strong to withstand an internal pressure of $p = \rho g z_d$, where ρ is the density of water (1000 kg/m^3).

The theoretical power required for pumping water at flow rates from 0.1 to 1000 kg/s (0.36 to 3,600 tonnes/hour) is shown in Fig. 16.24. A single man might be expected to generate about 50 W and thus he can lift 1 kg/s through a height of 5 m or 0.1 kg/s through a height of 50 m. A water wheel driving the pump might generate 10 kW and thus it is theoretically capable of lifting 10 kg/s through a height of 100 m, or 100 kg/s through a height of 10 m.

In practice the power supply to the pump must also overcome frictional losses, which are generated by the motion of water in the suction and delivery pipes and the motion of the piston in the cylinder. In addition to these there may be leakage losses from the delivery pipe or passed the seals of the suction and delivery valves or passed the piston. Assuming that the pipes and the piston do not leak then the main pressure loss is that due to friction as the water flows through the pipes. This can be expressed as a friction head, z_f, by the well-known relation:

$$z_f = \frac{4f(z_s + z_d)v^2}{2gd} = \frac{4f(z_s + z_d)}{2gd}\left(\frac{4m}{\pi\rho d^2}\right)^2 \qquad 16.11$$

f is the friction factor (about 0.001 for very smooth metal pipes, and about 0.02 for moderately rough pipes), d is the pipe diameter, and v is the average velocity in the pipe. The second equation is found by substituting the continuity equation $m = \rho A v$, where ρ is the density of water, A is the cross-sectional area of the pipe ($\pi d^2/4$). Here it is assumed that the suction and delivery pipes have the same diameter and friction factor, and that the pipe lengths are identical to the vertical heights, z_s and z_d. Thus the power required for pumping and to overcome friction is

$$W_f = mg(z_s + z_d + z_s) = mg(z_s + z_d)\left(1 + \frac{32fm^2}{g\pi^2\rho^2 d^5}\right) \qquad 16.12$$

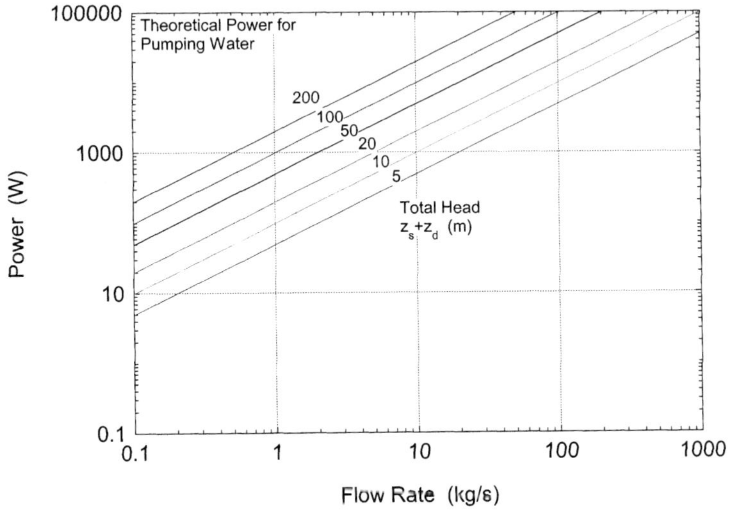

Fig. 16.24 Theoretical power required to lift water through a total height (z_s+z_d).

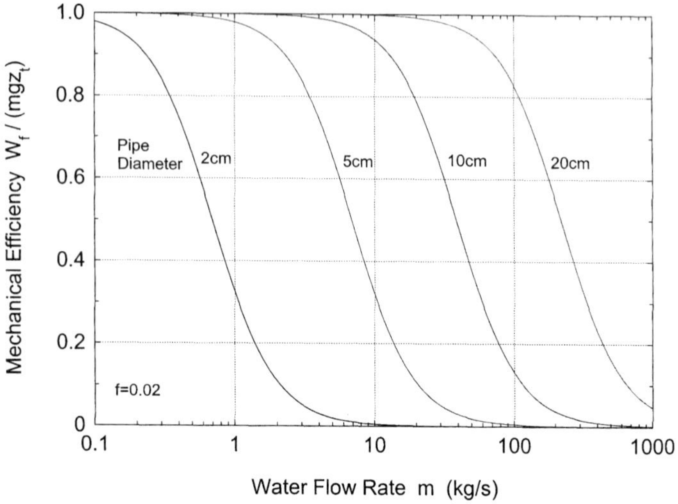

Fig. 16.25 Influence of water flow rate and pipe diameter on the efficiency of reciprocating lift pumps.

It is convenient to define the mechanical efficiency of the pump as

$$\eta_m = \frac{W}{W_f} = \frac{1}{1 + \dfrac{32 fm^2}{g\pi^2\rho^2 d^5}} \qquad 16.13$$

The most important variables here are the diameter of the suction and delivery pipes, which is raised to the fifth power, followed by the mass flow rate, which is squared, and the pipe friction factor f, which is linear. The variation of a pump's mechanical efficiency with mass flow rate and pipe diameter is illustrated in Fig. 16.25. Clearly, 2 cm bore pipes are good for flow rates up to 0.5 kg/s, 5 cm bore pipes up to 5 kg/s, 10 cm bore pipes up to 20 kg/s and 20 cm bore pipes up to 100 kg/s. If these values are exceeded than frictional losses rise very quickly and the required power rises sharply.

Example
It is required to pump 50 tonnes of water per hour from a well 6 m deep to a reservoir 15 m above the pump. If the pipes are 10 cm bore and the friction factor is 0.03 determines the power required.

$\rho = 1000$ kg/m^3. $g = 9.81$ m/s^2.
If the pump operates at 10 cycles/min determine the volume of the cylinder.

$m = 50$ tonnes/hour $= 13.88$ kg/s
From Equ. 16.12 $W_f = 3.18$ kW

The volume flow rate is 13.88/1000 = 0.01388 m^3/s
= 0.8328 m^3/min
Swept volume of cylinder is 0.8328/10 = 0.08328 m^3 = 83.28 litres.

CHAPTER SEVENTEEN

MEDIEVAL, RENAISSANCE, AND EARLY MODERN MINING

Introduction

We have seen, in Chapter Fourteen that a fundamental change in attitude towards manual work occurred in the Middle Ages and was fostered by Christianity and by the monastic movement. Benedict, the founder of western monasticism, emphasised poverty, chastity, obedience, communal living, physical labour, a minimum of conversation, and the distribution of alms to the poor. His monks were supported by their own labour and the duty of labour was strenuously recommended not only as a means of securing their daily subsistence but also as a penance and a physical exercise. In the ancient world technology was seen as undermining faith in the Gods or in nature, but medieval man considered that God instituted technology for and on behalf of humanity and that man's technical achievements greatly honoured Him. Technical processes, that were once considered to threaten political stability, were now regarded as morally beneficial because they satisfied humanity's physical needs, increased sociability, and thereby stabilised society. Monastic labour and their interest in manufacturing embraced mining and the smelting and working of metals to make objects of beauty and utility for church worship. The monks, travelling and corresponding throughout Europe, were at the forefront of technology and their endeavours provided a model that stimulated lay manufacture and the demand for metal.

The earliest English mines of which we have any record are the Cornish tin mines. These mines were known and worked by the Romans but seem to have been run down, if not abandoned, during the Dark Ages. By the end of the twelfth century they were worked again, Fig. 17.1, and production was 483 tons per year. Fig. 17.1 displays tin output from 1200 to 1800. Until about 1680 output never exceeded 1000 tons per year and usually fluctuated between 400 and 800 tons/year. Although there were fluctuations in output there was no long-term growth. Even in the thirteenth-century

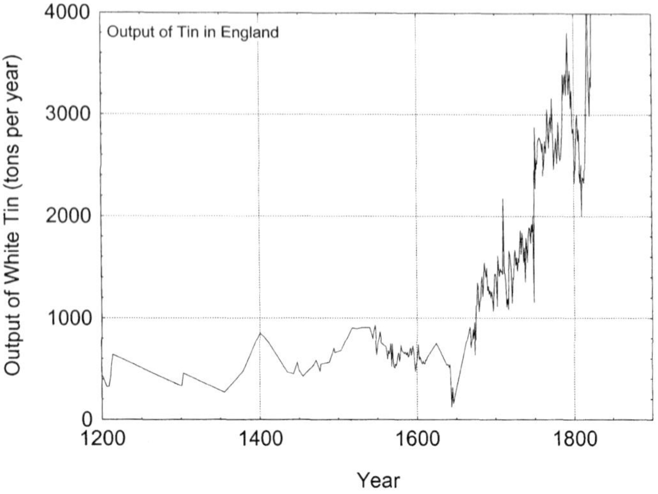

Fig. 17.1 Output of tin in England, 1200 to 1800.

output sometimes exceeded 600 tons/year and was much the same as in the seventeenth century.

According to the sixteenth-century traveller Leland, see Chandler (1993, 69–70, 82–6), the tin mines were very extensively worked when he passed by on his itinerary of England in the period 1539–45 and they seem to have been the most notable mining activity that he observed. He tells us that from Mr Godolphin's house to Lelant is four miles, and in all Cornwall there are no greater tin workings, but in fact, he observed tin mines and workings all over the known mining districts of Cornwall and Devon. Highston, near the Tamar in east Cornwall, "would be a barren waste were it not for its rich workings of tin, which is both mined and streamed." He found similar conditions as he passed through Bodmin, Redruth, Lelant, St Just, and Newlyn. There were no workings between Dodman point and Falmouth and all along the south coast from Looe to the Tamar, he found that "the soil is fertile for corn and grass but there is no tin working".

Leland was apt to blame the tin workings for silting up the Cornish rivers and estuaries. He says, "There was a time before the harbour [at St Erth, about a mile above Lelant] was blocked up by the tin workings, when good tall ships used to come to this place." He made similar charges concerning the bridge at Lostwithiel that was silting

up, and thought that in the course of time it would affect the whole of the Fowey estuary. Similarly, he thought the tin workings were silting up the river Dart and its estuary. The Torey Brook in Devon he notes "was always stained red, from the sand which it flows over and carries with it from the tin mines."

Cornish mining, in comparison to continental practice, was comparatively backward until about 1650 when there was a marked growth in output that continued until the end of the nineteenth century. This growth was attributable to improved technology that enabled deeper, more waterlogged, veins to be drained and worked. Improved ventilation, better means to transport the ore to the surface, and more mechanised surface processing, also helped to increase productivity and reduce the miner's burden.

The earliest period of coal mining in England after the Roman period also has its roots in the twelfth and thirteenth centuries when opencast mining was done along the banks of the rivers Tyne and Wear. Salzman (1964, 6) says that by the end of the thirteenth century practically all the English coalfields were being worked to some extent, for the most part by open-cast works, but pits were emerging. Bell-pits were small vertical shafts cut to the seam and exploited by mining a little way from the shaft base. The shaft was abandoned when the air supply became foul, or it became too difficult to drag the coal to the shaft, or the workings became waterlogged. Exploitation of deeper seams required greater sophistication and was not undertaken until the surface or near surface seams were exhausted. Bell-pits, survived until recently in the neighbourhood of Leeds, Oldham, and elsewhere. They quickly became unsafe. One Piers le Graver was killed by the collapse of the pit in which he was working alone at Silkstone in 1290, and there are numerous instances in Derbyshire in which miners were killed by the fall of their pits. When as much coal as could be safely removed had been obtained, the pit was abandoned and as a rule it was filled up.

An interesting case that suggests that mines were being drained is recorded in Derbyshire in 1322, when "Le Dampe" killed Emma, daughter of William Culhare, while drawing water from a "colepyt" at Morley. Besides suggesting that pits now needed to be drained it is one of the few early references to choke damp, or "stithe", as it was often called. Another indication of the need for drainage at Morley Park occurred 40 years later when it was said that a certain pit was drowned, "for lack of a gutter". Salzman also says that in 1354 the monks of Durham, when obtaining a lease of coal mines at Ferry, had leave to place pits and water gates where suitable and ten years later a lease at a mine at Gateshead stipulated provision of timber for the pits and water gate. In 1486 the monks of Finchale, ever in the forefront of medieval technology, erected a pump worked by horses at Moorhouse. Such pumps did not become more common in Britain until the middle of the sixteenth century.

Coal, mined by these simple methods, was shipped down the coast from Newcastle to London where it was, in consequence, known as sea-coal. In London, it was burnt mainly for industrial rather than domestic use, for coal was often considered unhealthy and was thought to release poisonous fumes when burnt. It does, of course, release carbon monoxide when burnt with an insufficient supply of air, as does any carboniferous fuel, including wood and charcoal. However, wood and charcoal were usually preferred because they were considered to have a better odour. Coal production did not expand greatly until the sixteenth century when wood supplies around London were scarce and sea-coal became cheaper than wood.

Leland in about 1540 saw coal pits in many places. He says that a Mr Bradshaw, of Haigh near Wigan, had recently "discovered much cannel, a substance like sea-coal, on his land, and this has proved profitable for him." In Northumberland, he found that "Coquet Island stands on a very good seam of sea-coal, and at low tide the men dig for it on the seashore by the cliffs, with very good reward." "From Redden Burn along the Tweed to Berwick there is virtually no woodland, but coal is dug for fuel at Murton, a small village in Glendale, two miles from Berwick". Leland also found coal pits at Gatehead and near the river Went in Yorkshire, at Bolton in Lancashire, at Wombridge in Shropshire, at Walsall and Wednesbury in Staffordshire, and in the Mendips. At Cawood and Rotherham in Yorkshire Leland found that there was "a good supply of wood, yet the inhabitants burn predominantly coal, which is found in great abundance there, and is sold very cheaply. There are excellent coal pits a mile from Rotherham." Coal, unlike tin, needs little or no surface treatment, so Leland made no complaints about pollution.

Mining, of course, is a dangerous occupation, and the engineer needs to ensure that conditions in a mine are as safe as possible. Safer working conditions usually developed only after serious accidents had occurred. According to Salzman (1964, 4), the first recorded victim of coal mining was Ralph, the son of Ulgar, who in 1243 drowned in an open cast mine. Bowman (1990, 137) found several deaths related to seventeenth-century mining activities in the parish registers of Ashton-under-Lyne, in the Lancashire coalfield, namely:

4th July 1627: "Thomas Slater slayne in a colepitt with the fall of the roof."

31st October 1662: "Mary Winterbottom Bestow, widow, found dead in ye Bancke under a load of coles."

21st February, 1663: "Joseph Taylor, Alt, found in a Cole Pitt off Lowside and fell in the 17th day att night as was sayd."

6th February, 1673: "Elisha Knott killed in a coale pitt at Fairbothom."

No doubt many other, and earlier, records exist and certainly many fortunate escapes occurred. Butterworth (1823, 18) records that a certain man by the name of Turner, having finished work for the day wished to ascend the mineshaft. He jumped into a coal bucket, with other miners, and had not ascended halfway when a sudden jerk of the winding rope threw him out. However, the buckle of his shoe caught hold of an iron strip surrounding the top of the bucket. "He was suspended thus in so perilous a situation with his head downwards, so that no one could or durst try to relieve him, and he was wound up the other half-way in this frightful manner." At the top he was rescued from his dangerous position and boasted that "I have out-done you all, I have done what none of you dare copy." Mine accidents did not usually have such a fortunate outcome.

The most serious mining accidents are those associated with dust or gas explosions in collieries. Fig. 17.2 shows the number killed and the frequency of such accidents in British collieries during the eighteenth, nineteenth and early twentieth centuries. Only the most serious accidents, those killing more than twenty miners, are shown. About 150 such accidents occurred over the period, although not all are recorded. The frequency of these accidents and the large numbers killed, generated great concern and many experiments were made to eliminate them or reduce their severity. Faraday investigated dust explosions as early as 1840.

If the dangers were severe in British mines they were worse elsewhere. Explosions were the most dramatic and frequent cause of death in mines but there were numerous other and more important sources. The total number of deaths in collieries in several countries during the early 1920's is compared in Table 17.1 and Fig. 17.3, Peele (1927, 1594–6).

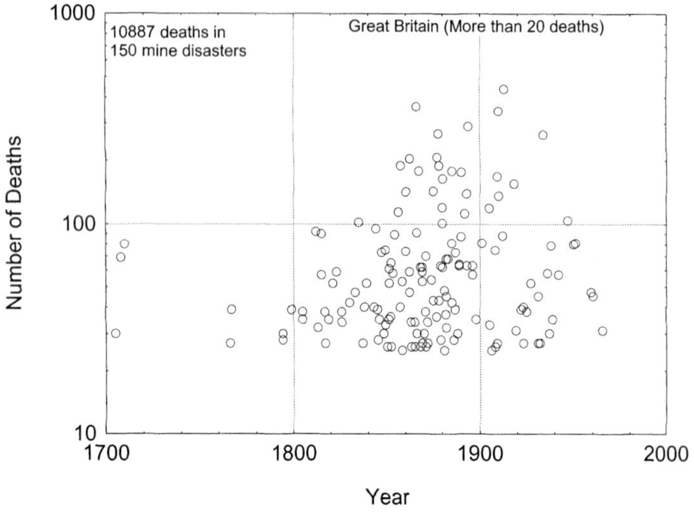

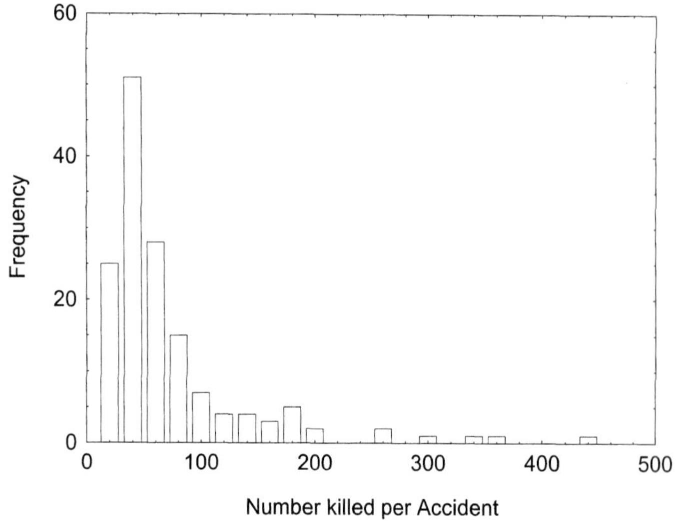

Fig. 17.2 Major mine explosions in Great Britain involving 20 deaths or more deaths per explosion, 1830–1930.

Table 17.1 Numbers killed in Collieries from all causes.

Country	*Deaths per 1000 man-years**			*Deaths per million tons mined*		
	1920	*1921*	*1922*	*1920*	*1921*	*1922*
United States	3.78	4.20	4.89	3.45	3.94	4.15
Great Britain	1.11	1.08	1.09	4.11	4.02	3.86
France	1.13	0.99	0.87	7.22	6.42	5.37
Belgium	1.12	0.91	0.91	7.33	6.09	6.07
Germany	2.16	1.86	1.89	11.53	10.25	10.30

* One year is of 300 working days duration.

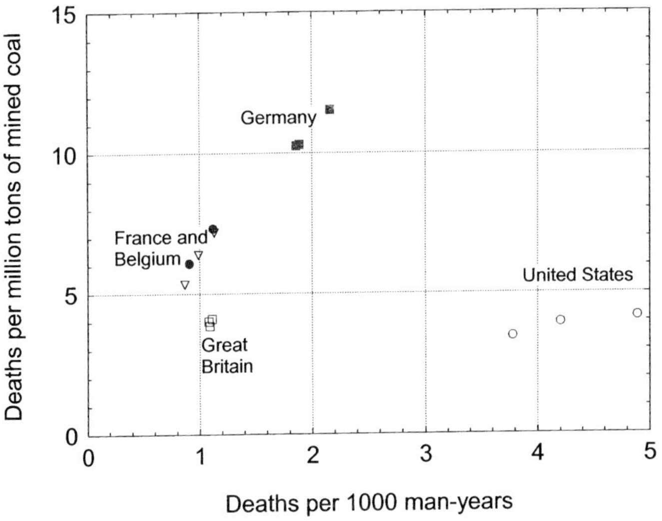

Fig. 17.3 Deaths and productivity in collieries, 1920–22.

Roof falls were the most common cause of accidental death followed by haulage accidents, gas and coal-dust explosions, other underground explosions, shaft accidents, and lastly surface accidents. Metal and mineral mines were safer than collieries but all mines were dangerous. As far as the miner was concerned the important statistic was the number of deaths per thousand man-years and in this respect Belgium, French, and British collieries were very good, having about one death per thousand man-years. This appears to be a very small risk but it is not so. If a miner worked for fifty years in a coalmine, he would have a 1:20 risk of accidental death. In Germany, the risk was twice this and in the United States, it was four times greater. If the same data are expressed in terms of the number of deaths per million tons of output then the United States and Great Britain were the safest, having about four deaths per million tons. In France and Belgium the deaths per million tons was about four and in Germany it was about ten. Even in comparatively modern times coal mining was a dangerous occupation and at earlier periods, it must have been much worse. The productivity, it should be noted, was about 1000 tons/man-year in the USA, 250 tons/man-year in Great Britain, and about 180 tons/man-year in Europe.

We have to wait until the Renaissance before any real knowledge of improved mining technology emerges, and then we have a flood of information. In 1540, Biringuccio published his classic account of metals and metallurgy, "Pirotechnia". He was a practised and observant engineer writing in his own language and he evidently intended his work to be read by others working in his field. Sixteen years later, in 1556, a German physician, Georg Bauer (commonly known by the Latinized name of Agricola) published "De Re Metallica". Being written in Latin it was aimed at a different, more scholarly, readership. Overall "De Re Metallica" is better known and better illustrated. In addition, there were numerous other texts relating to mining and taken together they give a clear description of mining problems and the methods adopted to overcome them. The remainder of this chapter addresses these problems.

Drainage

Biringuccio (1540, 18–19) gives sensible advice on where mines should be started to ensure good drainage and points to the difference between medieval and ancient practice. "Take care that you always begin your excavation low down and as nearly as possible at the base or root of a mountain, in such a manner that the mine proceeds in a straight line to cross the vein of the ore in the shortest or surest way that is indicated.—The ancients used another method, by which, as is seen in old abandoned mines, instead of beginning from the bottom at the roots of the mountains (as the moderns do), they began to mine in the upper part of the mountain where the ore appeared in the light of day at the surface, and digging down as in wells they followed it to the bottom, now here, now there, as it appeared.—Above all, the moderns have a better understanding because of the impossibility [in the old method] of being able to dispose of the waters which often are so abundant that they multiply the owner's expense and labours through the great number of helpers required in order to make wheels, pumps, pipes, pistons, and other similar contrivances for drawing out the water.—It is a much better and surer method to begin excavating at the root of the mountain rather than at the top or upper ridge and to penetrate within little by little. In order to make it easier for the waters to run off and for the workmen to carry, there should be a gradual rise of half a braccio every ten canne."

An early mine is described by Taylor (1859, pl. 1) and is illustrated in chapter 16, Fig. 16.13. It takes the form of an adit, which drained water from the work-face and along which coals were transported. Men or boys carried the coals from the workings, either on their backs or on sledges. A vertical shaft provided natural ventilation and a winch mounted above it provided additional means of removing the coal. The vertical shaft was usually shorter than the adit and it was usually easier to remove the coal by the winch than by dragging a sledge along the adit. From the mine exit, the coal was carried on the backs of horses or donkeys. A typical load weighed about 125 kg.

Biringuccio (1540, 20) confirms that mines were worked on a very large scale and in the most arduous of conditions. Speaking of a mine in a valley between Innsbruck and Halle in Austria, he says that "—as they mined they advanced a little less than two miles by my estimate before they saw the sparkle and shadow of ore—when I was there looking I saw a wall of the hardest limestone forming a very large open space where more than two hundred men stood working at one time, both above and below, and having no other light than that of lanterns.—Furthermore I wish to tell you that there was in the middle of the mine a channel which collected all the waters that fell continuously from the various openings and issued in such quantities that I surely believe it would have satisfied every demand of a large mill. In going in and out of the mine I remember that because of water above and water below I was soaked as if I had passed through a heavy rain—"

Conditions in pre-industrial mines are well illustrated in an eighteenth century drawing prepared for Diderot's encyclopaedia; Fig. 17.4. This illustrates the problems, dangers, and discomforts of mining: drainage, ventilation, and lighting that uses up the available oxygen. Water was pumped off or, preferably, drained through a drainage gallery out of the hillside. The darkness, cramped space, awkward digging, crude hoists, simple tools, and dim torches are exemplified.

Biringuccio's advice on positioning the entrances and the use of adits to drain off the water was sound but it was not always possible to put into practice. Pryce (1778, 150) remarks "With all the skill and adroitness of our miners, they cannot go any considerable depth below the [lowest] adit, before they must have recourse to some contrivance, for clearing the water from their workings." Frequently the lowest part of the mine was a long way beneath the lowest possible entrance and so the water had to be pumped up to a suitable drainage adit. Taylor (1859, 17) reports that as a rule, a greater weight of water than of coal must be raised to the surface of the Northumberland and Durham collieries. At the Percy Main Colliery near Newcastle, the weight of water raised has been nearly thirty times that of coal and in other cases seven or eight times that of coal. This presented tremendous problems before the invention of fire engines (steam engines) for pumping and, no doubt, made it impossible to work in many seams. Taylor says that in 1610, Sir George Selby stated in parliament that the Newcastle coalfield would be worked out within a period of twenty-one years. He was no doubt referring to the seams that could be drained by the methods then available. Two centuries later the Newcastle coalfield was still producing 16 million tons of coal annually, largely due to the use of steam driven pumps to drain the levels.

Agricola (1556) describes several types of water pump, a good account of which is found in Wolf (1968). In Fig. 17.5, we see a simple continuous chain with dippers attached. However, the power available from one man is very limiting and the system needs a large reduction ratio if the man is to develop sufficient force to move the dippers. Two reduction gears, each of 6:1 reduction are used to give an overall reduction ratio of 36:1 so although the load applied by the man is quite small his cranking speed must be quite high and even then the speed of the dippers is very low.

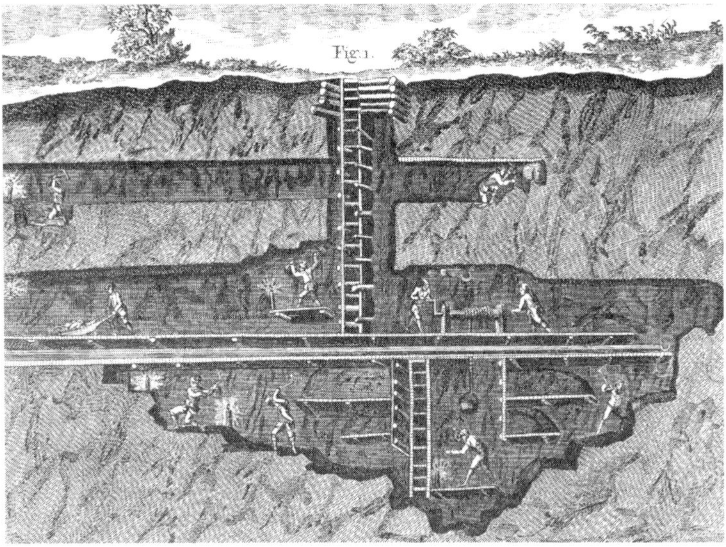

Fig. 17.4 Mining. Diderot and d'Alembert (1751–72).

The performance of a dipper wheel may be estimated by using Equ. 17.1 to 4, set out below, and some reasonable assumptions. A man turning a winch might develop a power of 75 W for relatively long periods of time (several hours). If he exerts an average force of 100 N on a crank of 0.5 m radius then he can rotate the crank at a steady speed of 14.3 rev/min. With an overall gear ratio of 36, a mechanical efficiency of 0.6, and assuming the radius of the chain drum is 1.0 m, then he can lift a load of 1080 N (about 110 kg) and the lifting speed of the dippers is 2.5 m/min (about 4 cm/s). This is a very slow lifting speed. If the dippers contain about 6 kg of water each then about 18 dippers may ascend on the chain as shown in Fig. 17.6. To hold 6 kg of water the dippers would need to be about 16 cm diameter and 32 cm high. Thus, 18 dippers fitted reasonably close together need about 8 m of chain. The amount of water lifted depends on the height through which it must be raised. If we take this to be 6 m then the pumping rate is about 0.76 kg/s (about 2.8 tonnes/hour). The pump is quite effective if rather slow.

These simple calculations corroborate the geometry shown in Fig. 17.6. Agricola remarks that "It cannot be constructed without great expense, and it carries off but little water, and is somewhat

Fig. 17.5 Simple pump consisting of a chain of dippers driven by one man through a large reduction ratio. Agricola (1556, 173).

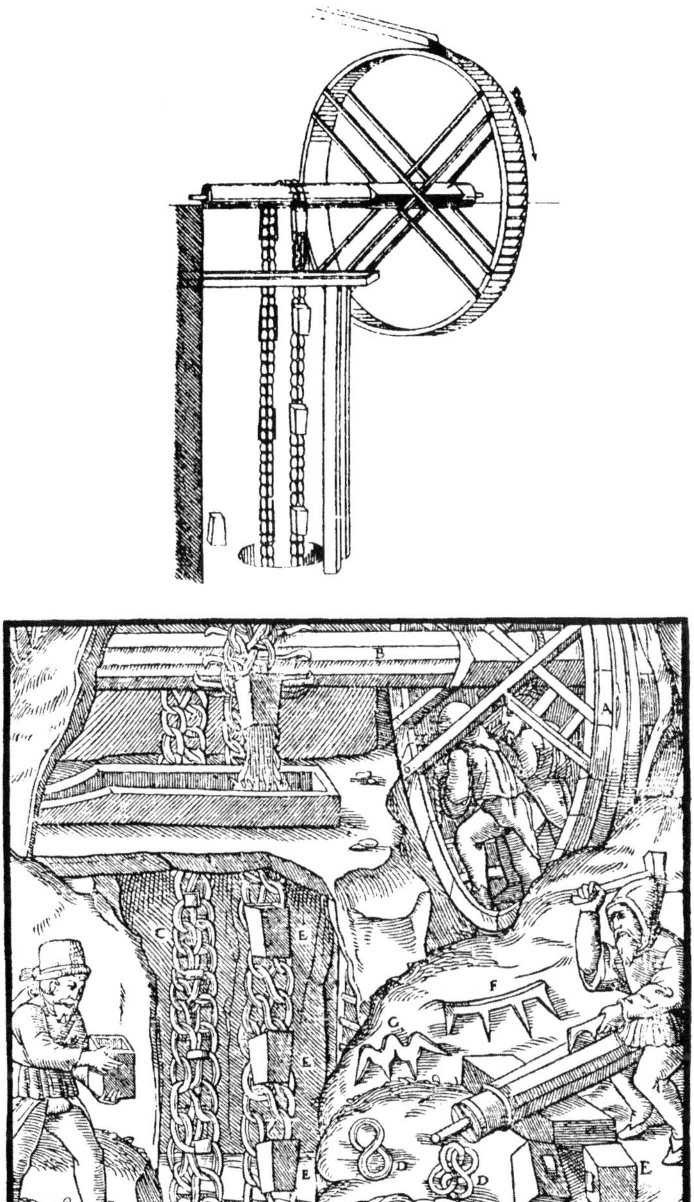

Fig. 17.6 Top: chain pump driven by a water wheel, about 1700. Taylor (1859, Fig. 2). Bottom: chain pump driven by a two-man treadmill. Agricola (1556, 174).

slow, as also are all other machines, which possess a great number of drums [i.e., a large overall gear ratio]". The construction of the pump seems to be very sound. The frame, wheels, pinions, and shafts are all of iron and owe much to the design of clock reduction gears. The main horizontal shaft carrying the chain of dippers rests in iron rollers.

Such devices for raising water continued in use for many centuries and were improved by replacing the manual winch by one driven by a water wheel, as illustrated in Fig. 17.6. The anonymous author of the "Complete Collier" of 1708 says that "in some places we draw the water by water, with water wheels or long axle-trees, but there is not that convenience of water everywhere". The long axle-tree is the drive shaft from the wheel to the chain of pots. Clearly, a water wheel generates much more power than does a man and it can lift the pots more rapidly.

Fig. 17.7 shows a simple suction pump fitted with clack valves and directly driven by a man. As we have seen, the Romans used such pumps. In the figure, suitable wooden tree trunks, commonly elm, are brought from the forest to be bored out to a diameter of about 125 to 150 mm. The ends are tapered to form male or female spigots so they can be more effectively joined. Scattered around are boring tools, bored-out logs for use as water pipes or piston cylinders, pistons and rods, and various valve fittings. The piston is a flexible leather bag that contracts on the down-stroke and expands to fill the cylinder on the up-stroke. As Fig. 17.7 indicates, such a pump operated by one man could scarcely produce a lift of 2 m. Nevertheless, from Equ. 17.4, the mass lifted per unit time would be perhaps three times greater than the man-powered dipper. Sometimes a simple rocking bar was used to drive the piston rod instead of the direct drive illustrated. Alternatively, cams driven by a waterwheel could lift the piston rod and the pistons could fall back under their own weight.

A much more sophisticated pump is illustrated in Fig. 17.8. This pump is suitable for lifting water from deep mines, and is one that Agricola says was "invented ten years ago, is the most ingenious, durable, and useful of all, [and] can be made without much expense". It was powered by a water wheel about 4.5 m diameter through a crank of about 1.2 m radius. The rotational motion of the crank was converted into a reciprocating motion by a connecting rod, one end of which was constrained to move in a vertical plane by means of guide levers. Wear at these points must have been a great problem. A small working model, built according to the data specified by Agricola, is displayed in the Science Museum, London. The great advantage of this pump is that the power input is considerably increased by the application of a water wheel, and the use of pumps in series reduces the pressure applied to the pipes.

A duplex type of pump driven by a man-powered crank is illustrated in Fig. 17.9. Two suction pumps discharge into a common crankcase hollowed out of a solid block of beech wood, 1.5 m long, 0.75 m wide, and 0.5 m thick and split along the axis of the iron drive-crank. A simple seal was formed around the crank by means of iron and leather washers on both sides of the hole. "Then the upper part of the box is placed upon the lower one, and properly fitted to it on every side, and where they join they are bound by wide, thick, iron plates, and held with small, wide, iron wedges, which are driven in, and are fastened with clamps." Agricola warns that "since a wooden box frequently cracks open, it is better to make it of lead, or copper, or brass." The box is fitted with two suction pipes, one for each pump, and a common delivery pipe. The system is still limited to the power of a single man and the height that water can be lifted is inadequate. The maximum height of a suction pipe is confined to about 8 m to avoid the formation of air bubbles in the water. The strength of the wooden/lead/copper/brass box and the efficiency of the seals restrict the maximum height of the delivery pipe. The crank is fitted with a flywheel, which helps to maintain a constant speed but was mistakenly thought to reduce the work of the driver.

Another type of water pump described by Agricola is the rag and chain system; see Fig. 17.10. This is a very ancient system. The chain ran in a vertical pipe and at about 2 m intervals a ball was fitted. The balls are "made of the hair of a horse's tail sewn into a covering to prevent it being pulled out by the iron clamps on the driving drum: the balls are such a size that one can be held in each hand". When water was available to drive it, Agricola tells us, a 7 m diameter water wheel lifts water from a shaft 62 m deep and a 9 m diameter water wheel from a shaft 72 m deep. Where water was not available, a horse whim or even, as in the illustration, men in a treadmill could provide the power.

The hoist shown in Fig. 17.11 could be used, presumably, for lifting the ore from the mineshaft or for lifting water. Agricola says it was used for raising water. A timber wheel 11m in diameter was

Fig. 17.7 A simple suction pump. Agricola (1556, 177).

Fig. 17.8 A series of suction pumps driven by a water wheel and crank mechanism. Agricola (1556, 185).

Fig. 17.9 A duplex type of mine pump with two suction valves. Also see Fig. 17.8 which Agricola says was invented about ten years ago. Agricola (1556, 180).

Fig. 17.10 Rag and chain pump driven by men in a treadmill. Agricola (1556, 197).

Fig. 17.11 Reversible water wheel for raising water from a mine. Agricola (1556, 199).

fitted with a double row of buckets. These were supplied with water from two water troughs and arranged so that the wheel could be reversed. A man controlled the direction and speed of the wheel by lowering or raising the two water gates that fed the troughs. The main horizontal axle was 11 m long and 0.6 m square. The winding drum fitted on this axle consisted of four hubs, 1.3 m diameter at 1.3 m intervals, covered in wooden strips. The haulage chain is wrapped around this drum. A brake drum of about 2 m diameter was fitted so that if the controller "cannot close the water gate quickly enough, and the water continues to flow, he calls out to his comrade, and bids him raise the brake on the drum and stop the wheel." The bucket suspended in the air before the controller looks rather odd but it is connected to a float in the sump and indicates to the controller the level of water in the sump.

Pumping water in British mines still used the dipper wheel, similar to that shown in Fig. 17.6, but driven by horses. John Taylor, the water poet, writing in 1618 describes a visit he made to the mine of Sir George Bruce, at Culross, near Dunfermline, Scotland, Chandler (1999, 34). The mine could be entered either from the land or from the sea. Taylor entered from the sea and says that at high tide the sea was prevented from flooding into the mine by "a round circular frame of stone, very thick, strong, and joined together with glutinous or bituminous matter so high, that the sea—[could not]—overflow the height of it." Within this protective cage they dug more than 40 feet [12.2 m] down. "At last they found what they expected, which was sea-coal, they following the vein of the mine, did dig forward still, so that in the space of eight and twenty, or nine and twenty years they have digged more than an English mile under the sea, that when men are at work below, an hundred of the greatest ships in Britain may sail over their heads." "The sea at certain places doth leak, or soak into the mine, which by the industry of Sir George Bruce, is all conveyed to one well near the land; where he hath a device like a horse mill, that with three horses and a great chain of iron, going downward many fathoms, with thirty six buckets fastened to the chain, of which eighteen goes down to be filled, and eighteen ascends up to be emptied, which do empty themselves (without any man's labour) into a trough that conveys the water into the sea again; by which means he saves his mine, which otherwise would be destroyed by the sea—."

German mining technology led the world in the sixteenth century and their methods were certainly in advance of British practice. Taylor (1859, 21) tells us that the double water wheel, shown in Fig. 17.11, was not adopted in the Newcastle coalfield until the eighteenth century. Pumps, similarly, were not used until the eighteenth century. A report made by his manager to the Earl of Mar in 1709 stated that mines in the Newcastle coalfield were drained by means of water wheels and horse engines with chain pumps. The common depth of pits was from 40 to 60 m and a few going down to 100 m. Steam engines were not mentioned. The Earl called in an engineer from Derby, George Sorocould, who recommended replacing the chain and buckets with a water pump. No one could be found to do the work. Eventually a millwright from Montrose, John Young, who had worked in Holland, and "the mechanical priest of Lancashire", succeeded, and the water wheel and buckets were replaced with cranks and beams working force pumps; see Taylor (1859, 19).

Pryce (1778, 150), writing of the Cornish tin and copper mines, says that the hand pump or force pump will do well for small depths. The water was pumped into barrels and lifted from the mine up a shaft or run off along an adit. When the water flow was very large, they erected a horse whim at the surface to lift the barrels up the shaft, see Fig. 17.12. Using two horses, they lifted barrels containing 60 gallons [272 kg] and with four horses they lifted 120 gallon barrels [544 kg]. Such whims were also used to raise coal or ore. Taylor (1859, 20) states that for a shaft, 100 m deep, eight horses were required, two at a time, and 200 kg of coal was lifted in wooden containers. The weight lifted was 84 tons per day [85 tonnes per day], or about 10.5 tonnes per day per horse. Water would be lifted at the same rate.

Another "water engine is the rag and chain which consists of an iron chain with knobs of cloth stiffened and fenced with leather, seldom more than nine feet [3 m] asunder [apart]". The chain is lifted by a wheel of 0.7–1 m diameter, furnished with iron spikes, to "inclose and keep steady the chain", so it may pull the knobs through a wooden pipe of 8 cm to 13 cm bore and from 3.5 to 7 m long. As the leather knobs move up the pipe they form an effective seal and lift the water. Pryce states that "they are usually worked by hand like those in our navy" and the men "suffer much in their health and strength from the violence of the labour, which is so great that I have been witness to the loss of many lives by it." Such

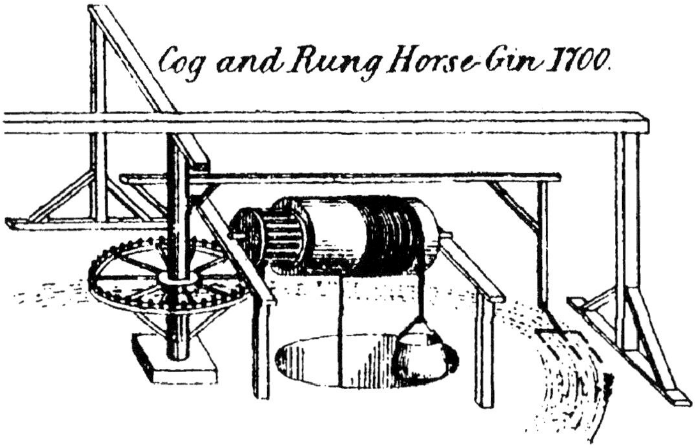

Fig. 17.12 Cog and rung horse whim, about 1700. Taylor (1859, Fig. 17.3).

a pump of 10 cm diameter "requires five or six fresh men every six hours to draw twenty feet [6 m] deep". To keep it going constantly required the employment of twenty or twenty-four men. Pryce says that these devices "are generally laid aside on account of the great expense, and the destruction of the men." They were, nevertheless, very effective in removing water.

The answer to this problem was to replace manpower by water-power. "The water wheel with bobs, is yet a more effectual engine; whose power is answerable to the diameter of the wheel and the sweep of the cranks fixed in the extremities of the axis [see Fig. 17.13]. Over them two large bobs are hung upon brass centre gudgeons supported by a strong frame of timber, and rise and fall according to the diameter of the sweep of the cranks". To each crank is fixed a straight balk timber, the connecting-rod, that causes the bobs to reciprocate as the water wheel rotates. One end of the bob is over the shaft and is connected by a large iron chain to a perpendicular rod of timber that works a piston in an iron or brass hollow cylinder. Pryce (1778, 151) reports that the water engine wheel at Cooks Kitchen Mine, is 48 feet [14.6 m] diameter, and works a tier of pumps of nine inches [23 cm] bore, divided into four lifts, and draws eighty fathoms [146 m] under the adit. If the stream of water was sufficient to fill the buckets it would draw forty fathoms [73 m] deeper. The power of this wheel was said to be the same

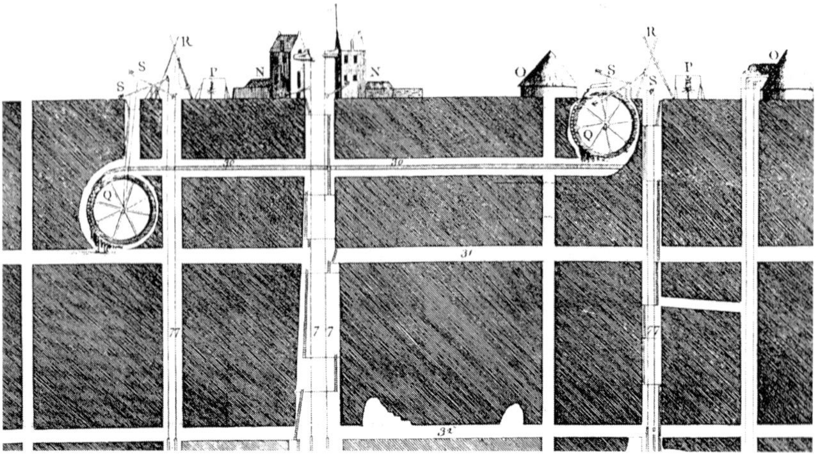

Fig. 17.13 Water wheel driving a reciprocating force pump through a set of bobs. Pryce (1778, pl IV).

as that of a 47 inch [1.2 m] "fire engine house cylinder". Where the water supply was adequate and was available for a small rent, the water driven pump was preferred to the fire [steam] engine. This remained the case until well into the nineteenth century.

Ventilation

Oxygen in mines is consumed not only by the miners but also by their candles and lamps. It is usually reckoned that a candle requires as much air as a man and the need for fresh air was recognised long before there was any scientific understanding of combustion. "—it is necessary to have great quantities of oily liquids that may be burned", says Biringuccio, "such as oil of olives, nuts, linseed, hempseed, resin of trees, or extracts of the fat of land animals or fish, for it is impossible to work underground without the light of fire. This cannot be kept alive unless the mine has some hole for air, either a wooden pipe or some other opening." More generally, we can say that mines need to be ventilated because both miners and their lamps consume the oxygen and some mines, particularly collieries, emit dangerous or poisonous fumes. Explosions in mines are usually caused by poor ventilation that fails to dilute and remove the explosive gases.

Fig. 17.14 Pivoted barrel over a ventilation duct. Agricola (1556, 203).

Natural currents arise because the air inside a mine is often hotter than that outside. These currents are enhanced if two or more shafts or passages are connected and their entrances are at different heights. Fires in the up shaft, or chimneys built over the up shaft, further enhance natural ventilation. Mass transfer by diffusion also refreshes the air in a mine but this is usually a slow process. To enhance ventilation further, some form of forced convection is required. Agricola records three techniques that were used in the sixteenth century to force the ventilation of mines: cowls, fans, and bellows.

Fig. 17.14 illustrates a cowl ventilator. A barrel with an opening in the side is fixed onto a ventilation pipe so it is free to rotate about the pipe. A tail fin is fitted opposite to the opening allowing the wind to rotate the barrel such that the opening always faces upwind. The wind entering through the opening is retarded thus creating a slight rise in pressure inside the barrel. This rise in pressure drives air into the ventilation pipe and hence into the mine. Fitted over the down shaft this simple device improved the ventilation, at least on a windy day.

Fig. 17.15 Centrifugal fan for ventilation of a mine. Agricola (1556, 204).

The second ventilator described by Agricola is a simple centrifugal fan, shown in Fig. 17.15. They were made either as cylindrical drums or as rectangular boxes. As Agricola notes, "The drum is far superior to the box; for the fans so fill the drum that they almost touch its side, and drive into the conduit all the air that has been accumulated." The operator rotates the vanes at high speed and the air is drawn in and is accelerated radially by the vanes into the delivery ducts.

Before the nineteenth century British practice in mine ventilation lagged behind that of Germany. Probably this was because British mines could be ventilated successfully and inexpensively by natural convection. Pryce (1778, 147) describes the process by means of Fig. 17.16. An adit was first driven from point *A* into the hillside. A little further on they dug a shaft to remove the ore from the workings and to supply air to the work face. The next shafts *C* and *D* were sunk for the same purpose. The working shaft is *D* and a new shaft *E* is not yet sunk to the work face *F*, at the end of the adit. "For want of the shaft *E* being holed upon the end *F*, the air is very close and suffocating; nay, the adit end [*F*] must be deserted for want of air." To remedy this they build a close stage of boards, *G*, about 30 cm high from the bottom of the adit. This is called a "saller" and is continued to about 2 m of the end at *H*. These boards are sealed with soil or clay and the space under the boards conveys air to the work face. The air then discharges back along the adit to the working shaft *D* where it meets a door at *K* and it therefore ascends the shaft. This system works because the candles, the miners' bodies and the surrounding rock heat the air making it less dense than the atmospheric air. The hot air in the mine thus ascends the shaft because it is lighter than the atmospheric air.

Pryce goes on to explain that another way of forcing down air is by the pipe in shaft *C*. The top of this pipe can be turned towards the wind and receives the air which is forced down through the funnel *M* into the adit at *N*, whence it circulates back again through the workings to shaft *D*. This, of course, is identical to the cowl ventilator described by Agricola over two centuries earlier. Pryce (1778, 147) says that "This air pipe is seldom used in adits, because the saller is more cheap and easy, the difference in expense in the air pipe being considerable where an adit shaft is thirty or forty fathoms [55 to 75 m] deep." Nevertheless, he admits that an air pipe is indispensable in sinking a shaft that is void of good air and it is

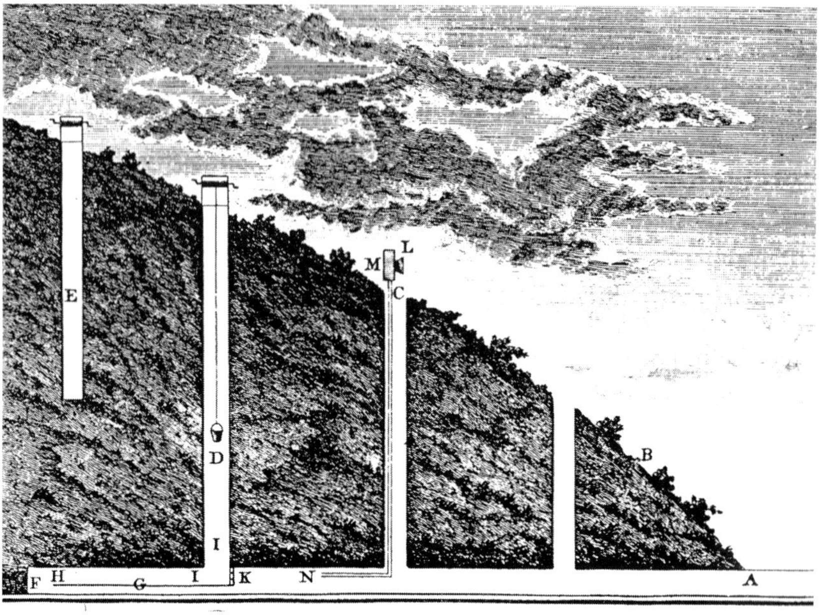

Fig. 17.16 Natural and forced ventilation of Cornish mines, after Pryce (1778, 147).

seldom that a shaft can be sunk to a depth of 55 to 75 m without an air pipe. A serious disadvantage of the cowl ventilator is, of course, that it does not work when the wind is insufficient whereas a saller always works so long as there is a temperature difference between the mine air and the open air.

Natural ventilation, although independent of working parts that could fail, was not always dependable. Hinsley (1969, 25–39) quotes Gabriel Jars who wrote in 1764 "I have remarked for a very long time and have heard it said by all miners, that the air circulates with difficulty in the mines at the budding and at the fall of the leaf, that is to say at spring and autumn; it even happens that some works are suspended at that time for want of air." Such stagnation occurs when there is insufficient temperature difference between the mine air and the open air, as may occur in spring and autumn.

Sir Robert Moray, Pryce (1778, 148) informs us, described a method practised at Liege for ventilating adits. They erect a chimney, thirty feet [9 m] high, at the entrance to the adit and lay a pipe to the work face. A fire at the foot of the chimney draws the air along the pipe from the work face and expels it up the chimney. The pipe needs to be quite long. Pryce tells us that in Cornwall, adits seldom exceed 800 m in length although there are some that are two or three times that length and at Freiberg in Saxony the passage underground is about 8 km in length.

The frequency of explosions and the large numbers of miners killed in the nineteenth century compelled colliery owners to adopt more positive ventilation measures. These measures took the form of chimneys and furnaces, to heat the air in the up shaft, or mechanically driven air pumps. The latter, however, were subject to breakage and until mechanical ventilators became reliable, they were not popular. As late as the 1850's the 10 m thick coal seams of South Staffordshire depended on enhanced natural convection, Gibbons (1851, 9). Gibbons was induced to publish his experiences "in consequence of a frightful explosion taking place at Oldbury" in December 1846. It was not one of the more serious explosions plotted in Fig. 17.2 but it was enough to prompt those in responsible positions to reduce the risk of future explosions by publishing accounts of safe practice.

Gibbons thought that the existing practice of sinking two shafts, an up-shaft and a down-shaft, of about 2 m diameter and perhaps 180 m in depth, and relying on natural convection, was dangerous. The shafts were usually started at the same level so they relied entirely on temperature difference to drive the air current. Such airflow is trifling and subject to change by atmospheric wind and temperature variations. The air frequently becomes quiescent and alternately ascends or descends in the same shaft. The miners say that the pits "fight". When the airflow reverses the dangerous emissions from the coal-face, instead of being removed from the mine, return to it, mix with more gas, and may issue into the gate-road, reach the fire point, and cause an explosion. This danger is seriously increased when the up-cast shaft, which removes the gas from the mine, is also used as the working shaft. The ascent and descent of the skips disturbs the airflow. Worst of all, when a loaded skip arrives at the top of the shaft a runner is wheeled out over the shaft, and under the skip, whilst the coal is landed. This completely stops the airflow and when the empty skip descends it drives the air and gas back into the mine.

Fig. 17.17 shows the system Gibbons used. One pit shaft is sunk, instead of two, and in the side of the shaft, a smaller shaft of about 1 m diameter is cut to form an air chimney and is afterwards

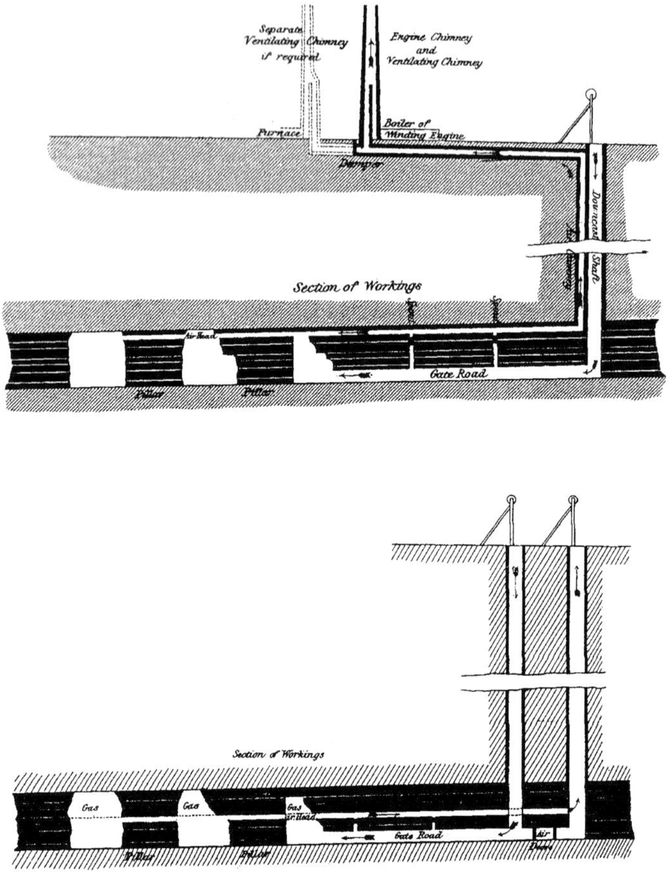

Fig. 17.17 System of mine ventilation adopted by Gibbons (1859, pl. 24 and 26) in the 10 m coal seams of South Staffordshire.

separated from the main shaft. Both shafts are bricked. The efficiency of this air chimney was demonstrated when the shafts were sunk because the time required to clear the fumes from gunpowder explosions was very greatly reduced. The fumes normally clear before a man can descend to the bottom of the shaft so there is no delay. The gate-road is driven in the usual way from the bottom of the shaft but an airhead is driven from the air chimney at the top of the coal seam, or higher if possible. The gate-road and airhead are driven in parallel and are connected by a vertical spout every 15 m. When a new spout is opened the old one is bricked up. The fresh air thus descends the main shaft, travels the length of the gate-road to the work face, then ascends to the airhead and returns to the air-chimney and to the surface. Even a spout of 10 cm diameter is sufficient to clear gas from the work face, but normally the spout is large enough to pass a man and about 6 m long in the 10 m coal seam. Gibbons found that the airflow is sufficient "to ventilate the gate-road and to enable the men to work with candles in the spout with perfect safety." As the coal is mined and the light mine-gas is released, it rises to the top of the workings and is drained off by the airhead and the airshaft. The ascending air is lighter than the descending air because it:

1. mixes with and carries away the lighter gases issuing from the mine face,
2. is heated by contact with the warm walls of the mine, and
3. is heated by the activities of the numerous workmen and horses in the mine and, of course, by the many candles.

The airflow, however, is often quite feeble and needs to be enhanced by erecting a ventilation chimney, of sufficient height, over the air-shaft. This may be aided, occasionally, by a small furnace or steam jet, to further heat the ascending air to make it sufficiently strong to defy wind and weather. This additional heating may make use of the exhaust gases from the boiler if a steam engine is in use at the mine, but before the use of steam engines it had to be provided by a separate furnace as shown in Fig. 17.17.

Although they are reliable and run on coal mined from the colliery, separate furnaces to heat the ascending air are nevertheless expensive to run. Taylor (1859, 31) says that at the Haswell colliery near Newcastle the temperature in the up shaft was 73°C and the down shaft was 10°C. The height of the shaft was 285 m so the pressure head driving the air was 52 m. The observed quantity of air passing up the shaft from the mine was 2,689 m^3/min or 3,359 kg/min. Thus the useful power developed by the furnace was 28.3 kW. The coal consumed in the furnace was found to be 218 kg/hour so the coal consumption rate was 7.7 kg/kWh and the thermal efficiency of the system was only 1.7%. A water wheel driving a mechanical ventilator would not consume any coal and a steam engine driving a mechanical ventilator would consume much less than 7.7 kg/kWh so it is not surprising that mechanical ventilators should be preferred when they became reliable. The ventilation chimney at the Seghill

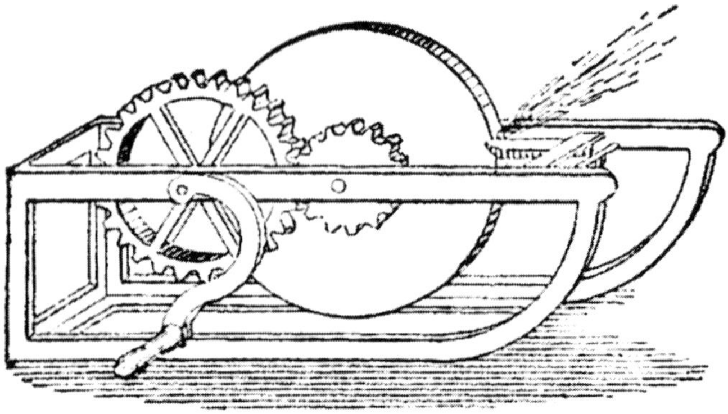

Fig. 17.18 Miner's steel-mill lamp, Knight (1884, 1314).

Colliery near Newcastle was said to be three times more efficient, having a thermal efficiency of about 5.3% and using only about 2.5 kg/kWh. Taylor reports that by 1859 mechanical ventilators were in use in the South Wales mines, at the Eaglesbush, Dyffryn, Risca, and other collieries. On the other hand, a coal consumption of about one tonne every five hours is not excessive in relation to the quantity mined in that period so the case for mechanical ventilators was not too strong.

Gibbons (1851, 9) gives some idea of the necessary quantity of air needed in a colliery. If a mine produces 100 m^3/min of gas then 30 times as much air (3,000 m^3/min) must be passed through the workings in order to dilute it to the point of safety, plus a further 500 m^3/min to supply the workmen, horses and candles. Before the introduction of safety lamps, in 1816, a steel mill was often used to illuminate mines where gas emissions were high, Fig. 17.18. This consisted of a steel that was rapidly rotated against a flint to produce a continuous series of sparks sufficient to give a feeble light. Very small points of flame of short duration, it appears, could not ignite the gas.

Driving Tunnels and Shafts

Biringuccio (1540, 24) says that "For excavating and breaking rocks strong and powerful tools are required, like large hammers and iron picks, long thick crowbars, mattocks and strong spades, picks both

Fig. 17.19 Miners' tools, ore barrows, and baskets. Biringuccio (1540, 24).

with and without handles, and similar iron tools, all of fine and well-tempered steel." Fig. 17.19 illustrates a selection of Biringuccio's tools, ore barrows and baskets.

Until the seventeenth century, tunnels continued to be driven through rock by the manual work of two or three miners hammering drills (iron picks) into the rock and levering or splitting lumps from the rock face. The method was sometimes called the "stook and feathers". One miner would hold the drill or pick while one or two others would strike the drill with hammers. When the hole was drilled the feathers, which were two thin plates of iron thickening towards the lower end, were placed in it and a wedge was driven between them until the rock split and fell away. Although gunpowder had been in use since the fourteenth century, it was not applied to blasting rocks until more than two hundred years later. Gaspar Weidl at Chemnitz in Hungary first used gunpowder for blasting in mines at the beginning of the seventeenth century. The method spread to Germany by 1627, and Prince Rupert of Bohemia, who had interests in both German and English mines, brought the technique to England. Earl (1994, 42) describes the use of gunpowder in the Cornish tin mines. At first gunpowder was loaded into holes of about 75 mm diameter drilled into the rock, but by 1720 it was found that holes of about 40 mm diameter were sufficient. The holes to hold the gunpowder still had to be driven by hand but the labour in loosening the rock from the rock face was greatly reduced. The rate at which holes can be drilled using hand-hammer drilling varies

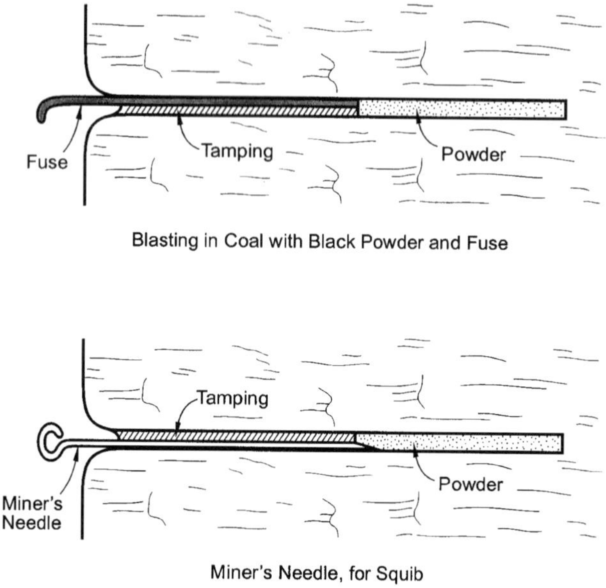

Fig. 17.20 Gunpowder, tamping, and a miner's needle in a drilled hole. The needle is removed and a fuse inserted into the hole.

from 150 to 500 mm/hour, depending on the hardness of the rock and the size of the hole. Fig. 17.20 illustrates a typical arrangement for blasting with gunpowder.

The powder was tamped in so that when it exploded, generating great heat and pressure, the rocks at the side of the hole were burst away. The loading and tamping had to be done carefully and the hole was scraped clean and dry. Gunpowder is rapidly destroyed by water so a wet hole was sealed by ramming down a ball of clay with a "claying bar", so driving the clay into the water bearing cracks. Next, the correct amount of gunpowder, which was always a matter of careful judgement and experience, was loaded into the bottom of the hole using a charging spoon or was wrapped in a paper cartridge if the hole sloped upwards.

A problem, which took some time to resolve, was how to ignite the charge safely. The earliest fuses consisted of small pipes made up of quills or straws and filled with gunpowder. Usually, a number of quills or straws were joined to form a long thin tube by pushing the narrow ends into the wider ends. Gunpowder was crushed to a fine powder on a dry shovel, and then poured into the tube until it was full. A long tapering rod, the miner's needle, was pushed down the drilled hole into the powder, and ramming with clay sealed the hole. The needle was removed carefully leaving a small hole in the rammed clay suitable for the quill or straw fuse. When ready to fire, the end of a candle was fixed to the fuse emerging from the shot hole. The candle was lit and burned for sufficient time to allow the miners to retreat to a place of safety before the powder flashed off down the tube and the charge exploded. Firing gunpowder in this way was always risky for it frequently failed and there was then a delay until it was considered safe to re-enter the mine.

A more reliable safety fuse was not invented until 1831 when William Bickford built a factory for their manufacture in Tuckingmill, Cornwall. Gunpowder (or blackpowder) is usually made from saltpetre (potassium nitrate, KNO_3), charcoal (carbon, C), and sulphur (S), mixed together in various proportions. Its preparation is discussed in the section on weapons. When gunpowder explodes, the reaction is:

$$\underset{\text{solid}}{20\ KNO_3} + \underset{\text{solid}}{30\ C} + \underset{\text{solid}}{10\ S} = \underset{\text{solid}}{6\ K_2CO_3} + \underset{\text{solid}}{K_2SO_4} + \underset{\text{solid}}{3\ K_2S_3} + \underset{\text{gas}}{14\ CO_2} + \underset{\text{gas}}{10\ CO} + \underset{\text{gas}}{10\ N_2}$$

As can be seen, the products of reaction contain a significant proportion of the poisonous carbon monoxide (CO) and this means that mine ventilation is particularly important. Gunpowder is not made in very different strengths but the blasting effect is controlled by the size of the grains. Large grains burn slowly whereas small grains burn quickly. Slow burning grains are often considered to have a "heaving" effect and remove the rock in large lumps. This is particularly useful in collieries for large lumps of coal are usually preferred. Quick burning grains produce a shattering effect giving smaller lumps, and this too is useful if the ore is to be ground or milled. Leaving an air space between the gunpowder and the tamping clay was also used to control the rate of burning.

Transport and Hoisting

After the ore was won it needed transport from the work face to the shaft where it could be hoisted to the surface. Shovelling was the most elementary method of moving ore from the immediate vicinity of work face. Usually one man can shovel material at a rate of about 1 to 2 m^3/hour. This declines with time as fatigue sets in and as the distance carried or thrown increases. In ancient mines, as mentioned above, the ore was often placed in a leather bag, dragged or carried to the shaft, and hoisted by rope, perhaps aided by a winch. Wheeled barrows or trolleys, which require horizontal galleries in all parts of a mine, were not known until the Renaissance, but baskets and leather bags continued in use. Biringuccio tells us that "—there must be many baskets, large and small, sacks made of untanned skins, and barrows with and without wheels to carry the broken pieces out of the mine."

Wheelbarrows are now considered economical for short distances only and usually one man can move 1 or 2 m^3/hour a distance of 30 m. Manoeuvring a barrow in the tight confines of a mine was quite difficult and soon small trucks, with a pin to engage in a grooved plank, came into use and were described by Agricola. Sometime around 1530 wooden rails, to ease the passage of heavily laden trucks, were in use in German mines, see Fig. 17.21, and made their appearance in English colliery accounts of 1597–8. A horse drawn coal truck running on rails is shown in Fig. 17.22 and they were in common use around Newcastle upon Tyne in the late eighteenth century.

Mules, ponies or horses are much more effective than are men, although they can be used underground only where there is sufficient room. A good mule, pony or horse can pull 10% of its body weight constantly for up to 10 hours and for short periods he can pull 40% of his body weight at a speed of about 1 m/s. A 700 kg horse thus develops about 700 W, or slightly less than one horse-power. For two horses it is commonly assumed that the power increases by 1.9 times, for three horses by 2.55 times, and four horses by 3.2 times. If the horses exert a joint pull of 1400 N at about 1 m/s, and the mechanical efficiency is 70%, then they could hoist a load of about 600 kg at a speed of ascent of 0.17 m/s (10 m/min).

The coefficient of rolling resistance (the ratio of horizontal tractive force to the weight pulled) for wheeled vehicles depends on the nature of the road surface and is specified in Table 17.2. Granite tramways, it should be noted, have a very low coefficient of rolling that makes it relatively easy to move loads. A 700 kg horse can pull a 5 tonne load along a level granite tramway at 1 m/s but over dry, packed earth he can only pull 1.4 tonnes, and over loose sand his load is reduced to about 0.44 tonnes. These figures refer to steady

Fig. 17.21 Railway truck, from *Ursprung gemeyner Bergrecht*, *c.* 1530.

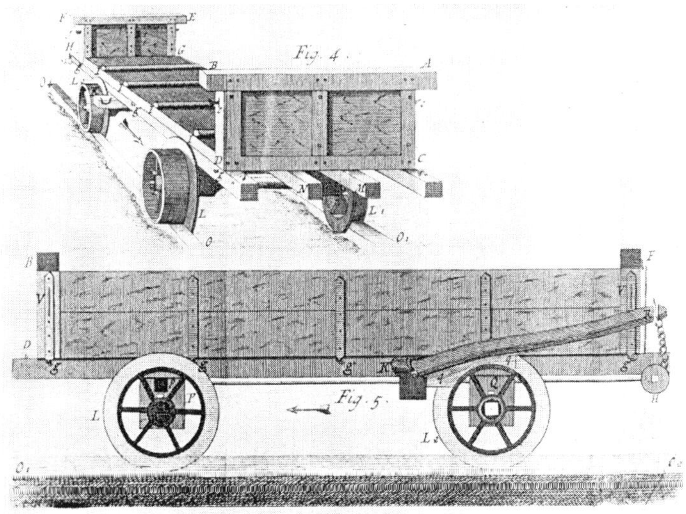

Fig. 17.22 Allen's Railway truck, 1731, from Desagulier's *Experimental Philosophy*, 1734.

Table 17.2 Coefficient of Rolling Resistance for various road surfaces.

Road surface	*Coefficient of Rolling*	*Road Surface*	*Coefficient of Rolling*
Loose sand	0.160	Compact gravel	0.026
Mud	0.085	Worn wood block	0.026
Loose gravel	0.074	Average macadam	0.023
Dusty earth	0.053	Asphaltic concrete	0.020
Dry packed earth	0.050	Sheet asphalt	0.019
Packed cinders	0.046	Good macadam	0.019
Dry, oiled road	0.031	New wood block	0.017
Cobble stones	0.027	Granite tramway	0.014

work over long periods. A horse can always pull greater loads for shorter periods. A 70 kg man pulls a sixth of that of a horse.

For hauling ore up the mineshaft a two-man, two-bucket winch in which the buckets ascend or descend in opposition, was often used. Such a winch has a small capacity but is useful for shafts up to 30 m deep. The performance of a winch driven water wheel is described below (Equ. 17.2) and that analysis applies equally to a winch driven hoist. For a simple winch, the crank radius might be 0.4 m and the drum radius 0.15 to 0.2 m. Thus the mechanical advantage might be about 2.5 and, for steady work over a long period, a man might lift a 25 kg basket of ore at a speed of about 0.25 m/s (15 m/min).

The weight that could be hoisted from a mine was greatly increased by the use of a horse whim. Such a whim was described by Agricola and is illustrated in Fig. 2.3 of chapter 2. Two horses drive the hoist and a whipple-tree is used to share the load equally between them. The power is transmitted through a crown and lantern gear to a horizontal axis main shaft fitted with a large brake. The shaft is also fitted with a drum around which is the winding chain. The drum is liable to heavy wear by the winding chains and is made, therefore, of replaceable wooden rundles. A large hook and chain are provided so that the load may be taken off the horses whilst the ore is being unloaded from the hoist. The circular horse track is about 15 m diameter and the vertical axle is about 12 m long and about 0.4 m square. The shafts and axles are fitted with iron pins running in iron journals. The velocity ratio between the horses and the buckets is about six.

Ore Dressing

At the surface, the ore needed to be processed to concentrate the metal and remove as much of the gangue as possible before it was sent to the smelters. Tin ore, for example, may contain less than 1% of metal and copper perhaps 3%. After dressing, the metal concentration may be raised to about 60% for tin and 70% for copper. The techniques and principles described by Renaissance engineers to achieve this remain in use to the present day although, of course, the techniques are now mechanised. Lump ore, if hard, was sometimes roasted to make it lighter, more friable, and therefore easier to crush. Roasting often had other benefits too in driving off unwanted elements. The ore was hand picked and broken to suitably sized pieces with a hammer. It was then crushed to a fine powder to enable the metal to be separated from the gangue. In ancient times this material, as we have seen, was ground by hand, but by the late medieval period and the renaissance water wheels were used to drive a stamp through a camshaft and a set of cams.

Stamps were rated on the weight of the falling part. In twentieth-century mills they weigh 600–700 kg (USA), 700–900 kg (South Africa), and 400–500 kg (Old Californian practice). However, old Cornish stamp heads were much lighter, weighing perhaps 65 kg, and this, perhaps, is similar to the earliest stamps. A typical stamp mill was illustrated by Agricola and is reproduced in Fig. 17.23. A water wheel turned a shaft fitted with cams that engaged with tappets, lifted the timber columns (the stamps), and allowed them to fall on the ore. Stamp heads weighed perhaps 65 kg and, after many drops, crush the ore to a powder, if the mortar ran dry, or to a pulp if it was filled with water. The pulp discharged through a plate fitted with suitable sized holes, perhaps "the bigness of a moderate pin".

Shoes and dies were eventually made of chilled cast iron or steel. The average life of the dies was only about 60 days and a typical wear rate was about 0.1 kg per tonne of ore crushed (0.01%). For the shoes, life was about 60 to 90 days. Cams were equally spaced on the camshaft and usually lifted 4, 5, 6 or 10 heads. The drop sequence was important and had a marked effect on the work done. The rules governing drop sequence were

i. no two adjacent stamps should fall in succession
ii. when one stamp is falling its neighbour should be rising

Fig. 17.23 Water driven stamp mill to crush mined material before separation from rock. Agricola (1556, 284).

Common sequences that satisfy these rules are 1, 3, 5, 2, 4; 1, 4, 2, 3, 5; 1, 5, 3, 4, 2; and 1, 5, 3, 2, 4. The first sequence best satisfies the rules but the other sequences were also used. It was held that the second sequence distributes the pulp better on the dies and facilitates discharge through the screens. For a 6-head stamp, the falling sequence was 1, 6, 3, 4, 2, 5 and the angle between the cams was 60 degrees. When a ten stamp battery was used the sequences were 1, 7, 3, 9, 5, 2, 8, 4, 10, 6 or 1, 6, 3, 10, 2, 7, 4, 9, 5, 8, and the angle between the cams was 36 degrees.

Stamps continued in use into the twentieth century but they are now used only for processing small quantities of ore or to process ore to a very fine slime. They were replaced by rolls and later by jaw crushers. In 1796, John Taylor at the Wheal Friendship mine in Devon improvised a roll-crushing machine using two discarded pipes from a pump. His roller crusher was not particularly success-

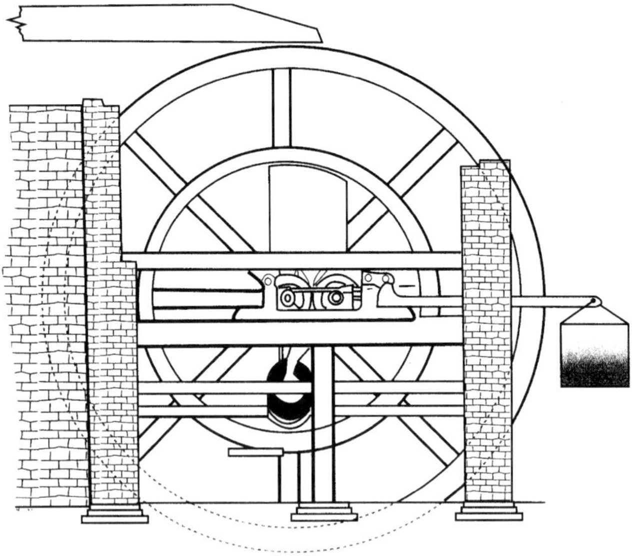

Fig. 17.24 Roll type ore crusher worked by a water wheel. After Earl (1994, 81).

ful for it did not produce as high a proportion of fines as conventional stamps. However, it was soon improved into an efficient machine consisting of two cast iron rolls of about 60 cm diameter, see Fig. 17.24. One roll was driven by a water wheel and the other was an idler that was pressed against the driven roll by a loaded lever. This allowed the idler to open thus permitting the safe passage of any particularly large or hard lumps of ore. The crushed ore then passed through a revolving sieve and the large lumps were collected and re-cycled through the crusher.

After crushing in stamps or rolls the ore was sieved into various sizes and the oversize pieces were returned to the crushers. Concentration of the crushed ore was effected in one of three ways, namely, hand sorting of larger pieces, jigging of smaller pieces, and washing on inclined surfaces for the smallest pieces.

Jigging the crushed ore is illustrated in Fig. 17.25. It remains, even today, a common way of treating small pieces of ore from 1 mm to 15 mm diameter. A sieve full of material, of about the same size, is jigged up and down under water. The heavy material containing the metal falls to the bottom and the lighter material rises to the top. The lighter material is removed and replaced with new ore and

Fig. 17.25 Jigging ore. Agricola (1556).

the process repeated until a sufficient depth of heavy material is formed to justify its removal. The whole process is repeated using finer and finer sieves to extract the maximum amount of rich material from the crushed ore.

Wet material from the stamps formed sands or slimes that were separated by washing in a thin flow of water on a slightly inclined table. This technique was very ancient and had been used, for example, in the ancient Greek mines at Laurium. In all such devices, the grains of material rest on the sloping surface under their own weight and by friction, but they may be dragged along the surface by the flow of water if the drag force is sufficient to overcome friction. Thus the water removes the lighter grains and the heavier, metal bearing grains remain on the table. The method is more effective if the grains are approximately the same size because the water flow tends to move both large, light gains and small heavy grains. The crushed ore may be washed like this several times to extract the maximum amount of metal.

Fig. 17.26, which is also an illustration from Agricola's book, shows a canvas strake for treating large volumes of finely crushed material. The sloping surface was covered with canvas and the sands or slimes from a set of stamps provided the raw material that was spread over the canvas. A thin sheet of water was allowed to run down the inclined surface at a suitable speed and the lighter material was removed. After washing for a sufficient time, the canvas-sheets were lifted and the fine, metal-bearing material was removed.

In addition to these mechanical methods of concentrating the metal in the ore chemical methods were also used, particularly in the extraction of gold from its ore. Gold, as it is mined, is usually mixed with quartz and other rocks and Diderot (1751) illustrates the method of its extraction my means of mercury, Fig. 17.27. The ore is separated in this mill by washing (Fig. 2) and then crushed under a set of stamps operated by a water wheel (Fig. 1). Mercury absorbs gold on contact and this is used to extract the gold from the ore. Under the stamps, the crushed ore is mixed with mercury to form an amalgam with the gold to separate it chemically from the rock. Gold is retrieved from the amalgam by distillation. The amalgam is heated to evaporate the mercury, which is then condensed for re-use, after which the remaining gold is melted and skimmed in a crucible.

Fig. 17.26 Canvas strake for separating sands and slimes. Agricola (1556).

Analysis

Performance of a Winch Driven Dipper Wheel

A man, a horse or a water wheel, may be used to drive a dipper wheel (chain of pots) but in each case the power input is

$$W = 2\pi R_1 F_1 N_1 = F_1 v_1 \qquad 17.1$$

R_1 is the crank radius, F_1 is the crank force, N_1 is the crank rotational speed, and v_1 is the crank velocity. A proportion of this power is lost in mechanical friction and the remainder lifts the pots, thus

$$\eta W = 2\pi R_2 F_2 N_2 = F_2 v_2 \qquad 17.2$$
$$N_2 = N_1 / G$$

Fig. 17.27 Gold Processing. Diderot and d'Alembert (1751–72).

η is the mechanical efficiency of the mechanism, R_2 is the chain-drum radius, F_2 is the lifting force, N_2 is the chain-drum rotational speed, v_2 is the chain velocity, and G is the overall gear ratio. The number of dippers lifted may be estimated as

$$n = \frac{F_2}{m_d g} \qquad 17.3$$

m_d is the mass of water held in each dipper and g is local gravity. Finally, the mass of water lifted per unit time may be determined from

$$\eta W = mgz \qquad 17.4$$

m is the mass of water lifted per unit time, and z is the height of the lift.

Analysis of a Cowl Ventilator

Fig. 17.28 shows the simplified air passages for a cowl ventilator. Wind at ambient pressure p and velocity v blows directly into the inlet pipe of the ventilator. Section 1 is the inlet pipe, section 2 is the interior of the barrel, and section 3 is the barrel exit pipe that connects to the mineshaft, section 4. It is assumed that the mine-

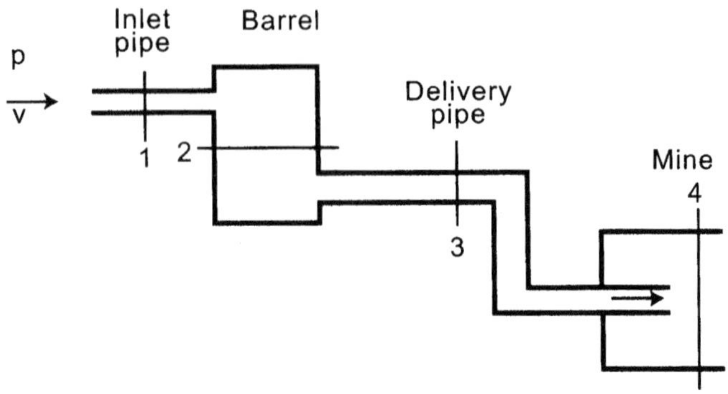

Fig. 17.28 Air flow through a cowl type ventilator.

shaft has other openings to the atmosphere and thus the pipe pressure at section 4 is atmospheric pressure, p. Pressure loss at entrance to the pipes is ignored but pressure loss at exit from the pipes is not ignored, consequently there is no pressure recovery at exit from a pipe and thus $p_1 = p_2$ and $p_3 = p_4 = p$. Changes in the vertical height of the pipe are ignored. Bernoulli's equations between the atmosphere and the mine is

$$\frac{p}{\rho} + \frac{v^2}{2} = \frac{p_4}{\rho} + \frac{v_4^2}{2_l}\left(\frac{v_1^2 - v_2^2}{2}\right) + \left(\frac{v_3^2 - v_4^2}{2}\right) \quad 17.5$$

The bracketed terms are the exit losses for pipe *1* and pipe *3*. Substituting the assumptions then from Equ. 17.5 we have

$$v^2 = v_1^2 - v_2^2 + v_3^2 \quad 17.6$$

If Q is the volumetric flow rate then the volumetric continuity equation is

$$Q = A_1 v_1 = A_2 v_2 = A_3 v_3 \quad 17.7$$

A_1, A_2, A_3 are the cross-sectional areas of the pipes. Substituting Equ. 17.7 into 17.6 gives

$$v_1 = \frac{v}{\sqrt{1 - (A_1 / A_2)^2 + (A_1 / A_3)^2}} \quad 17.8$$

From Equ. 17.7 the volumetric flow rate of fresh air into the mine is

$$Q = \frac{A_1 v}{\sqrt{1 - (A_1 / A_2)^2 + (A_1 / A_3)^2}} \qquad 17.9$$

Thus, the airflow rate into the mine depends only on the wind velocity, v, and the pipe areas A_1, A_2, and A_3. If the areas at sections *1*, *2*, and *3* are equal (a simple parallel inlet pipe conveys the wind into the mine) then the volumetric flow rate is

$$Q = A_1 v \qquad 17.10$$

This is somewhat greater than the flow rate given by Equ. 17.9 so we conclude that retarding the inlet flow by fitting a barrel reduces the flow (it introduces an extra pressure loss at the enlargement of flow into the barrel). If the barrel area is very large and the inlet and outlet areas of the barrel are equal then the loss of flow is $Q = 0.707 A_1 v$. The reduction of about 30% would not be noticed unless careful measurements of wind velocity and flow velocity could be made.

Mine Ventilation

The oxygen content of mine air is diminished by the presence of miners, animals, and naked flames (lighting), all of which consume oxygen and emit carbon dioxide and steam. Dust from mining operations, the products of explosives used in blasting (particularly carbon monoxide), and gaseous emissions from the rock surfaces pollute the mine air. In mineral workings, gas emissions from the rock surface are not usually significant unless the rock or the mine contains decomposed biological material that may form methane, but in coal mining, methane and carbon monoxide commonly forms and emissions are very dangerous. They must be removed by adequate ventilation. In wet mines the air is usually saturated and approaches 100% relative humidity.

Table 17.3 Explosion Ranges for Some Gases in Air. Haddock (1924, 79).

Gas	*Symbol*	*Lower Limit %*	*Upper Limit %*
Methane	CH_4	5.6	14.8
Hydrogen	H_2	5.0	72.0
Carbon Monoxide	CO	13.0	75.0
Ethane	C_2H_2	4.0	22.0
Coal Gas	—	5.0	28.0

The dangerous levels of commonly occurring gases in air are specified in Table 17.3 and are taken from Haddock (1924, 79). Miners use other terms for theses gases. What is usually called "firedamp" is mainly methane with a small amount of hydrogen and other hydrocarbons. Blackdamp is the name given to any gas that extinguishes a candle. Usually it is a residual mixture of gases left after the removal of methane and oxygen and consists of 1% to 15% carbon dioxide, the remainder being nitrogen. Whitedamp is simply carbon monoxide. What is sometimes called afterdamp is the result of an explosion of firedamp, coal dust and air. Hydrogen sulphide, which sometimes occurs in mines, is often called "stonedamp".

Carbon dioxide commonly occurs in mines but is not toxic unless exposure is very long. In amounts greater than 1% or 2% it is tolerated but at 3% to 4% breathing is laboured and at 6% there is a marked panting, increase in pulse rate, headache and flushed face. Unconsciousness occurs at 11% but death does not occur unless exposure continues for several hours.

Carbon monoxide, on the other hand, is very toxic. It commonly occurs: in mines having weak ventilation, in those where gunpowder (blackpowder) is used, and in mines where fires are allowed to smoulder. It also occurs in fire fighting. For continued exposure anything above 0.03% is unsafe, levels above 0.15% are considered dangerous, and above 0.4% death may occur. Small amounts of carbon monoxide cannot be detected by the miner's safety lamp and so canaries, white mice, or other small animals, which are very sensitive to small amounts, were formerly taken down mines to give early warning of its presence. Canaries proved so sensitive and certain that they were once officially approved for rescue work in the UK and in the USA, Table 17.4.

Table 17.4 Effects of Carbon Monoxide on Canaries.

%CO	*Effect*
0.09	Very slight distress after 1 hour
0.12	Weaker than above after 1 hour
0.15	Distress in 3 minutes. Fell from perch in 18 minutes
0.20	Distress in 1.5 minutes. Fell from perch in 5 minutes
0.29	Fell from perch in 2.5 minutes

Hydrogen sulphide, H_2S is considered more acutely poisonous than carbon monoxide. Air containing as little as 0.07% will cause death to man and many animals after an exposure of one hour. Its occurrence in mines is, fortunately, rare and easily detected by its obnoxious odour. Much less rare, however, are deaths from ingestion of siliceous dust, silicosis. Formerly this caused more deaths to miners than accidents of all kinds and it is, perhaps, surprising that the problem was not detected until about 1903. The remedy is to prevent dust getting into the atmosphere by using strong ventilation, water sprays, and withdrawing miners until dust levels have settled.

Pollution from men and animals arises mainly because oxygen is converted to carbon dioxide by breathing; see Table 17.5. In normal breathing a man expires about 7 l/min but this may rise to about 15 l/min or more during periods of heavy exercise. Expired air is always saturated. A horse or mule usually requires three to five times more air than does a man. Pollution by a candle flame or small lamp produces about the same effect on the atmosphere as a man.

Table 17.5 Volumetric Analysis of Fresh and Expired Air.

Constituent	*Fresh Air*	*Expired Air*
Oxygen, O_2	20.90	16.15
Nitrogen, N_2	79.06	79.90
Carbon Dioxide, CO_2	0.04	3.95
Total	100.00	100.00

Natural Convection

Most mineral mines and small collieries relied on natural convention currents and diffusion to ventilate underground workings. Fig. 17.29 shows the layout of a very simple mine consisting of two shafts and a work shaft. The lower shaft may be an adit to drain the mine whereas the upper shaft may be the main shaft for access or exit from the mine workings. If the adit relies on natural drainage then its exit from the mine will be lower than the main level and the height of the adit is then a negative. None of the shafts need be either horizontal or vertical but they are, for convenience, considered to be of a constant cross-section. Air motion in these shafts arises because mine air is normally hotter than atmospheric air and

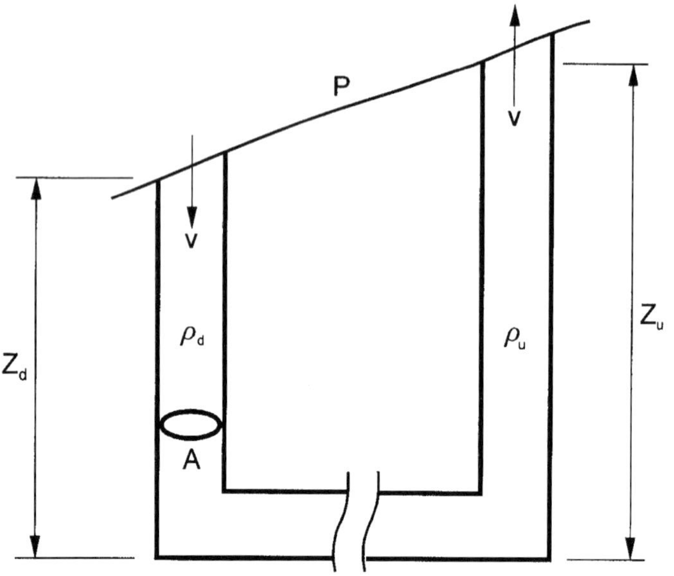

Fig. 17.29 Layout of a simple, two-shaft mine.

increases more or less linearly with depth. Such temperatures do not vary with time and have neither a daily nor a yearly cycle at depths greater than about 10 m. Generally, the temperature near the entrance is close to the temperature of the ambient air entering the mine. However, the air temperature quickly attains that of the surrounding rock, the temperature of which increases with depth at an average rate of about 1°C per 35 m. However, this figure is subject to great variation. Depending on local conditions it may vary from 1°C per 100 m in the very deep gold mines of South Africa to 1°C per 18 m in some collieries. The rock temperature very often limits the depth to which a mine may be worked for it is very arduous to work with rock temperatures exceeding 40°C. This means that the average depth of mine should not exceed about 700 m although some modern gold mines exceed 2000 m. Ancient mines, however, rarely exceeded about 60 m.

Under normal conditions air enters at the lower shaft, descends (or ascends) to the work-face and rises through the higher of the two shafts. Even if a mine has but a single shaft a weak air current will form; usually the air descends at the outer edge of the shaft and rises in the centre. The pressure difference driving the air motion is caused by the density differences in the columns of air and is determined by summing the pressure changes round the mine.

$$\Delta p = \Sigma \rho_i g z_i = g[\rho(z_u - z_d) + \rho_d z_d - \rho_u z_u] \qquad 17.11$$

Δp is the pressure difference between the entrance and exit to the mineshaft, ρ is the ambient air density, g is gravitational acceleration, ρ_u and ρ_d is the air density in the up-shaft and down-shaft, and z_u and z_d is the vertical height of the up-shaft and down-shaft respectively. Pressure changes in the mine have a very small effect on density but temperature changes are more significant and we may assume that the density of the outside air, the up-shaft, and the down-shaft is p/RT, p/RT_u, and p/RT_d, thus Equ. 17.11 may be written as

$$\frac{\Delta p}{\rho g} = z_u \frac{\Delta T_u}{T_u} - z_d \frac{\Delta T_d}{T_d} \qquad 17.12$$

ΔT_u is the average temperature rise in the up-shaft and ΔT_d is the average temperature rise in the down-shaft, relative to the outside temperature. If $\Delta p > 0$ then the flow direction is as indicated but if $\Delta p < 0$ then the flow reverses. Such flow reversals, when they occur, are usually caused by changes in the outside temperature and may be due to either annual or daily fluctuations.

The pressure difference round the mine generates air motion and a steady motion occurs when the driving pressure is equal to entrance pressure loss and the frictional pressure loss. The losses round the mine are an entrance loss ($0.5\ v^2/2$), a friction loss ($fSv^2/(2A)$), and an exit loss ($v^2/2$,) thus

$$\Delta p = \frac{\rho v^2}{2}\left(1.5 + f\,\frac{S}{A}\right) \qquad 17.13$$

v is the air velocity, f is the friction factor (see Table 17.6 below), S is the surface area of the mine airways, and A is the cross-sectional area of the mine airways. For circular or square section mineshafts the ratio $S/A = 4L/D$ where L is the total length of the mineshafts and D is the diameter (if circular) or the height (if rectangular). The friction factor, f, depends on the roughness of the rock surfaces over which the air must flow, on the degree of curvature of the shafts, and on whether they are clear or partly blocked. The friction factors shown in Table 17.6 are derived from data in Peele (1927, 1199).

Equating Equ. 17.12 and 13 gives the air velocity through the mine, v

$$v = \sqrt{\frac{2g(z_u \Delta T_u / T_u - z_d \Delta T_d / T_d)}{1.5 + fS/A}} \qquad 17.14$$

Table 17.6 Friction Coefficients for Airways in Mines.

Type of Airway	*Friction Factor f*					
	Straight		*Slightly Curved*		*Highly Curved*	
	Clear	*Partly Blocked*	*Clear*	*Partly Blocked*	*Clear*	*Partly Blocked*
Smooth Lined	0.003 to 0.006	0.008 to 0.01	0.006 to 0.009	0.011 to 0.014	0.011 to 0.014	0.015 to 0.019
Sedimentary Rock	0.009 to 0.021	0.014 to 0.026	0.012 to 0.025	0.017 to 0.03	0.017 to 0.03	0.022 to 0.034
Timber	0.024 to 0.03	0.095 to 0.037	0.028 to 0.036	0.032 to 0.04	0.032 to 0.04	0.037 to 0.045
Igneous Rock	0.027 to 0.06	0.032 to 0.065	0.031 to 0.063	0.036 to 0.068	0.036 to 0.068	0.04 to 0.072

Example: Air Velocity in a Mine
Determine the air velocity in a mine if the temperature in the shafts is 30°C (303 K), the outside temperature is 27°C (300 K), the up-shaft is 100 m deep, the down-shaft is 80 m deep, and local gravity is 9.81 m/s^2. The friction coefficient is 0.015, and the ratio of surface area to cross-sectional area is 700. What is the pressure difference if air density is 1.2 kg/m^3?

From Equ. 17.14 the velocity is 0.57 m/s
From Equ. 17.12 or Equ.17.13 $\Delta p = 2.34$ N/m^2.

Clearly, even quite small temperature increases in the mineshaft are capable of generating significant air movement. The pressure difference, in this case, is 2.34 Pa which is about 0.0023% of the normal atmospheric pressure, and the volumetric flow rate, assuming a cross-

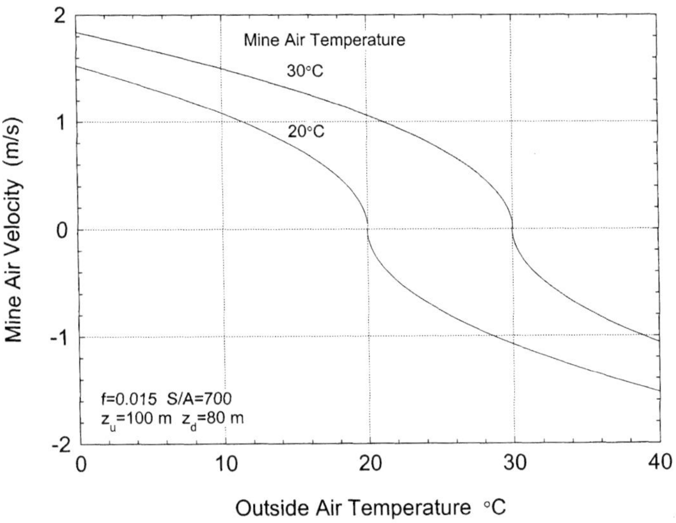

Fig. 17.30 Influence of outside air temperature on the air velocity in a mine.

sectional area of 2 m^2 is 1.16 m^3/s (7000 l/min). In this example, the outside temperature was taken to be 300 K but if this increased to 303 K (the same as the mine air temperature) then the flow velocity would be zero. The variation of mine air velocity with outside air temperature is illustrated, for this example, in Fig. 17.30. The reversal in airflow that occurs when outside air temperature exceeds the mine temperature is clearly seen.

An acceptable air velocity is usually less than 1 m/s. An air velocity of 0.5 m/s is barely perceptible, at 1 m/s the air movement is pleasant, but at 1.5 m/s, and at the usual mine temperature, it is uncomfortable for perspiring miners. At about 1.2 m/s candle flames flicker inconveniently and at 2 m/s they are likely to be blown out unless protected. Of course, airflow velocity may be controlled, if excessive, by doors (regulators). The pressure loss when air passes through a door is

$$\Delta p_1 = \frac{\rho v^2}{2}\left[\left(\frac{A}{a}\right)^2 - 1\right] \qquad 17.15$$

v is the velocity of the air approaching the door, A is the area of the passage and a is the area of the door opening. When the door

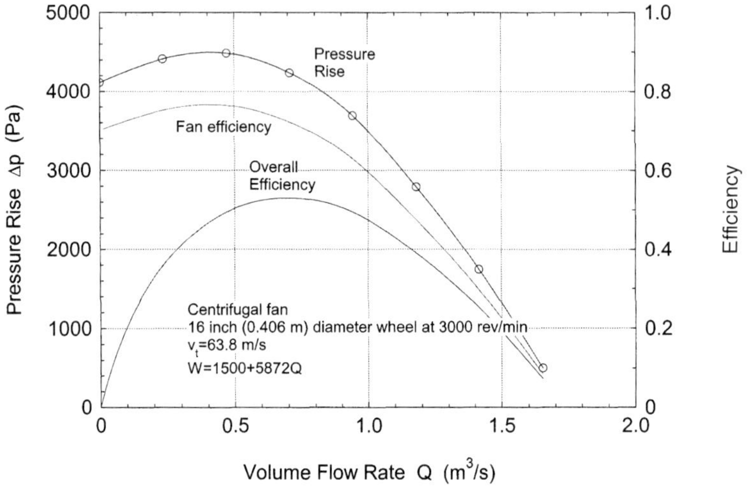

Fig. 17.31 Characteristics of a modern centrifugal fan.

is fully open ($a = A$) there is no pressure loss but if $a << A$ then the pressure loss is very great and air flow may cease.

The airflow may be insufficient if $z_u \, \Delta T_u / T_u = z_d \, \Delta T_d / T_d$ in Equ. 17.14 or if the quantity $f\, S/A$ is large. Under these conditions, the airflow may be improved by artificially heating the air in the up-shaft, or by cooling it in the down-shaft. Fires lit in the up-shafts were often used in mines.

Forced Convection

The efficiency of a simple centrifugal fan, of the sort illustrated in Fig. 17.15 is defined as

$$\eta = Q\Delta p \, / \, W \qquad 17.16$$

Q is the volumetric flow rate of air (m^3/s), Δp is the pressure rise across the fan (N/m^2), and W in the power required at the drive shaft (W). From this the efficiency is zero when the flow rate is zero or when the pressure rise is zero (power input cannot be zero since some power is required to overcome friction). At some intermediate flow rate the efficiency is a maximum, usually not more than about 60%. Equ. 17.13 gives the pressure rise necessary to drive air around a mine and the volumetric flow rate is $Q = Av$, where A is the cross-sectional area of the mineshaft, thus Equ. 17.16 may be written as

$$\eta W = (1.5A + fS)\frac{\rho v^3}{2} \qquad 17.17$$

For centrifugal fans with radial vanes the relation between pressure rise and rotor tip speed, v_t is

$$\Delta p = \rho v_t^2 \qquad 17.18$$

Thus from Equ. 17.16 the power is also given by

$$W = Q \rho v_t^2 / \eta = m v_t^2 / \eta \qquad 17.19$$

m is the mass flow rate

Example: Forced Convection of a mine
A man develops 64 W when turning the input shaft of a centrifugal fan. If the fan efficiency is 25%, the cross-sectional area of the mineshaft is 2 m^2, the surface area of the mine is 1400 m^2, and the friction factor is 0.015, determine the air speed in the mineshaft. The density of air in the mine is 1.2 kg/m^3. What is the volume flow rate, what is the pressure rise and what is the rotational speed if the rotor diameter is 1.3 m?

From Equ. 17.17 v = 1.035 m/s Volume flow rate is $Q = Av$ = 2 × 1.035 = 2.07 m^3/s
Pressure rise from Equ. 17.13 is 7.71 N/m^2.
From Equ. 17.18 the rotor tip speed is v_t = 2.54 m/s.
Rotational speed is $N = v_t/(\pi D)$ = 0.62 rev/s = 37.2 rev/min

The example shows that a man-powered fan such as that shown in Fig. 17.28 could supply sufficient air to a reasonably small mine. Modern mines, and many ancient mines, are much longer and more complex than that assumed in the example, consequently the pressure drop is much greater and they usually require much more powerful fans. Fig. 17.31 shows the characteristics of a modern centrifugal fan. The pressure rise and power input are very much greater than in the example. The efficiency remains at about 70% at low and medium flow rates but as the flow rate exceeds the design value the pressure and efficiency drop towards zero. The overall efficiency includes the mechanical efficiency, both of which are zero for zero flow rate.

Air Temperature Distribution in a Mine Shaft

Atmospheric air enters a mineshaft at atmospheric temperature T_a and is heated as it flows along the shaft by a hot wall at temperature T_w. The energy equation for heat transfer between the air and the shaft wall is

$$\rho D^2 v c_p \frac{dT}{dx} = 4Dh(T_w - T) \qquad 17.20$$

ρ is the air density, D is the shaft diameter or height (for a square section), v is the air velocity, c_p is the specific heat of air. T is the air temperature, x is the distance along the shaft from the entrance, h is the heat transfer coefficient between the air and the wall, and T_w is the wall temperature. Using Reynolds' analogy the heat transfer coefficient may be estimated from the friction coefficient, f, using

$$h = \frac{f}{2} \rho c_p v \qquad 17.21$$

The friction coefficient, for different mines, is specified in Table 17.6 above. Substituting into Equ. 17.20 gives

$$\frac{dT}{dx} + \frac{T}{x_c} = \frac{T_w}{x_c} \qquad 17.22$$
$$x_c = \frac{D}{2f}$$

The solution of this equation gives the air temperature in the mine as a function of the distance, x, from the entrance to the mine. Two solutions are of interest, namely that for a constant wall temperature and that for a linearly increasing (or decreasing) wall temperature.

If the wall temperature is constant, and the air temperature at entry to the mineshaft is T_a, then the solution to Equ. 17.22 is

$$T = T_w - (T_w - T_a)\exp(-2fx / D) \qquad 17.23$$

This relation is plotted in Fig. 17.32 and illustrates the rise in air temperature from the outside air temperature at the shaft entrance to the wall temperature at some distance into the shaft. As an approximation, the air temperature and the wall temperature are equal at a distance of about 2D/f from the mine entrance. If the friction factor from Table 17.6 above is 0.02 then the air temperature rises to the wall temperature at a distance of 100D from the entrance.

If the wall temperature increases linearly with distance so that $T = T_0 + kx$ then the solution of Equ. 17.22 is

$$T = T_0 + kx - \frac{kD}{2f}(1 - \exp(-2fx / D)) \qquad 17.24$$

This result is also plotted in Fig. 17.32. At large distances from the entrance, the air temperature is equal to the wall temperature minus

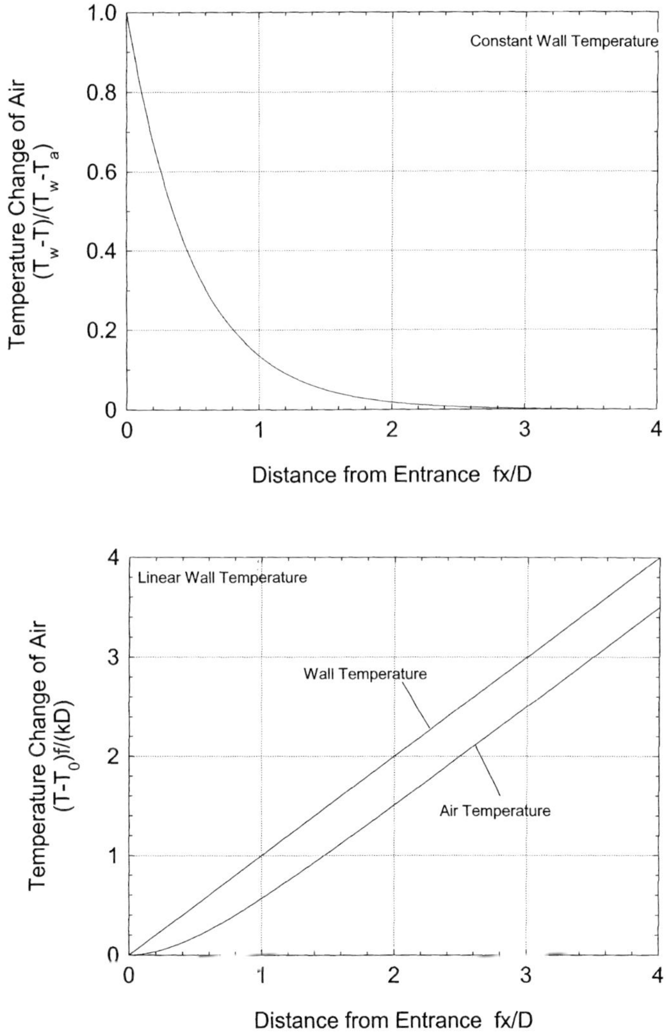

Fig. 17.32 Air temperature in the entrance region of a mineshaft. Top: if the wall temperature is constant. Bottom: if the wall temperature increases linearly with distance.

kD/2f. If the airflow direction was reversed so the air was leaving the mine rather than entering, then its temperature would be greater than the adjacent wall temperature by *kD/2f*. Thus, the temperature difference between air in the entrance shaft and air in a parallel exit shaft would be *kD/f*. For a typical pair of shafts we may take $k = 1/35$°C/m, $D = 2$ m, and $f = 0.02$, thus the temperature difference between the shafts would be 2.86°C. This is sufficient to give a degree of ventilation even if the entrance and exit to the two shafts were at the same level.

Geometry of Rolls

The angle of nip is critical in the design of rolls. It is the angle formed by the tangents to the roll faces at their points of contact with the ore to be crushed, Fig. 17.33 (top). If the nip angle is too large then the ore will skid of *f* the surface of the rolls and will not pass through. It is, of course, related to the coefficient of friction between the rolls and the ore. In practice, the nip angle rarely exceeds 25 degrees. A little elementary geometry shows that the nip angle, θ, is related to the diameter of the rolls, D, and the size of the ore, d, and the reduction ratio, r, by

$$\cos\left(\frac{\theta}{2}\right) = \frac{D + d / r}{D + d} \qquad 17.25$$

Table 17.7 Power Consumption of Rolls.

Size of Product, h (mm)	*Quantity Crushed, (tonnes/kW-hour)*
< 6	0.55
6 to 18	0.8
8 to 25	1.0
25 to 38	2.1
38 to 50	3.8

The reduction ratio is the ratio of the initial size of the ore to the minimum distance between the rolls. This relation is plotted in Fig. 17.33 (bottom). The theoretical capacity of rolls, Q, is the volume of a continuous ribbon of ore, the length of which is the velocity of the periphery of the rolls, v, the thickness is the minimum distance

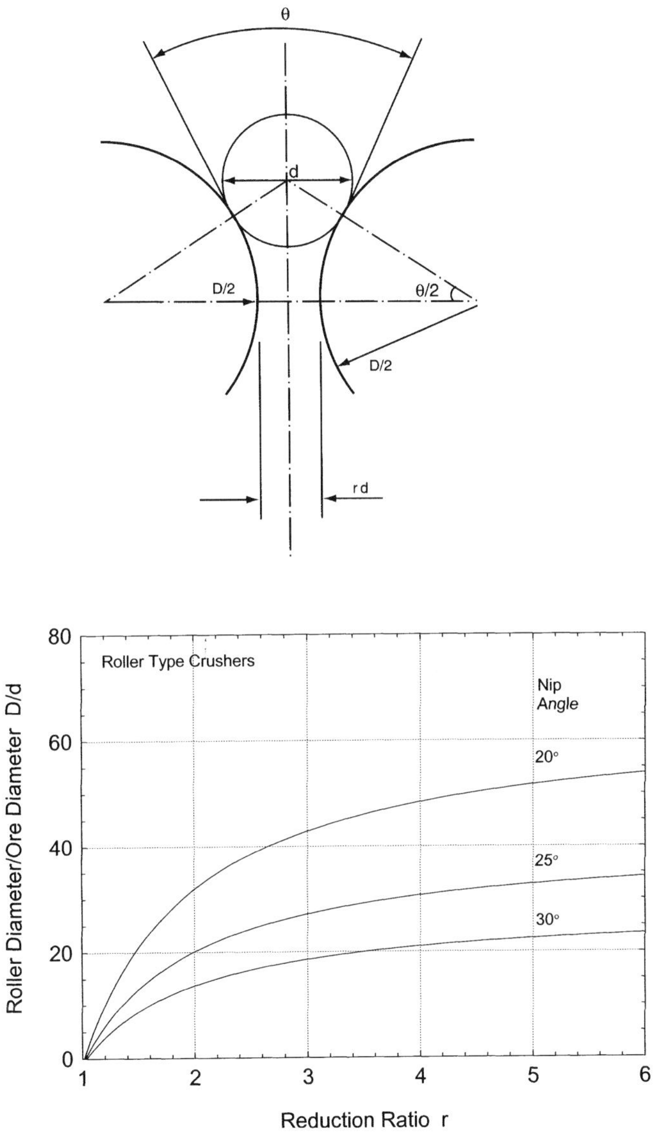

Fig. 17.33 Top: geometry of a pair of rolls. Bottom: ratio of roll diameter to ore diameter for various reduction ratios with a nip angle of 25 degrees.

between rolls, h, and the breadth is the length of the rolls, b. Thus

$$Q = bhv \qquad 17.26$$

For rolls set corser than $h = 25$ mm the actual capacity is about 5% of the theoretical capacity. For 6 mm $< h < 25$ mm the capacity is about 15% to 20% of the theoretical value, and for h < 6 mm an average performance is about 20% to 30% of the theoretical. Very small distances between the rolls may result in capacities exceeding 100% of the theoretical value because the ore may force the rolls apart.

The power required to drive the rolls and crush the ore is based on the amount of new feed introduced, per kW-hour, on the type of rock, size reduction, size of product, and the re-circulating load. Average values are given in table 17.7.

Separation on an Inclined Water Plane

The movement of grains of material down an inclined surface depends on the frictional force, which resists motion, and the drag force, which causes the grain to move if it is greater than the frictional force. Fig. 17.34 shows these forces. If the grain moves then the drag force generated by the water is greater than the frictional force between the grain and the surface, that is

$$C_d \rho_w A v^2 / 2 > f(\rho_s - \rho_w)Vg \qquad 17.27$$

C_d is the drag coefficient, ρ_w is the density of water, A is the cross-sectional area of the grain, v is the velocity of the water, f is the

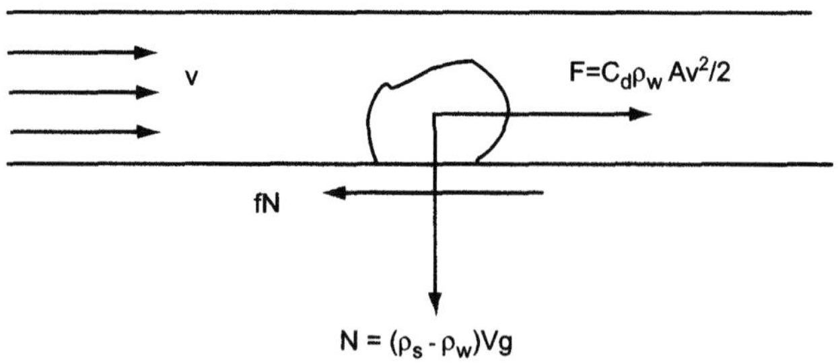

Fig. 17.34 Forces acting on a grain when water flows along a slightly inclined plane.

coefficient of friction, ρ_s is the density of the solid material (grain), V is the volume of the grain, and g is local gravity. In Equ. 17.27 it is assumed that the grain size is less than the depth of water and therefore the grains tends to float. If the grains are spheres of diameter D then Equ. 17.27 may be written as

$$v > \sqrt{\frac{4fgD}{3C_d}\left(\frac{\rho_s}{\rho_w} - 1\right)} \qquad 17.28$$

Clearly, small dense grains will move at the same water velocity as large low-density grains and for the system to work satisfactorily only grains of a similar size should be washed. If the drag coefficient and the friction coefficient are equal, which is approximately true, then the estimated velocity at which grains will move is shown in Fig. 17.35. For example, if the grain size is 5 mm then a water velocity of 0.5 m/s will remove all grains having a specific gravity less than about 5 but all those having a specific gravity greater than about 5 will be retained on the surface. In practice the volume flow rate of water may be about 10 times to 20 times the flow rate of crushed ore and the incline may be 4° to 6°. The crushed ore may be dragged back to the top of the incline by wooden handled rakes and several washings may be required to concentrate the ore.

Another method of separation is flotation. If the velocity $v = 0$ then the force acting on the grains is

$$N = (\rho_s - \rho_w)Vg \qquad 17.29$$

Thus the grains will sink to the bottom of the liquid if the density of the solid grains is greater than the density of the liquid. On the other hand, if the density of the solid grains is less than that of the liquid then $N < 0$ and the grains float to the surface. By a judicious choice of fluid the lighter and heavier grains can be separated.

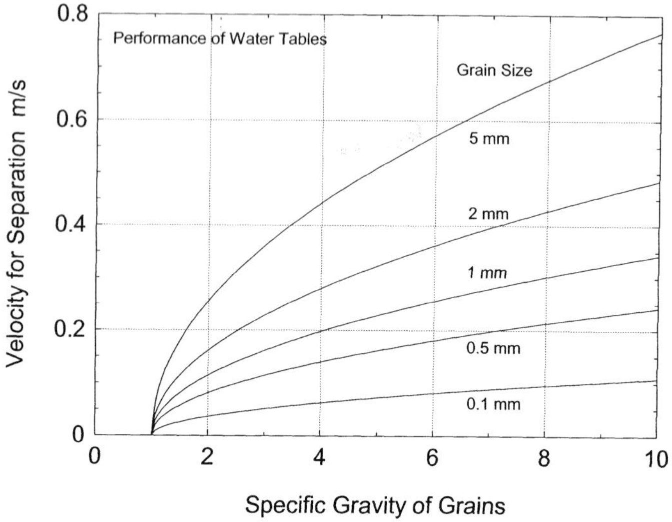

Fig. 17.35 Performance of water tables showing the effect of specific gravity and grain size on the velocity at which grains will move along a surface.

CHAPTER EIGHTEEN

ANCIENT AND MEDIEVAL METAL WORKING

Introduction

Metals have provided poets and writers with an almost inexhaustible supply of metaphors and figures of speech for hardness, strength, toughness, value, and beauty, and the circumstances of their manufacture have often been compared to Hades, Stromboli, Vesuvius, and in some cases to Paradise. That sensible Victorian writer Samuel Smiles, in *Men of Invention and Industry (1884)*, comes closest to the true value of metals when he says of iron that it is "the soul of every manufacture, and the mainspring perhaps of civilized society." Civilized society, however, needs many such springs. Metal production has always been difficult, demanding as it does: knowledge, strength, wealth, and determination. In ancient Greece, if Aristotle is to be believed, the smith was despised because his work caused bodily deformation and the long hours left insufficient time for other duties, consequently he did not make a good citizen. In the 19th century the blacksmith's calling had not changed significantly but "the muscles of his brawny arms" were more admired, as, for example, Longfellow's poem *The Village Blacksmith* illustrates. By 1839, when this was written, the local blacksmith was no longer the sole provider of metal products and he was viewed more romantically, but love it or hate it, metalworking is indispensable to civilized life.

Metal articles and metalworking are the products and processes most closely associated with engineering. Metalworking, in the period of interest, comprises mainly of smelting, casting, forging, rolling, slitting, wiredrawing, and machining operations such as filing, turning, drilling, boring, and screw cutting. This chapter, therefore, examines the historical development of these processes. Smelting is of fundamental importance and the ancient and medieval processes are significantly different from modern processes; so different, in fact, that they must be understood if medieval mechanical technology is to make sense.

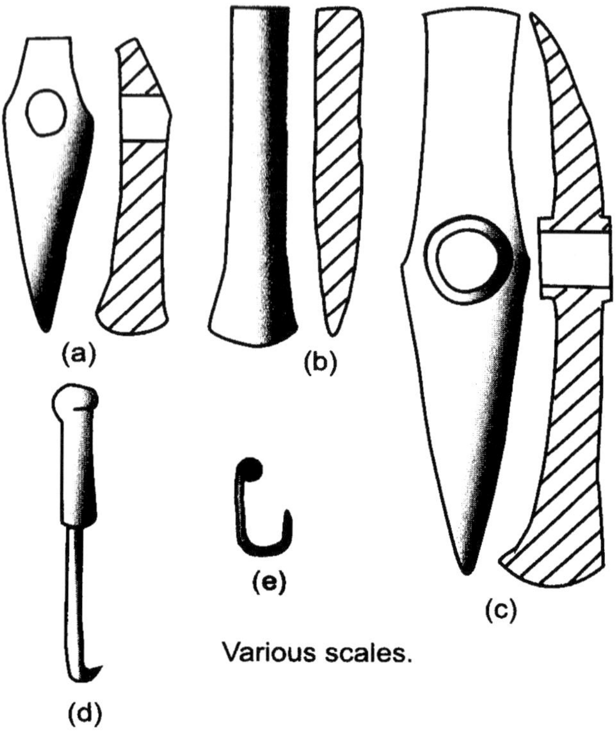

Fig. 18.1 Early Bulgarian and Romanian copperware of the fifth and fourth millennia B.C. a: hammer-axe, b: Chisel, c: axe-adz, d: awl in antler handle, e: fish-hook. After Champion et al. (1984, 145).

Prehistory

Copper was one of the earliest metals worked by man and, as illustrated in chapter 16, several extensive copper mines have been found in Britain and dated to the second and third millennia B.C. But copper metallurgy had started in the fifth millennium in the Balkans and in the fourth millennium in Iberia, Fig. 18.1. Gradual development from the use of native copper to smelting copper ores, casting and hammering occurred. Native copper is frequently found, sometimes in very large deposits, as around Lake Superior. The North American Indians continued to find and use native copper for centuries without ever discovering how to smelt it from ore but in Europe native copper did not occur in such large quantities. Cyprus was the major source of copper ore in the ancient world and may have given its name to the metal. Pure copper melts at a moderate temperature (1080°C) and is relatively soft and easily worked. Small amounts of impurities, such as arsenic, occur naturally and give a harder, more useful metal. Kilns, used to fire pottery, were adapted for smelting and could reach temperatures in excess of 1000°C and from the earliest times charcoal was used as a fuel. Charcoal is wood with the volatile components driven off by raising its temperature to about 900–1000°C by partial combustion in a restricted air supply. The weight is reduced to about half as the volatile components are driven off and the resulting fuel is light, porous and nearly pure carbon. Theophrastus (*Enquiry into Plants*, 5.9–4,6) in the fourth century B.C. described the properties of various woods for making charcoal, Humphrey et al. (1999, 41–2). In the silver mines, for example, they prefer charcoal made from pine whereas smiths prefer it made from fir. He also describes the method of manufacture. Charcoal's great advantage over wood and coal is that it is free of any impurities, such as sulphur, which seriously affect the properties of metals, and it burns evenly.

The first ores used were the oxides and carbonates of copper mixed with earthy material (gangue). Every effort was made to reduce the gangue and to separate it from the mined ore before it went into the furnace but pure ore free of gangue could not normally be obtained. The oxide and carbonates of copper are easily smelted in a furnace, the CO, obtained by burning charcoal in a restricted air supply, stripped away the oxide and left the metal, which melts and flows to the bottom of the furnace where it may be extracted when cool. However, sulphide ores were also used in the late Bronze Age and these needed a more complicated process.

The discovery of smelting was probably accidental, as was the discovery of bronze, which occurred shortly after the discovery of copper. The Copper Age (Chalcolithic period) was quite brief in the Middle East and in Europe and bronze soon came to be preferred. Bronze is copper alloyed with 5% to 15% tin and the addition of tin resulted in a harder metal that was able to take a sharper edge and was more resistant to wear. Tin ore (cassiterite) is not very widely distributed and normally contains less than 3% tin so it is likely that the first bronze was cast from ores in which both copper

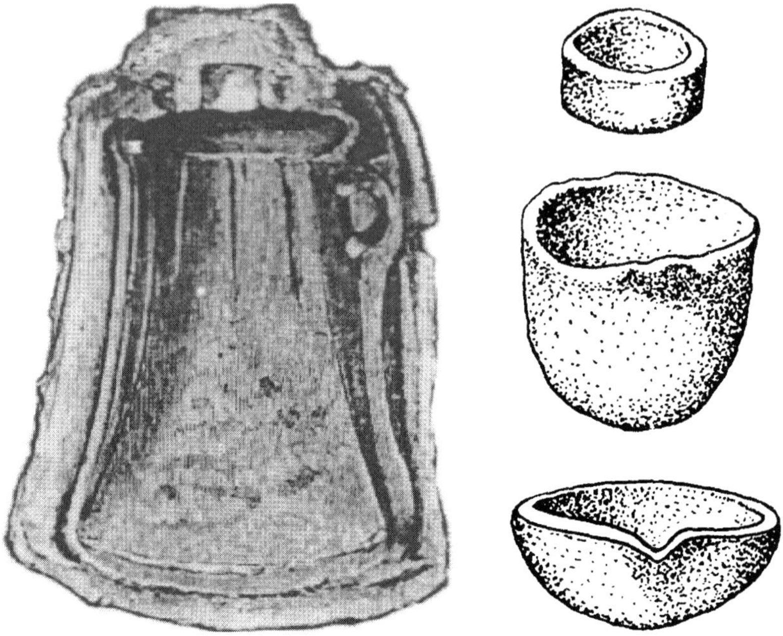

Fig. 18.2 Left: an English mould made of bronze and used to cast bronze celts. Right: Prehistoric crucibles.

and tin naturally occur. Be that as it may, bronze articles were made and have survived in some quantities although the source of the tin ore is not now known.

Casting implies the existence of furnaces, crucible, moulds, and waste products (slag), and these have been found in pre-historic contexts. Fig. 18.2 (left) shows a bronze mould that would have been used to produce bronze celts (a prehistoric metal implement with a chisel edge) and Fig. 18.2 (right) shows prehistoric crucibles made from clay or clay and sand. Open casting was used first, and is usually indicated by a difference in the oxygen content of the copper between the top and bottom surfaces. Enclosed casting was soon developed and was commonly used.

Pre-history ends with the invention or introduction of writing. In the Middle East this occurred about 3000 B.C. but in Britain writing was not used until contact with Roman civilisation in the first century B.C. Thus the use of iron preceded the use of writing in Britain whereas further east this was reversed. One of the earliest

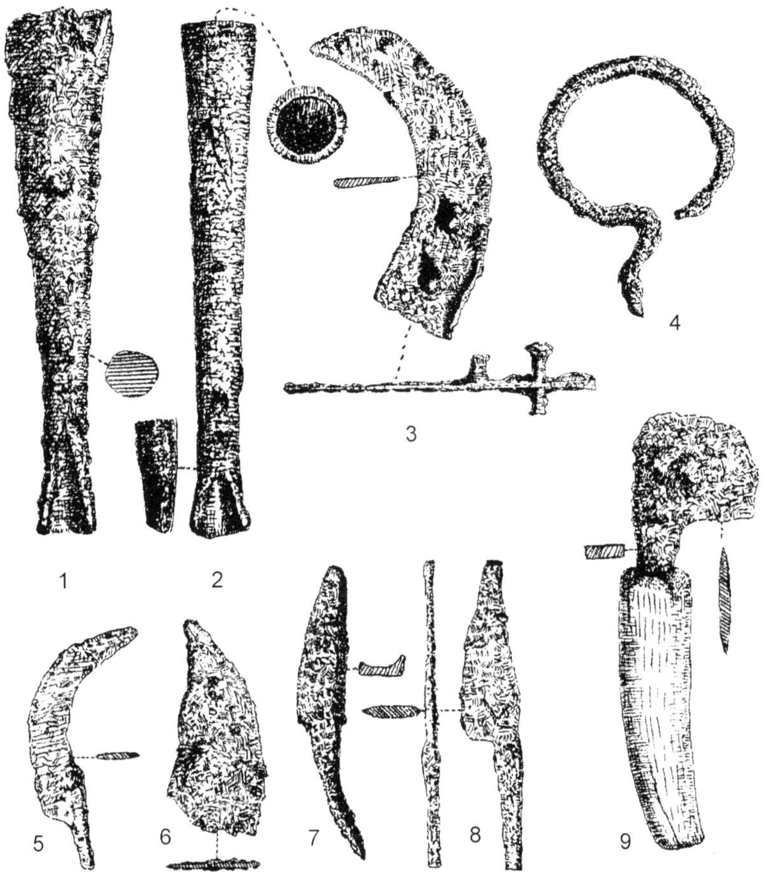

Fig. 18.3 Iron objects from All Cannings Cross, Wiltshire. 1, 2 Socketed iron gouges, 3 curved iron blade with two rivets, 4 iron ring, 5, 6, 7, 8 tanged iron blades, 9 iron implement in socketed deer horn handle.

indications of iron working in Britain is supplied by a hoard from Llyn Fawr, Glamorganshire, which may date from the sixth century B.C.; see Schubert (1957, 3). Amongst many bronze objects were three wrought iron items: the hilt of a sword, a spearhead, and a socketed sickle. But it was not until about 450 B.C. that iron was regularly worked in Britain.

Celtic invaders introduced it and a typical early settlement, at All Cannings Cross, Wiltshire, was occupied from 450–300 B.C.; a selection of their ironwork is shown in Fig. 18.3. Two later Celtic invasions brought more advanced techniques. Trade with the continent developed and iron bars were used as money. Currency bars were half finished iron bars of standard weight, much used in the west of the country but not used in the south-east where the last Celtic invaders, the Belgae, introduced coin in about 75 B.C. Production of iron, tin and lead was well established by the time of the Roman invasion and the need to secure the available ores was a prime reason for Roman occupation.

The earliest furnaces were shallow holes dug in the ground and known as bowl furnaces. The most ancient bowl furnace for smelting iron in the British Isles was discovered at Round Pound, near Chagford, Devonshire, in a pre-historic settlement founded sometime after 400 B.C. The settlement consisted of 27 huts and the furnace was in the centre hut, possibly the chief's dwelling. The hut was divided into two compartments, one being sleeping and living quarters with a small hearth and the other a small workshop with the furnace, Fig. 18.4. The furnace was a small, roughly circular, clay-lined pit 38 cm × 30 cm and 15 cm deep. It was filled with soil made black and sooty with charcoal, and there was a large mass of iron slag adhering to one side, which identified the ore as local Devonshire haematite ore. The five stones close to the furnace mouth may be the remains of a low circular reinforcement wall. About 3 cm from the iron slag is a granite stone, rising a little above the ancient ground level and probably a support for the pipe through which air was forced into the furnace using bellows (the tuyeres). Similar bowl furnaces of the third or second century B.C. have been found elsewhere and iron blooms or ingots have been found.

Ancient Empires

As early as the Bronze Age casting techniques became quite advanced and the lost wax (*cire perdue*) process was used. One of the oldest known lost-wax castings is a silver model of a reclining ram, attached to a seal in the Uruk style, Fig. 18.5 (left), dating from about 3100 B.C. In the lost-wax process, Fig. 18.5 (right), a wax model of the required object is made and it is then covered with a soft refractory material such as clay. The wax model also has wax feeders and risers. The clay and wax is baked in an oven, the clay hardens, retaining the shape of the wax model, and the wax melts and escapes

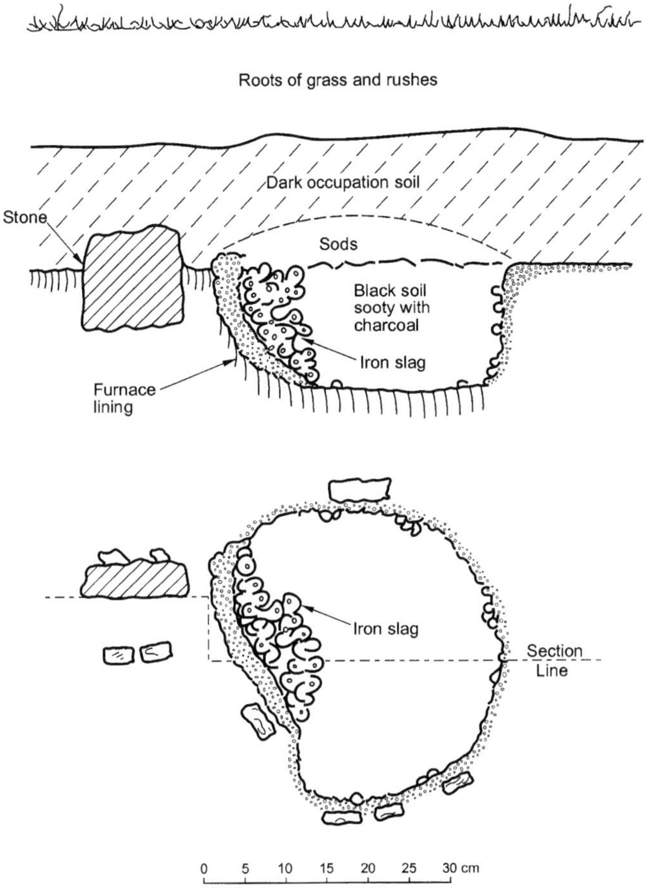

Fig. 18.4 Bowl furnace at Round Pound, Kestor Rock, Devon. a: Section. b: plan. After Schubert (1957, 20).

through a hole in the clay covering, or through the feeder and riser. The hardened clay is then used as a mould and molten metal is poured in at the top and allowed to solidify. When cold the clay is then broken away revealing a metal casting of the original wax model. The feeder and riser need to be removed and the casting polished and finished. The method was very effective and was used throughout antiquity and down to the present time.

Gold and silver were also commonly worked in the third millennium B.C. Dishes dating from 2600 B.C., now in the British Museum, were made of gold and silver and have handles and spouts brazed

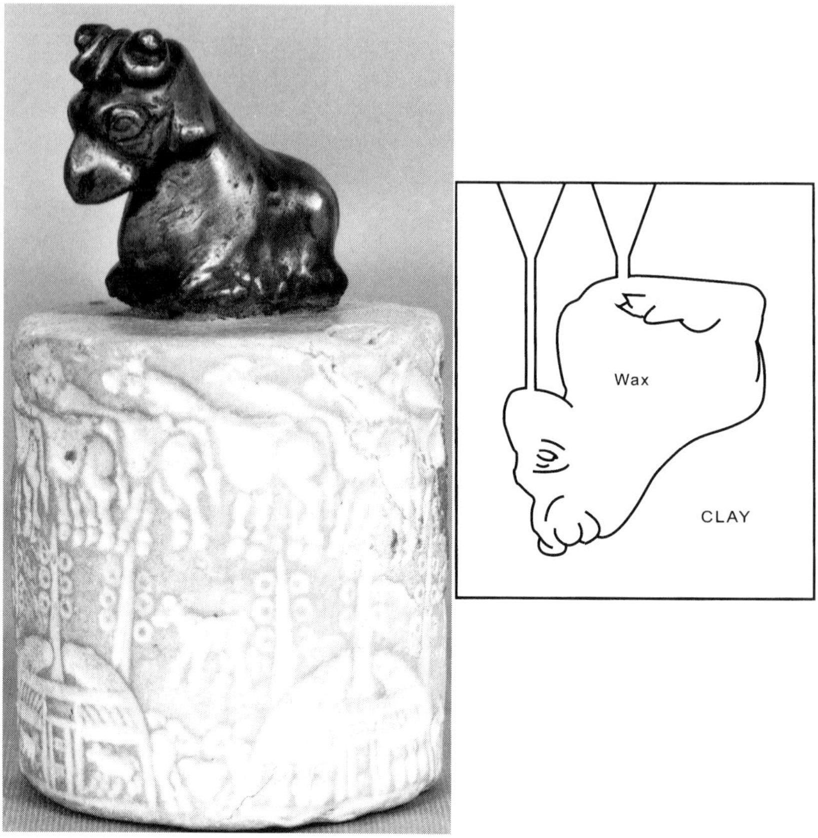

Fig. 18.5 Left: silver figure of a reclining ram cast by the lost-wax process, Uruk style, 3500–3100 B.C. Ashmolean Museum, Oxford. Right: possible mould for the lost-wax process.

into place using a gold and copper solder having a lower melting point than electrum (gold with 25% silver). The overlap between the Bronze Age metalworking and the Neolithic Age is clearly illustrated in items such as the pre-dynastic flint knife from Egypt, Fig. 18.6 (top). A beautifully worked flint with a curved blade is fitted with a sheet gold handle ornamented in repoussée designs showing serpents, flowers, goats and other animals. As mentioned in chapter 16, the Egyptians of the Old Kingdom organised expeditions of miners and quarrymen to Sinai to obtain copper ores and a surviving wall painting depicts a metalworkers' workshop, Fig. 18.6 (bottom). In the

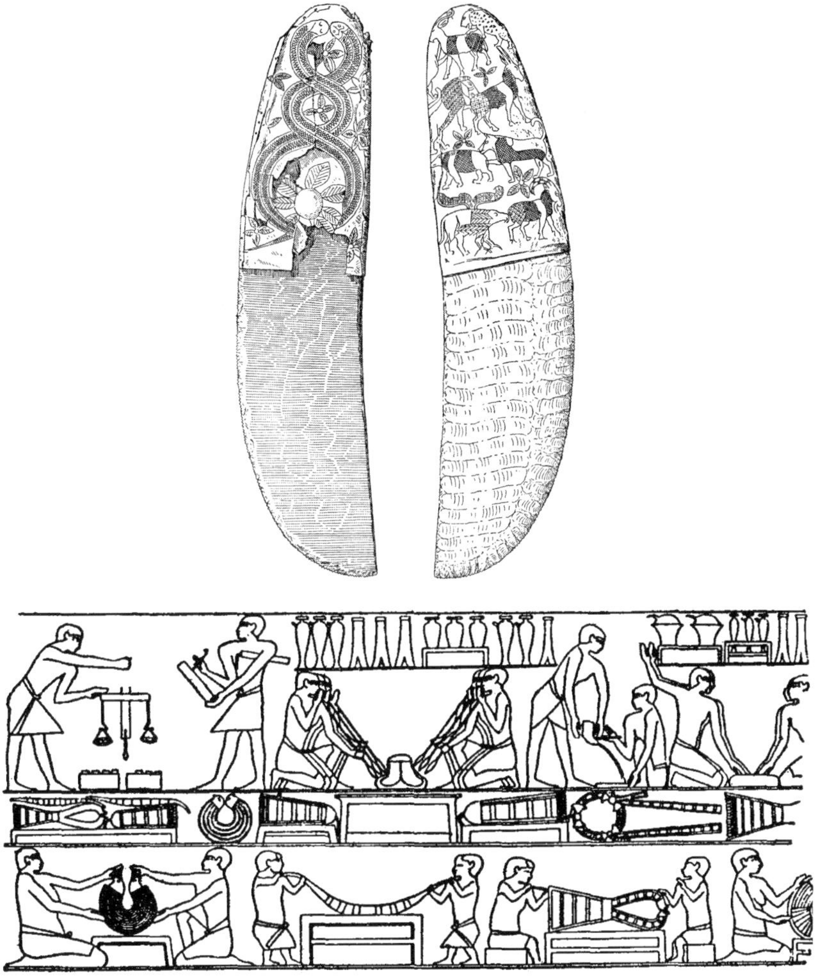

Fig. 18.6 Top: pre-dynastic Egyptian flint knife with sheet gold handle in repoussée. Bottom: metalworkers' workshop in the Old Kingdom.

upper register at the left, workmen are weighing precious metals and malachite; in the centre is a simple furnace with six men encouraging the fire with blowpipes; on the right workmen are casting metal and hammering (forging) the cooled castings. Bellows, evidently, were not in use in the Old Kingdom and even today primitive peoples use blowpipes to increase the temperature of simple furnaces. In the lower register what appear to be dwarves are employed in making necklaces and other costly ornaments. Bronze was not in

Fig. 18.7 Egyptian foot bellows from the time of Tuthmosis III, 1504–1450 B.C. Knight (1884, 269).

use in the Old Kingdom and copper was used for tools, bolts, nails, domestic vessels, and weapons. Silver was imported, possibly from Cilicia in Asia Minor, and gold was mined in the granite mountains along the Red Sea coast. In the time of Tuthmosis III, in the eighteenth dynasty, the king of Cyprus paid yearly tribute with 108 blocks of copper weighing about 1.8 kg each (about 200 kg), in addition to lead, and costly stones, Breasted (1912, 313). Bronze was certainly in use by this time but it is not certain from where the necessary tin was obtained but probably it was imported from Persia. From the 4th dynasty onwards the court officials took charge of metal-working activities and they may even have been the responsibility of the Pharaoh's sons.

During the time of the New Kingdom, foot bellows, Fig. 18.7, replaced blowpipes as a means of encouraging the fire. A pair of leather bags or cylinders attached to a disc were alternately inflated or compressed and the air expelled through blowpipes.

Presumably the operator's foot acted as a valve by covering a hole, thus sealing the bellows during the compression and allowing the entry of air during expansion. Apparently the bellows were expanded using strings so the bellows were attached to a heavy stone or otherwise anchored to the ground.

The metal most closely associated with ancient Egypt is gold and certainly they had mines in the mountains between Thebes, on the Nile, and the Red Sea, but Egypt did not acquire large amounts until the famous eighteenth dynasty and the start of the New Kingdom. Once the foreign invaders, the Hyksos, were defeated in the Delta, using the very weapons that the invaders had introduced into Egypt (horse drawn chariots and the composite bow) the Pharaohs turned their attention to Nubia in the south. *Nubia* means *land of gold* and their gold mines, see Fig. 16.10 of chapter 16, were the main attraction. Gold mines, such as that discussed in chapter 16, occur everywhere between the Nile and the Red Sea and there is no doubt that they were ruthlessly exploited, probably using slave labour, prisoners of war, or convicts. Among recent discoveries is the location of the town of Berenice Panchrysos that was mentioned by classical writers as a gold working centre. Excavations continue but already it seems to have been about 2 km long with a population of about 10,000.

Davies and Friedman (1998, 138) quote from a cache of diplomatic correspondence known as the 'Amarna Letters', to illustrate Egypt's great wealth of gold. The King of Assyria wrote to Amenhotep III (1386–1349 B.C.): "Great King of Egypt, my brother—Gold in your country is like dust; one simply picks it up. Why are you so sparing of it? I am building a new palace. Send me as much gold as is required for its adornment." In a similar vein the King of Mittani wrote to Amenhotep III: "May my brother send me very great quantities of unworked gold—and much more gold than he sent to my father. In my brother's country gold is plentiful as dust—Whatever my brother wants for his house, let him write and take it.' Egyptian Gold may have appeared as common as dust to foreign royalty, but the conquest and retention of Nubia and the winning of ore in the hot, dry mountains was not easy. In return for this gold the Egyptians imported the many commodities they lacked.

Iron was known in Egypt from a very early period although it did not become common until about 800 B.C. Early iron was of meteoric origin and was referred to, both in Egypt and in Mesopotamia, as 'the metal of heaven'. It was valued greatly although very few iron objects have survived for so long a period. An exception is an iron tool from the masonry of the Great Pyramid at Giza. Although iron ore is found in the gold bearing regions there are no workings. As mentioned in Chapter Sixteen, iron was not used to any great extent in the third millennium and throughout the second millennium it was treated as a precious metal, small articles of iron being suitable as gifts for kings, as for example, when a Hittite king sent Ramesses II (1279–1212 B.C.) an iron sword and promised a shipment of the metal. The Hittite kings seemed, at this time, to have had a monopoly. The ore was plentiful and available in most places but neither the technique of producing the high temperature required for smelting it nor the method of extracting it at lower temperatures was generally understood. Iron melts at 1530°C but almost pure iron can be extracted from the ore in the solid state at temperatures of about 1000°C. The Hittites are usually credited with the discovery of iron smelting and working by about 1200 B.C. Iron ores are abundant in the Anatolian Mountains and iron swords, iron writing-tablets, and iron statues of gods and animals are mentioned in Hittite texts, but only a few iron objects of the Hittite period have been found. Their monopoly did not last. Iron has been used in the Austrian Alps since about 1000 B.C. and iron products have been discovered in thousands of graves dating from 1000 to 700 B.C., near the salt-mines of Hallstatt in Upper Austria. As mentioned above, iron was being worked in Britain, at the very edge of the civilised world, by 450 B.C.

Greece and Rome

Homer's account of how Hephaestus, at the request of Thetis, made a complete set of armour and a marvellously decorated shield for Achilles is poetic rather than technical but, as stated in chapter 14, the story reveals the reverence accorded to smiths at this time. The scene was frequently depicted. Fig. 18.8 shows a seated Hephaestus (the smith-god) handing the completed armour, which consists of helmet, greaves, and shield, to Thetis. The quality of the armour illustrated is confirmed by surviving examples of Greek armour. Despite Homer's poetic intentions his account does contain interesting technical information. We learn, for example, that "Hephaestus left her [Thetis] and went back to his forge where he turned the bellows on the fire and bade them get to work. The bellows—there were twenty of them—blew on the crucible and gave a satisfactory blast of varying force, which increased at critical moments and subsided at others", (Iliad, XVIII). Clearly, by the eighth century B.C., when the myths were first written down, bellows were in use and had replaced the blowpipes illustrated in earlier Egyptian paintings. Homer, in fact, uses two or three different words for *bellows*, which may indicate different types, such as the Egyptian foot bellows mentioned above, that were tested before a common type emerged to become the accepted design. Ausonius (*Moselle*, 267–269), writing in the fourth

Fig. 18.8 Hephaestus presenting armour (helmet, shield, and greaves) to Thetis. The shield is a cross between a normal hoplite shield and the heroic Boeotian form. Attic red-figure cup from Vulci.

century A.D. describes the operation of the valves: "Just as when the blast of air drives on the smithy fires, the woollen valve plying in the beechwood hollow [of the bellows] takes in and confines the winds now by this hole, now by that one", Humphrey et al. (1999, 222). A pair of bellows is illustrated on a Roman oil lamp, Fig. 18.9, which amusingly shows them being used by a man to encourage the lamp's flame.

A description of an ancient furnace has survived in a passage from Dioscorides (*Pharmacology*, 5.75), which may be quoted from the translation of Humphrey et al. (1999, 222). Writing in the first century A.D. Dioscorides says: "In a house with a double roof, build a chimney and close to it, towards the top, a suitable window opening upwards. For the bellows, break through the wall of the room next to the chimney a small opening that leads into the furnace itself.—Joined to this room is another one for the bellows and the man

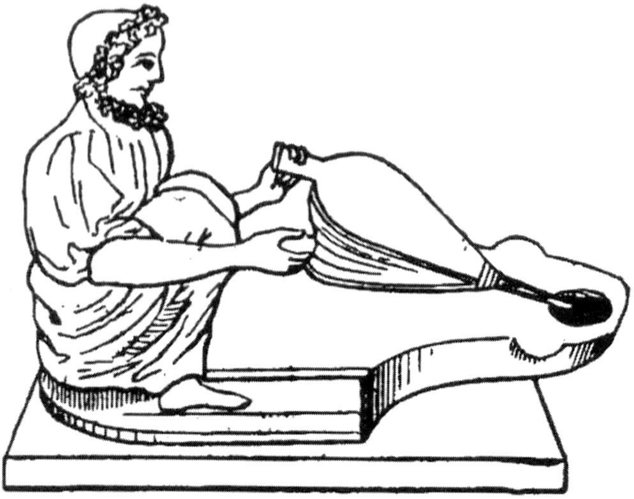

Fig. 18.9 A Roman oil-lamp showing a man encouraging the flame with a pair of bellows. Knight (1884, 222).

working the bellows. Then the coals are put into the furnace and are kindled. Finally the workman, standing nearby, sprinkles on the *cadmia*, which has been beaten into small pieces, from a place above the head of the furnace—".

By the fifth century B.C. armour and weapons were manufactured on a large scale in specialist workshops. For example, in 404 B.C. Cephalus, the father of the orator Lysias, had a workshop that was seized by the Thirty (the rulers of Athens at that time). They took 700 shields, a large quantity of gold, silver, jewels, and furniture, and 120 slaves. (Lysias, *Against Erathosthenes*, 19). But this seems to have been exceptional and smithies, in Greece, remained small workshops rather than industrial sized factories. A scene in a Greek Smithy of about 500 B.C. is shown in Fig. 18.10. The smith's tools and his products are hung up on the walls. Visitors, shown on the right, were quite common, especially in winter when they came in from the cold. Hesiod (*Works and Days*, 493) advises the farmer. "Pass by the smithy and its crowded lounge in winter time when the cold days keep men from field work—for then an industrious man can greatly prosper his house".

Greek skill in casting, and in particular in casting bronze statues, reached heights, both technically and artistically, that have hardly

Fig. 18.10 Greek smithy, about 500 B.C., from an Attic amphora. Left, the master smith; centre the smith's slave; right, spectators.

been bettered. The most famous is the Colossus of Rhodes, a statue of the Sun god and one of the Seven Wonders of the Ancient World. It was a hollow, bronze statue, cast in several pieces and riveted or welded together. Pliny (XXXIV, 41–46), writing in the first century A.D. says, "No statue has commanded greater admiration than the Colossus of Rhodes made by Chares of Lindos, a pupil of Lysippus. It was about 32 m high. Sixty-six years after its erection the Colossus was toppled by an earthquake, but even lying on the ground it is amazing. Few people can make their arms meet round its thumb, and the fingers alone are larger than most statues. Where the limbs have been broken off, there are huge, gaping holes, and inside you can see the large, heavy pieces of rock that the sculptor used to steady the Colossus when set up. Records show that it took 12 years to complete, at a cost of 300 talents; this money came from the sale of the siege engines abandoned by King Demetrius when he became tired of the long, drawn out siege of Rhodes [305–304 B.C.]".

Rome, of course, was not to be outdone, and Pliny goes on to say, "Large-scale statues also used to be made in Italy. We can see the Tuscan Apollo in the library of the Temple of Augustus. It is debatable whether this statue—which measures 15 m from head to toe—is more remarkable for the quality of its bronze or for its beauty. After he had defeated the Samnites, in a war fought under a solemn oath [293 B.C.], Spurius Carvilius made the statue of Jupiter that stands in the Capitol from the captured breastplates, greaves, and helmets. The figure is so large it can be seen from the Temple of Jupiter Latiaris on the Alban Mount.—But of all such colossi the Mercury made by Zenodorus for the Averni, in Gaul, has pride of place. It took ten years to complete, at a cost of 40 million sesterces. When he had given sufficient proof of his skill in Gaul, Zenodorus was summoned to Rome by Nero and there he cast the gigantic statue—about 32.5 m in height—intended as a likeness of Nero, but now dedicated to the Sun after the condemnation of Nero's crimes; it is a breathtaking sight." Pliny seems to have visited Zenodorus' workshop in Rome. He says, "In his studio we used to admire not only the extraordinary verisimilitude of the clay model but also the framework of small timbers that formed the armature for the work in hand. This statue shows that erstwhile expertise in bronze casting has not gone by the board." Casting, assembling, and finishing of a large bronze statue is illustrated on an Attic black figure cup, now in the Berlin Museum, Fig. 18.11. Pliny tells us that the bronze used for statues and for sheets of the metal, is of copper, a third of which is copper scrap, and tin that is alloyed in the proportion of one part tin to eight parts copper. To impart the colour known as *Grecian* a tenth part of lead and a twentieth part of silver is added and it is suitable for moulding. Pot bronze, named from the vessels that are made from it, is composed of three or four parts of silver-lead for every hundred pounds of copper. Lead added to Cypriot copper makes the purple in the borders of robes on statues.

The fine results of such methods are illustrated in Fig. 18.12, which shows the Chimera of Arezzo. This mythical beast has the head of a lion, the body of a goat, and the tail of a serpent. The bronze was found in 1554 A.D. and caused a great sensation amongst the Florentine artists of the time. It once belonged to the Medici collection and it was said that its restoration was placed in the hands of Cellini, but he does not mention it in his autobiography. It subsequently underwent a 'freely interpreted' restoration, now barely recognisable. The inscription TINSCVIL on the right paw is a dedication, as a votive offering, to TIN, the Roman god Jupiter. The chimera is an Etruscan hollow bronze from the fifth or fourth centuries B.C., and was probably part of a group in which it hurled itself upon the hero Bellerophon, mounted on the winged horse Pegasus, who eventually killed it. It is, despite some restoration, a fine example of hollow casting by the lost wax process.

Fig. 18.11 Bronze worker's foundry. Top: casting and assembling a bronze statue. Bottom: finishing the statue. Attic black figure cup, Berlin Museum.

A good illustration of a Roman goldsmiths' workshop was found on the walls of the Black Room in the House of the Vettii at Pompeii. Aulus Vettius Restitutus and Aulus Vettius Conviva, two affluent freedmen, owned the house. Perfumers, fullers and flower dealers are also illustrated on the walls of this room and instead of slaves the workers in these pictures are depicted as rather sweet cupids. In the picture portraying goldsmiths, Fig. 18.13, a large furnace is on the right. The pictures are not too clear but Rostovtzeff (1945, 273) thought the cupid on the right was intently occupied in chiselling a large metal bowl, probably a bronze bowl which was being prepared for inlaying with silver, and he thought the other cupid was busy keeping the furnace going by means of a blowpipe and at the same time heating a piece of metal which he holds with a pair of tongs. However, it seems more reasonable to say that the cupid on the

Fig. 18.12 The Chimera of Arezzo. A fine example of Etruscan bronze casting of the fifth or fourth centuries B.C.

right is operating a pair of bellows whilst his colleague in front of the furnace is heating a piece of metal that he holds with tongs. A third cupid is hammering a small piece of metal, presumed hot, on an anvil. Near him stands a counter with three open drawers and two pairs of scales. The drawers probably contain standard weights for weighing the precious metals to be melted or forged. Further left a lady customer discusses with the proprietor the weight of a jewel and beyond them two cupids are hammering a larger piece of metal on an anvil.

Iron ores, says Pliny (XXXIV, 144–150), are found almost everywhere and he recognised many different kinds: "some lands produce a soft iron akin to lead; others, a brittle, coppery kind especially to be avoided for wheels and rails—for this purpose the soft iron is suitable. All the different kinds of iron are called 'edging ores', a term not used in the case of other metals. This term I derived from

Fig. 18.13 Goldsmiths depicted in the Black Room of the House of the Vettii at Pompeii, first century A.D.

the expression 'to draw out a sharp edge'." The brittle kind of iron, especially to be avoided for wheels and rails, seems to be cast iron, which implies that the Romans were able to raise the temperature of their furnaces sufficiently to melt iron. Pliny confirms this when he says, "It is remarkable that when ore is smelted the iron liquefies and subsequently becomes spongy and brittle". If iron liquefies then cast iron containing about 3–4% carbon is produced and this is certainly brittle. Some cast iron objects have been found and as we will see, almost pure iron was produced in Roman furnaces at Wilderspool, Cheshire, and was probably made from re-melted cast iron. Steel, which was difficult and expensive to produce at this time was used sparingly and weapons were usually made of a soft iron and edged with steel. The Romans also knew how to harden and temper steel by quenching in oil or water although Pliny's ideas on the usefulness of the different waters were mistaken. He says: "There is also a great difference in the way furnaces are used: by one special process the iron is smelted to give hardness to a blade; by another to give solidity to anvils and hammer-heads. But the chief difference is the water into which the red-hot metal is at intervals plunged. The water in some districts is more beneficial than in others and has made places famous for their iron, as, for example, Bilbilis and Tarragona in Spain, and Comum in Italy, although there are no iron mines in these places." Pliny thought that Chinese iron, which was exported to Rome along with fabrics and skins, takes the first prize and the iron of Parthia the second, and he was impressed by the high tenor of the ore from Noricum, in the Austrian Alps. Smaller forgings were usually quenched in oil "lest they should be hardened by the water and become brittle" and in this he was correct. He notes that iron can be protected from rust by white lead, gypsum, and pitch.

The island of Elba was famous for the production of iron. Diodorus of Sicily (*History*, 5.13.1–2) says that it takes its name from the abundance of sooty clouds close about it. "They have a great abundance of the iron ore, and the men working the ore pound the rock and burn the broken stones in cleverly designed furnaces. Then smelting the stones by means of a great fire in these furnaces, they cut the product into moderately sized pieces resembling large sponges in appearance"; see Humphrey et al. (1999, 219). This spongy material was shipped to the mainland (Pozzuoli) where it was processed further and forged to make armour, double-pronged forks, sickles and other tools.

A Roman blacksmith working at his anvil is illustrated in the tomb relief shown in Fig. 18.14 (top). Behind him is a furnace at which an assistant is operating bellows and protects himself from the heat by a simple shield. The blacksmith appears to be making a lock, illustrated at the bottom right, so perhaps this was his specialisation. Hammer, tongs and a piercing tool are also illustrated. The tomb relief in Fig. 18.14 (bottom) shows an ironmonger's shop with his wares (scissors, knives, chisels, hammers, pliers, saws, etc) on display.

The sites of furnaces used by the Greeks and Romans in smelting ore and fusing metals have, in many cases, survived. The older type of furnace was similar to the prehistoric bowl furnace shown in Fig. 18.4 and remained in use for a long time, even to the end of the Roman period. Usually it consisted of a hole, dug in the ground, lined with clay and perhaps fitted with a small chimney. The more advanced type is known as the shaft furnace and is similar to that illustrated in Fig. 18.27 in which the ore and the fuel were heated together. Often these furnaces could be employed only once, which explains why there are so many on some sites. However, more permanent shaft furnaces were developed.

In Britain, according to Julius Caesar, small quantities of iron were produced along the coast and it was considered a precious metal because of its rarity. Iron bars were used among the inhabitants as money, and a hoard of iron currency bars averaging 0.625 kg each, was found in 1919 near Winchester. Roman ironworks were in production at Maresfield, Sussex by the time of Vespasian (d. 69 A.D.), and near Battle fifty years later. More important workings developed on the Wye and in the Forest of Dean.

Fig. 18.14 Top: a Roman blacksmith at work, second century A.D. Bottom: an ironmonger and his wares. Isola Sacra, Ostia, first century A.D.

The earliest and most complete of the Roman iron-smelting furnaces in Britain were found by May (1905, 427–432) at Wilderspool, Cheshire, Fig. 18.15. It was found in conjunction with about 100 Roman coins dating from the 29 B.C. to 337 A.D., and with quantities

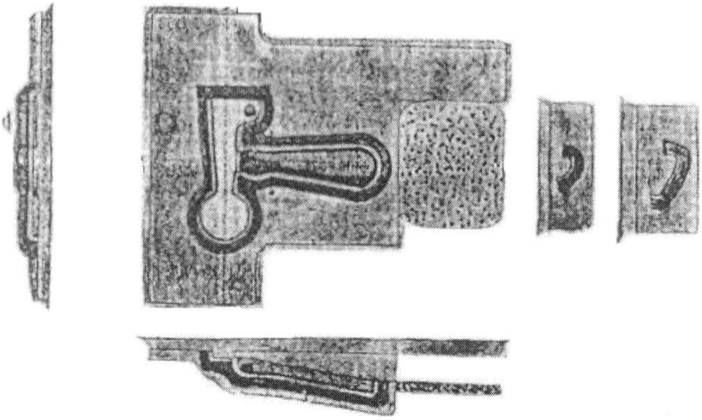

Fig. 18.15 Roman iron smelting furnace found at Wilderspool, May (1905, 427–432).

of Samian ware pottery. The site also had a wide range of other Roman industrial workings such as pottery, tile and glass making, and lead and bronze working. The iron-furnace is a roughly T shaped cavity cut in a large platform of boulder clay. The circular pit in the crosshead of the T is 70 cm diameter and 7.5 to 10 cm deep. It is fitted with a fan-shaped flue or stoke hole 130 cm long and of depth decreasing from 10 cm to 3 cm outwards. At right angles to the flue is a sloping trough 200 cm long and increasing in width from 30 to 60 cm and its depth increases from 15 cm to 45 cm and is the means for collecting the molten metal from the circular furnace. An unusual feature is a covered pipe that connects the outer end of the flue with the top of the sloping trough and May thought it acted as a return pipe to convey molten metal into the trough that may have run past its open mouth. The interior of these cavities was calcined to a depth of 5 cm and the clay was reddened by long continued heat for more than 15 cm beyond. The size and solidity of the clay platform indicate that it was a permanent structure and it is assumed that the circular pit was surmounted by a chimney, perhaps 100 cm high or more, to provide a natural draught. The site also yielded a length of corroded metal pipe, thought to have been a tuyere fed from a bellows. The flue and trough were also overarched.

May concluded that the iron ore was smelted in a continuous process in which the products of the circular hearth were run at intervals into the sloping trough and the lighter slag removed by skimming it over the lip. This was confirmed by the presence, just outside the trough, of large quantities of spent slag mixed with charcoal, cannel and ordinary coal, and nodules of clay ironstone and red hematite. The cast-iron left in the trough was then ladled or tapped into clay moulds.

Analysis of the ore was 90–96% ferric oxide with very little siliciceous matter. The slag had very little lime and considerable amounts of silica, ferrous oxide and ferric oxide. The large amount of iron in the slag was necessarily consumed in removing the silica. Analysis of the furnace lining showed that it contained large amounts of ferric oxide, which confirmed its use in smelting iron. The evidence uncovered by May confirmed that the Romans made and used cast iron and this was also confirmed by the composition (3% carbon) of a iron block also found in his excavations and by a Roman cast iron statuette found in a slag heap at Beauport Park, Hastings. In the base of another furnace May found a much-oxidised strip that was 99.9% iron and had only 0.06% carbon. This was a pure variety of soft iron, made from magnetic ore or red hematite. It must, he thought, have been purified from cast iron because such a low proportion of carbon would have been practically impossible to make directly from ore. However, cast iron objects from the Roman era remain rare and if cast iron was regularly produced it must have been mainly for the purpose of converting into steel. May also concluded that cannel coal was the main fuel.

Furnaces for roasting ore prior to smelting were also found at Wilderspool and elsewhere and also smiths' furnaces for re-heating the smelted metal so that it may be forged. Schubert (1957, 50) also notes that a complete novelty to Britain was the welding furnace, Fig. 18.16. This was necessary because the size of bloom was usually quite small and to make larger masses the small blooms must be hammer-welded together. Such a furnace, dated to 340 A.D., was found at Corstopitum, along Hadrian's Wall. Circular in shape, it had an internal diameter of about 1.8 m and a probably height of 1.5 m. No slag was found either in or around the furnace. The figure shows how the large mass, found in the furnace, was built up from small blooms that had been produced by direct reduction. The blooms were heated to welding temperature (tuyeres of baked clay were found in close proximity to the furnace), and then removed and hammered to form a cohesive mass. This process was repeated until the required size was attained. Although a little charcoal was

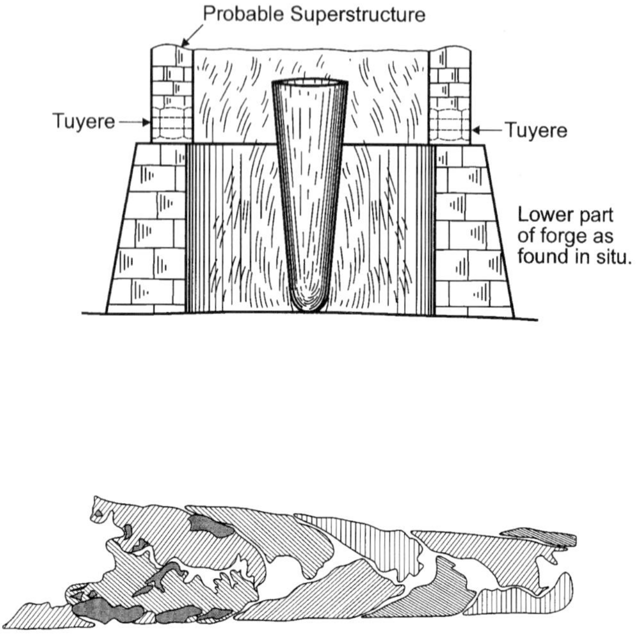

Fig. 18.16 Top: section through a Roman welding furnace, Corstopitum. Bottom: 156 kg mass of iron found in the furnace and built up of smaller blooms by hammer welding. After Schubert (1957, 51).

found the main fuel was ordinary coal. Similar blocks of welded iron have been found at Chedworth Roman Villa, Gloucestershire.

Medieval

Mining and metallurgy in Europe seem to vanish for several centuries after the fall of the Roman Empire and it is not clear how much continued or in what areas. Salzman (1964, 22) notes that Oswy, King of Kent, in 689 granted to the abbey of St Peter of Canterbury land at Liminge "in which there is known to be iron-ore"; which suggests that the monks at least had revived the craft. There is also a legend that about 700 A.D. Alcester, Warwickshire, was a centre of iron working. Apparently, the pagan smiths refused to stop their work to listen to St Egwin, Bishop of Worcester, when he visited the town and he could not be heard for the noise of their

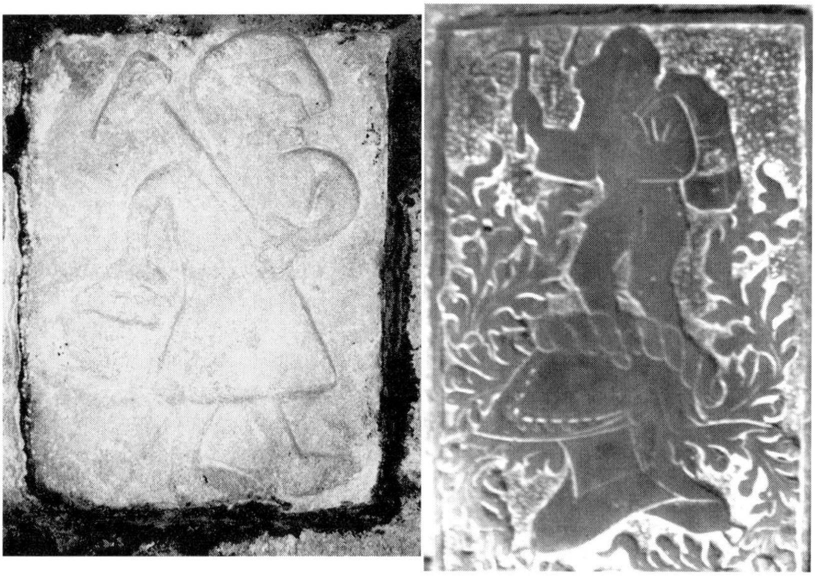

Fig. 18.17 Left: *T'owd Man*, with his pick and kibble is a carved figure of a medieval, or perhaps Anglo-Saxon, lead miner found at Bonsall but removed to Wirksworth church in about 1876. Right: a fifteenth century free-miner from a brass in Newland church, Forest of Dean.

anvils and consequently the earth swallowed them up. The legend was probably invented to explain the Roman remains below the ground, and to enhance the cause of Christianity, but the association of Alcester with smiths may have some validity. It is not history but it is what survives from the period, relating to iron, and we must rely mostly on archaeology for more reliable data. Nevertheless, tribute must be paid to the work of the monks, for whom manual labour was a requirement of their rules and a check against idleness.

The Romans certainly mined for lead and silver in Britain and pigs of lead have been found stamped for Brittannicus (44–48 A.D.) and Claudius (49 A.D.), but the next reference is 835 A.D. when lead mines at Wirksworth, Derbyshire, were leased by the Abbess of Repton to Duke Humbert. The parish church at Wirksworth is very fine with many surviving Anglo-Saxon carvings, amongst which, though not originally from Wirksworth, is a carved relief showing a lead miner with his kibble and pick, Fig. 18.17 (left), and known locally as "T'owd man". Fig. 18.17 (right) is a somewhat similar

Fig. 18.18 Left: Saxon and Saxo-Norman iron smelting sites. Right: references to iron working in the Domesday Survey. After Crossley (1981, 30–31).

depiction in a memorial brass of a free miner of the fifteenth century from Newlands in the Forest of Dean.

Crossley (1981, 29) notes that towards 1000 A.D. there is growing evidence that districts once active were again producing iron. In East Anglia there are three examples, Aylesham, Mucking, and Witton, of early Saxon iron smelting, see Fig. 18.18 (left), and at Ramsbury, Wilts, is the best-preserved site of the period, dated by radiocarbon to the late eighth century. The remains of low shaft furnaces, first introduced by the Romans, were found here. But despite this, references to iron tools in ninth century monastic inventories are rare and it is not until the eleventh century that iron goods and smiths are mentioned frequently. At West Runton, Fig. 18.19 (left) is an eleventh century bowl-hearth, 500 mm in diameter, set on natural clay from which the slag was tapped during smelting, and two bellows, with tuyeres set at right angles to the slag notch, had been used. At Stamford, Fig. 18.19 (right), operations were on a larger scale and ore roasting and smelting furnaces have been excavated. Crossley (1981, 30) thought the eleventh or twelfth century shaft furnace excavated in 1964 was the best early medieval example in England, being over 1 m high and 400 mm diameter.

The Domesday Survey contains many references to iron working,

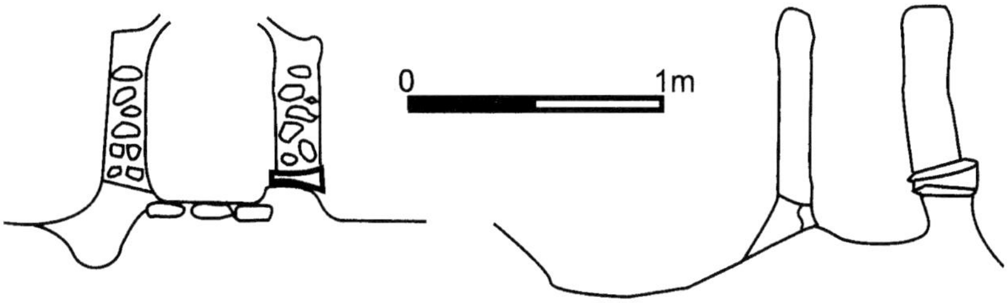

Fig. 18.19 Saxo-Norman smelting furnaces. Left: West Runton. Right: Stamford. After Crossley (1981, 31).

Fig. 18.18 (right). Salzman (1964, 23) says that Gloucester rendered as part of its farm 36 "dicres" of iron, probably in the form of horseshoes, and 10 rods suitable for making bolts for the king's ships, while Pucklechurch paid 90 "blooms" of iron yearly. There were six smiths in Hereford, each of whom made 120 horseshoes per year for the king; at Hessle, in the West Riding, were six ironworkers, and the survey mentions iron mines at Rhuddlan, Cheshire (now in Wales), in Sussex, and elsewhere. Schubert (1957, 78) found 25 such references and although this does not seem large he notes that English knives at this time had a high reputation in France and that William of Poitiers, chaplain to William the Conqueror, states that German skilled workers often resided in England because of the high level the Anglo-Saxons had attained in all forms of craftsmanship. Schubert (1957, 89) also notes that two mills, at Lexworthy, Somerset, rendered two blooms of iron each as their customary rents, which suggests, but does not prove, that they were proper iron mills in which waterwheels were used to drive the bellows. Lead mines and workings are also referred to in the Domesday survey. In the time of Edward the Confessor the Derbyshire mines at Bakewell, Ashford, and Hope yielded £30 per year and five wainloads of lead, but by 1086 the yearly value had fallen to £10 6s, probably because of the devastations caused by the conquest.

An Anglo-Saxon smith of the eleventh century working at his forge is illustrated in Fig. 18.20 (top). On the right of the picture the smith is working at an anvil and forging a piece of iron with a hammer. To the left of the anvil stands a man holding tongs in his right hand, ready to lift the hot iron. In his left hand he holds a tool bag. Neither furnace nor bellows are apparent. Three centuries later, in

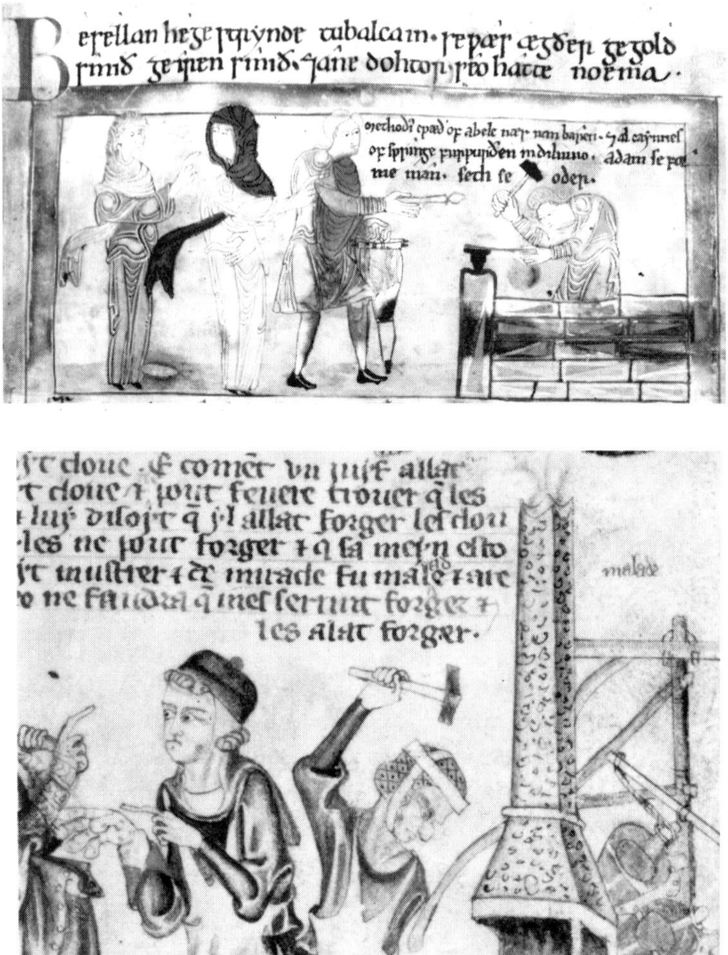

Fig. 18.20 Top: Anglo-Saxon smith working, early eleventh century. From the BM; Cotton MS, Claudius B IV, fol. 10. Bottom: blacksmith's forge with bellows, anvil, and an effective chimney, 1325–30. The Holkam Bible, BM, London.

1325–30, some progress is evident in the form of hand-operated bellows and a substantial chimney; see Fig. 18.20 (bottom). The illustration is a scene from the "road to Calvary" and it is not the smith who forges the iron but his wife. The smith was ordered to make nails for the cross but he refused under the pretext that his hand was injured. When he was told to show his hand his pious falsehood was made retrospectively true.

The position in continental Europe, after the fall of the Roman Empire was similar to that in England. Rents in blooms of iron were paid at the Abbey of Lorsch in Hessen, in 788, and in the early tenth century at the Abbey of Fulda. As in England iron objects, particularly arms, were held in high regard and the traditions of pattern welding in the manufacture of swords, see chapter 22, probably using ores from several sources, were maintained. There have been important excavations of large groups of shaft furnaces dated to the eighth and ninth centuries. At Zelechovice are 24 shaft furnaces set into banks, with bellows blowing air through tuyeres sloping downwards into the furnace. An interesting feature was that the hearths were built with rear chambers into which the blooms were pushed in the later stages to lessen the risk of oxidation and to give the iron a higher carbon-content. By the twelfth century the ores of eastern France were in use and there were important centres in Champagne and in the valley of the Blaise and Marne. The Cistercians, as described in chapter 14, were the main producers and workers of iron ore in Champagne from 1250 to the seventeenth century and were as active in Europe as they were in northern England and in the Forest of Dean.

In 1161, according to Salzman (1964, 23), Robert de Busli had given the monks of Kirkland a site in Kimberworth for four forges, two for smelting and two for working iron, with rights to dig ore and to take dead wood for fuel. At Egremont an iron mine was granted to the monks of St Bees, and at Denby a similar grant was made about 1180 by William FitzOsbert to the Cistercians of Byland. In Derbyshire, towards the end of the following century, Sir Walter de Abbetoft gave to the monks of Louth Park a wood at Birley in Brampton and two smithies, namely one bloomery and one forge, with rights to take beech and elm for fuel. But it was in the southwest that the greatest development took place. During the twelfth century the Forest of Dean was the main centre of the iron industry. The accounts of the sheriffs of Gloucester record a constant output of rough and manufactured iron: bars, nails, pickaxes and hammers sent for Henry II's building work at Woodstock, Winchester, and Brill; horseshoes, arrows and other warlike material dispatched to the army in France; spades, pickaxes, and other miners' tools provided for the Irish expedition of 1172; iron bought for the Crusade which Henry projected but did not live to perform; and 50,000 horseshoes made for the Crusade of Richard I. Throughout the thirteenth century the Forest of Dean retained a near monopoly of the English iron trade and in 1282 there were as many as 72 forges in the Forest leased from the Crown by various iron-smelters.

The Weald of Sussex and Kent was also developing. As early as 1254 the sheriff of Sussex was called upon to provide 30,000 horseshoes and 60,000 nails, and in 1275 the king's master smith, Master Henry of Lewes, purchased 406 iron rods "in the Weald" for £16 17s 11d. A year or two later he obtained another 75 rods from the same source, and paid £4 3s 4d to a certain smith of the Weald for 100 iron rods. The Weald had the great advantage of nearness to London and soon grew to a dominant position. Schubert (1957, 139) discloses that the output from bloomery furnaces at Tudeley, near Tonbridge, in the period 1330–54 averaged 166–167 blooms per year. Similarly a forge at Tindale, Northumberland produced approximately 167 blooms per year in 1333–1335. If a bloom weight of 15 kg is assumed then the level of output at Tudeley and at Tindale was about 2.5 tonne per year. Such a small output, if typical, may explain why so little evidence of medieval iron smelting has survived. Its scarcity persuaded Parliament, in 1354, to enact that no iron, either wrought or unwrought, should be exported, and for the next two centuries the English market was supplied with iron and steel from Spain and Germany.

Reasonable evidence of medieval lead smelting has survived, particularly in Derbyshire. Blanchard (1981, 72–84) states that during the first phase of development, 1290–1360, smelting took place in a 'bole', which was a simple bowl furnace, using as fuel a mixture of brushwood and charcoal and having a draught hole facing the prevailing wind. Alternating layers of charcoal and crushed and washed ore were used. Kirkham (1969, 119–138) says such boles were used throughout the medieval period but they were not particularly efficient and the waste slag, known as *blackwork*, remained rich in metal. After about 1360 separate ovens, located next to the bole, are found. These are circular stone structures about 1.5 m diameter and 60 cm high, Fig. 18.21. Initially they were used to process newly mined ore but later they were used to re-smelt blackwork and were known as blackwork ovens.

In a blackwork oven, Fig. 18.21 (top) the floor is of puddled clay with a bowl-shaped hollow in the centre in which the crushed and washed ore (*a*) was laid and charcoal (*b*) was packed around it. The hearth was sealed with a layer (*c*) of clay and once ignited the charcoal was maintained at a high temperature by an intense blast of air from bellows whose nozzles (*d*) passed through the side of the hearth. The molten lead, extracted from the slag, ran off into a mould (*e*) outside the hearth. Power for the bellows was provided either by water or manpower. The latter required two men alternately depressing two bellows (*B*), which then returned to their original position by the spring-action of two saplings (*C*). In Fig. 18.21 (bottom) the saplings are weighted, rather than sprung, see Kirkham (1969, 127). The use of the screw is not clear, but if the saplings were sprung against a fixed support then it could be used to adjust the tension.

The impact of these new ovens was impressive. Initially it permitted 45% of the metal remaining in the slag to be recovered. Later this was raised to over 60%. Blanchard (1981, 119–138) reports that the prevailing extraction rate with the richest ore was 2.2 tons of ore per ton of lead, but with the depletion of the ore this increased with time to 3.7 tons of ore per ton of lead from the deeper more fragmented veins, and by 1528–36 even the richest ores required 9 tons of ore per ton of lead and consequently production costs had increased significantly. To make matters worse the dissolution and destruction of over 650 monasteries in the 1530's released large quantities of roofing lead onto the market, forcing down prices. But the industry survived and the introduction of the 'Durham' oven opened up a new era and enabled ores of much lower metal content to be smelted. In this method, devised in Weardale in 1460, ore and blackwork were mixed in the ratio of 1 of ore to 2.4 of blackwork and the charge in the oven was reduced. The waste tips of centuries, containing much un-smelted metal, suddenly became available and freed the industry from reliance on mines nearing exhaustion. Other smelting arrangements were developed and are discussed by Kirkham (1969, 119–138).

In the later Middle Ages metal working in Europe underwent two

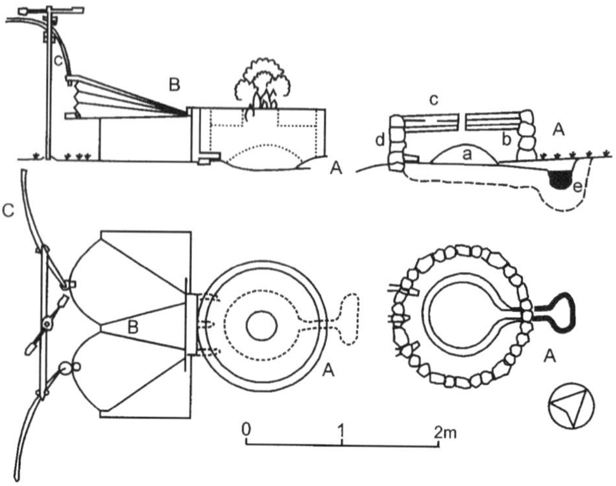

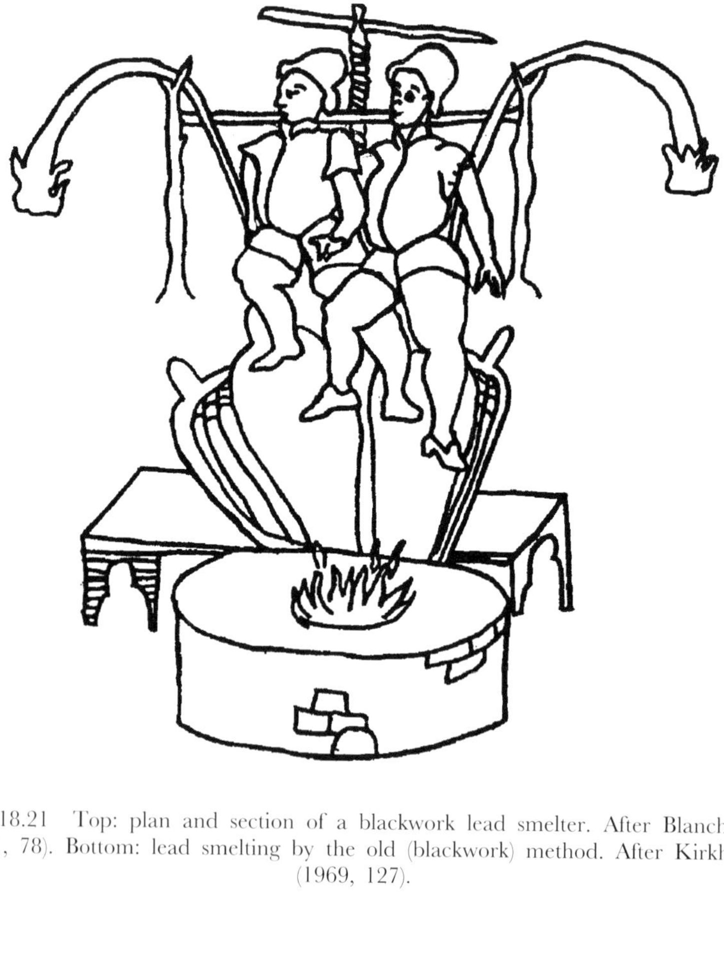

Fig. 18.21 Top: plan and section of a blackwork lead smelter. After Blanchard (1981, 78). Bottom: lead smelting by the old (blackwork) method. After Kirkham (1969, 127).

major changes. Firstly the smith's lot was eased considerably by the introduction of water-powered hammers in the forge and water-powered bellows for smelting and re-heating. Secondly, the blast furnace producing cast-iron was introduced. As already mentioned, there is an implied use of waterpower in the Domesday survey where two mills at Lexworthy, Somerset, were expected to pay customary rents in iron blooms. Crossley (1981, 29–49) states that in France in 1116 an iron mill, which is assumed to have been a powered hammer, replaced a tannery. In Spain, around Barcelona, some 14 iron mills had been built before about 1200 and during the early thirteenth century they appear in France, Germany, Moravia, and Hungary. A Hungarian reference of 1227 suggests the use of both hammers and powered bellows. Progress in Flanders and in England appears to have been slower. Schubert (1957, 133) thought it likely that the monastic orders, especially the Cistercians, would have introduced water-powered bloomeries into England. Certainly, the earliest example for an English forge hammer is at Chingley, Kent, on the lands of Boxley Abbey, Fig. 18.22 (left). It dates from the early fourteenth century, a time when European industry was generally in recession. The main surviving structure is a timber wheel race and a massive frame, thought to be the foundation for a hammer. Large quantities of charcoal were present, for reheating the blooms in hearths, which, Crossley suggests, may have been blown by water-powered bellows. There were no smelting furnaces on the site and it is assumed that iron blooms came from powered bloomeries in the surrounding woods.

The labour shortage, that followed the Black Death in the middle of the fourteenth century, no doubt accelerated industrial mechanisation. An early drawing of a water-powered bellows is from a book by Taccola, "the Archimedes of Sienna" Italy, written in 1438–41; see Fig. 18.22 (right). It shows an overshot wheel, supplied by a chute from a pond, driving a shaft held in bearings and fitted with three cams. Two of the cams lift a pair of large bellows blowing air into a furnace through a set of replaceable tuyeres. The bellows, after being lifted by the cams, fell under their own weight. Such bellows were fitted with air inlet valves and their fall was often aided by additional weights, none of which are shown. After reviewing the scanty documentary evidence Schubert (1957, 137) concluded that waterpower was applied earlier to working the bellows than for lifting the hammer, but the recent archaeological evidence, as we have seen, does not support this.

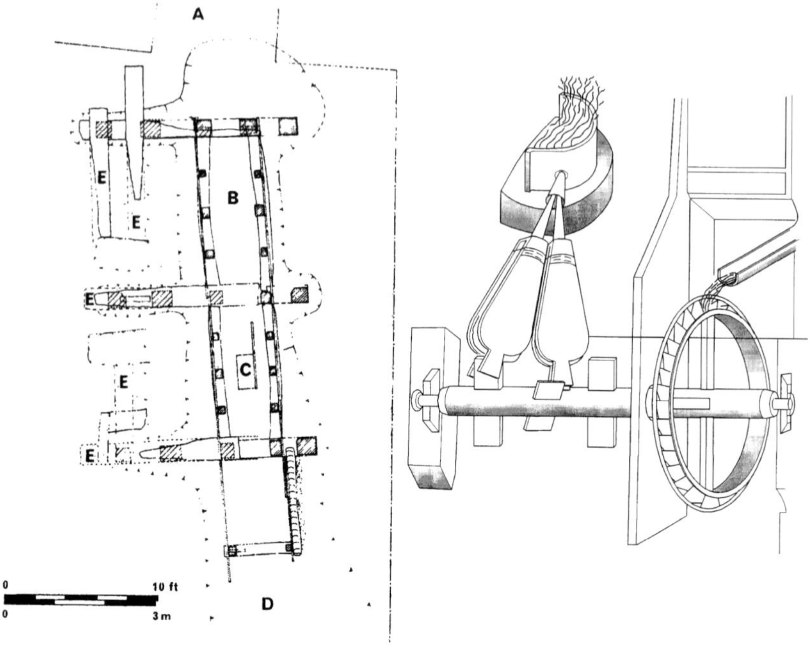

Fig. 18.22 Left: remains of a fourteenth century timber wheel-race and hammer frame, Chingley, Kent. A dam, B wheel-race, C wheel fragment, D tail-race, E hammer-base timbers and beam-slots. After Crossley (1981, 38). Right: bellows operated by waterpower, *c.* 1440. After Schubert (1957, 135).

As mentioned above, output from medieval bloomeries was about 2.5 tonnes per year but production was ten times higher after water-power was adopted. At Byrkenott, Schubert reports an output of 278 blooms per year at 88.5 kg per bloom (24.5 tonnes per year). Moreover, this increased output was attained with fewer men. At Treeton, near Sheffield, a bloomery equipped with waterwheel and a pair of bellows per furnace required only three workers for two hearths, whereas at Llantrisant, Glamorgan, a newly erected bloomery operated without waterpower in 1530 required nine workmen for the two hearths, four of whom operated the bellows.

After the introduction from Europe of cast iron from the blast furnace in the sixteenth century the powered bloomery for the production of iron died out in southern England but in the north and the west, where local needs did not justify the greater output of a blast furnace, it was retained. As late as 1636, a complete new bloom-

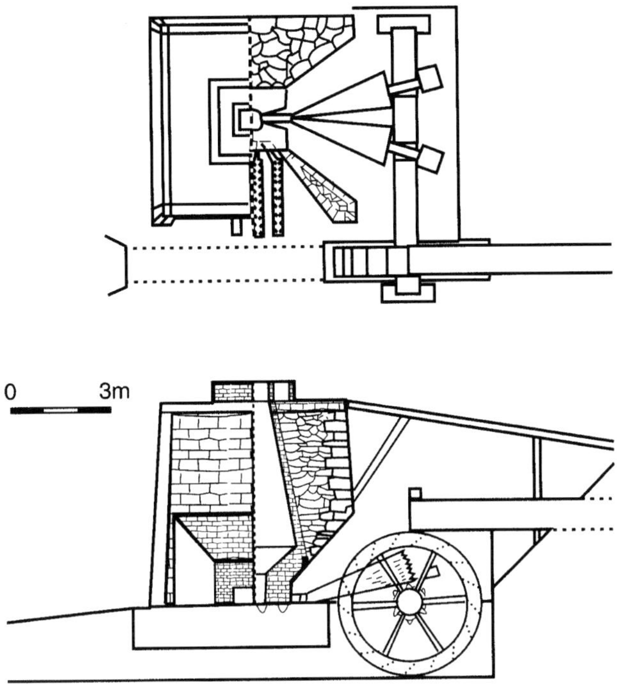

Fig. 18.23 Charcoal blast furnace, based on the Chingley furnace. After Crossley (1981, 39).

ery, with powered hearth and hammer, was built on the Esk at Muncaster Head, Cumberland, and a bloomery at Stoney Hazels, Furness, was not converted to a finery forge until the early eighteenth century, Crossley (1981, 37).

One of the first descriptions of a blast furnace is a diversion in an untitled treatise on architecture written by the Florentine sculptor-architect known as Filarete; see the translation by Spencer (1963, 201–06). He describes a trip to an iron mine in March 1463 and the method he observed was not to change significantly until the eighteenth century. The ore was first roasted with limestone, probably to reduce the sulphur content, and was ground, sifted and prepared for the charge. Layers of charcoal and ore-limestone alternated and were loaded through the top of the furnace from an upper floor and were blown by means of a two bellows through a common tuyere on the floor below. Fig. 18.23, based on the Chingley furnace gives some idea of what it looked like. There was an iron notch near the tuyere but no slag notch. The process appears to have been continuous and molten iron was tapped from the furnace every twelve hours to cast into pigs, and 20 to 25 new charges per day were added at the top. He notes that the process smelled strongly of sulphur.

Evidently the use of powered bellows had increased the temperatures in the furnace to the point where iron, unlike the bloomery process, melted, flowed to the bottom of the furnace, and could be tapped into moulds to produce pig iron. Filarete says "when they [the bellows] blow, they make a sound or a note so loud that it sounds like a stormy sea". Apparently they were much larger than those he had seen previously. The idea that you could melt iron and cast it in moulds was so novel that Filarette emphasized the point, "As it came out it ran exactly like bronze or bell metal—there is no doubt that the iron had no form whatsoever when it was underground. Thus when the iron came out of the furnace, it took whatever form was imposed on it as is the nature of metals. An additional proof of this, there is a cast bombard in the Castello of Milan in the form of a lion. It looks like it was lying down."

An important difference with the bloomery, which produced almost pure iron, was that the higher temperature of the blast furnace caused some dissociation of the carbon monoxide gas, which produced solid carbon particles that diffused rapidly into the molten iron. Thus the molten iron contained about 3–5% carbon and the resulting cast-iron was very brittle and quite unlike the malleable product of the bloomery. Filarete continues: "They take the iron to another furnace where it is melted again. Then they beat it out with hammers as they think best." This, clearly, is the finery process by which the excess carbon is removed from the iron by a second melting in contact with air so that the oxygen could remove the carbon by combining with it to form carbon monoxide. The result is almost pure iron, similar to that produced by the bloomery, and capably of being forged, as Filarete makes clear. This is known as the indirect process. Filarete also notes that the iron is forged by hand "This is done, because they had not yet arranged or completed a hammer-mill here" and he goes on to describe a water powered hammer-mill he had seen at Grottaferrata Abbey, near Rome, Fig. 18.24 (top).

Water powered tilt hammers for forging hot metal proved to be very effective and remained in use well into the twentieth century, although Nasmyth's invention of the steam hammer early in the nineteenth century replaced them for all but the lightest work. The

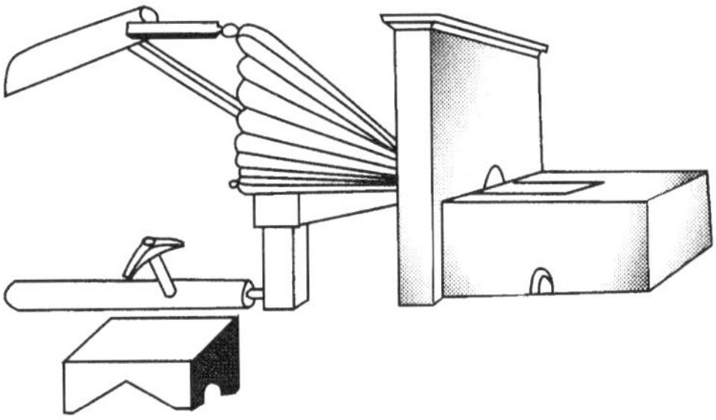

Fig. 18.24 Top: water-powered smithy bellows, forge, and tilt hammer at Grottaferrata, near Rome, 1461. After Spencer (1963, 202). Bottom: water-powered tilt hammer at Abbeydale, Sheffield, built in 1775 and in use until about 1940 for forging scythes.

hammer shown in Fig. 18.24 (bottom) was in regular production, until about 1940, hammering out scythe blades in Sheffield. A water-wheel turned the main shaft through step-up gearing and the camshaft was fitted with perhaps 24 cams per revolution so the hammer operated at a high frequency of about 160 stokes per minute. To describe someone as going 'full tilt' when they work hard is thought to derive from the speed of such tilt hammers.

It is unclear exactly where the blast furnace first developed but it seems likely to have been in Flanders where there are references to what might be blast furnaces from about 1400 and there are references to cast-iron ammunition in the early fifteenth century, particularly in eastern France. The workers who first brought iron casting to England at the end of the fifteenth century were said to be French and the process was always known as the 'Walloon' process. The first English blast furnace was built at Newbridge, Sussex in 1496, almost a century after its first use on the continent.

It was some time before cast iron came to be used as a material in its own right and initially the majority was converted into bar iron by removal of its carbon content in finery and chafery furnaces. The process is described below in some detail but essentially a furnace, lined and filled with charcoal and cast iron was ignited and blown until the cast iron melted. The carbon monoxide resulting from the partial combustion of the charcoal reacted with the carbon in the molten iron leaving pure iron. Pure iron, however, has a higher melting point than cast iron and so it became pasty and solidified as the reaction proceeded so the metal was extracted by collecting the pasty material on the end of a pole. A forge, for the conversion of cast iron into malleable iron is illustrated in Fig. 18.25 and gives some idea of its size. The main forge, *A* in the figure, has four water-wheels for driving a finery, a chafery, and two tilt hammers. The feed pond, partly shown in the figure, could cover several acres, and the coal store and charcoal yard, *D*, were larger than the forge. Seven worker's cottages and stables surrounded the forge yard, *G*, and cinder waste from the furnaces was dumped at *F*.

Very little is known about the manufacture of steel in medieval England and it seems that much was imported by the Hanseatic League through the Steelyard in London. Steel was used mainly for tipping both tools and weapons that required a hard or sharp cutting edge. This was done by hammer welding steel onto the softer iron at a high temperature. Schubert (1957, 120) found an entry in

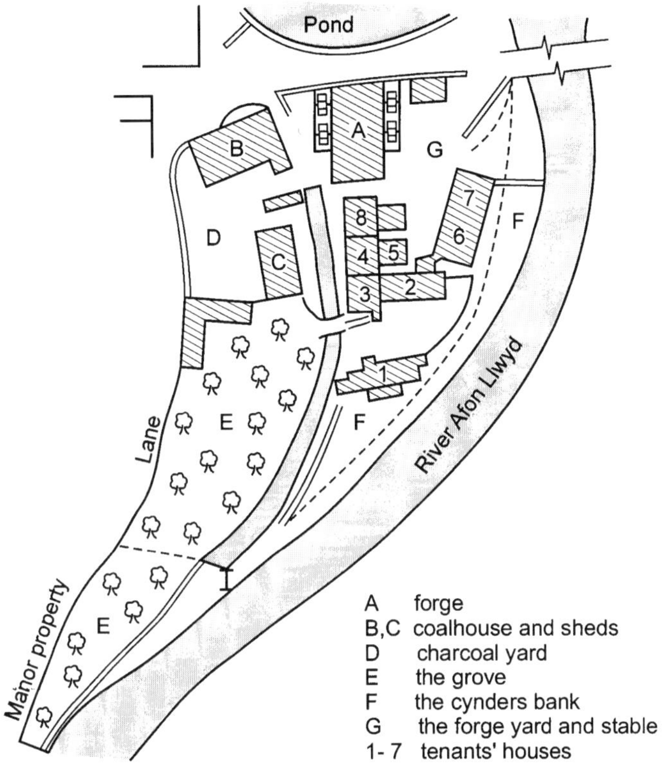

Fig. 18.25 The Forge at Pontypool. After Schubert (1957, 275).

the accounts of the Abbey of Durham in 1368 that refers to about 30 kg of iron being converted into steel. The method used was probably cementation (surface carburisation), which was attained by packing charcoal around wrought iron and heating for several hours to allow the carbon to diffuse into the surface of the iron. The method was known to the Romans and practiced in Roman Britain, and the German monk Theophilus described it in the twelfth century. At the end of the process the hot, carburised iron was quenched in water to harden the carburised surface.

Steely iron could also be produced directly from the clay ironstones of the coal measures in Yorkshire, Shropshire, and Staffordshire, which were suitable because of their manganese content. Similarly, the brown haematites of the Forest of Dean could produce a steely iron. The steel helm of the Black Prince is famous but its origin is not known. Equally fine is the English steel helm in Hereford Cathedral of Sir Richard Pembridge who died in 1377. It was said that no penknife could scratch its polished surface. It was made in three pieces welded together so well that the seams are invisible. Long periods of hand hammering of the surface are thought to account for the helm's hardness and resistance to corrosion. A famous maker of swords was Andrea de Ferrara who produced swords, some still surviving, in the Scottish Highlands and comparable in quality to those of Toledo or Milan. His swords, it was said, were tempered so that the point could bend to touch the hilt and spring back uninjured. Schubert notes that in a test in 1464 arranged between an English and a foreign artificer to produce four steel dies, two engraved and two embossed, the English artificer was judged the more skilful and was the winner. The supply of English or Scottish steel, however, was always small until the end of the eighteenth century and supplies were usually imported.

Casting and forging were the two main processes used to manufacture metal articles. After forging, half-finished bars were sold on to other smiths some of whom specialised in making particular articles. One such product, wire, was needed for many purposes and has, in fact, a very long history. An early reference in the Bible briefly describes how fine gold wire was made for weaving into linen "And they did beat the gold into thin plates, and cut it into wires" (Exodus 39, 3). Similar wire has been found in ancient Egypt, Greece, Rome, and China. No doubt the same method was tried with other materials but they were more difficult to beat into plate and cut into strips. Wrought iron could be forged into long rectangular bars about 20 mm wide but smaller sizes were difficult to handle. Iron wire was required to make chain mail and rivets in medieval armour, and for nails.

England's main industry from the fourteenth century was wool production and this required wire for making wool cards. At some point, probably late in the first millennium A.D., it was discovered that pulling a metal rod through a slightly smaller round hole in a drawplate reduced the diameter and increased the length. If the wire was drawn through a succession of smaller holes it could be reduced to any required diameter and length. Such a drawplate, with three rows of holes of varying diameter, is shown in Fig. 18.26 (bottom). The plate was found in Norway and is from the Viking age (800–1050 A.D.). An eleventh century German monk, Theophilus, has describes

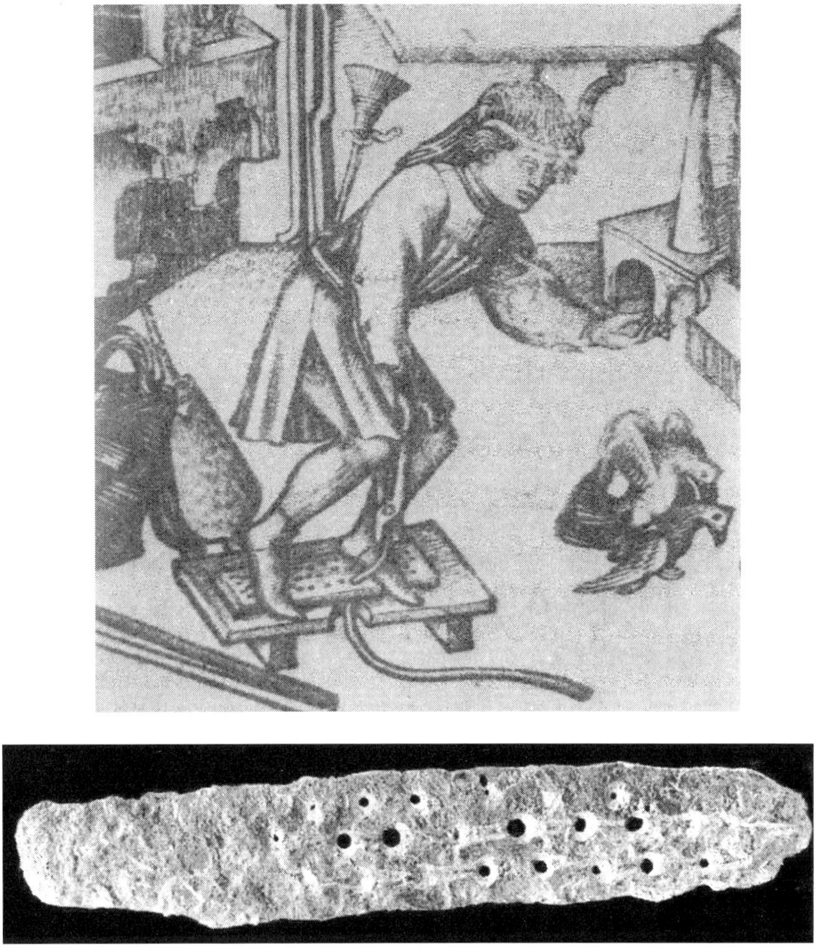

Fig. 18.26 Top: drawing wire through dies by hand. Detail from an etching of St Eligius in the Goldsmith's Shop, by the Master of Bileam. Bottom: early medieval draw-plate of the Viking Age found in Norway.

the use of a four row draw plate to produce wire, and Fig. 18.26 (top) shows an etching by the Master of Bileam in which St Eligius, working in a goldsmith's shop, is pulling wire through a die manually. The force required is approximately 70% of the product of the material's yield stress and the cross-sectional area of the hole. Thus direct pulling is satisfactory for soft materials and small diameters but for harder materials and larger diameters some form of mechanical advantage is required. Rudolf of Nuremberg, in about 1350, is credited with being the first to develop a method of wire drawing with the aid of waterpower. Most materials are not as ductile as gold and cannot be drawn out very far before they break. Thus if the material is not very ductile it must be annealed frequently, by heating it to a temperature above the re-crystallization point, which is about half the melting point of the metal.

Analysis

Smelting

Smelting is the extraction of metals from their ore, usually by heating to a high temperature, with a carboniferous fuel such as charcoal or coke. Typically the metal melts and flows downwards under its own weight to form a pool of liquid metal at the base of the furnace. A hole at the base is then opened and the liquid metal flows into suitable ladles and moulds where it cools to the desired shape. Earthy material not removed at the mine must be removed in the smelting and this is usually done by means of limestone which reacts with the earthy materials to form a molten slag that is lighter than the metal and thus floats on top of the molten metal. Although relatively simple processes suffice for most common metals, iron is very different and the processes used in and before the Middle Ages were quite unlike those used and understood by modern engineers. The reason for this is that pure iron melts at about 1530°C, much higher than most common metals, and the earliest furnaces were unable to reach this temperature, thus iron and steel had to be produced without melting in what is called the bloomery process.

Wood was the first and most obvious fuel for a smelting furnace but it was abandoned at quite an early period and replaced by charcoal. When charcoal burns in air it can reach a temperature of about 2600°C if just sufficient air is supplied for complete combustion, that is, to convert the carbon into carbon dioxide. This occurs when 11.1 kg of air is supplied per kg of charcoal and is called the stoichiometric ratio. However, the combustion products have to heat up the other contents of the furnace, and the air supply is usually much less than the stoichiometric ratio, consequently temperatures in simple furnaces rarely exceed about 1200°C. When wood became scarce and coal came to be used instead the impurities in raw coal (particularly the small amount of sulphur) produced a very poor quality iron. To remedy this the procedure adopted for wood, namely removal of the volatile constituents, was espoused and coal was converted into coke by heating it in a restricted air supply to about 1000°C. This produced a light, porous material that was nearly pure carbon and similar to charcoal. Iron made with such a fuel proved to be satisfactory and its discovery was one the important factors in eighteenth century industrial evolution.

Copper Smelting

Copper pyrites, found in Cornwall, is 35% copper in the pure state, but because of impurities the average yield was not more than 12%. The main impurities are quartz, iron, sulphur, and arsenic. The iron and quartz is separated as a fusible slag and the sulphur and arsenic is dissipated by conversion into sulphurous and arsenious acids, through oxidation in a furnace. At Swansea, the seat of copper smelting in Britain, reverberatory furnaces (calciners and smelters) are used. These are furnaces in which the fuel is burnt in a separate hearth and the hot gases are passed to an adjacent chamber to heat the ore. The more usual furnace arrangement is for the ore and fuel to be mixed in the same chamber.

Oxide and carbonate ores are simply roasted (heated in air) to form copper oxide and either carbon dioxide or water:

$$CuCO_3 = CuO + CO_2 \text{ or } Cu(OH)_2 = CuO + H_2O$$

The copper oxide is then roasted with coal to form copper and carbon dioxide:

$$2CuO + C = 2Cu + CO_2$$

Sulphide ores are more difficult to reduce. These are first roasted in a reverberatory furnace with the added silica, when the iron in the ore is oxidised and forms a slag of ferrous silicate with added silica (SiO2), whilst the copper remains in the combination with the sulphur as cuprous sulphide Cu2S, which fuses and sinks below the slag:

$$2CuFeS_2 + 4O_2 + SiO_2 = Cu_2S + 2FeSiO_3$$

The cuprous sulphide is separated and again heated in air. Part of the cuprous sulphide is converted into oxide, and this reacts by "self-reduction" with the remaining sulphide to form crude metallic copper:

$$2Cu_2S + 3O_2 = 2Cu_2O + 2SO_2$$
$$2Cu_2O + CuS = 6Cu + SO_2$$

Copper is an easily wrought metal and useful for many purposes. Alloyed with zinc it forms brass, which was extensively used for ornaments and alloyed with tin it formed bronze, which was a harder and more useful metal.

Tin Smelting

Cornwall, in the nineteenth century, produced about half the world's tin and the remainder was from southern Asia, Bolivia, and Peru. The Romans knew Cornish tin and, as a major element in bronze, it was an important metal. The early Egyptians and Babylonians must have known it, since they too used it to make bronze, but where they obtained their supplies remains a mystery. Tin ore (cassiterite) occurs as the dioxide, SnO_2. This is stamped to a powder and repeated washing separates the earthy impurities (gangue). Metallic impurities of similar density (copper, arsenic and tungsten) cannot easily be separated by washing and are removed by roasting and washing. Roasting is done in a reverberatory furnace, where sulphur and arsenic are evaporated, oxides of other foreign metals are formed, and copper sulphide is converted into sulphate. A second washing removes these oxides mechanically and dissolves the copper sulphate. The tin ore is then smelted at high temperature in a reverberatory furnace. One tonne of charge is mixed with about 170 kg small coals and perhaps with a little slaked lime to serve as flux for the siliceous impurities. After 6 to 8 hours the tin is reduced and is run into cast-iron pans from which it is ladled into ingot moulds. A liquation process may be used to refine it further. Besides being a major constituent of bronze, tin came to be used for coating other metals such as iron or copper vessels (tin ware) for domestic use. It was also used for silvering looking glasses and is a major constituent of pewter.

Iron Smelting by the Bloomery Process

According to Chambers (1842, 745) the normal charge for blast furnaces in the early nineteenth century was 2.75 tons of roasted ore per ton of iron, 1.25 tons of limestone per ton of iron, and 7.85 tons of coal per ton of iron when using cold air blast (that is, before about 1830). This amount of coal requires 87 tons (78,000 m^3) of air for complete combustion and needs, therefore, large blowing engines driven by steam engines or by water wheels. At earlier times much smaller furnaces had to be used because the means to blow the necessary air was limited. To make a 15 kg bloom (a 12 cm cube) of iron requires about 41 kg of ore, 19 kg of limestone, 118 kg of coke and 1310 kg (1170 m^3) of air for complete combustion. This is equivalent to about 2 m^3 of air per minute for 10 hours. The power required to blow air is about 70 W for a flow rate of 1 m^3/min therefore the requirement is about 140 W. Thus complete combustion would require the efforts of about 2 or more men operating bellows continuously for 10 hours. This illustrates the problem of supplying sufficient air to a furnace, even for a small bloom of 15 kg. One or two men operating bellows supplied a very small proportion or what was needed. However, a small chimney built over the furnace would normally supply significantly more air. If less air than stoichiometric is supplied then combustion is incomplete and the temperature is lower. This, in fact, proves to be advantageous because incomplete combustion of charcoal produces carbon monoxide, CO, which is a reducing gas, that is, it combines readily with oxygen. As iron ore is largely iron oxide the carbon monoxide easily reduces the ore to produce almost pure iron without melting the iron. This can occur at temperatures as low as 800°C but at higher temperatures, about 1200°C, the slag formed from the gangue melts and is more easily separated from the metal.

The temperature distribution measured by Tylecote et al. (1971, 349) in a small shaft furnace, 30 cm diameter and 1 m high, is shown in Fig. 18.27. The charge was hardwood charcoal of 6–25 mm size and a mechanically driven pump delivered air at rates of 0.2 m^3/min and 0.3 m^3/min. A single 2.5 cm diameter tuyere was placed 20 cm above the bottom of the furnace and about 3 cm from the front wall. At 0.2 m^3/min the temperature reached about 1100°C at the tuyere level but 15 cm above this it was only 700°C. The hot area near the wall at the top was caused by combustion of some

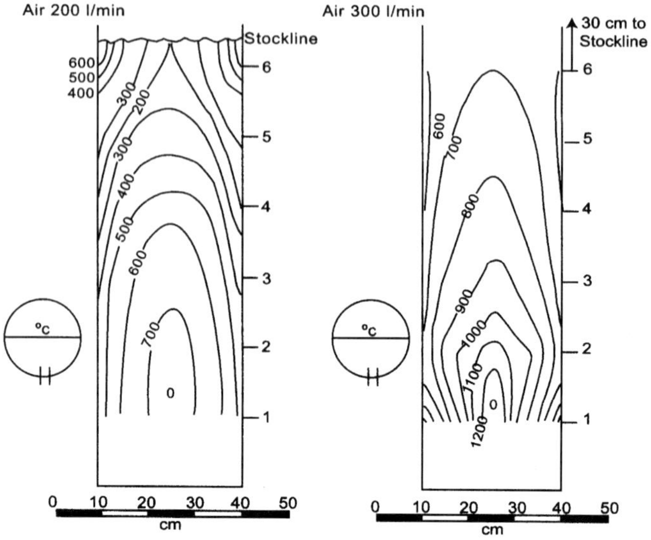

Fig. 18.27 Left: Temperature distribution in a charcoal bed with an air flow rate of 0.2 m^3/min. Right: temperature distribution in a charcoal bed with an air flow rate of 0.3 m^3/min. After Tylecote et al. (1971, 349).

of the CO by air leaking down from the top surface. The carbon monoxide burnt on leaving the top surface as it mixed with the air. At 0.3 m^3/min the temperatures were much higher and reached over 1200°C at the tuyere and falling to 700°C about 70 cm above the tuyere. The pressure drop through the charcoal bed was quite small, about 3 Pa for a flow rate of 0.23 m^3/min. This may be determined with good accuracy from Ergun's equation, see for example, Bird et al (1960, 200), but the pressure drop is very small and may be ignored. The natural draft caused by a 1 m high shaft acting as a chimney was found to be almost, but not quite, sufficient to generate the necessary airflow. At 0.1 m^3/min the CO_2 in the furnace was found to vary from 1.0% at the bottom of the furnace rising to 2.7% at the top and CO varied from 30% to 26.2%.

The iron ore used in this furnace was mainly FeO or Fe_2O_3 so the hot CO gas generated by the charcoal reacted with the ore, stripping it of oxygen to form gaseous CO_2 and leaving a solid lump of nearly pure iron in the solid state. Tylecote et al found that a successful smelting required air flowing at 0.3 m^3/min and ore crushed to a fine size. The process is illustrated in Fig. 18.28. After the fur-

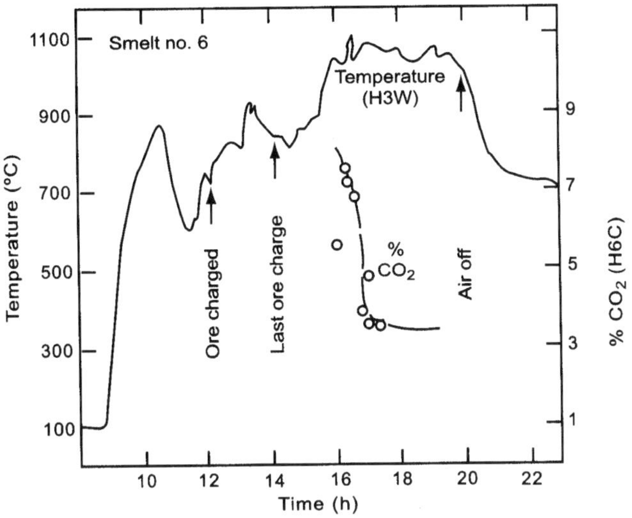

Fig. 18.28 Operating condition during a successful smelt. After Tylecote et al. (1971, 350).

nace had been warmed by burning charcoal for about 2.5 hours with a flow rate of 0.3 m^3/min the top gas could be ignited. Charging with cold ore started after 3 hours and a bed depth of 1 m was maintained by adding a 2:1 charcoal to ore ratio. The temperature in the centre reached 1070°C and 845°C by the wall at the time when the last ore was added. The mass of ore used was 10.5 kg and a further 11 kg of charcoal was added after the last charge of ore. The contents were extracted 13 hours later in a single lump weighing 5.8 kg and consisting of 2.05 kg of slag attached to 3.65 kg of porous iron (35% of the ore) containing very little slag. The carbon content varied from 0.1% to 0.8%. The iron bloom (steel in this particular case) was found immediately below the tuyeres.

Thus steel or nearly pure iron had been formed without melting and the furnace temperature was always well below that required to melt pure iron (1530°C). As the bloom cannot be melted and cast like other metals, at least in simple furnaces, it must be forged to the required shape. This forging is also necessary to remove the bloom's porosity and expel any entrained gangue. The earthy material in the ore was mainly silica and during the smelt it reacted with some of the iron oxide to produce a silicate (slag) that melts at about

1200°C and floats free of the iron if the temperature is sufficient. This is a very convenient way of removing the gangue from the iron but it is wasteful for the slag may contain over 50% iron. The ferro silicate, however, protects the iron from carburisation and helps to remove harmful elements such as phosphorus. Eventually it was discovered that adding limestone to the charge removed the gangue without using the iron and this, of course, increased the yield of the ore. The lime, CaO, unites with the silica, SiO_2 and alumina, Al_2O_3, of the ore to form a fusible slag of calcium silicate and aluminate that melts and separates from the iron because it is lighter. This is the bloomery process that was used to produce iron, in Europe and elsewhere, from pre-historical times until about 1500 A.D.

Lead Smelting

The ore (galena) is generally distributed but the largest supplies were in Britain and Spain, the British ores being obtained from Cornwall, Derbyshire, Shropshire, Yorkshire and Wales with a little from Scotland and Ireland. Galena is 87% lead, 13% sulphur when pure and it is relatively easy to reduce to pure lead. Lead was known in ancient Babylon and Egypt and is mentioned in the Bible (Job XIX, 24). It was often confused with tin but Pliny recognised the difference and called tin 'white lead'. The Romans carried out lead extraction in England and smelting in Derbyshire was active in the medieval period

In the nineteenth century the fine galena from the stamps was smelted in a reverberatory furnace, sometimes called a Flintshire furnace. This was generally about 2.5 m long, 2 m wide and 0.6 m high at the centre. Fire was placed at one extremity and separated from the smelting part by a wall that was built about half the height of the furnace. The hearth slopes from the wall that separates it from the fire to the other end of the furnace, and was hollowed from the sides to the centre. This was enclosed by an arched roof fitted with a small aperture for admitting ore from a hopper above. About one tonne of ore was put into the furnace and spread over the hearth with a rake. For the first two hours a gentle heat was generated by putting small coals on the furnace, the doors of which were kept shut. This was called the roasting process and was to dispel all sulphurous vapours from the ore.

$$2PbS + 3O_2 = 2PbO + 2SO_2$$
$$PbS + 2O_2 = PbSO_4$$

The smelter stirs the ore at intervals and a shovel full of quicklime was thrown in to liberate the oxide of lead from the ore and allow it to react with any sulphur that may have resisted the roasting. The temperature was then raised and the remaining lead sulphide reacts with the two oxide products to form the metal:

$$PbS + 2PbO = 3Pb + SO_2$$
$$PbS + PbSO_4 = 2Pb + 2SO_2$$

In about 4 hours the furnace was at its greatest heat, after which a tap hole was opened and the lead ran into an outer basin from which it was cast into semi-cylindrical moulds (bars). Lead sheet was cast on large tables, and smoothed and thinned by rollers. Lead was also much used to produce pipes and lead shot was formed by dropping molten lead in a shower from a great height into water. The addition of 1% arsenic helped the lead to form spheres. Such towers were conspicuous in London in the early nineteenth century

The Blast Furnace

The main zones of a blast furnace are illustrated in Fig. 18.29. The charge is introduced at the top of the furnace and consists of ore, charcoal, and limestone. In the eighteenth and nineteenth centuries the ratios used at the Calder Iron Works were 2.75 tons of roasted ore per ton of iron, 1.25 tons of limestone per ton of iron, and 2.05 tons of coal per ton of iron when using a hot air blast, invented by Neilson at the Clyde Ironworks in 1829, Chambers (1842, 745). Before Neilson's invention, the ratio was 7.85 tons of coal per ton of iron and this illustrates the great saving in fuel introduced by the hot air blast. Air, whether pre-heated by the waste gases from the furnace or unheated, is blown in at the bottom of the furnace at the tuyeres. Combustion occurs in the hearth immediately above the tuyeres where the temperature reaches about 1400°C. The reaction is:

$$2C + O_2 = 2CO$$

The gaseous combustion products consist mainly of carbon monoxide and nitrogen, which is mostly inert, and at this high temperature some of the carbon monoxide dissociates into carbon dioxide

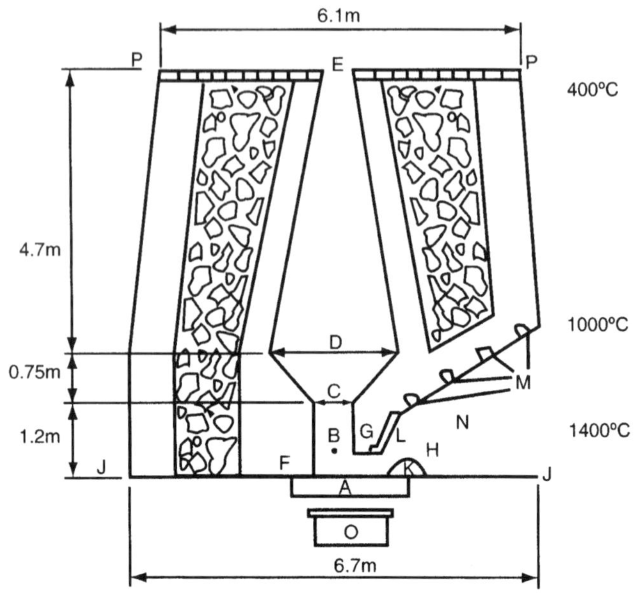

A = large hearth stone B = tuyere hole C = top of hearth D = top of boshes
E = top of shaft F = back wall G = timp stone H = dam plate
JJ = ground level K = dam L = timp plate M = iron lintels
N = working aperture O = ground moisture chamber PP = tiled platform.

Fig. 18.29 The main zones of a blast furnace, *c.* 1620. Schubert (1957, 202).

and very fine solid carbon particles, similar to those forming in a candle flame.

$$2CO = C + CO_2$$

From here the gas ascends to the top of the furnace, gradually reducing in temperature as indicated in Fig. 18.29 and increasing the temperature of the descending charge. The top part of the furnace is shaped so that the descending charge is less likely to adhere to the walls. In this part the ore is dried of its water content and is reduced, that is, the oxygen in the iron oxide ore reacts with the carbon monoxide of the ascending gas to give carbon dioxide, leaving a solid of almost pure iron and, of course, the earthy ingredients of the ore. The reaction is:

$$FeO_3 + 2CO = 2Fe + 3CO_2$$

Also in this upper region the limestone added to the charge is decomposed into quicklime and carbon dioxide.

$$CaCO_3 = CaO + CO_2$$

At a lower level the lime, CaO, unites with the earthy constituents of the ore (silica, SiO_2, and alumina, Al_2O_3) to form a fusible slag of calcium silicate and aluminate. When this melts it trickles down to the hearth where it floats on the surface of the molten iron.

The mass of gas leaving the top of the furnace reduces the volume of the charge near the boshes and allows the higher material to descend steadily under its own weight. The lower part of the furnace, called the boshes, converges because in this part the slag and the iron begin to melt and liquid droplets fall under their own weight and are guided by the taper into the hearth at the very bottom. The molten iron collects in the hearth and the lighter slag floats on its surface, but below the level of the tuyeres. The iron, in the region below the boshes, absorbs the very fine carbon particles from the combustion gases and the molten iron gathering in the hearth normally contains about 3–5% carbon.

To observe the smelting process the iron-founder looks into the furnace through a small opening between the nozzles of the bellows, called the 'founder's eye'. Through this he was able to judge the process by the colour and appearance of the iron droplets trickling down from the ore and by the appearance of the slag floating on the molten iron. Dark coloured slag and iron indicated a low temperature and thus the fuel supply was too small. In this case the iron was insufficiently reduced and much remained in the slag and the iron that was reduced was of a white colour, intermixed with particles of slag, and liable to great loss in forging. This form of iron is called white cast iron and the carbon is in the form of cementite, which makes it hard and brittle. If the iron droplets and the slag were of a bright colour then the reduction was complete and the iron produced was darker and less intermixed with slag. It is called grey cast iron because the carbon is in the form of free graphite and results in a cast iron that is softer, more tenacious, less fusible and less liable to loss when it was forged.

When the molten iron collecting in the hearth reaches the upper edge of the dam stone the blast is turned off and the clay sealing the tapping hole is pierced to run off the molten metal into moulds. Normally this occurred once or twice each day and continued throughout the period of blowing, termed the campaign, without being interrupted either on Sundays or festival days. The full capacity, in the sixteenth century, might be about one tonne at one tapping but this was not reached until about eight weeks after the campaign started. The campaign ended because the hearth and the lower parts of the furnace crumbled and burnt-out and the yield of iron decreased markedly. Initially three or four campaigns per year were necessary to keep the furnace in good repair but this re-building cost money and production was lost, consequently there was commercial pressure to increase the length of a campaign. Improvements in the design and construction of blast furnaces eventually decreased this to one per year.

The tapped iron ran into a bed of sand that was moistened with a little water to make it cohere and yet retain sufficient porosity to permit the free escape of gases generated by the heat of the molten iron that evaporated the water and organic matter in the sand. To control the flow rate from the furnace the founder used an iron hook called a 'stopping hook'. When the iron had been cast, sand was thrown over its surface to reduce the heat lost by radiation and to prevent the iron at the surface from being chilled and thus turned into white cast iron instead of grey. Cast iron as it comes from the blast furnace is full of impurities, such as: silicon, sulphur, phosphorus (the last two being harmful impurities), fragments of charcoal, and earthy matter, which requires to be removed by a refining process. There are three varieties of cast iron. Grey cast iron is of a dark colour, open in the grain and melts at a lower temperature. White cast iron is excessively brittle and hard and melts at a higher temperature. An intermediate grade is known as mottled iron. Schubert (1957, 244) says that the output attained at the end of the sixteenth century was one ton in 24 hours and was similar to that attained on the continent. By 1650 an output of two or three tons per day was reached, although in the Weald it never exceeded about 1.25 tons per day.

Occasionally, an un-smelted residue called a 'bear' collected in the hearth. It was caused, for example, by pieces of stone detaching from the hearth as it became worn out, or by un-melted slag. Another cause, known as ore 'tumbling', was when a quantity of ore adhered, 'getting together', as the ironmaster at Heathfield worded it, "which will not readily melt, or perhaps not at all, and then falls down in a lump". It was a serious matter for if a bear was too large to extract through the tapping hole then the furnace had to be blown out. After each casting the blast was stopped and the hearth was cleared of slag adhering to the walls using long iron hooks. When the last casting had been finished and the furnace was blown out, the interior required a complete overhaul and was a constantly recurring expense.

The Forge

Instead of obtaining malleable iron directly, as in the medieval bloomery, the blast furnace first produced a highly carburised cast of pig iron (2–5% carbon). Strictly it was a new metal, an alloy of iron and carbon that was too brittle for the forge and had to be decarburised by re-melting and oxidation in separate furnaces known as finery and chafery furnaces. The operation in the finery had three stages: melting down the pig iron; refining; and working the refined iron into a bloom. The process is described, for example, in Schubert (1957, 272–291).

The bottom and sides of the furnace were lined with charcoal dust and then the fireplace was filled with charcoal and the pig iron positioned for smelting. One end was covered with charcoal and placed slanting downwards towards the tuyere. The fire was lighted, the blast started, and the pig gradually pushed forward over rollers and exposed to the strongly oxidising blast. This melted the cast iron surface at about 1200°C and the oxidising blast removed the carbon. Removal of the carbon raises the melting point towards about 1500°C and thus the decarburised iron became pasty and could be collected by stirring and working with long iron bars known as ringers. The semi-solid metal on the end of the ringers was broken up with an iron crossbar and those parts not sufficiently decarburised were exposed to the oxidising blast and the process repeated until all parts were sufficiently refined. Finally, the metal was again melted near the tuyere and formed into pasty lumps, which were left in the bath of molten slag at the bottom of the furnace before being gathered and kneaded into a bloom of malleable iron. The whole process took about an hour and produced about 10 or 15 kg of malleable iron.

The iron bloom was taken out of the furnace and beaten with heavy hammers to remove the surface crust and slag and was then put under a heavy water powered hammer where it was consolidated into a thick square shape. The bloom was returned to the finery where many of the impurities were 'sweated' out and on returning to the hammer it was elongated further. The hardest and most carbonaceous of the particles left in the iron were removed in a chafery furnace. This was a separate furnace operated at a higher temperature generated by a stronger blast. In it the bloom was heated and consolidated under a power hammer and forged to the final bar shape.

Performance of Bellows

If air is treated as incompressible then the power required to operate a bellows is:

$$W = Q(p_d - p_s) \qquad 18.1$$

Q is the volume flow rate, m^3/s, p_d is the pressure during the delivery stroke, and p_s is the pressure during the suction stroke. From Bernoulli's equation the pressure difference between the bellows and the atmosphere is related to air velocity in the delivery valve, v_d, and the air velocity in the suction valve, v_s by:

$$p_d - p_a = \rho\frac{v_d^2}{2}$$
$$p_a - p_s = \rho\frac{v_s^2}{2} \qquad 18.2$$

Hence the power to drive the bellows is

$$W = \frac{Q\rho}{2}\left(v_d^2 + v_s^2\right) \qquad 18.3$$

The velocity in the suction and delivery valves is related to the cross-sectional area of these valves by the continuity equation

$$Q = \frac{C_d A_s v_s}{2} = \frac{C_d A_d v_d}{2} \qquad 18.4$$

C_d is the discharge coefficient. This equation assumes that the suction valve is open for half the time and the delivery valve for the other half. Thus the power required is also expressed as

$$W = \frac{2\rho Q^3}{C_d^2}\left(\frac{1}{A_d^2} + \frac{1}{A_s^2}\right) \qquad 18.5$$

Example
A bellows has a delivery tube of 3 cm diameter and a suction valve of 9 cm diameter. If a man uses 60 W in driving the bellows determine the theoretical flow rate of air assuming atmospheric air density is 1.2 m^3/s. The discharge coefficient is 0.65.

$A_d = 3.142/4*.03^2 = 0.000707\ m^2$ $A_s = 3.142/4*0.09^2 = 0.00636\ m^2$
From Equ. 18.5 $Q = 0.0173\ m^3/s = 1.04\ m^3/min$.
From 18.4 $v_d = 75.48 m/s$ and $v_s = 8.39\ m/s$.

CHAPTER NINETEEN

RENAISSANCE AND EARLY MODERN METAL WORKING

Introduction

Many of the metal forming process discussed in the previous chapter were continued and developed in the Renaissance and early modern period. These include the blast furnace, the forge, and the wire drawing mill. At the beginning of this period England lagged behind continental Europe and foreign workers were frequently invited to teach their methods. Iron casting was boosted by the need for ever heavier and stronger cannon but was hampered in England by a growing shortage of wood for making charcoal. This prompted many attempts to use coal rather than charcoal to fuel the furnaces and eventually the success of Dud Dudley, Sir Clement Clerke, and Abraham Darby, in using coke rather than charcoal, reduced the cost of iron and made it readily available. The need for coke was further and dramatically reduced by the introduction of the hot air blast. Steel production in England was always difficult because local ores were unsuitable, so throughout most of the period England relied on imports either of steel or of appropriate ores. Nevertheless, the development of fuel-efficient methods of production made steel cheap and readily obtainable. Metals continued to be cast and forged to their final shape and metal cutting operations were avoided whenever possible, consequently machine tools received relatively little attention and the only machines regularly used before the nineteenth century were lathes, screw-cutting machines, and boring machines.

Wire Drawing, Battery, Slitting and Rolling

Biringuccio in 1540 described various methods of drawing wire of gold, silver, iron, copper and brass and said that wire was drawn of every metal except tin and lead and could be made in any thinness and length; Biringuccio (1540, 378). In Fig. 19.1 (top) Biringuccio

Fig. 19.1 Top: wire drawing with capstan, windlass, and drums. Bottom: drawing heavy wire by means of waterpower. Biringuccio (1540, 379–80).

illustrates three draw benches suitable for most metals except iron. He says the metal bar must be reduced under the hammer as round and long as possible and then it must be annealed and taken to the draw bench. The drawplates are about 15 cm long with several rows of holes of successive sizes and firmly fixed to a block of wood. The tongs must have flat, serrated mouths and open legs; they are attached to a stirrup-shaped iron ring on a belt or rope wrapped around a windlass. The tongs close and grip the wire automatically when pulled and release when the tension is relaxed. The windlass may be horizontal or vertical. The wire must be well lubricated with new wax placed by the craftsman into one of the holes of the drawplate. For small diameter wire such large devices were not convenient and the wire was wound directly from drum to drum through the drawplate. Biringuccio also describes how gold is hammered and soldered over copper bars before being drawn and comments that "if it is not done for fraud, this process is a beautiful thing—The wire may be drawn out so fine that the eye can scarcely perceive it, yet it is always very well gilded all over".

For drawing iron wire a stronger water-powered machine was required, Fig. 19.1 (bottom). The waterwheel rotates a crank in two bearings and a short distance away a drawplate is fixed in a heavy wooden frame. Between them a pit is made to the depth of a man's knee in which the craftsman sits on a suspended seat. No great effort is demanded of him and "His only care is to seize with the tongs the end of the wire that issues from the drawplate with every return that he makes". The stirrup on the tongs ensures that the grip on the wire is proportional to the drawing force and the man is only required to release the tongs and reposition them at the end of each stroke. The suspended seat makes it easier for him to push himself backwards or forwards relative to the wire.

The first English wireworks operated by waterpower were built in 1566–7 near to Tintern Abbey, Monmouthshire, although, of course, the monasteries had been dissolved by this date. William Humfrey, Assay master at the Royal Mint, was the prime mover and later it was taken over by a company called the "Society of the Mineral and Battery Works". Schubert (1957, 292) says that the wireworks was about 15 m long and 9 m wide and was equipped with four waterwheels, two furnaces for annealing, and four hammers for initial shaping of the wire rods. It was built in response to concerns to make England more self-supporting in manufactured goods for, as the unknown author of a report stated, wire was a product which "God hath ordained in other countries" but not in England. Christopher Schütz, a follower of Agricola, and many of his workers were persuaded to come to England. They were skilled in the production of brass, the manufacture of which was new to England, and this was their initial objective at Tintern. Shortage of equipment delayed the work and they were directed instead to set up a wire drawing workshop on the continental style. They had not the experience to do this but nevertheless it was attempted. However, it did not succeed until after another foreign worker, Barnes Keyser, who was a skilled wire-drawer, was obtained and had re-trained the workforce. The local iron ore proved inappropriate for wire drawing and success was not achieved until Osmond ore and an experienced smith were obtained from Westphalia. Osmond iron was made from a particular ore from Sweden and was famous throughout the Middle

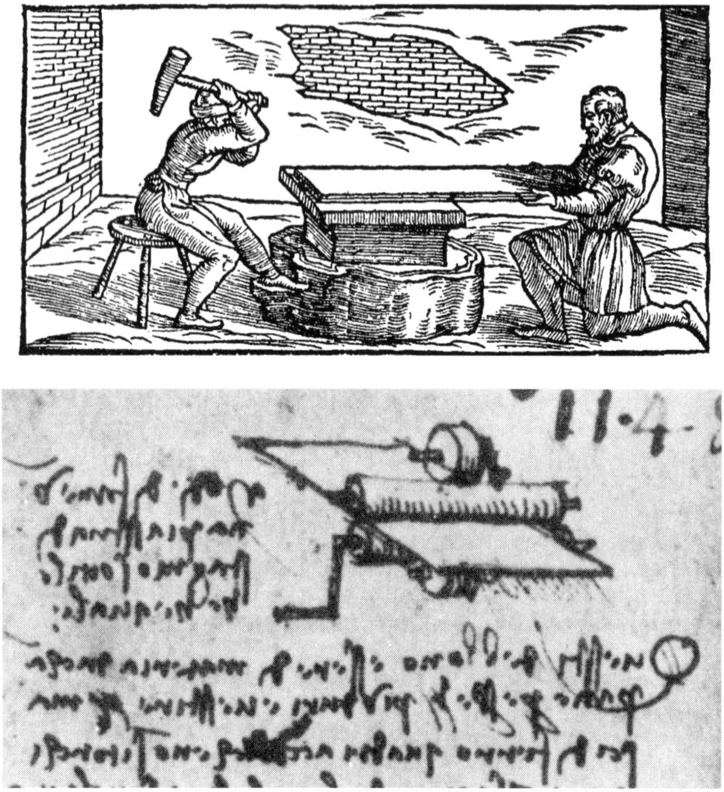

Fig. 19.2 Top: goldbeaters working on a duplex plate of gold and silver. Bottom: design for a rolling mill by Leonardo da Vinci (1452–1519). Biringuccio (1540, 382).

Ages for its great ductility and strength, qualities that made it ideal for wire drawing. The Tintern wire works proved, eventually, to be very profitable. By 1597 over 5,000 people were employed on the manufacture of wire goods in England and in 1607–08 a second wire works was opened at Whitebrook, Monmouthshire.

In addition to wire, iron and other metals were sold on to other smiths, remote from the centres of production, in the form of plate. Before the invention of rollers, metal sheets had to be hammered out at the forge, as illustrated, for example, by Biringuccio, Fig. 19.2 (top). The process was carried out in a battery mill. In this case a silver plate is being coated with gold and hammered out very thinly by using flat hammers. The resulting plate was so thin it could be cut with scissors and used to coat thread to produce spun gold. Battery by flat hammers was a tedious process but it continued in use until the seventeenth century. But even before the time of Biringuccio, Leonardo da Vinci had made the first designs for a rolling mill, Fig. 19.2 (bottom). The note on the drawing reads: "The art of making a thin and even tin-plate. These rollers should be made of bronze in order that they may be harder; and they are supplied with iron axles so that they cannot wear out. While one turns the other round, they stretch out the plate whose width is half a yard". It sounds as if he had seen or made such a machine. A hand crank turned the lower roller and he recognized that friction would draw the plate through the rollers and rotate the upper roller. He also recognized the danger of the rollers deflecting under the high loads when rolling very thin plate and to counter this he proposed a pair of back-up rollers, similar to modern rolling mills. Such rollers may have been used in rolling lead sheet for roofing or for making lead pipes. It is the earliest drawing of a rolling mill. A century later, in 1607, Zonca illustrated a small rolling and slitting mill for making lead strips for stained glass windows, see Keller (1964, 51). Despite this, battery with flat hammers continued.

Such manual methods were used to produce all kinds of plate and sheet. The first battery mill to produce plate in England was set up at Greenwich by Henry VIII in 1511 and it was solely for the production of chased and inlaid decorative armour, an example of which is shown in Fig. 23.10 of chapter twenty-three. This too was essentially a foreign art, particularly strong in Milan, Germany and Belgium. It was not until the erection of the wireworks at Tintern, and the formation of the Society of the Mineral and Battery Works, that the method of flattening iron into plates and sheets by means of water powered hammers was made generally available in England. Such battery or plating mills were usually small, having a single waterwheel driving a lighter hammer with a large flat face and operated at higher speeds. They became more common in the seventeenth century and were used to make "frying pans, dripping pans, salt pans, fire shovels, armour for back, breast, and head, and all sorts of iron plate", see Schubert (1957, 303). Battery mills continued in use even after the invention of rolling mills, which could produce plate faster and more accurately.

Forging by means of tilt hammers, discussed in the previous chapter, was supplemented by the drop hammer, Fig. 19.3. This is simply a

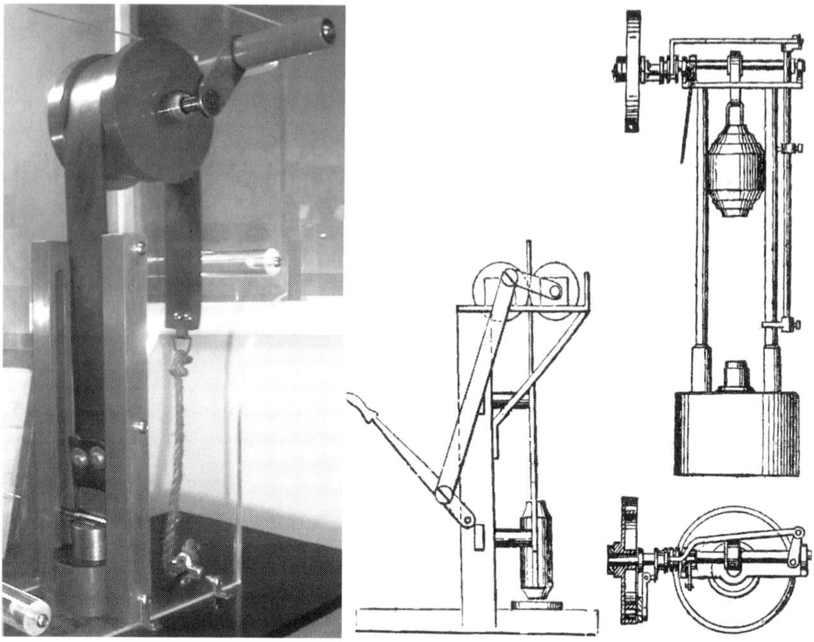

Fig. 19.3 Left: model of a simple drop hammer (Finch Foundry, Devon). Right: industrial drop hammers, Knight (1884, 754).

heavy weight (the hammer), raised by machinery, and allowed to fall on the hot metal to relieve the blacksmith of much of his labour. A model drop hammer, Fig. 19.3 (left) consists of an overhead shaft, turned clockwise by hand in the model but normally driven by a water mill. A heavy hammer on the anvil has a leather strap wrapped around the overhead shaft and the other end of the strap is allowed to dangle at a convenient height. By pulling lightly on the free-end it tightens around the rotating overhead shaft, which thus lifts the heavy hammer. The gain is typically about five, so a pull sufficient to raise 10 kg will raise about 50 kg. When the weight reaches the desired height it is released and falls on the heated metal. The industrial drop hammers illustrated in Fig. 19.3 are of a more sophisticated design. In Fig. 19.3 (centre) the weight is attached to a strap, which is drawn upwards between two pulleys that are brought together to compress the strap. The driving pulley is fast on its axle while the other is loose. A lever arrangement moves the axle of the loose pulley to compress the strap and thereby lift the hammer. In Fig. 19.3 (right) a strap, which winds onto the main shaft, raises the hammer when a loose pulley is engaged by means of a clutch. The clutch is operated by an arrangement of levers.

A process that was very common at one time but is now discontinued is slitting. Slitting mills were used to cut metal plates into a large number of thin, rectangular bars. Such bars were not easy to forge in long lengths of less than about 20 mm thickness because the long, thin, hot bars were too flexible and cooled too rapidly. There was, of course, a great demand from the smiths for such products in order to make nails and the numerous small items of everyday life. Today, rectangular, circular and hexagonal bars, and bars of other cross-sections, are supplied from the draw mill or the extrusion press but before these processes became universal, after about 1850, the combined slitting and rolling mill was used; see Jenkins (1918, 445–46 and 487–89). Such mills seem to have appeared first in the sixteenth century around Liège, Belgium and in Germany and the first license in England was granted to the Elizabethan mining engineer Bevis Bulmer in 1588. Jenkins quotes from *Stow's Annals* of 1631, which says "The cutting of iron bars in a mill for the ready use of smiths to make long rods and all sorts of nails, was first brought to England in the year 1590 by Godfrey Box of the Province of Liège: who set up the first mill for this purpose near Dartford, Kent." It seems that Bulmer obtained the license but used Box to set up the mill. Jenkins found that a Godfrey Box died at Dartford in 1604. The process was described in Plot's *Staffordshire* in 1686: "—those [plates] they intend to cut into rods, are carried to the slitting mills, where they first break or cut them cold with the force of one of the wheels into short lengths; then they are put into the furnace to be heated red hot to a good height, and then brought singly to the rollers, by which they are drawn even, and to a greater length: after this another workman takes them whilst hot and puts them through the cutters, which are of diverse sizes, and may be put on and off, according to pleasure: then another lays them straight also whilst hot, and when cold binds them into faggots, and then they are fitting for sale." Elsewhere, Plot's account suggests that slitting and rolling mills were fairly common, at least in Staffordshire.

The earliest drawings of slitting and rolling mills that Jenkins could find were by Swedenborg, in 1734, Emerson, in 1758, and Diderot

Fig. 19.4 Top: a slitting, rolling and shearing mill of 1758, Emerson's *Mechanics*. Bottom: a slitting and rolling mill, Diderot and d'Alembert (1751–72).

in 1751–65. The latter two are reproduced in Fig. 19.4. Emerson's drawing gives a better idea of the mill's design whereas Diderot's drawing gives a better idea of its operation. In Fig. 19.4 (top) the slitting rollers are marked *AB* and the plate rollers are *CD* and *SP* is the clipping mill. The water wheels, *E* and *F*, rotate in opposite directions because the stream flows towards the viewer on the right, turns through 180 degrees (not shown) and flows away from the viewer on the left. The wheel on the right drives the lower rollers of both mills and the wheel on the left drives the upper rollers, thus the top and bottom rollers rotate in opposite directions. A disadvantage of this arrangement was that the top and bottom rollers might rotate at different speeds. At a later date they were geared together so they rotated at the same speed. The perspective of the clipping shears is slightly askew because the plane *LPR* should be vertical. An iron bar *TQ* may be cut in two by the steel edges *SP*, *RP* which are operated by a cam *V* on the main drive shaft. Diderot's drawing, Fig. 19.4 (bottom) shows iron bar being removed from the heating furnace and passing through the slitting rollers. Again, two independent water wheels (not shown) are used; one to drive the upper rollers and one for the lower rollers. The bearings of the roller appear to be water-cooled. A clipping mill is not illustrated.

By 1785 there were 16 slitting and rolling mills in England and 8 in Ireland, many of which have been located. The Irish mills slit 1,700 to 2,000 tons a year in total whereas the English mills slit from 800 to 1,500 tons each, say 18,000 tons per year in total. Jenkins says that the Irish iron was obtained from Russia through London, and the cost of transport from London to Dublin was much higher than that from Russia to England, largely because it could be brought to London as ballast in ships that would otherwise return un-laden.

Iron bar was required not only in square section from the slitting mill but increasingly as circular section suitable for machining on a lathe. Medium and small diameters could be obtained from a wire drawing mill but for larger diameters a bar mill was often used. Fig. 19.5 shows a water powered bar mill at Newburn on Tyne. Each pair of working rolls has semi-circular grooves, reducing in diameter along the barrel, so that larger bar could be reduced gradually by passing it through successive holes until the desired size is reached.

Casting

Biringuccio in 1540 described the method of making cast iron in a blast furnace and Agricola copied his description some years later. Biringuccio (1540, 63–4) writes "First (supposing that you have found the ore and have mined a goodly amount and have roasted, sorted, well re-sorted, and washed it) you must have one or more furnaces of the type called large blast furnaces which can hold great quantities of charcoal and are constructed on the inside like the one I shall show you drawn here [Fig. 19.6]. Near these is a large pair of bellows, very close to the wall of the furnace, like a great pair of wings.

Fig. 19.5 A water powered bar mill of 1845, Newburn on Tyne.

Fig. 19.6 Blast furnace for refractory ores. Biringuccio (1540, 64).

These are usually 6 or 8 braccia [3.5 or 4.7 m] in height and are set in motion by a strong water wheel to which they are connected so that they have a large extension for inflation. With their powerful blast, which enters the furnace by means of a tube about two and a half braccia [1.5 m] from the bottom, the ore is smelted after the furnace has been filled with charcoal. Depending on the kind, some ores are smelted once and others twice before they are disposed to become iron good enough to give to the forge to be drawn out by the forge hammer. In spite of all this, it often happens that, however much care the masters have taken, it has not been possible to bring the metal to such a softness that it can be worked. This is due to the malignity of its companions, which melt with it and became inseparable from it". The malign component, of course, is carbon. The first smelting produces cast iron containing perhaps 3–4% carbon and is very brittle and the second smelting is needed to reduce the carbon content to make wrought iron, although Biringuccio is ambiguous on this point and does not indicate that the second melting is any different from the first. However, his drawing, Fig. 19.6, makes it clear that large and powerful bellows were required to heat the metal to its melting point and also illustrates the relatively small size of the furnace and the relatively small amount of iron that could be cast.

A large variety of items could be cast in iron from the blast furnace but, in fact, very little was produced because its main use was in the manufacture of bar iron (wrought iron) to supply the forge. Usually less that 5% of cast iron was used for casting, except at the furnaces that produced guns and shot. Biringuccio (1540, 235) describes the manufacture of suitable gun moulds at great length; see Fig. 19.7 (bottom). The mould consists of a main body *A* closed at one end by a fitted breech mould *B*. A core *C* is built up of clay on an iron rod and is held centrally in the mould by an iron support (a chaplet) *D* and by a clay disc *E*. The trunnions are made of wood and taper towards the barrel to ease their subsequent removal; they are nailed onto the barrel mould. Bricks *F* close the trunnion cavities. The whole mould is reinforced with an armature of heavy iron bands and iron wire and it is buried upright in a pit (breech down, muzzle up), for casting.

The pattern on which the clay mould is built up is made either of wood or clay, but in either case the cornices and other decorative and projecting parts are applied over a layer of wax so that the pattern can be driven lengthwise out of the mould, leaving these parts to be removed later by hand. The wooden trunnions are removed before the pattern is driven out of the mould. The ends of the pattern are extended to act as journals that rest and rotate on appropriate trestles, as shown in Fig. 19.7 (top). A layer of soft clay is applied as the pattern is rotated, and is shaped by a stickle board securely held on the trestles, rather like a lathe. Several layers are applied to build up a proper thickness and each is allowed to dry before the next is added. The first layers are allowed to dry naturally, to avoid disturbing the wax, but later layers are dried by some application of heat. The core is built up on an iron bar to give rigidity and straightness, and this is covered with layers of soft

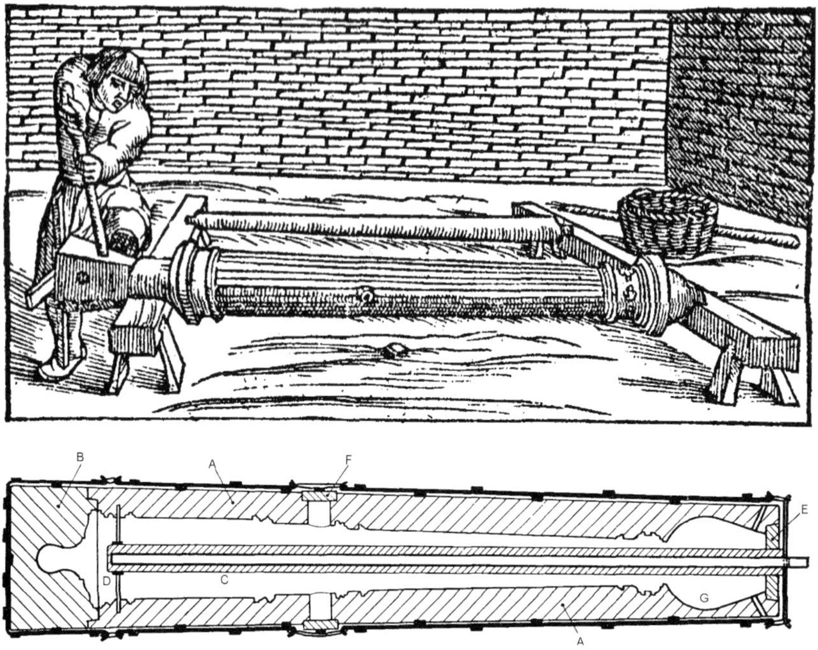

Fig. 19.7 Top: the pattern from which gun moulds are made. Biringuccio (1540, 235). Bottom: a completed gun mould, after Smith and Gnudi in Biringuccio (1540, 452).

clay, turned on trestles and baked in the same way as the main pattern. It is located in the centre of the main pattern by means of chaplets at *D* and the clay disc at *E*.

Despite a late start in the production of cast iron, England, by the sixteenth century had attained a great reputation in the manufacture of cast iron cannon. Schubert (1942, 137) has found evidence that Frenchmen employed in the Tower of London introduced iron casting into England at the end of the fifteenth century, and several ironworks were set up in Ashdown Forest, Sussex. At Newbridge cast iron guns with detachable chambers were made (mentioned in accounts in 1509–10) and these are the first cast-iron cannon made in England. A new impetus came when Henry VIII re-launched English expertise by inviting foreign gun founders to live in England and teach their methods. These included Peter Bawde, a Frenchman, Peter van Collen, probably from Cologne, and members of the Arcana family from Casena in Italy. Bawde, in 1528, was "Gunner in the Tower" and he initiated Ralph Hogge, an English ironworker, into the mysteries of gun making. Gunners at this time were not only soldiers but also had to make their own guns and shot. It seems that Bawde had the expertise in casting bronze guns and Hogge the expertise in making cast-iron and the result, in 1543, was the first successful casting in one piece of an iron gun, which was then bored out in the same way as existing bronze guns. Such guns, of cast iron, were very much cheaper than comparable weapons cast in bronze. The main technical advance was in the amount of metal that could be cast at one time. A large cannon weighed about 4 tonnes, although those first cast were demi-culverins weighing about 2 tonnes, and to supply this, double furnaces were used, that is, two furnaces in a single structure. The new guns were also improved by casting the barrel and the chamber in one piece; a process that must have been impossible until the double furnace was adopted. Ralphe Hogge's blast furnaces were at Buxted, in Ashdown Forest and the Weald was quickly established as the premier iron-working district.

The superiority of English naval cannon was one of the factors in the defeat of the Spanish Armada in 1588. By the 1580's cast iron cannon had displaced bronze cannon in northern Europe; the Swedes, Dutch, and Germans having bought cast iron guns and learned the technology from the English. Mediterranean countries, however, remained loyal, for a time, to bronze guns and rowed galleys. Cast iron was similar to bronze in terms of strength and inferior in terms of corrosion resistance, but its great advantage was that it cost about a third as much as bronze, Guilmartin (1974, 175). By 1574 the fear that Wealden guns may be exported to England's enemies prompted the government to enquire into the number of the nation's furnaces and forges, and to control their exports. This revealed that there were 107 ironworks of which 49 were furnaces and 58 were forges, although Schubert (1957, 174), has found references to others working at this time that are not on the list. Only seven furnaces were outside the Weald, namely four in Wales and three in the Midlands. John Norden, in his "Surveyor's Dialogue", 1607, says that there were near 140 hammers and furnaces for iron in Sussex, but as he was the crown surveyor of woods in several counties he was concerned at the rate they were being consumed for fuel and perhaps he exaggerated the number of hammers and furnaces. Schubert's figures are plotted in Fig. 19.8 (top) and show the dominant position attained by the Weald in the sixteenth century. Instead of importing iron it was now exported in the form of

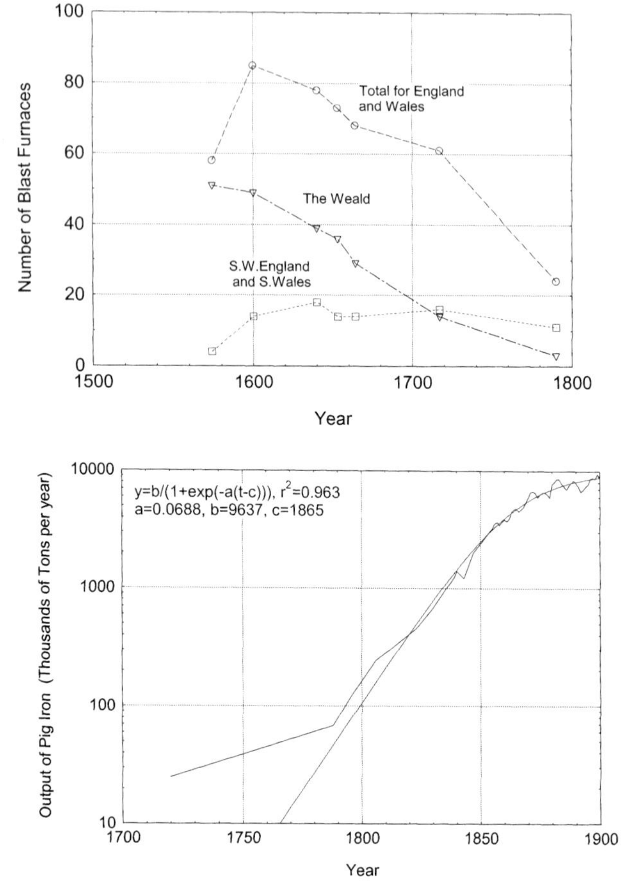

Fig. 19.8 Top: number of blast furnaces in England and Wales, 1574–1800. Schubert (1957, 175). Bottom: output of pig iron in Great Britain, 1720–1900. Mitchell and Deane (1962, 131–3).

cannon. For example, Sir Thomas Leighton and Sir Henry Neville obtained separate patents from Elizabeth I to send their ordnance abroad, and this prompted Sir Walter Raleigh to complain to the Commons "I am sure heretofore that one ship of Her Majesty's was able to beat ten Spaniards, but now, by reason of our own ordinance, we are hardly able to match one to one", Smiles (1882, 38). Proclamations against export were again issued but they did not stop the trade.

John Norden's concerns about diminishing woodlands were certainly valid. Smiles (1882, 40), states that at Lamberhurst 5 tons of iron was produced per week at the expense of 200,000 cords [720,000 m^3] of wood per year. Although this seems to be an exaggeration it works out to be about 2,770 m^3 of wood per ton of iron, and compares favourably with the figures used by Constantinou (1992, 70–71) who found that in Cyprus it required about 6,000 m^3 of pine to produce a ton of copper. Much of this wood was consumed in the manufacture of charcoal. Woods near to London were the main source of domestic fuel and their destruction caused great concern. To protect this fuel supply an Act was passed in 1581 prohibiting the use of woods within 14 miles [22 km] of the Thames for making iron and forbidding the erection of new ironworks within 22 miles [35 km] of London. Similar legislation by later Parliaments forced the Sussex iron masters to look elsewhere for their fuel supplies with the result that many of them migrated from the Weald to the Forest of Dean and other places. During the English Civil War Parliamentary forces destroyed many Royalist ironworks, which were never subsequently re-opened. The result was that iron production in the Weald had virtually ceased by 1800 and Wales, southwest England and the Midlands took over production. In 1674, after the Restoration, all the royal ironworks in the Forest of Dean were demolished to conserve timber for shipbuilding. This need to conserve woodlands was a general problem and the steady decline in the number of furnaces during the seventeenth century became more rapid in the eighteenth century so that only 25 furnaces were active by 1790. This is not the whole picture, however, because as Fig. 19.8 (bottom) indicates, the total production of pig iron increased during this period because the output per furnace increased.

The need to conserve woods for shipbuilding and for domestic fuel turned men's thoughts to the use of coal rather than charcoal to heat their furnaces. In 1612 Simon Sturtevant, a German miner, took out the first patent to smelt ores of all kinds, and iron ore in particular, using sea-coal, pit-coal, earth coal, and brush fuel. As was usual at this time the patent specified nothing (it was considered to be secret) and although he conducted many experiments he failed and his patent was cancelled. The following year it was granted to John Rovenzon but he too failed, having succeeded only in reinventing a reverberatory furnace, but blown by a windmill. The first successful use of coal, since the Roman period, was by Dud Dudley, 1599–1684, the illegitimate son of Lord Dudley. Smiles (1882, 47) says that there were at this time some 20,000 smiths and workers in iron living within 10 miles (16 km) of Dudley Castle and the town of Dudley was a great centre of iron manufacture. If this were so then the need to protect local woodlands would have been great. In 1619, at the age of 20, his father removed Dudley from his university studies to take charge of one of his furnaces at Pensnet, Worcestershire, which consisted of one furnace and two forges. Being restricted by want of charcoal Dudley experimented with pit-coal and had immediate success. He found his iron "to be good and profitable, but the quantity did not exceed three tons per week". After a second trial he wrote to his father and through him he obtained a patent for 14 years in 1621. It is not known how he succeeded with pit-coal when others had failed but it seems likely that he removed the volatile components by partial combustion to form coke and that he used coke to smelt his iron ore. Within a year the "Great May Day flood" destroyed his furnace, to the joy of his competitors, but it was rebuilt and produced iron goods cheaper than could be done using charcoal. After complaints from rivals the King ordered samples of Dudley's bar iron to be sent to the Tower to be tested independently, where it was found to be "good merchantable iron".

Being sued by the local ironmasters, Dudley moved to Himley in Staffordshire but he was without a heavy forge and was forced to sell his cast iron to local forge masters for conversion to bar iron. These forge masters, he claims, decried his iron and gave him a poor reputation, so he set up a new furnace and forge at Askew (Hasco) Bridge, Sedgely, Staffordshire. His bellows and furnace were larger than usual and he was able to make 7 tons of iron per week, then a British record, using coal from a newly opened pit having a seam 3 m deep and lying immediately over ironstone. All looked well until a mob of rioters, apparently incited by rivals, destroyed his bellows and other machinery and he seems to have been given no peace, being harried by lawsuits and eventually imprisoned for debt. His patent ran out in 1638, but in the face of great opposition and with three financial supporters, he obtained another patent. Before much could be done he was engaged in Scotland, on the King's behalf, and then along came the Civil War, during which Dud Dudley was a Royalist Colonel and was at many of the major battles. He was captured in 1648 and condemned but not beheaded and his estates sequestrated. During the Commonwealth Cromwell and Parliament did not support Dudley's work but granted the patent to a

Captain Buck, Major Wildman, and others. They erected coal-burning furnaces in the Forest of Dean but failed to produce good iron. Dudley was invited to inspect these workings but condemned them without revealing his own secrets and Cromwell's furnaces were eventually abandoned. After the Restoration, Charles II refused to extend Dudley's patent, despite his loyal services, and consequently he took no further interest in coal-fired blast furnaces. In 1665 Dudley published an account of his work in "Metallum Martis", but he did not reveal his methods, and whatever they were, they died with him in 1684.

In a recent paper, King (2001) has investigated the adoption of coal during the late seventeenth and early eighteenth centuries in smelting iron, copper and lead. In particular he has established firm links between Dudley and Sir Clement Clerke, who was subsequently very active in the introduction of coal fired reverberatory furnaces for smelting these metals. Clerke's earliest industrial enterprise was a furnace built by Dud Dudley at Dudley "for making iron or melting down ironstone with charcoal made of wood and pitcoal together to be blown and set on work by the strength of men and horses". Horsepower rather than water power was used to blow the furnaces. King believes that Dudley's methods did not die with him but were passed on to Sir Clement during this period of co-operation. Certainly, Clerke and his son were enthusiastic advocates of the use of coal rather than charcoal for smelting metals and although he did not invent the reverberatory furnace he was responsible for its successful application to smelting.

However, it was not until the first half of the eighteenth century that Abraham Darby at the Coalbrookdale Ironworks perfected the method of smelting iron ore in a blast furnace by using coke rather than charcoal as the fuel. But his methods were not taken up immediately. By the mid-century they had not spread outside Shropshire but by the end of the eighteenth century they were general throughout England. The remarkable (hundredfold) increase in pig iron production engendered by the needs of industrialisation and use of coke rather than charcoal as the main fuel, is clearly demonstrated after this date, see Fig. 19.8 (bottom).

In addition to cannon, other items were cast in iron, such as firedogs, grave-slabs, railings, gates, and a range of items for industrial and agricultural use. Cast iron rollers for crushing sugarcane were made in 1663, garden rollers in 1671, and rollers for slitting and rolling mills. Pots and pans for domestic use were cast at Buxted

Fig. 19.9 An English blast furnace and ironmaster illustrated in a cast iron fireback dated 1636.

as early as 1543. Formerly, fireplaces and chimneys were used only in grand houses, monasteries and castles but in the Elizabethan and Jacobean ages they became common in ordinary cottages and houses, and created a demand for cast iron firebacks to protect the brickwork of domestic fireplaces. The fireback shown in Fig. 19.9 is not particularly early, it is dated 1636, but it is interesting in that it shows the iron founder, Richard Lenard, in the centre and his furnace, with flames emerging from its top, at the bottom left. A wheelbarrow, various tools and products together with his initials and shield containing a hammer and pincers, are also depicted. A wheelbarrow is appropriate because a blast furnace required charging with large quantities of ore and charcoal many times per day. The fireback was cast at Brede furnace and Richard Lenard was also the tenant of the Sackville furnace at Little Udimore, Sussex. The demand for cast iron pots, kettles, and pans led to improved methods of casting,

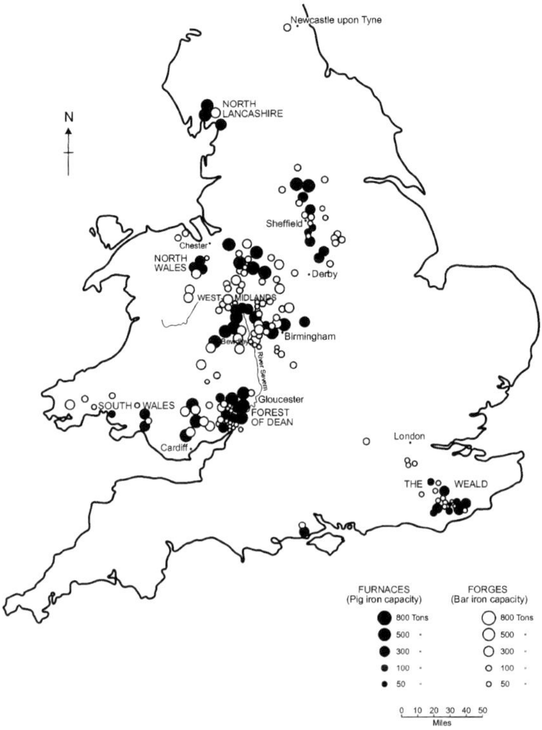

Fig. 19.10 Furnaces and forges in England and Wales, 1717. After Schubert (1957, 193).

and in particular to the method of casting in dry sand moulds instead of moulding with loam or clay. Abraham Darby is credited with this invention in 1707, together with the introduction of multi-parted moulding boxes. The dry sand, of course, did not generated steam when the hot iron was poured and consequently the mould was less likely to rupture.

The distribution of ironworks in England and Wales in 1717 is shown in Fig. 19.10 and is based on the list compiled by John Fuller, of Heathfield furnace, Sussex, and the forgemaster at Monmouth, William Rea. Their list gives a total of 176 ironworks consisting of 60 furnaces (55 in production), 114 forges, and some surviving bloomeries in North Lancashire, which were not counted. This industry was scattered and small-scale and the distribution was dictated by the need for large quantities of charcoal and ore and the impracticality of transporting either over any distance in the absence of good roads or canals. The distribution confirms the diminishing importance of the Weald and the dominant position of South Wales, the Forest of Dean, and the West Midlands. Most of the Wealden losses occurred, as discussed above, during and after the Civil War.

Cannon manufacture continued to be important and Crossley, (1975, 1–37) describes it at a manufacturing site he excavated at Pippingford Park, Sussex. Two blast furnaces produced iron here in about 1717 and their remains and the remains of a gun-casting pit and cannon-boring mill were found. The plan of the western furnace, Fig. 19.11 (top), shows the positions of the tailrace, bellows, hearth, and the casting pit for the guns. Fig. 19.11 (bottom) shows a cross-section through the casting pit, which is 4.5 m deep, lined at the base and the sides with oak timber and set on a base of puddled clay. The timber was arranged so that external pressure forced their edges together to ensure water-tightness. The felling date of one of the timbers was 1698±5. The space between the mould table and the double bottom was used as a sump to collect water seeping into the pit and a 5 cm diameter lead pipe ran vertically down one side to enable the sump to be pumped out. The remains of an unfinished, faulty cannon were found in the east furnace wall. It weighed about 280 kg. The un-machined casting, complete with its casting head, was discarded because it contained many inclusions and shrinkage voids, which were detected by X-ray photographs but in the eighteenth century were detected by the gun founder tapping the casting and listening for any deadness or ringing. The mould had been fitted with a wire wound core and was remarkable similar to that illustrated by Smith and Gnudi (1990, 452) and shown in Fig. 19.7. From Table 4 in chapter 23 it appears that it was intended to be one of the smallest cannon, namely a falcon of about 6.4 cm bore. Cannon were bored out after casting and the site of the cannon-boring mill near the east furnace is described below.

Casting hot metal, it should be remembered, was potentially dangerous and accidents frequently occurred. Hogg (1963, 35–6) records that in 1716 two 24 pounder brass cannon were being cast at Mr Bagley's foundry and were witnessed by a distinguished gathering including the Surveyor-General of the Ordnance, the Chief Firemaster of England, the Clerk of the Ordnance, and Mr George Harrison, afterwards superintendent of His Majesty's Foundries. Alas the moulds were still damp and "—the burning metal no sooner sunk down to

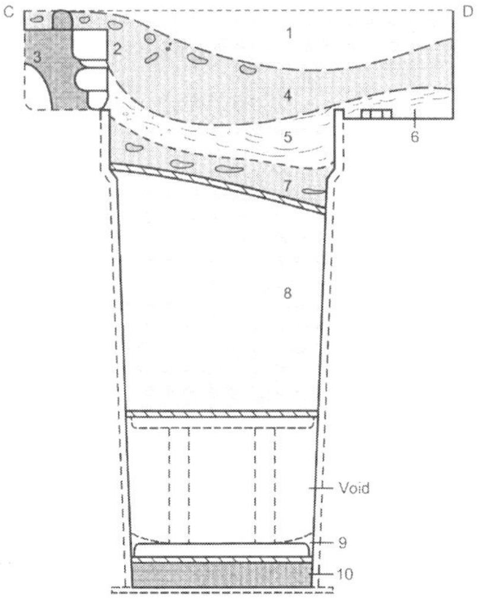

Fig. 19.11 Top: plan of Pippingford West Furnace. Bottom: cross-section of gun casting pit, 4.5 m deep. After Crossley (1975, 4, 5). 1 = black floodsilt, 2 = rubble and soil, 3 = clay and stones, 4 = course blackened sand, 5 = silt, woodchips and clay lumps, 6 = clay and plank fragments, 7 = silt and woodchips, 8 = sand filling casting pit, 9 = silt, 10 = puddle clay.

the bottom of the mould, but with a noise and force equal to that of gunpowder, it came pouring up again, blowing like the mouth of a volcano, or a little Vesuvius." Seventeen persons out of twenty five, including Mr Bagley, were killed and the remainder were badly injured and burned.

An important development of the early eighteenth century was the introduction of foundries, often remote from the centres of cast iron production, which took pig iron from the producers and re-melted it for casting into moulds. This was necessary because the blast had to be turned off when the blast furnace was being tapped and casting the iron into many small moulds rather than into large pigs took a long time with the risk that the furnace might cool too much and the process cease. Schubert (1957, 252–3) quotes a Major John Wade who was well aware of the problem in 1653: "in the casting of shot [for cannon] there is much delay in the works, for the furnaces must stand still for a season whilst metal is taken out with ladles and put into moulds and therefore, if it should continue still casting of shot, it would cool and at last blow out." If it was necessary to re-melt cast iron pigs in a separate furnace before casting then this could be done almost anywhere and thus foundries sprang up even in parts of the country, remote from the iron ore deposits. They commenced in 1702 with the introduction of an improved reverberatory furnace, attributed to a London physician and chemist, Dr Edward Wright. Initially it was developed for smelting lead but was eventually adopted in iron foundries with excellent results. In a report to the French Government in 1775 Marchant de la Houilière stated that: "during the twenty years which have elapsed since the English began to adopt this method of first smelting iron ore with coke in a blast furnace and then passing the cast iron so obtained through a reverberatory furnace before casting cannon, not one cannon so made had burst aboard their ships". He contrasts this with French manufacture, perhaps with some exaggeration, and says: "our sailors fear the guns they are serving more than those of the enemies".

Wrought Iron and Steel

Cast iron, which is brittle and lacks the ductility required for forging, and wrought iron, which is very ductile but lacks the strength and hardness of cast iron, were the two forms of iron that were readily produced by the medieval blast furnace and bloomery. There are numerous processes used to make cast iron more malleable or to make wrought iron harder and stronger. This is usually done by reducing the carbon content of cast iron or by increasing the carbon content of wrought iron. In either case, steel is produced.

Biringuccio (1540, 69) describes a process of making steel from wrought iron by breaking the iron into small pieces and immersing them in a bath of molten cast iron. The wrought iron does not melt because its melting point is higher than that of cast iron. However, the iron absorbs carbon from the molten cast iron and is transformed to steel. Biringuccio says: "Having previously made under the forge hammer three or four blooms weighing thirty to forty pounds [13 to 18 kg], each of the same iron, they put these while hot into this bath of molten [cast] iron.—They keep it in this melted material with a hot fire for four to six hours, often stirring it with a stick as cooks stir food.—taking out one of the masses and bringing it under the forge hammer to beat it out, and then, throwing it into the water while it is as hot as possible, the temper [quench] it; and when it has been tempered [quenched] they break it to see whether every part has changed its nature and is entirely free from every layer of iron." The pieces of steel are then withdrawn from the bath and hammered into bars.

Prince Rupert, one of Charles I's most able generals during the Civil War and a man of many talents, is said to have invented another method. In 1671 he was granted a patent, part of which concerned "softening cast or melted iron so that it may be filed and wrought as forged iron is". The process consisted of melting cast iron with iron ore (FeO_2). The oxygen in the ore reacted with the carbon in the cast iron to produce carbon dioxide and thus partially decarburising the iron. In 1676 the patent was transferred to John Brown of Brenchley, Kent, but it is not known to what degree the discovery was used. Samuel Lucas rediscovered the process in 1804 and subsequently used it at various foundries and forges.

The common method by which wrought iron was made from cast iron using finery and chafery furnaces to reduce the carbon content of the cast iron was described in the previous chapter. It would be useful if these processes could be stopped at an apt point to give steel having a carbon content of 0.1% to 1%. This, however, is not easy and the finery and chafery usually produced iron having a very low carbon content. To produce steel the carbon content had to be increased slightly to the range 0.1% to 1%. One way to do this was to introduce new cast iron into the finery to mix with the almost pure iron with the result that the forged product would be steel that could be hardened by quenching the hot product in water. Such steel was imported into England in the seventeenth century and was known as German steel or sometimes as Cullen (Cologne) steel, although it was usually made in Austria.

Yet another method of carburising wrought iron to give steel was by the cementation process. This process is very ancient. Derry and Williams (1960, 121) say that the Chalybes, a subject tribe of the Hittites, invented the process in about 1400 B.C. It consisted of steeling wrought iron bars by heating them in direct contact with charcoal, which diffuses into the surface of the iron, and then hammering to distribute it more evenly. The process was repeated many times until a satisfactory result was obtained and a great deal of charcoal was consumed in the process. The best-known Roman steel was Seric iron, which was supposed to be Chinese but was actually wootz, a high carbon crucible steel that reached Rome through Abyssinia from southern India.

Steel could also be made directly from certain ores. This too was an ancient method that survived into modern times. Attempts were made to find such ores in Britain but most British ores contained small amounts of phosphorus, which weakens the steel and makes it brittle. However, small quantities of suitable ores were found and used. For example, raw steel was obtained directly from ore in a blast furnace near Tongwynlais, Glamorgan, in about 1565. German workers were used and the work continued for about 10 years but was eventually abandoned, probably because the local ore was not entirely fitting.

Moxon (*Mechanic Exercises*, 1677) was the first in England to describe the medieval process of cementation. The iron was covered with a mixture consisting of dried and powdered cow horns, hooves, salt, and stale urine or wine vinegar. It was then encased in loam to seal it and heated to a red heat, about 900°C, for a period of time. The iron, when recovered, could be hardened by quenching in water. This, of course, produced a hard surface without affecting the strength and ductility of the core. The depth of the hard case depended on the length of time the iron was maintained at the red heat. The process was found to be adequate for items requiring a polished surface, such as gun barrels, tobacco boxes, buttons, and the heads of walking sticks.

In 1614 William Ellyott and Mathias Meysey, probably in conjunction with Sir Basil Brooke, were granted a patent for 21 years to convert iron into steel in a reverberatory furnace. Wrought iron and other ingredients (not specified in the patent but probably charcoal) were placed inside a sealed container or pot and this was put into a reverberatory furnace where it was heated by burning coal. The temperature attained was probably around 900°C, and after several days, to allow the carbon to diffuse from the charcoal through the metal surface, blisters formed on the surface and the resulting steel was known as blister steel. The use of mineral coal rather than charcoal was a distinct advantage and the process must have been both technically superior and considerably cheaper than earlier methods, especially when several pots were processed in the same furnace. The process was developed successfully before the patent expired although only the best quality wrought iron made from Swedish or Spanish ores could be used. English ores were unsuitable though Russian ores were used at a later date. This, in fact, was the process used in England for most of the following two centuries and England seemed content to rely on imported ores for making steel.

Fig. 19.12 shows the type of cementation furnace that came to be used in Sheffield and elsewhere. It consisted of a conical chimney, 10 m to 20 m high inside which were two pots or chests made of sandstone or firebrick and set on either side of a central flue. On the bottom of the chests was placed a layer of charcoal dust; and over this alternating layers of charcoal and iron filled the chest and the whole was covered with clay to exclude air. Beneath these chests was a long grate containing coal or coke. The flues were carried around from the furnace, so as to distribute the heat evenly around the chests. The fire was kept up for 8 or 10 days during which the temperature of the chests was maintained at about 1100°C. By withdrawing a test bar from an aperture in the side the degree of cementation was determined, and when the conversion of iron into steel was complete the fire was extinguished and the whole left to cool for 6 or 8 days longer before it was removed. Initially such furnaces could handle only 3 or 4 tons at a time but this increased during the nineteenth century to 40 tons.

The blister steel produced in this way did not have a uniform composition because the carbon was not diffused evenly through the

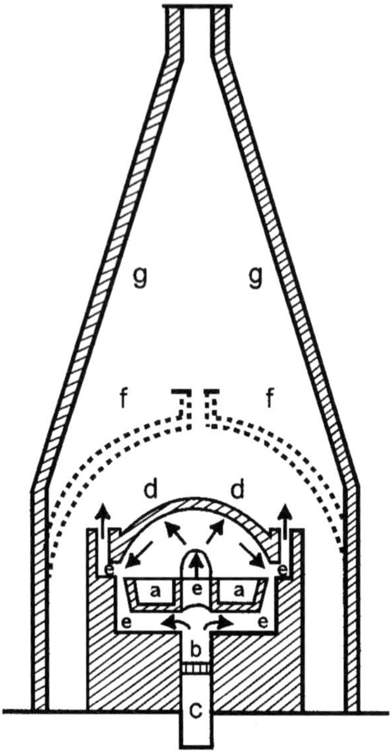

Fig. 19.12 A cementation furnace. Arrows show the direction of the flue gases. a,a = pots containing iron bars embedded in charcoal; b = fireplace with grate; c = ashpit; d,d = arched roof; e,e = fire flues; f,f = older type of domed roof and chimney; g,g = late eighteenth century tall chimney. After Schubert (1957, 327).

metal but was richer at the surface than in the centre. If the carbon at the surface was too great it became hard and brittle and unfit for the forge. However, a second cementation process to reduce the carbon could correct this. Blister steel was never of uniform composition and it varied in toughness and ductility depending on the amount of carbon it had absorbed. It was then heated to a bright red and either forged or rolled into bars to consolidate it and convert it to a more convenient form. Often, long bars were broken into shorter lengths, bundled together, tied with wire, heated to a welding temperature and then forged into a single bar to produce what was known as shear steel. This gave the carbon a more uniform distribution but even so there were alternating layers of high and low carbon steel. To improve it still further it could be cut in half and the pieces welded again under the hammer to produce double shear steel. This steel combined flexibility and strength and was capable of taking and holding a sharp edge and was for a long time the symbol of quality in Sheffield cutlery.

Shear steel, however, was not proper for all applications. A Doncaster watchmaker, Benjamin Huntsman, thought himself greatly hindered by the poor quality of steel available for watch springs and he turned his attention to finding a better. Smiles (1882, 104) says his first experiments were made at Doncaster but he moved to Handsworth, near Sheffield, in 1740 where suitable fuel was more readily available. There he seems to have conducted numerous experiments over many years before he finally hit on the solution. Essentially Huntsman produced a uniform distribution of carbon throughout the iron not by forging it but by melting the blister steel from the cementation furnace in a clay pot and then casting it into ingots. His clay pots were called crucibles and the steel he produced was called crucible steel.

Fig. 19.13 (top) shows a crucible steelworks of 1797, which would not be unlike that originally set up by Huntsman. Crucibles were made to hold about 12 kg of blister steel broken into 0.5 kg lumps. The crucibles were filled with the broken steel to which a flux of carbonaceous and vitreous substances was added and the contents were sealed in with lids. They were placed in a melting furnace and raised to a white heat by a coke fire where the metal melted in about three hours, the coke being replenished as required. The completion of the melting process was indicated by the clear surface of the melted metal, which had the dazzling brilliance of the sun. The pots were then lifted from the furnaces and the liquid steel poured into ingots of the required shape and size.

The steelworks of 1797, Fig. 19.13 (top), has a peaceful rural setting. The large and airy cellar provides the necessary air supply for the combustion of the coke, and the high rectangular chimneys provide the draught for the fire and carries the hot gases through the melting holes on the upper floor. A man in the cellar is preparing the charges by breaking up blister steel into 0.5 kg lumps and filling clay crucibles. On the upper floor a crucible is being drawn from a furnace hole and on the left an ingot is being cast. At this time, 1797, it appears that the charge weight was about 12 kg, roughly

Fig. 19.13 Top: a Sheffield crucible steel melting shop, 1797. Bottom: five furnace holes of a surviving crucible steelworks at Abbeydale, Sheffield.

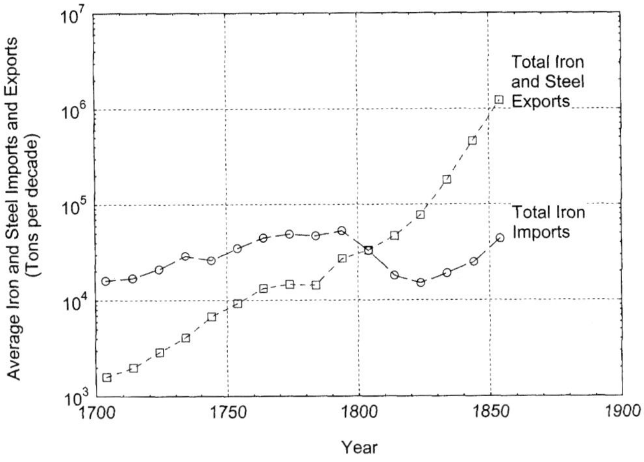

Fig. 19.14 Average imports and exports of iron and steel. Mitchell and Deane (1962, 140–47), England and Wales 1700–1791, Great Britain 1792–1814, United Kingdom 1814–1860.

double that of thirty years earlier. Steel production, it will be noted, was measured in tens of kg whereas cast iron production was measured in tonnes. Steel continued to be produced in this way until the 1940's in many small steelworks, and Fig. 19.13 (bottom) shows the five furnace holes of a surviving, but presently disused, crucible steelworks at Abbeydale, Sheffield. These processes became very important. By the middle of the nineteenth century Sheffield produced 90% of England's steel and over 50% of European steel. These methods were, however, replaced by the Bessemer and open hearth processes later in the nineteenth century. The average yearly imports of iron and steel from 1700 to 1860 are illustrated in Fig. 19.14. The imports and exports of iron are mostly in the form of bar iron (wrought iron). As can be seen, England relied heavily on imports during the eighteenth century and did not become a net exporter of iron and steel until the early years of the nineteenth century. Imports of iron declined drastically during the Napoleonic Wars but home production made up the shortfall and exports continued to grow. Iron imports did not start to increase again until about 1820 when peace was established.

One of the reasons for the rise in exports of wrought iron bar was due to Cort's invention in 1784 of an improved method of

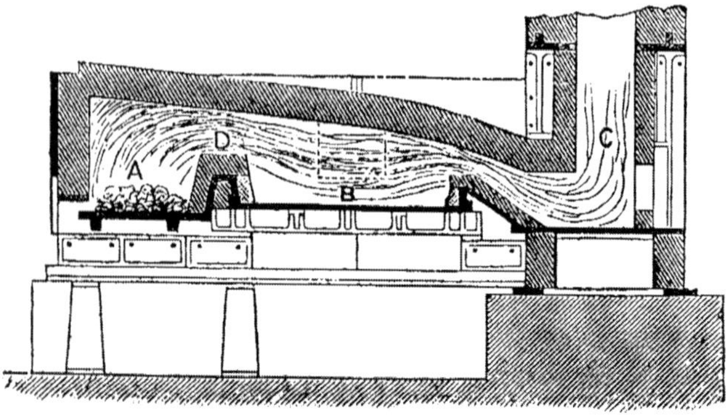

Fig. 19.15 Puddling furnace. A fireplace, B puddling hearth, C chimney, D fire-bridge. Chambers (1874, 409).

producing wrought iron appropriate for rolling and forging from cast iron. Although none of the particular processes were completely original Cort's method impressed contemporaries and they quickly adopted them and he is usually considered the inventor of the puddling technique. Fig. 19.15 shows a puddling furnace. Essentially it is a reverberatory furnace having a fire at *A*, and a puddling hearth at *B* separated by a fire-bridge *D*. The draught of the exhaust gases ascending a chimney at C creates a large air supply. The charge is typically about 0.25 tonne of which 20–50% is refined iron and the rest is broken pig-iron with a variable proportion of hammer slag and iron scale. After about half an hour the charge is melted and is stirred until it begins to "boil" by the formation and escape of carbon monoxide which forms a blue flame over the surface as it reacts with the excess oxygen in the combustion gases. As the carbon in the pig iron is reduced by oxidation the melting point rises and pasty masses of malleable iron separate and are gathered together into balls weighing about 40 kg on the end of a long iron rod. The whole process takes about an hour or more and is very laborious and the airflow rate has to be regulated as the process advances.

The iron ball is a spongy lump of malleable iron mixed with cinder and required to be welded into a coherent mass and at the same time to squeeze out the cinder. This is called *shingling* and is usually done by a tilt hammer. After hammering to a solid oblong mass it is formed into bars by means of the grooved rollers of a rolling mill. The puddle bar so created is cut into short lengths bundled together, raised to a welding heat and then hammered or rolled to form any required form of sheet, plate or bar. Sometimes the bundling operation is done twice. The result is a very malleable and ductile iron containing a very small amount of carbon, which may be used as it is or converted into steel by the cementation process.

Hardening Steel by Quenching

The most remarkable and useful property of steel is its ability to change its hardness when quenched from a proper temperature. It was once thought that no other metal possessed this property but many alloys are now known to behave in this way. The method is very ancient. Homer (*Odyssey, Book IX*) describes how he and his colleagues blinded the Cyclops by hardening an olive branch in a fire and plunging it, hissing, into his single eye. He says: "I was reminded of the loud hiss that comes from a great axe or adze when a smith plunges it into cold water—to temper it and give strength [hardness] to the iron. That is how the Cyclops' eye hissed round the olive stake." Several hundred years later the technique had developed so that Pliny (*Natural History XXXIII, 146*) was able to write: "It is usual to quench smaller iron-forgings with oil lest they should be hardened by water and become brittle." If steel is heated to redness and plunged into cold water it becomes extremely hard but at the same time it is too brittle for use. On the other hand if it is cooled very gradually it becomes soft and ductile but deficient in strength and hardness. The process of tempering gives steel instruments intermediate qualities of brittleness and ductility, hardness and softness, which ensure the proper strength and hardness for the intended use. For this purpose, after quenching in water from a high temperature has hardened the steel, it is partially softened by heating to a lesser temperature and plunging again in cold water or some other quenching medium. The severity of the quenching media is: brine, water, oil, air. The least severe medium to give the required hardness should be used to minimise distortion and cracking. If oil is the quenching medium it should have a high flash point; ordinary lubricating oil is usually unsuitable.

Various methods have been used to indicate the proper temperature for tempering but the most common is to observe the change in colour that occurs on the surface of the steel as the temperature

Table 19.1 Colour of steel at various temperatures.

Temperature °C	*Colour*
500–600	Very dull red
600–700	Dark blood red
700–800	Cherry red
800–850	Bright red
850–900	Pinkish-red
900–950	Pinkish-orange
950–1000	Orange
1000–1100	Yellow
1100–1200	Yellowish white
1200–1300	White
over 1300	Brilliant white

increases. Biringuccio (1540, 371) was aware of this. He writes: "Because the first colour that is shown by steel when it is quenched is fiery white, it is called silver; the second which is yellow like gold they call gold; the third which is bluish and purple they call violet; the fourth is ashen grey. You quench them at the proper stage of these colours as you wish them more or less hard in temper. If you wish it very hard, heat your iron well and quench it rapidly in the tempering baths that you have prepared or in clear cold water." When making a drill for drilling out the ignition hole in a gun barrel he says "The point should be made like a little chisel with its edge obtuse and level, and it should be tempered at the beginning of the violet." Table 19.1 gives the colour of steel at various temperatures and Biringuccio's simple classification in descending order of temperature (silver, gold, violet, and grey) is seen to be valid. Such colours are useful indicators of temperature and are commonly used in simple forging operations.

The modern practice in hardening a steel cutting edge is to heat it to about 800°C (bright red) and to quench by plunging the edge into cold water. When the quenched steel is removed from the water part of the tool shank remains hot and this heat conducts back towards the cutting edge. This is quickly polished and watched carefully. As the temperature of the cutting edge increases its colour changes, as specified in Table 19.2, and when the desired temperature is reached the whole tool is quenched in water. Articles such as watch springs need to be hardened and tempered throughout and

Table 19.2 Tempering temperatures and colour for typical steel products.

Temperature °C	*Colour*	*Typical Products*
230	Pale straw	Best razors, surgical instruments
245	Full yellow	Common razors, penknives
255	Brown	Small shears, scissors, cold chisels, hoes
265	Brown with purple spots	Axes, plane-irons, pocket knives
275	Purple	Table knives, large shears
285	Bright blue	Swords, watch springs, bell-springs
295	Full blue	Fine saws, daggers, augers
315	Dark blue	Hand and pit saws

in this case the entire article is quenched in water from about 800°C and the entire article is reheated to the tempering temperature before being quenched again.

Machine Tools

By machine tools we mean mechanical devices for turning, boring, planning, cutting screw threads and cutting gears. Machine tools are closely associated with industrialisation in the nineteenth and twentieth centuries but a few are much older. The bow-drill, for example, was known in ancient Egypt and many ancient artefacts appear to have been machined on the lathe, although the lathes themselves have not survived either as objects or in pictures. The scene in Homer's *Odyssey*, mentioned above, in which the Cyclops was blinded by driving a heated olive pole into his eye not only tells us that iron was commonly hardened by quenching but it also describes a primitive drilling operation. Homer (*Odyssey, Book IX*) writes "Seizing the olive pole, they drove its sharpened end into the Cyclops eye, while I [Odysseus] used my weight from above to twist it home, like a man boring a ship's timber with a drill which his mates below him twirl with a strap they hold at the other end, so that it spins continuously."

The earliest illustration of a machine tool dates from 1390, Fig. 19.16, and is a boring machine, operated by a bow spring and used for drilling holes in pearls prior to stringing them as necklaces, Klemm

Fig. 19.16 A bowstring operated boring machine. From the Mendelschen Zwölf-Brüder-Stiftung, 1390.

(1959, 64). From the same source comes a foot-operated lathe, Fig. 19.17 (left). A spring pole tensions a rope that is wrapped around the work piece and connected to a foot pedal. The work piece is fixed between centres in a headstock and tailstock and as the foot pedal is operated the work piece rotates one way then the other. A tool rest is clamped to the bedplate to support a cutting tool and as the work rotates the turner cuts the outside to the desired cylindrical shape. These devices, of course, produce an oscillating rotation so the cutting process was intermittent and irregular, but nevertheless some very fine work was fashioned. Such lathes continued in use for a very long time. Fig. 19.17 (right) shows a similar lathe from around 1800 and they did not die out in Europe until later in the nineteenth century and they are still used in parts of Asia.

Leonardo da Vinci, Fig. 19.18, designed a smoother and more even drive for a lathe, Hart (1961, 225). In this case the foot treadle, instead of being connected to a sprung pole over the lathe, was connected to a crank and flywheel on the main shaft. This device had recently been introduced to drive spinning wheels but its precise date and inventor is unknown. The tailstock was fitted with a screwed dead-centre to support the work piece and the large disc

Fig. 19.17 Left: a foot operated pole lathe. From the Mendelschen Zwölf-Brüder-Stiftung, 1390. Right: A foot operated pole lathe of about 1800. Science Museum, London.

was used as a flywheel to reduce the speed oscillations caused by the varying cutting load. Fig. 19.18 also shows Leonardo's design for a treadle-operated sawmill. This is based on a crank, connecting rod, and crank-pin arrangement similar to that in modern reciprocating engines. In this way the saw reciprocates vertically although driven by a rotating crank.

The lathe was also adapted for spinning hollowware, particularly pewter. Fig. 19.19 shows a pewterer at his lathe in 1568 and an assistant turning a crank on a large flywheel, which also appears to drive the lathe spindle by means of a belt. Clearly this provides a much smoother drive to the lathe shaft than either a treadle or a sprung pole, and is preferable to Leonardo's treadle operated crank. It is a comparatively early example of a belt or rope drive, although such drives had been in use on spinning wheels for spinning yarn for almost a century. However, it does require an additional man. Metal spinning is an old technique that was commonly used to make hollow shapes of cylindrical section, such as those on display in Fig. 19.19 (left). Usually, a wooden former was fitted to a lathe and rotated with the main shaft. A sheet of pewter was clamped against

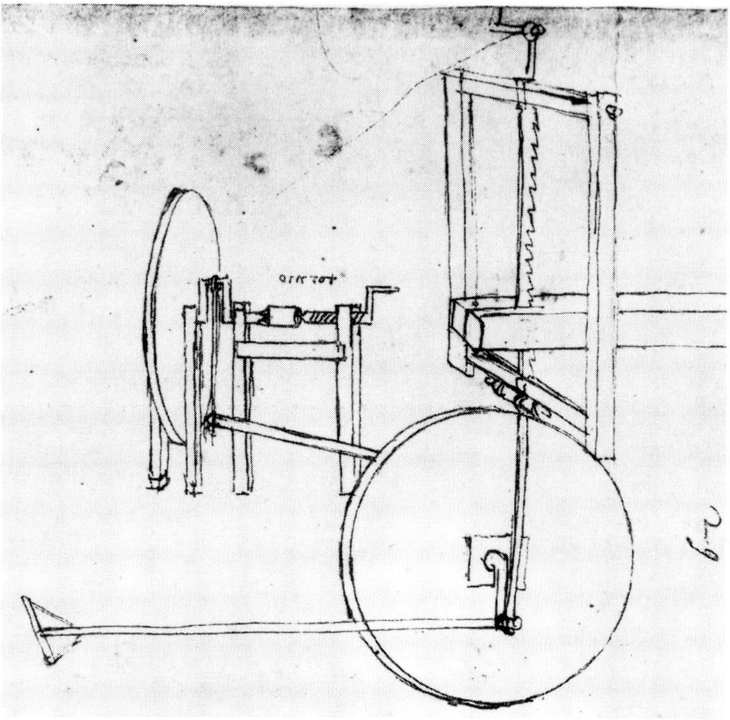

Fig. 19.18 Leonardo's designs for a treadle driven lathe (left) and a treadle driven sawmill (right).

it using the tailstock and the lathe rotated. A long, wooden spinning tool, pivoted on a pin, was used to deform the pewter sheet and press it into the former. With a collapsible former, made of several pieces, articles with a neck at the open end were spun. Pewter is mostly tin with up to 15% lead and it was often used for domestic hollowware in the medieval and early modern periods and some Roman examples survive. It was commonly made in England, where tin was readily available, but was less common elsewhere. With more than about 15% lead there was a danger of lead poisoning, particularly as pewter was used mostly for tableware.

Screw cutting was one of the most difficult processes so screws for clamping were avoided whenever possible by the use of wedges or other devices. Nevertheless they were made. Dickinson (1941) has investigated the origin and manufacture of wood screws, of which Fig. 19.20 is an example from about 1620. He also found examples

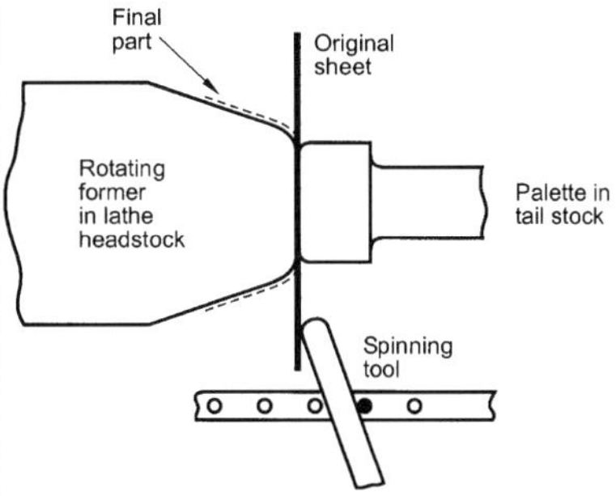

Fig. 19.19 Left: the pewterer spinning hollowware at his lathe. Schopper, 1568. Right: method of spinning in a lathe.

Fig. 19.20 Wood screw of about 1620 from an Italian snaphaunce.

of a fifth-century A.D. Roman wood screw, now in Reading Museum, although he expressed doubts about its authenticity, and a fifth-century German wood screw from Dalheim, near Metz. All appear to have been well made in their original condition. Scribing the crest or root of the thread on a circular bar and then chiselling and filing it by hand was the only way to create screw threads. Such threads were of quite poor accuracy but once produced they could be copied. Heron of Alexandria designed a simple machine to cut an internal thread in a drilled hole; see Fig. 1.14 of Chapter One. Essentially the tool copied an existing external thread and this idea of copying a master thread to create a new thread was commonly used. An example is the mandrel lathe.

Fig. 19.21 (top) shows such a lathe but with the rest for the cutting tool removed and shown below. The headstock on the right was clamped to the bedplate and contained the master screw thread fitted with a handle. The work to be screwed protruded from the master screw, to which it was fastened, and was supported in a tailstock. As the master screw was turned by its handle the work piece was pushed towards the tailstock and if the tool rest and cutting tool was clamped in position then a screw thread, having the same pitch as the master screw, was cut on the work piece. Of course, several passes, using a small depth of cut, were necessary to form a complete screw thread. If the master screw had a left-hand thread then a right-hand screw thread was cut on the work piece. A disadvantage was that a different master screw was needed for each pitch since the master can only reproduce its own pitch.

Leonardo da Vinci (1452–1519) invented an improved screw-cutting lathe, Fig. 19.21 (bottom), which could cut threads of different pitch from the same master screw. His note accompanying the sketch says, "This is the way you make a screw. You turn the middle wheel, which rests on the screw you wish to make. If you wish to make screws with greater or lesser-inclined threads [pitch], then remove the wheels *s* and *f* and replace them with the wheels *a* and *b* or the wheels *c* and *d*. Accordingly you push the stirrups *h*, *e* and *t* farther apart or closer together; also those at the plane *K* [the bearing carriage or saddle] and the block *g*. The plane is that part which contains the screw-cutting tool which, in being pushed forward, cuts the thread of the screw." The work piece is in the centre and is shown threaded to the cutting point on the saddle and plain beyond. In Leonardo's design the tool rest moves with the saddle on two mas-

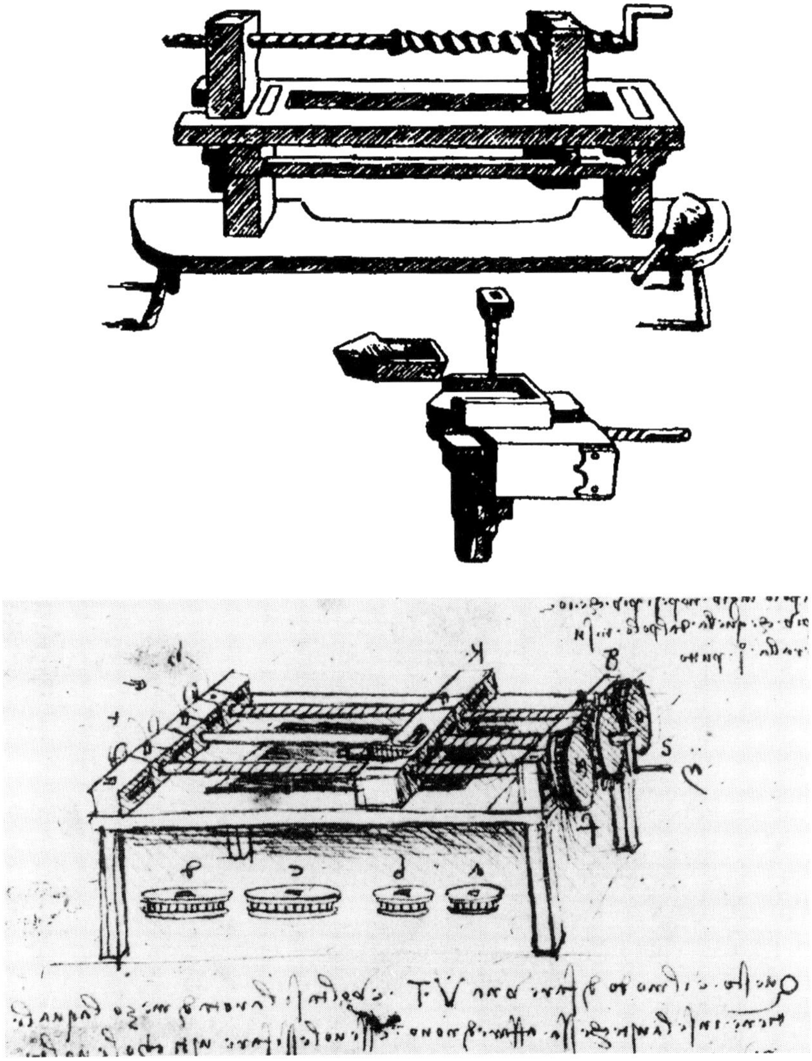

Fig. 19.21 Top: screwcutting lathe of 1480 with a tool rest having a screw feed. Bottom: Leonardo da Vinci's screw-cutting lathe.

ter screws (lead screws) and changing the gear ratio between the lead screw and the work piece varies the pitch of the cut thread, as on a modern lathe. Errors in the pitch of the lead screw were reduced when copied onto the workpiece. The design, clearly, could be got to work but it is not known if it was ever brought to a successful

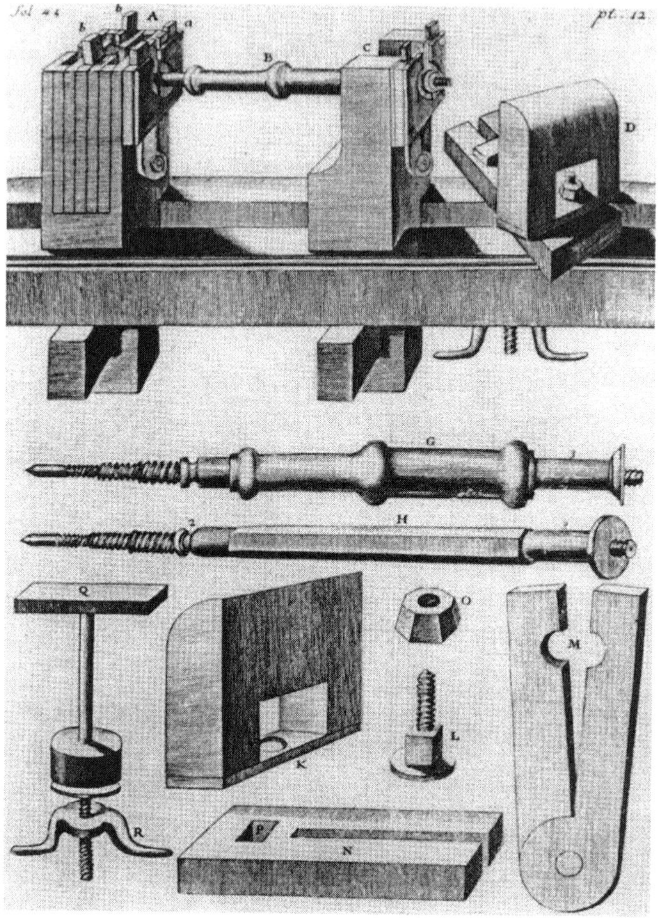

Fig. 19.22 Mandrel lathe, *c.* 1700, Plumier.

conclusion and, like most of Leonardo's designs, it does not seem to have influenced later designers.

Mandrel lathes for copying screw threads continued in use and an example from Plumier in about 1700 is illustrated in Fig. 19.22. The method of rotating the lathe is not illustrated. Two units, *A* and *C* are clamped to a bedplate and a mandrel *B* connects them. This mandrel is also illustrated below in *G*. Three master threads are at the left hand end next to a groove. Item *M* below is a scissors clamp, four of which are normally positioned in the unit *A* so they may engage with the groove or one of the three master threads. When one of the scissors is closed to locate in the groove of the mandrel the lathe may be used for ordinary turning work; the work piece being fastened to the thread at the right hand end of the mandrel. If a scissors clamp is engaged in one of the screw threads then when the mandrel rotates it moves to the right, pushing the work piece past the cutting tool and copying the master thread onto the work piece. The cutting tool is held in the tool holder, *D*, which is clamped to the bedplate. Similar machine tools for cutting internal and external screw threads were illustrated by Diderot and D'Alembert in their encyclopaedia and illustrated in Fig. 1.17 of Chapter One.

Ramsden, in 1766 invented a dividing engine for which, in 1777, he received a reward of £615 from the Board of Longitude, who were interested in promoting improved methods of clock making so that the degrees of longitude could be more accurately determined. He adapted his circular dividing engine to make accurate screw threads, Fig. 19.23. His engine consisted of a large wheel moved by a tangent screw and worm. The large wheel was 45 inches diameter and had 2,160 teeth so that six turns of the tangent screw moved the tangent wheel one degree. The guide screw *G* is turned by the handle, and at each revolution moves the larger tangent wheel one tooth. It also winds a length of watch spring onto the boss *P* and thereby pulls a slide on which the cutting tool *t* is fixed thus cutting the screw *C*, which at the same time is rotated by the gearing *c g* from the main shaft. The pitch of the thread was fixed by the diameter of *P*, which was chosen so that 600 turns of the handle gave a motion of 5 inches to the tool slide and thus the pitch was 20 threads per inch.

Lathes were the most commonly used machine tools and were used for turning all manner of articles in wood and metal. Invariably they were man-powered as the illustrations in Fig. 19.24 show. Here we have three different types of lathe. Two have a rope wrapped around the work piece that must be reciprocated, either by hand, as in the case of the lathe on the left, or by the foot pedal, as in the case of the lathe at the rear. In both cases the rope is tied to an overhead spring lath, which provides the necessary tension in the rope and also returns the rope to its starting position. Friction between the rope and the work piece causes it to rotate as the rope is pulled. A more continuous motion is generated when the work piece is rotated by means of a rope drive from a large wheel turned by an assistant. This rope drive has two wheels. The larger wheel is used

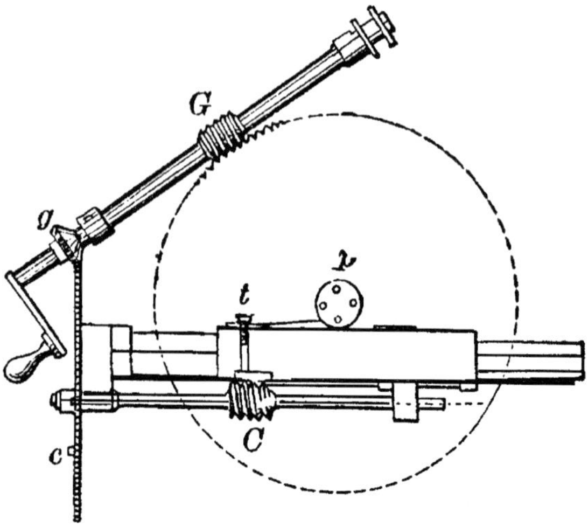

Fig. 19.23 Ramsden's screw cutting machine of 1766, Knight (1884, 712).

Fig. 19.24 Turner's workshop showing three different lathes. Diderot and D'Alembert (1751–72).

when a high rotational speed is required, as when cutting a small diameter work piece, and the smaller wheel is used when a slower rotational speed is needed, as when cutting larger diameter work pieces. The principle is that the cutting speed at the tool edge should be the optimum. The work piece is held between centres in the headstock and the tailstock. The rest on which the cutting tool is supported when in use, is fixed and clamped to the bedplate. For decorative work, where variations in diameter are relatively unimportant, there is no problem, but for engineering work, where a true cylinder of precise diameter is often required, considerable skill is necessary. This problem was not eased until the invention of the slide rest in the early nineteenth century. The various cutting tools necessary for the work are hung on the walls.

The greatest improvement in the lathe occurred at the end of the eighteenth century with the invention of the slide rest by Henry Maudslay, who made the invention shortly after starting in the workshops of Joseph Bramah and before setting up his own independent workshop; see Fig. 19.25. The cutting tool, instead of being guided by the hand of the workman, is attached to a holding frame or rest, which is made to move, or slide, by rotating a screw either by hand, or automatically using the same power which turns the lathe. With careful adjustment the cutting tool could be made to slide parallel to the axis of rotation of the work piece and thus a true cylinder could be cut to the required diameter. Similarly it could be made to move at 90 degrees to the axis of rotation of the work piece (along the face of the work) so the work may be faced-up in an accurate manner or, when turning a cylinder, the exact diameter may be set before the cut is made. This simple device revolutionised the accuracy of work that could be produced by the turner.

Gear wheels were often made from wood and on large wheels the teeth were made individually and fitted into slots cut into the wheel. Alternatively they could be cast from wooden patterns. For clock making, machined gears of greater precision were required and the gear cutting machines clockmakers developed were later adapted to heavier engineering. A gear-cutting machine is illustrated in Fig. 19.26. The disc to be cut is fixed on a spindle, one end of which has a graduated dividing wheel consisting of a number of equally spaced holes drilled on a common pitch circle. The spindle is located by means of a stop pin in one of the holes. A cutter on a slide rest is rotated; usually by hand but later by machinery, and advancing the

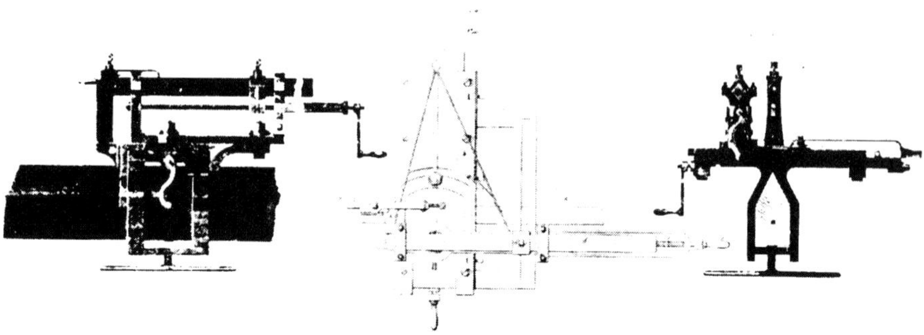

Fig. 19.25 Bramah's (i.e. Maudslay's) original slide rest lathe, 1794, from Buchanan's *Machinery and Millwork*, 1841.

slide rest enables the cutter to remove the metal between a pair of gear teeth. When the cut is completed the slide rest is re-wound and the dividing head advanced by a number of holes, depending on the pitch of teeth to be cut.

The face of the dividing head has various pitch circles containing different numbers of holes, which can be used to cut wheels having almost any number of teeth between 6 and 360. The number of holes per circle and the corresponding numbers of teeth that can be cut are shown in Table 19.3. Of course the cutter must be ground and sharpened to the tooth profile but the same cutter may be used for different diameters of gear so long as the ratio of the number of teeth to the diameter is constant.

Special purpose machines were often developed and in a survey of this nature it is not possible to mention even a small proportion of them, but some mention must be made of file cutting machines. This is because they have considerable similarity to the file-cutting machine invented by Leonardo da Vinci in the fifteenth century and illustrated in Chapter Seven Fig. 7.10. Such machines have a table to which a file blank is attached. A trip-hammer strikes a chisel to cut each groove and between blows the table and file are advanced by a small amount. Leonardo's machine was driven by falling weights but when file cutters were introduced in the nineteenth century they were usually driven from shafts powered by waterwheels. Knight (1884, 842) mentions that such machines were developed by Duverger, 1699, Fardouet, 1725, Thiout, 1740, Brachal and Gamain 1756, 1778, Raul 1800, Ericsson, 1836, and Robinson 1843, amongst others.

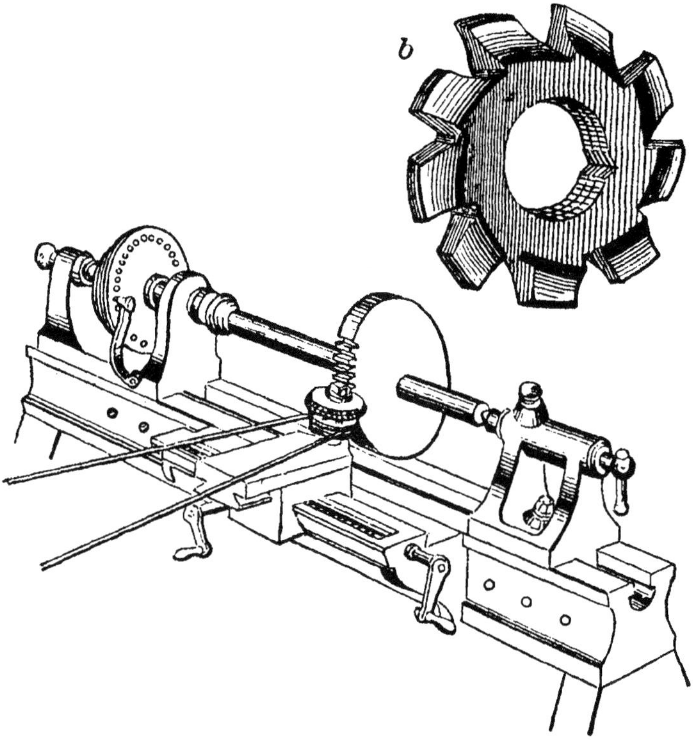

Fig. 19.26 Gear cutting machine, Knight (1884, 960).

File making by hand continued to almost the end of the nineteenth century in England and file making machinery was greatly resisted. Hand-cut files were often considered superior to cheaper machine cut files and this is, perhaps, an example of low price erroneously implying low quality.

Card and Studley's file-cutting machine, Fig. 19.27 may be a typical if late example. A feed screw and half nut that is automatically raised to stop the feed motion actuates the sliding head to which the blank is clamped. The anvil is a hemispherical block that rests in a socket for support. The anvil and feed movement are supported on a turntable, by whose adjustment the inclination of the teeth may be set. The chisel is supported on a flexible rod connected to the hammer handle by a spiral spring. The hammer is attached to a rocker shaft, which has an adjustable arm acted on by a cam on the main shaft.

Table 19.3 Number of Holes on Graduation Circle and number of teeth that can be cut.

	Graduation Circles										
	192	200	208	216	220	224	252	256	272	304	360
	192	200	208	216	220	224	252	256	272	304	360
	96	100	104	108	110	112	126	128	136	152	180
	64	50	52	54	55	56	84	64	68	76	120
	48	40	26	36	44	32	63	32	34	38	90
	32	25	13	27	22	28	42	16	17	19	72
	24	20		18	11	16	36	8			60
	16	10		12	10	14	28				45
Number of teeth	12	5		9	5	8	21				40
	8			6		7	18				36
	6						14				30
							12				24
							9				20
							7				18
							6				15
											12
											10
											9
											8
											6

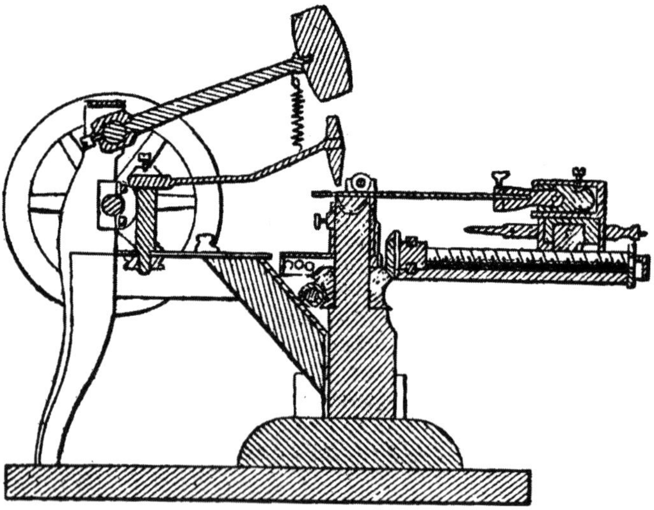

Fig. 19.27 Card and Studley's file cutting machine, Knight (1884, 842).

Fig. 19.28 Top: Boring machines for bronze cannon. The wheel on the left is a one or two man treadmill and that in the background is turned by hand. Bottom: treadmill boring machine with two boring tools, one driven by gears. Biringuccio (1540, 309–10).

Machine tools for boring gun barrels were used almost from the earliest guns and spread from the smaller handguns to the larger cannon. The need for a smooth bore in a cannon was not so critical because the clearance between the ball and the bore was typically about a centimetre. Biringuccio (1540, 309–10) illustrates two machines for boring out bronze cannon, Fig. 19.28 (top). He tells us: "For greater caution, for the beauty and safety of the gun, and to make sure that it achieves its purpose of shooting with perfect accuracy, gentlemen—soldiers or master gunners—began to desire that both large ones and small ones should be bored, as they do the arquebuses and iron muskets that are used for shooting by hand."

To meet this demand Biringuccio made a large double wheel such that one or more men could walk inside it to set it in motion, or if this was not possible he made use of a cannon-carriage wheel. "In the nave box I fixed a piece of wood, in the middle of which I put an iron spindle with a bent handle—. I supported this wheel on a journal and at the other end I made a good square-shaped socket. In the socket of this I fixed a thick spindle long enough to reach the end of the bore of the gun; and at the top of this I had a square piece of steel welded, with all its corners true, sharp, and well tempered, so that when it is put in the mouth of the gun and rotated it may make it exactly round. This is the usual method. In order to avoid the inconvenience given by that mass of steel and iron in tempering, adjusting accurately, and sharpening, I have also made them with little bronze castellations and then attached four steel corners so that it may be sharp."

"I have also fixed the gun—on a board of elm, walnut, or some other wood that was flat underneath and made a little cart. After I had fastened it well with iron bands or ropes—so that the borer should not move it while cutting, I then placed this on the surface of a platform made of several boards whose length is at least twice that of the gun and I made it strong and firm. Between the bed where the gun rests and the said surface I put three round rollers crosswise so that the gun could move easily as it was pulled. For this I arranged a little windlass crosswise in front, which could pull equally two pieces of cord attached to the edges of the bed with two small, good iron hooks. I also arranged another little windlass at the back in order to be able to pull it back when it runs too fast and in order to draw out the borer and the borings". Biringuccio, however, preferred the machine shown in Fig. 19.28 (bottom) which has a double treadmill driving two borers, one directly and the other through a set of gears. In this way he was able to bore two cannon at the same time. Biringuccio also describes the kinds of cutters and boring bars he used.

Crossley (1975, 1–37) has excavated the remains of a horizontal cannon-boring mill, somewhat similar to that described by Biringuccio; see Fig. 19.29. It consisted of a track for the carriage carrying the gun barrel and a circular foundation, which was thought to be for a windlass to pull the carriage. There does not seem to have been a second windlass to withdraw the gun carriage, as described by Biringuccio. The slots for the carriage track were 20 cm deep, 30–32 cm wide. One was 4 m long and the other 3.3 m. The oak beams,

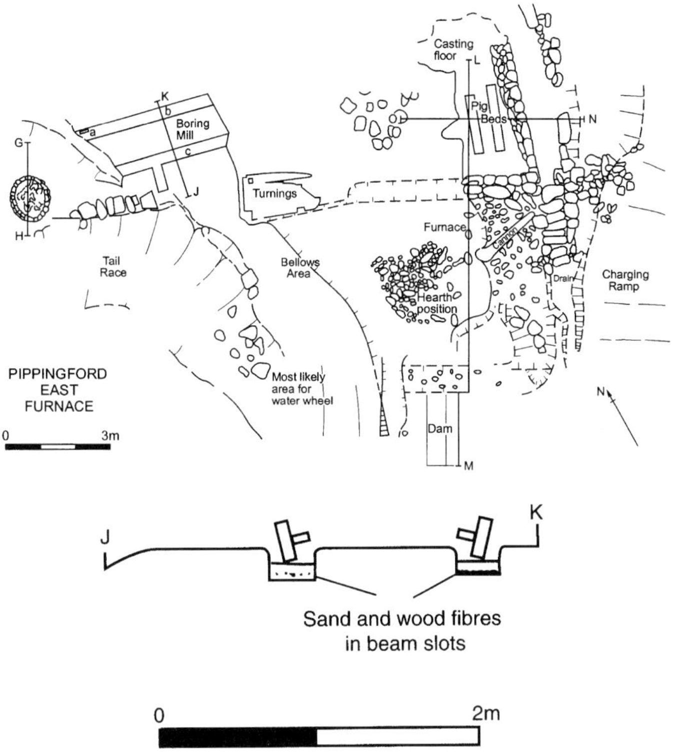

Fig. 19.29 Pippingford East Furnace 1712. Top: position of the horizontal boring mill. Bottom: tracks and wheels of the cannon-boring mill (section JK). After Crossley, (1975, 9, 11).

which originally occupied the slots, had rotted leaving only blackened sand and wood fibres. Three cast-iron wheels of 23 to 24 cm diameter were found in position on the track and lying between the tracks was a dome-shaped cast-iron artefact that was considered to be a chuck for holding a boring bar, although no boring bars were found. A pile of rusting iron turnings was found nearby. Despite the evidence for a waterwheel driving bellows for the eastern furnace there was no evidence that the boring mill had been powered by water.

Hendrik van Balen, 1575–1632, and Jan Breughel the Younger, 1601–78, illustrate an early example of a vertical gun-boring mill; see Fig. 19.30 (left). The painting is usually displayed at Kingston

Fig. 19.30 Left: vertical boring mill for cannon, from Jan Breughel the Younger's *Fire* (detail), *c.* 1620. Right: vertical gun boring mill driven by animal power. Diderot and D'Alembert (1751–72).

Lacy, Dorset (National Trust) and is one of a series illustrating the four elements, namely *Fire.* It was completed, apparently, before about 1620. The gun barrel is suspended, muzzle down, in a heavy wooden frame and a horse walking in a circle below rotates a boring bar. It seems likely that Breughel, or perhaps Balen, painted the boring mill from examples they had seen in The Netherlands. Behind the boring mill is a horizontal-axis treadmill that was used, evidently, to lift or lower the cannon. It is not clear in the painting but somehow the gun barrel is lowered slightly, after the horse completes each revolution, so the boring bar advances towards the breech cutting out a smooth and reasonably accurate bore. Great accuracy was not necessary because, as mentioned above, the clearance between cannon ball and bore was typically about 1 cm.

Steeds (1969, 5) says that a gun-boring machine was installed in the Royal Arsenal at Woolwich in about 1715 although no details of the machine are preserved and in 1713 a vertical gun-boring machine was invented by a Swiss engineer named Maritz. This gentleman later worked at the Netherlands State Gun Foundry at The Hague and the boring machine illustrated by Diderot in his encyclopaedia, Fig. 19.30 (right), is thought to be Maritz's machine. It is, however, remarkably similar to that painted a century earlier by Breughel or Balen. An animal, yoked at *T* on the lower left, rotates the central vertical shaft on which the boring bar and cutter is fastened. The cannon is shown at *H* and is suspended in a cradle that can move vertically, guided by the vertical posts *Ff* and moved by ropes passing over an arrangement of pulleys to a horizontal windlass. Rotating the windlass by hand raises or lowers the cannon and the reduction gearing gives the operators fine control of the cannon's movement. The vertical and rotary motions are independent and one imagines that each time the animal makes a turn, two workers operating the wheels marked *P* simply advance the wheels a fixed distance to impose the feed motion.

The first gun boring machine in which the barrel was rotated and the feed motion given to the cutting tool was in about 1758 when Jan Verbruggen and Jacob Ziegler designed and built a horizontal gun borer. A reconstruction is on display in the Dutch Military Museum, Leiden. Machines built later in the eighteenth century were adapted to increase the rate of production by having them bore cannons two or even four at a time and horizontal borers came to be preferred to vertical machines.

The main defect of these boring machines was that they were not particularly accurate. Pole (1970, 42) reports that Watt was in the greatest difficulty when boring the cylinders for his steam engines. His earliest cylinders were made at the Carron Works, where Smeaton had erected a machine for boring cannon, but they proved to be very inaccurate. The piston could not be kept steam tight in the cylinder despite the various expedients adopted. Watt complained of one cylinder of 18 inches (457 mm) diameter "at the worst place the longer diameter exceeded the shorter by three-eighths of an inch (9.5 mm)."

The cause of this defect was that the cutting tool of the boring machine was fixed at one end only, as if boring a cannon, and hence was not sufficiently rigid to guide the tool accurately when cutting a deep hole. John Wilkinson of Chester made the first accurate boring machine in 1775. He placed a strong bar completely through

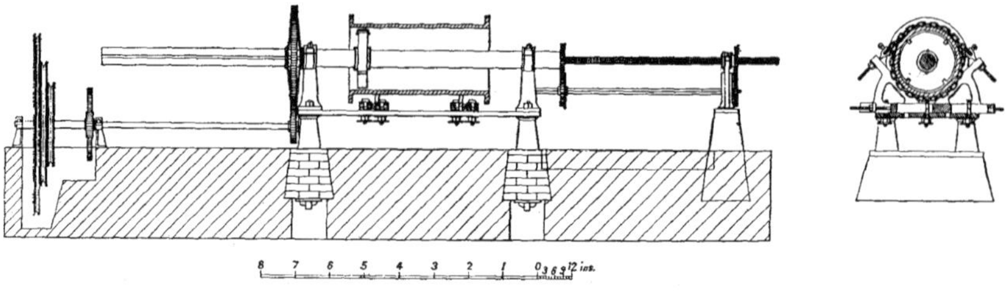

Fig. 19.31 Mathew Murray's cylinder boring-machine of about 1822 built for the engine works at Chaillot. After Steeds (1969, 38).

the cylinder, fixing it firmly at both ends on lathe centres. Cutters were fitted to this boring bar and when rotated and advanced slowly down the cylinder they were able to cut the metal with very considerable force without deflecting and thus they gave greatly increased accuracy. A horizontal cylinder boring-machine of about 1822 is shown in Fig. 19.31, and clearly illustrates Wilkinson's principle. The boring bar is supported by two bearings but can rotate and move axially. It is driven by the large gearwheel on the left through a sliding keyway, to enable the boring bar to move axially. The feed is controlled by the apparatus on the right, which is a screw driven by gears from the boring bar. As the boring bar rotates the screw pushes it along its own axis. The gear at the lower left of this device can slide along its square shaft.

A late development in machine tools was the planing machine. Essentially this is a machine for producing perfectly flat surfaces and is analogous to the operation of planing in woodwork. No similar operation existed for metal and the usual method of getting a flat surface was by chipping and filing; a tedious and expensive process, and not particularly accurate. There seems to have been numerous independent inventors of the metal planing machine. James Fox of Derby seems to have invented and used a metal planing machine in 1814 and according to information supplied by a workman to Smiles (1882, 49) it was similar to the planing machines that came into use at a later date but differed in detail, and the workman goes on to specify these differences. Similarly Richard Robert of Manchester has a claim, but the chief credit is usually given to Joseph Clement, another of Bramah's workmen who later set up an independent business. The cutters used were similar to those used on metal working lathes and not at all like those on wood planes, but unlike the lathe

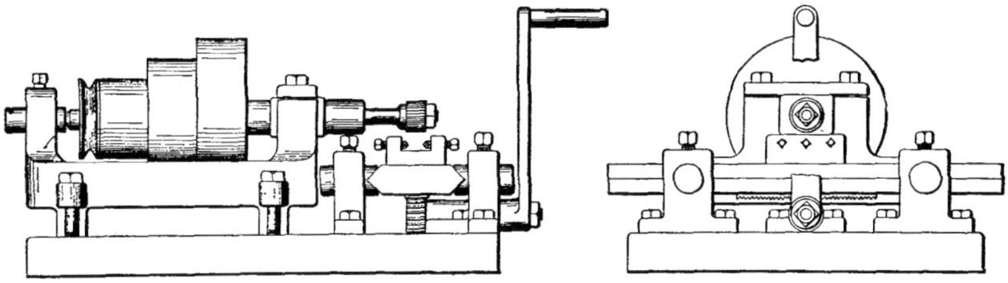

Fig. 19.32 The Middletown milling machine, *c.* 1818. After Steeds (1969, 43).

they worked in straight lines, the tool being stationary with the work moving under it and the tool advancing across the work a small amount between each cut. For ten years after it had been set up in 1825, Clement's planer, which he made available to others, was the only machine available for large work and it was an important source of his income.

It was, of course, but a small step to set the cutters into rotary motion to produce the milling machine. Probably the earliest successful milling machine was that of Robert Johnson of Middletown in the USA and shown in Fig. 19.32. Steeds (1969, 42) says it was of a primitive nature in that it did not have any vertical adjustment between the work piece and the cutter. The horizontal-axis rotating cutter was driven from pulleys on the left and the large handle on the right was rotated to move the work piece horizontally under the cutter using a rack and pinion.

Analysis

Hot and Cold Working of Metals

The greatest problem encountered in working metals is the magnitude of the forces needed to deform them and consequently the great size and strength of the equipment required. This was a severe problem before the advent of heavy industrial machinery and consequently the metal to be worked needed to be as soft as possible. The strength and hardness of pure metals is related to their melting point. High melting point metals are usually the strongest and hardest, see Fig. 19.33, and consequently the earliest metals that could be worked were the soft, low melting point metals such as

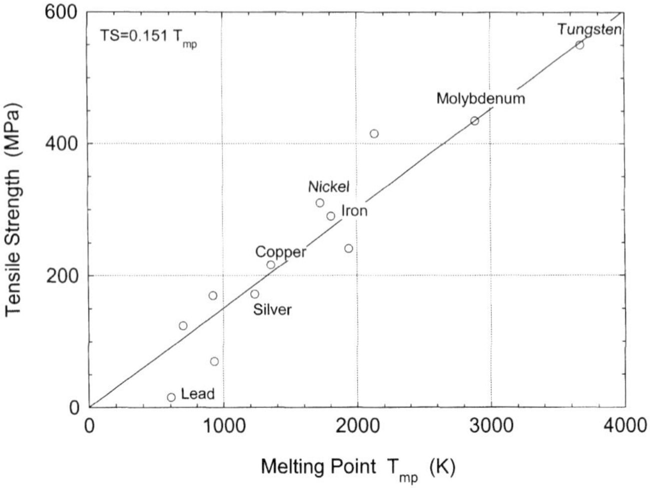

Fig. 19.33 Relation between melting point and tensile strength of metals.

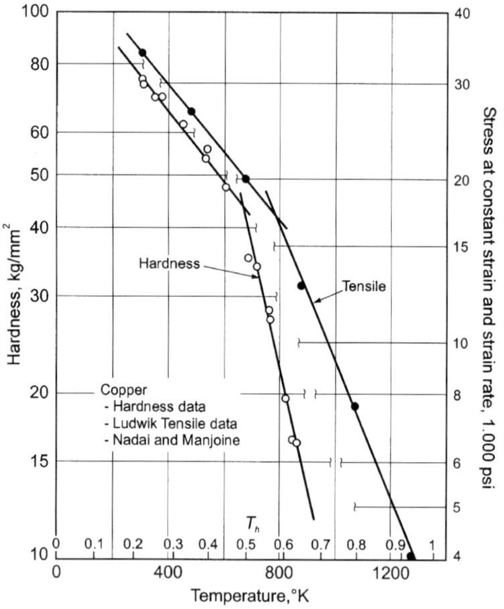

Fig. 19.34 Temperature dependence of the hardness and strength of copper. After Westbrook (1953, 221–47).

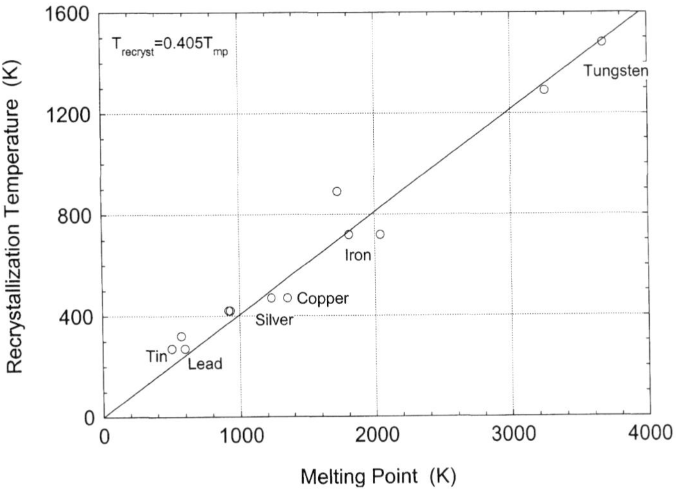

Fig. 19.35 Re-crystallization temperature and melting temperature for various metals.

lead, copper, gold, and silver. Iron, having a melting point of 1809 K and a tensile strength of 290 MPa was the hardest metal worked by pre-industrial methods.

Most metals retain their strength and hardness up to about half of their melting point but at temperatures above this the metal loses hardness and strength quite quickly and can be more readily worked. Fig. 19.34, for example, shows the variation of the hardness and strength of copper with temperature. In general the hardness and strength falls exponentially with increasing temperature but at temperatures above about a half of the melting point the rate of fall is quite rapid and the metal may have only 10% of its room temperature hardness and strength at about 80–90% of its melting point. This makes it possible to work even quite hard materials by forging, drawing through dies, rolling between rollers, or slitting with rollers.

The distinction between hot-worked and cold-worked metals is not precisely defined. When metals are worked at low temperatures the grains deform and stretch and the material often becomes harder to deform, that is, it work-hardens. At high temperatures, however, there is no work hardening and the grains re-crystallize continuously as they deform. The temperature at which re-crystallization starts for most metals is about 40% of the absolute melting point temperature, as shown in Fig. 19.35. Tin and lead re-crystallize at ambient temperature (300 K) and are, therefore, relatively soft and easy to deform, but because they are only a little above their re-crystallization temperature it takes some time for this to occur and they do work harden a little before softening again. Silver and copper are somewhat more difficult to work, and iron, which re-crystallizes at temperatures above 720 K (447°C) is the most difficult of the common pre-industrial metals. The re-crystallization temperature is the transition temperature between hot and cold working. But whether worked hot or cold a metal needs to be sufficiently ductile. Cast iron is much too brittle for excessive deformation at any temperature and some lead alloys melt at temperatures below 400 K and should be well into the hot working range at room temperature, but they also are brittle and unsuitable for working.

Two other important properties depend on whether a metal is worked hot or cold, namely the coefficient of friction and the yield stress. At high temperatures the coefficient of friction is also high and if the temperature is sufficient a metal can be welded to itself by hammering together its surfaces; a process that was familiar to ancient civilizations. At cold working temperatures the yield stress, at which a metal flows plastically, depends on the logarithmic or true strain and usually it increases somewhat with strain as the material work hardens. At hot working temperatures work hardening does not occur and the yield stress depends on the strain rate. Thus the equations for hot and cold work respectively are

$$
\begin{aligned}
Y &= K\varepsilon^{n} \\
Y &= C\left(\frac{d\varepsilon}{dt}\right)^{m} \qquad 19.1 \\
\varepsilon &= \ln(x / x_0)
\end{aligned}
$$

Y is the yield stress, ε is the true or logarithmic strain, x_0 is the initial length, x is the final length, K and C are the strength coefficients, n and m are the hardening coefficients. The first equation is applicable to cold working and the second to hot working. Typical values of K, C, n, and m are shown in Table 19.4 where it is apparent that hot working reduces the yield stress, and hence the necessary force, to 10–20% of the cold working force. Metals in pre-industrial societies were invariably worked hot for the simple reason that they did not have the powerful machinery to carry out cold working operations.

Many products could be cast to their finished dimensions directly when using, for example, the lost wax process to cast bronze, but for other metals this was not possible or not desirable. Furnaces could not melt iron until the late medieval period so the lost wax process was not possible. Also, one of the main reasons for working metals into their final shape rather than casting them directly was that worked metals usually have superior properties to cast metals. Most cast ingots require considerable hot working to refine and homogenise their crystal structure. Such hot working may reduce the cross-sectional area by 80% and improve properties such as impact strength and ductility. Cold working usually work hardens a material, increasing its tensile strength but reducing its ductility.

Table 19.4 Yield stress for hot and cold working.

	Hot Working			*Cold Working*			
Material	*Temperature °C*	*C MPa*	*m*	*K MPa*	*n*	*Reduction in Area %*	*Annealing Temperature °C*
0.08%C steel	1000	100	0.1	600	0.25	70	850–900
0.45%C steel	1000	120	0.13	950	0.12	45	790–870
Copper	900	41	0.2	450	0.33	78	375–650
Brass 30%Zn	800	48	0.15	500	0.41	75	425–750
Phosphor Bronze 5%Sn	700	160	0.35	720	0.46	60	480–675
Lead	100	10	0.1			100	20–200
Zinc	75	260	0.1			–	100

In forging, the material is compressed between metal plates to reduce its thickness and increase its area and to achieve this the applied stress has to exceed the yield stress of the material. In Fig. 19.36 the initial area is A_0 and length is x_0, and after forging the area is A and the length is x. The volume of material is the product $V = Ax$ and the work done, W, is then given by the integral

$$W = \int_{x_0}^{x} F dx = \int_{x_0}^{x} YA dx = \int_{x_0}^{x} YV \frac{dx}{x} \qquad 19.2$$

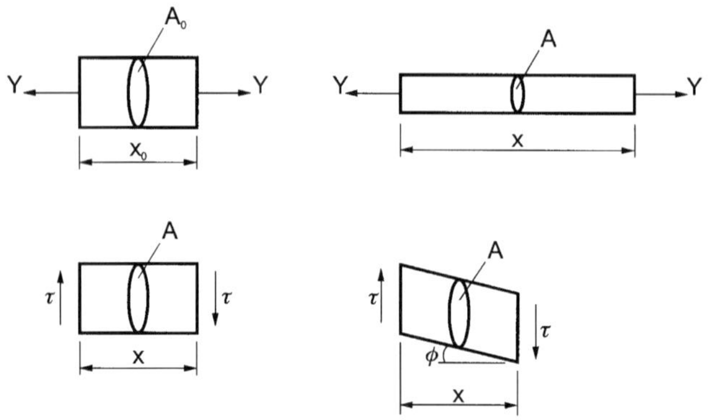

Fig. 19.36 Possible changes in shape during forging a simple bar. Top: homogeneous stretching. Bottom: shearing.

F is the applied force. If the volume and the yield stress are constant then the work done per unit volume in deforming the bar is

$$\frac{W}{V} = Y \ln\left(\frac{x}{x_0}\right) = Y \ln\left(\frac{A_0}{A}\right) \qquad 19.3$$

In practice the yield stress may not be constant, that is, the material may work-harden, and in this case an average yield stress may be assumed. Also, the yield stress may change with temperature so a value appropriate for the temperature of the material must be used. This type of deformation, in which rectangular elements remain rectangular as they change shape, is called *homogeneous*.

Equ. 19.3 is the lower limit to the work required to plastically deform the material and normally the required work exceeds this because additional work is done to overcome friction between the moving surfaces and also the material may experience other forms of deformation, such as shearing, which is also illustrated in Fig. 19.37. The work done in a plastic shearing process is

$$W = \int_0^{\varphi} Fx d\varphi = \int_0^{\varphi} \tau A x d\varphi = \frac{1}{2}\int_0^{\varphi} YV d\varphi \qquad 19.4$$

φ is the shear strain, τ is the applied shear stress and to create plastic deformation this, for many materials, is equal to half of the yield stress, $\tau = Y/2$. If the volume and yield stress are constant then Equ. 19.4 integrates to

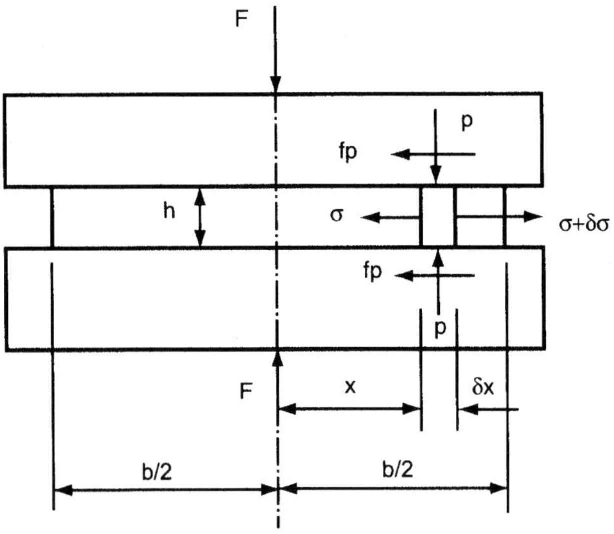

Fig. 19.37 Forging a metal slab between plates.

$$\frac{W}{V} = \frac{Y\varphi}{2} \qquad 19.5$$

If the purpose of an operation is to elongate a bar then such shear work is superfluous to the operation and if it occurs it is referred to as *redundant* work. Similarly work done against friction is redundant work and all forms of redundant work are sources of inefficiency.

Forging

Fig. 19.37 shows a thin metal slab, at some instant in time, being compressed between rigid plates. The geometry changes continuously during the process. The method of analysis is that described, for example, in Rowe (1977, 251). The pressure between the slab and the plate is p and varies over the slab face. As the slab deforms outwards it overcomes friction at the interface of magnitude fp. The stress in the x-direction is σ at distance x from the centre of the slab and is assumed to be uniform over the thickness of the slab. The slab thickness is h and slab breadth is b, both of which vary with time. The slab width is w (into the plane). For equilibrium in the x-direction of an element of length δx we may write

$$\sigma hw + 2fpw\delta x = (\sigma + \delta\sigma)hw \tag{19.6}$$

Thus, as δx and $\delta\sigma$ approach zero we can write

$$\frac{d\sigma}{dx} = \frac{2fp}{h} \tag{19.7}$$

The yield criterion is

$$p + \sigma = Y \tag{19.8}$$

Y is the average yield stress of the material and is constant, consequently any change in the stress σ results in a compensating change in the contact pressure p. Differentiating Equ. 19.8 and substituting from Equ. 19.7 gives

$$\frac{dp}{dx} = -\frac{d\sigma}{dx} = -\frac{2fp}{h} \tag{19.9}$$

Equ. 19.9 may be integrated to give the pressure distribution between the plate and the slab

$$p = A\exp\left(\frac{2fx}{h}\right) \tag{19.10}$$

A is a constant of integration, the value of which may be determined from the boundary condition, $\sigma = 0$ and therefore $p = Y$ at $x = b/2$, thus

$$\begin{aligned} \frac{P}{Y} &= \exp\left(\frac{f(b-2x)}{h}\right) \quad 0 < x < b/2 \\ \frac{P}{Y} &= \exp\left(\frac{f(b+2x)}{h}\right) \quad -b/2 < x < 0 \end{aligned} \tag{19.11}$$

This result is shown in Fig. 19.38 for four different values of $2fb/h$ (0.5, 1.0, 1.5, and 2.0). For the case of zero friction the pressure is constant and equal to the yield stress of the material. The contact pressure is equal to the yield stress at the outer edges of the slab and rises to a maximum value at the centre of the slab, the value of which depends on the coefficient of friction and the geometric ratio b/h.

If friction is low then the maximum pressure, p_{max}, at the centre of the plate and the average pressure over the width of the plate, p_m, may be approximated as

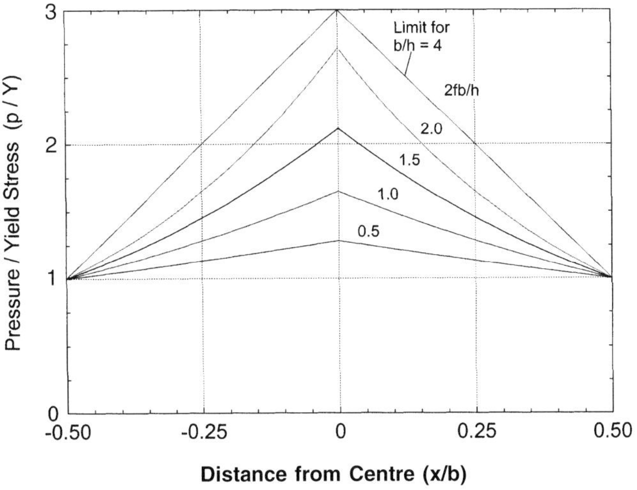

Fig. 19.38 Pressure distribution between plate and slab during compression (forging).

$$\frac{p_{\max}}{Y} = 1 + \frac{fb}{h}$$
$$\frac{p_m}{Y} = 1 + \frac{fb}{2h} \qquad 19.12$$

From this the applied force F is determined as the product of the mean pressure and the slab area, $F = p_m bw$.

This analysis is valid for low coefficients of friction ($f < 0.5$) but for high coefficients of friction, that often occur when hot working metals, the shear stress on the upper and lower faces of the slab, fp, may reach the critical value for shear failure, which for many metals is $Y/2$. If the friction is sufficiently high then this condition extends over the entire face of the slab and Equ. 19.9 is then written as

$$\frac{dp}{dx} = -\frac{d\sigma}{dx} = -\frac{Y}{h} \qquad 19.13$$

Integrating this and using the boundary condition that $p = Y$ at $x = b/2$ gives the result

$$\frac{P}{Y} = 1 + \frac{b}{2h} - \frac{x}{h} \qquad 0 < x < b/2 \qquad 19.14$$

$$\frac{P}{Y} = 1 + \frac{b}{2h} + \frac{x}{h} \qquad -b/2 < x < 0 \qquad 19.14$$

Thus the pressure distribution is now independent of the coefficient of friction and depends only on the yield stress and the ratio b/h. Equ. 19.14 is also plotted in Fig. 19.38 for the case of $b/h = 4$. As friction increases the centre of the plate reaches the limiting condition first and then it spreads out towards the edges of the plate until the entire plate follows the limiting, linear distribution. The coefficient of friction at which the limit is reached is found by equating Equ. 19.11 and 14 and putting $x = 0$. As an approximation, this occurs when $f = 0.5$. Thus Equ. 19.11 is valid if $f < 0.5$, and if friction is greater than this then it should be set equal to 0.5 or Equ. 19.14 should be used.

From Equ. 19.12 it is apparent that the contact pressure approaches infinity as the plate thickness approaches zero and thus there is a limit to the thickness to which a slab can be compressed. If the maximum pressure that can be applied to the plate is p_{max} then the minimum thickness is

$$h_{\min} = \frac{fbY}{(p_{\max} - Y)} \qquad 19.15$$

To produce a very thin slab by pressing between plates the coefficient of friction should be small, the plate length should be small, and the maximum pressure tolerated by the plates should be high in comparison to the yield stress of the slab material. The maximum pressure that can be applied to the plates is likely to be limited by their deflection, which may be comparable to the slab thickness.

Example

Determine the minimum thickness of plate that can be pressed if the coefficient of friction is (i) 0.5 and (ii) 0.05. The plate width is 0.2 m and the maximum pressure on the plates is 5 times the yield stress of the material being pressed.

From Equ. 19.15 h_{min} = 0.025 m (25 mm) if friction is 0.5 and 0.0025 (2.5 mm) if friction is 0.05.

Another simple case of considerable interest is compression of a circular disc. Rowe (1977, 257), for example, has shown that for plastic compression of a circular disc between overlapping rigid plates the pressure distribution is given by

$$\frac{P}{Y} = \exp\left(\frac{2f(r_0 - r)}{h}\right) \qquad 19.16$$

r is radius, r_0 is the outer radius of the disc, and the other symbols are defined above. The average pressure may be approximated as

$$\frac{p_m}{Y} = 1 + \frac{fr_0}{h} \qquad 19.17$$

Example

What load is required to forge a 0.6 m diameter steel disc to reduce its thickness from 0.3 m to 0.24 m between flat, rigid plates? The steel is at 900°C where the yield stress is 30 MPa and friction is 0.5.

The disc radius when compressed to 0.24 m increases to r_0 = 0.671 m, assuming the volume is constant. From Equ. 19.17 the mean pressure is p_m = 71.94 MPa, the disc area is 1.271 m^2, and the steady forging force is 91.43 MN (9320 tonnes).

Rolling

Fig. 19.39 shows the forces and geometry for simple rolling. A plate of width w and thickness b passes between a pair of rollers of radius R where it is reduced to thickness a. The reduction in thickness is $\Delta h = b - a$. The vertical force compressing the plate has magnitude F and the coefficient of friction is f. At the instant the plate enters the rollers the radial contact force is P, and fP is the frictional force dragging the plate into the rollers. If the plate is pulled through the rollers unaided then the horizontal component of the frictional force fP must exceed the horizontal component of the radial force P of the plate. This occurs when

$$f \geq \tan(\theta) \qquad 19.18$$

The horizontal distance over which contact occurs between the rollers and the plate is L and the angle of contact is θ. From the geometry of the roller, and assuming the change in plate thickness is small compared to the roller radius, we have

$$L = \sqrt{R\Delta h} \qquad 19.19$$

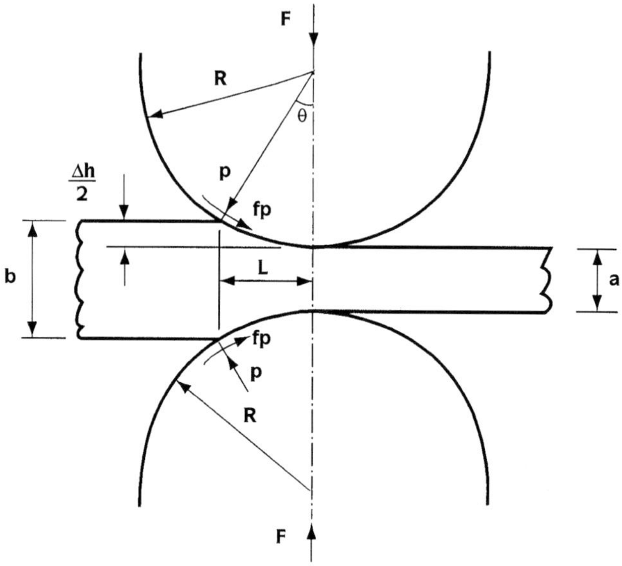

Fig. 19.39 Geometry and forces during rolling.

From the geometry we have

$$\tan(\theta) = \frac{L}{R} = \sqrt{\frac{\Delta h}{R}} \leq f \qquad 19.20$$

Thus the maximum reduction is $\Delta h_{max} = f^2 R$ and for a large reduction the friction and the roller radius must be large. The vertical force F required to restrain the rollers and compress the plate may be estimated as the product of the average contact pressure between the roller and the plate, p_m, and the area of contact between the plate and the roller, wL:

$$F = p_m w L \qquad 19.21$$

Now the average contact pressure for compression has been determined in the previous section (Equ. 19.12) and allowing for the different notation and taking the average depth as a+Δh/2 we have

$$p_m = Y\left(1 + \frac{fL}{(a+b)}\right) \qquad 19.22$$

From Equ. 19.18, 19.20 and 19.21 we obtain the simple result

$$\frac{F}{wRY} = \sqrt{\frac{\Delta h}{R}}\left(1 + \frac{f\sqrt{R\Delta h}}{a+b}\right) \qquad 19.23$$

$$\sqrt{\frac{\Delta h}{R}} < f$$

The radius of the rollers needs to be large to ensure that they "bite" and draw the plate through the rollers but this also increases the vertical force so it should not be excessive. Most cold rolling involves plate that is wide and thin and so it is more accurate to assume plane-strain conditions, for which the yield stress exceeds Y by about 15%, Rowe (1977, 208–50).

$$\frac{F}{wRY} = 1.15\sqrt{\frac{\Delta h}{R}}\left(1 + \frac{f\sqrt{R\Delta h}}{a+b}\right) \qquad 19.24$$

A source of error in this equation is the assumption that the radius of the rollers is constant when, in fact, the high loads normally encountered in rolling usually deform the rollers, thus increasing the roller radius somewhat. It is usually sufficient to replace R by a modified roller radius R_1, determined from

$$R_1 = R\left(1 + \frac{c}{\Delta h}\frac{F}{w}\right) \qquad 19.25$$

$$c = \frac{16(1-m^2)}{\pi E}$$

m is Poisson's ratio, E is the modulus of elasticity, and the other terms are already defined.

Equ. 19.23 works quite well but a more accurate allowance for friction is due to Ekelund (1933, 8–14). For cold rolling he found an equation similar to that derived above, namely

$$\frac{F}{wR_1Y} = 1.15\sqrt{\frac{\Delta h}{R_1}}\left(1 + \frac{1.6f\sqrt{R_1\Delta h} - 1.2\Delta h}{a+b}\right) \qquad 19.26$$

For hot rolling Ekelund (1933, 8–14) needed to take account of the influence of strain rate on the mean yield stress and, of course, for hot rolling the coefficient of friction is higher. He proposed

$$\frac{F}{wR_1 Y} = \left(1.15 + \frac{2v\mu\sqrt{R_1\Delta h}}{Y(a+b)}\right)\sqrt{\frac{\Delta h}{R_1}}\left(1 + \frac{1.6f\sqrt{R_1\Delta h} - 1.2\Delta h}{a+b}\right)$$

$$\mu = 1373 + 0.0983\Delta T \qquad 19.27$$

$$f = 0.84 - 0.0004\Delta T$$

$$Y = 100\mu(1.4 + C + Mn + 0.3Cr)$$

The temperature rise ΔT is °C, the rolling velocity v is mm/s, and the yield stress Y is kN/mm^2.

The maximum pressure that is exerted on the rolls is

$$p_{max} = Y\left(1 + \frac{2f\sqrt{R\Delta h}}{(a+b)}\right) \qquad 19.28$$

If the reduction in thickness is taken as the maximum possible, $\Delta h = f^2R$, then the minimum thickness of plate that can rolled is

$$a + b = \frac{2f^2RY}{p_{max} - Y} \qquad 19.29$$

Thus, to roll thin sheet it is necessary to use small diameter rollers, to reduced friction to the minimum, to roll materials with a low yield stress, and to use rollers that can withstand a high contact pressure. However, a small roller radius and a large roller pressure lead to excessive deformation and bowing of the rolls relative to the thickness of the strip, which eventually prevents further reduction. The phenomenon is discussed in more advanced treatments of rolling theory but it is sufficient to note that the minimum thickness depends on the elastic constant c defined in Equ. 19.25, the coefficient of friction, the roller radius, and the yield strength of the strip, thus

$$a_{min} \propto cfRY \qquad 19.30$$

For steel rolls a commonly used equation is

$$a_{min} = 35x10^{-3}fRY \qquad 19.31$$

Wire Drawing

Fig. 19.40 shows the geometry of a metal rod or wire being drawn through a set of conical dies. A force F draws the wire through the

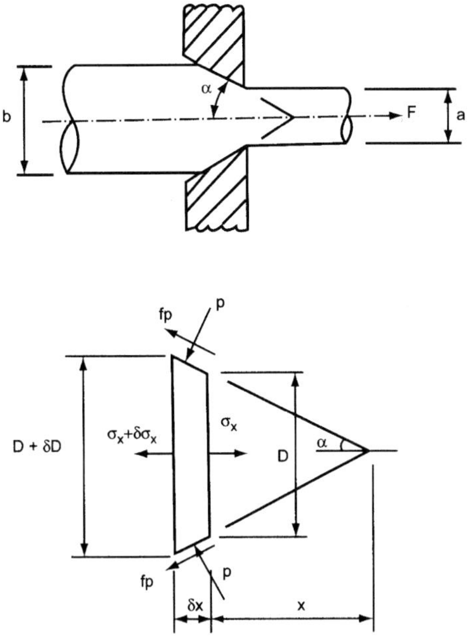

Fig. 19.40. Top: geometry of wire drawing. Bottom: the forces acting on an element.

dies having a cone angle α and reduces the wire diameter from b to a. The figure also shows the forces acting on an element of length δx, diameters D, and $D+\delta D$. The axial stresses acting on the opposite faces of the element are σ_x and $\sigma_x+\delta\sigma_x$, and the contact pressure at the die face is p. The friction between the die and the rod is fp. The shear stress acting on the faces is ignored. For equilibrium in the axial direction we have

$$\sigma_x \pi D^2 / 4 = (\sigma_x + \delta\sigma_x)\, \pi\, (D + \delta D)^2 / 4 + p\pi D\delta x \tan(\alpha) + fp\alpha D\delta x \qquad 19.32$$

This can be written as the integral equation

$$\int \frac{dD}{D} = -\frac{1}{2} \int \frac{d\sigma_x}{\sigma_x + p + Bp}$$
$$B = \frac{f}{\tan(\alpha)} \qquad 19.33$$

For radial equilibrium of the element

$$\sigma_r \pi D\delta x = -p\pi D\delta x + fp\pi D\delta x \tan(\alpha) \qquad 19.34$$

Thus

$$\sigma_r = -p\,(1 - f\tan(\alpha)) \qquad 19.35$$

As the values of f and α are small their product is very small in comparison to 1 and so the radial stress is approximated as $\sigma_r = -p$. If the material is assumed to deform plastically at a constant yield stress Y and the yield criterion is $\sigma_x + p = Y$ then Equ. 19.33 may be written as

$$\int \frac{dD}{D} = -\frac{1}{2}\int \frac{d\sigma_x}{(1+B)Y - B\sigma_x} \qquad 19.36$$

The solution of these integrals is

$$\ln\left[\frac{D}{[Y(1+B) - B\sigma_x]^{1/2B}}\right] = c \qquad 19.37$$

c is the constant of integration the value of which may be determined from the boundary condition $\sigma_x = 0$ when $D = a$. Thus

$$D = b\left[1 - \frac{B\sigma_x}{(1+B)Y}\right]^{1/2B} \qquad 19.38$$

The force F required to draw the rod through the die is the stress $\sigma_x = \sigma_a$ at diameter $D = a$; substituting this condition and re-arranging the equation gives the result

$$\frac{\sigma_a}{Y} = \frac{F}{\pi a^2 Y} = \frac{1+B}{B}\left[1 - \left(\frac{a}{b}\right)^{2B}\right] = \frac{1+B}{B}\left[1 - (1-r)^B\right]$$

$$r = 1 - \left(\frac{a}{b}\right)^2 \qquad 19.39$$

$$B = \frac{f}{\tan(\alpha)}$$

r is the reduction in area.

For the particular case when friction is zero ($B = 0$), Equ. 19.38 is indeterminate but a solution is obtained by integrating Equ. 19.36, with $B = 0$, to give

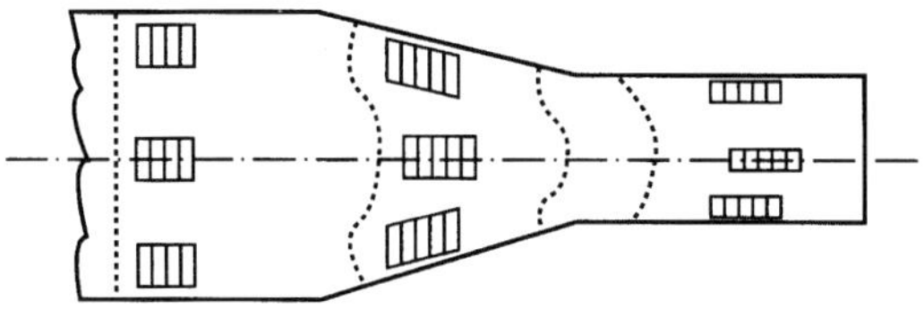

Fig. 19.41 Deformation of elements passing through the dies.

$$\frac{\sigma_a}{Y} = \frac{F}{\pi a^2 Y} = \ln\left(\frac{b}{a}\right)^2 = \ln\left(\frac{1}{1-r}\right) \qquad 19.40$$

$$r = 1 - \left(\frac{a}{b}\right)^2$$

The maximum drawing stress that can be applied is $\sigma_x = Y$, at which point the rod will break, thus the maximum reduction in area at a single pass is given by

$$r_{\max} = 1 - (1 + B)^{-1/B} \qquad 19.41$$

If $B = 0$ then $r_{max} = 0.632$ and if $B = 1$ then $r_{max} = 0.5$; in general the maximum reduction in area varies between these limits and a typical value ($f = 0.05$, $\alpha = 15°$, $B = 0.1865$) is about 60%.

In this analysis we have ignored the redundant work, which is related to the shear strains generated in the rod as it deforms, Fig. 19.41. Near the die face the metal rod experiences shear strain equal to the semi-angle of the die (α), but metal moving along the centre line experiences no shear strain, thus the average strain is $2\alpha/3$ and this has sometimes been used to allow for redundant work. However, the degree of reduction also affects the redundant work and Johnson, in Rowe (1977, 135) has used a correction φ, in conjunction with Equ. 19.39, as shown below. Johnson found this to apply to a large range of reduction ratios and die angles.

$$\frac{\sigma_a}{Y} = \frac{F}{\pi a^2 Y} = \varphi \frac{1+B}{B}\left[1 - (1-r)^B\right] \qquad 19.42$$

$$\varphi = 0.87 + \frac{1-r}{r}\sin(\alpha)$$

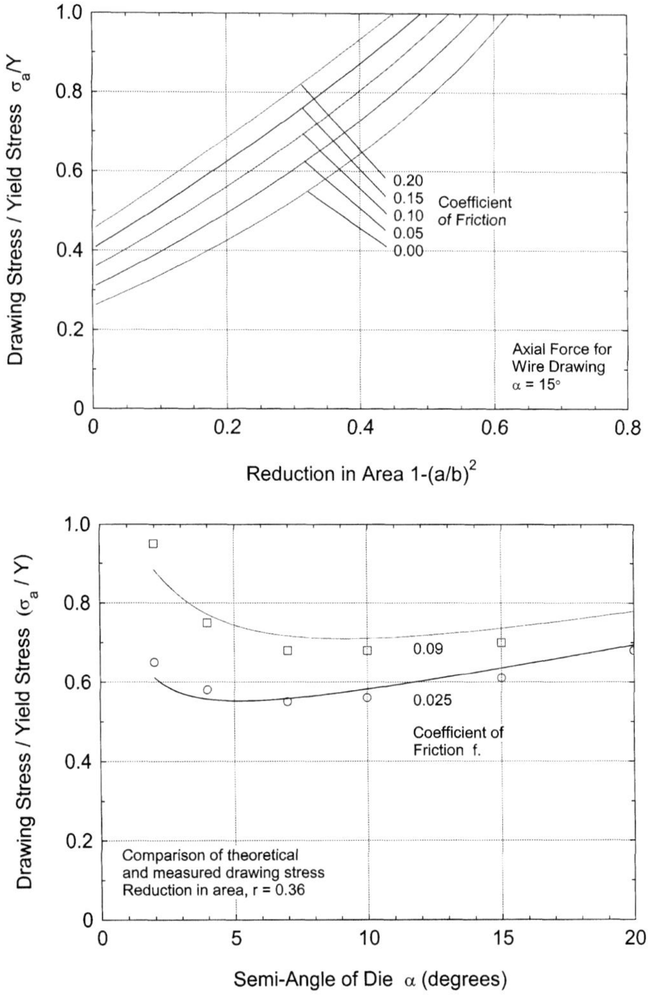

Fig. 19.42 Top: variation of drawing stress with reduction in area showing the influence of friction. Bottom: comparison of theoretical and measured drawing stress and the optimum die angle for a reduction in area of 0.36.

The variation of axial drawing stress with reduction in area is illustrated in Fig. 19.42 (top). The comparatively large stress required for very small reductions in area occur because the wire bulges slightly before it enters the die and for small reductions the bulge is comparable in size to the reduction. Friction, as can be seen, has a marked effect on the drawing stress and in practice lubrication is essential. The maximum reduction that can be attained with a single

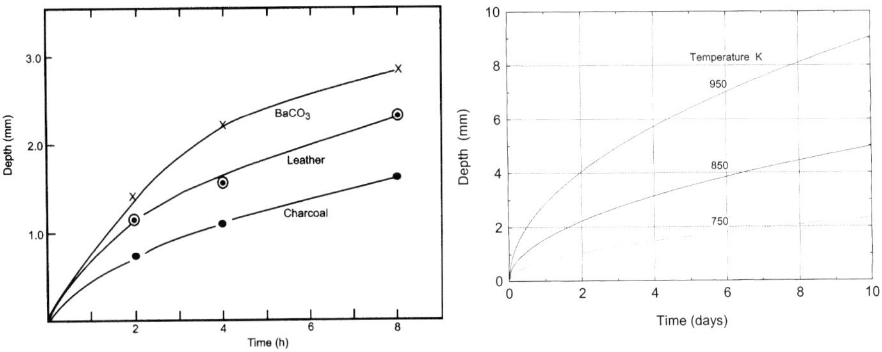

Fig. 19.43 Left: time required to carburise iron to 0.85%C when heated to 900°C and packed in various media. Right: depth of carburised layer for penetration of iron by carbon.

pass occurs when the drawing stress equals the yield stress and this occurs at reductions in the range 0.5 to 0.6 depending on the coefficient of friction. Experimental verification of Equ. 19.42 is shown in Fig. 19.42 (bottom) for a reduction of r = 0.36. Two types of lubricant, having coefficients of friction of about 0.025 and 0.09, were used. The figure also shows the variation of drawing stress with die angle and the optimum die angle for minimum drawing stress. Although there is not a marked variation in drawing stress it is nevertheless important to avoid either very low or very high angles.

Carburising

Fig. 19.43 (left) shows some data of Shaw Scott (1907) on the time required to carburise iron to various depths using different carburising media. This is a diffusion process and like all such processes the depth of the diffused surface layer can be calculated from

$$x = \sqrt{4Kt}$$
$$K = A exp(-B / RT) \qquad 19.43$$

x is the depth, K is the diffusivity of the carbon in iron, t is time, T is temperature, R is the universal gas constant (1.987 cal/mol/K), and A, B are constants. For carbon diffusing in iron $A = 6.2 \times 10^{-3}$ cm/s and B – 19200 cal/mol, Smithells (1976, 874). Equ. 19.43 is plotted in Fig. 19.43 (right).

Example
Determine the depth of the diffused layer when iron is held in contact with carbon at 950 K for a period of (a) 10 hours and (b) 10 days.

From Equ. 19.43 $K = 2.37 \times 10^{-7}$ cm/s and (a) $x = 1.85$ mm after 10 hours (b) $x = 9.05$ mm after 10 days. This is the maximum depth by which carbon penetrates and thus the carbon concentration is 0% at this depth. The depth for 0.8% C is less than this.

CHAPTER TWENTY

TEXTILE MANUFACTURE

Introduction

Textile production is a very ancient activity and during the pre-industrial era it was mostly carried on in the home or in small workshops. In Europe there were five main sources of fibre. Two sources were from animals, namely wool and silk and the other three sources of fibres were from plants, namely, flax and hemp, from which linen was made, and cotton. Silk and cotton, of course, were exotic products not native to Europe and were imported originally from the East. They were, for many purposes, superior to wool and linen and they were always in great demand. The burdens textile production placed on mechanical engineering were mainly in the manufacture of machines for preparing and spinning the thread, for weaving the cloth, and for fulling and other finishing processes. Progress in all these areas was slow but steady until the eighteenth century when a spate of new inventions in textile machinery augmented and then revolutionised the methods of production. These machines eventually required water or steam power to operate most productively and this, as is well known, led to the introduction of large-scale factory production with concomitant economic benefits and social disruption. It could be argued, and with some justification, that the demands of the textile industry, particularly the cotton industry around Manchester, created the engineering profession in the modern sense. Textile manufacture was by far the largest and most prosperous export industry in the United Kingdom in the nineteenth century, as it had been in the fourteenth century. It was larger than engineering, coal, machinery, iron and steel and most other industries.

It was, perhaps, inevitable that industrialisation should have been triggered by textile manufacture because, as described below, it escaped from its usual state of domesticity and developed to very large proportions on at least three occasions before it finally became firmly based in factories. The first occasion was in ancient Mesopotamia, in about 2000 B.C., the second was in ancient Greece, and the third occasion was in the Low Countries and Italy in the thirteenth and fourteenth centuries A.D. In this chapter the development of textile machinery in the pre-industrial era is briefly described along with those inventions that first ushered in factory production.

Prehistory and Ancient Civilisations

During the last ice age Neanderthal and modern man survived, in part at least, due to the warmth and protection afforded by fire, cave dwellings, and clothing. The earliest cave paintings, however, are mainly of animals and when humans are depicted they appear to be naked, although in a few examples a case can be made that some form of clothing is worn. Later postglacial paintings from Spain clearly show humans wearing feathered headdresses and although these paintings are mainly monochrome silhouettes they nevertheless are very lively and do convey the impression of clothing. Whitehouse, in Collins (1975, 103) says that some of the best evidence of the early use of skin clothing comes from two sites in southern Siberia. A statuette from Buret' seems to be fully clothed in a tailored Eskimo style skin, Fig. 20.1, and three smaller figures from Mal'ta have the fur indicated by small incisions. Carbon 14 dates for these objects is about 12,800 bc, say about 16,000–12,000 B.C. Finely made eyed-needles are found, in France, from about 17,000 bc and their survival is an indication that tailored clothing (animal skin) was in use.

During the Mesolithic and Neolithic periods the old hunter-gatherer economy gave way to one based on animal husbandry, grain production, animals used for draft purposes, permanent houses, pottery, and flint mining. Stern (1970, 245) is of the opinion that when an agricultural way of life replaced a hunting economy, animal skins become scarce, so a new kind of material for making garments was needed. But in this he is probably wrong because animal husbandry surely makes animals more readily available than does hunting, and therefore skins become more plentiful. It was not a shortage of skins that led to the invention and use of textiles but, perhaps, the desire for a superior product. Because some experience had been gained during the Mesolithic and Neolithic periods in weaving fishnets and baskets and in the manufacture of long cords, perhaps of sinew, Stern suggests that the way was ready for weaving cloth and for making felt by compressing wet fibres together.

Fig. 20.1 The clothed figure from Buret, 16,000–12,000 B.C. After Collins (1975, 104).

The most useful indication of when and where cloth was spun and woven is the huge number of stone or baked clay loom weights that were used from the earliest times to hold the warp taut. Although the cloth and even the wooden loom may have perished, the loom weights survive. In buildings that were destroyed suddenly, by fire for example, the loom weights are often found in a straight line where they fell. Similarly the existence of spindle whorls, Fig. 20.2, is a sure and surviving indication of spinning, even when the spindle and the thread have perished. They commonly occur on archaeological sites from the Neolithic period onwards. For instance, more than 8,000 were found at Troy alone.

Organic materials such as textiles survive best in the wet airless conditions found in bogs, or in the dry conditions found in Egypt, where they were woven from about 5,000 B.C. Linen, of course, was used from 3500 B.C. onwards in mummification and large quantities have survived. Many old European cloth specimens were found in the Swiss lakeside villages where undisturbed submergence in water has preserved them. Ancient fragments of clothing have also been found on the permanently frozen bodies of Scythian chieftains who were buried deep in underground tombs, but these are of more recent date.

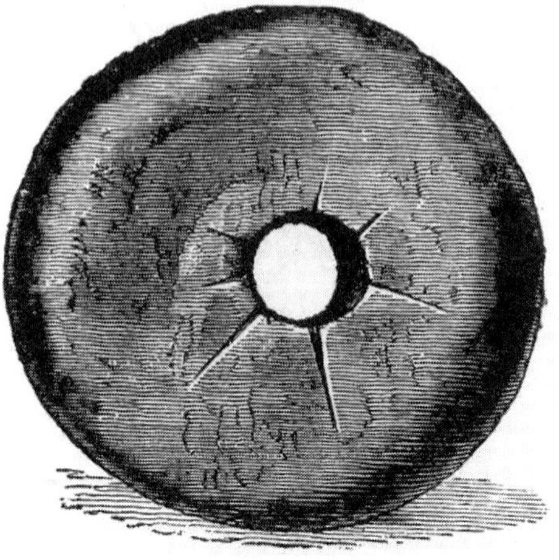

Fig. 20.2 A sandstone spindle whorl from Holyhead.

The earliest fully preserved and clothed human body was found in 1991 by German hikers at a height of 3,200 m on the Similaun glacier in the Ötztaler Alps in the south Tyrol, see Renfrew and Bahn (2000, 66). He is usually known as "The Iceman" or as "Ötzi". Fifteen radiocarbon samples give dates in the range 3365–2940 B.C., and average about 3300 B.C. Originally he was thought to have died from exhaustion, but recently an arrowhead has been found in his shoulder and his cause of death is not yet determined. His body was preserved in a depression in the glacier. He is unique in being the first natural, as opposed to interred or sacrificed, body to survive from prehistory and the 70 objects found with him constitute an invaluable "time capsule". His clothes included: a leather belt and leather loincloth, leather leggings covering three quarters of each leg, a long sleeved coat of tanned domestic goat hide, shoes with bearskin soles and deerskin uppers and filled with grass, a bearskin cap, and finally a long outer cape of woven grass or reeds, see Fig. 20.3. The emphasis here is on leather for only one woven item was found. This suggests that weaving was first applied to naturally occurring materials such as grasses and reeds before more appropriate spun fibres were invented.

In Egypt, during the Old Kingdom, Breasted (1912, 96) says that flax was plentifully cultivated, and a noble of high rank controlled

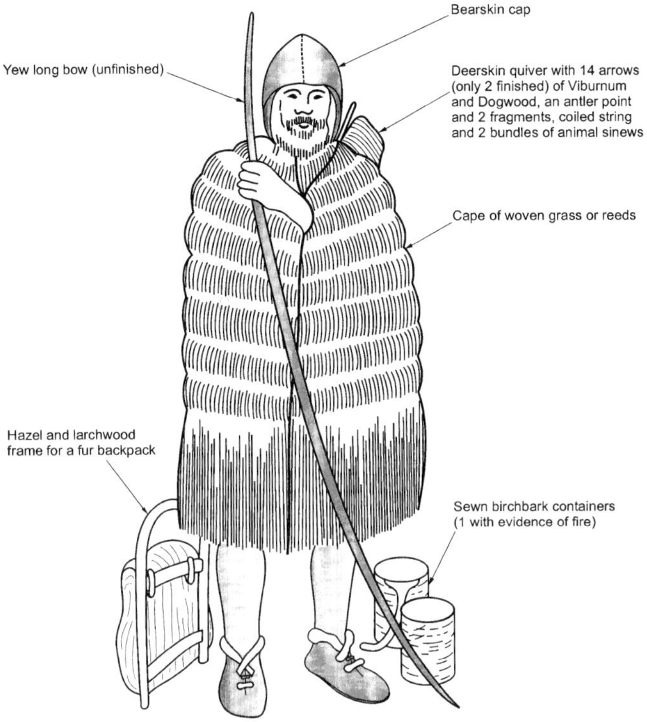

Fig. 20.3 The clothing of the "Iceman"; about 3200 B.C. After Renfrew and Bahn (2000, 67).

the Pharaoh's flax harvest. Spinning and weaving on the great estates was the work of women serfs. Coarse varieties of cloth were usually of good quality, but expensive clothes and fine cloths had already developed to the point where they had become the most immediate and obvious indications of wealth and importance. Skill in spinning and weaving fine flax had already developed to a very high level. The surviving specimens of the royal linens are of such exquisite fineness that the ordinary eye requires a glass to distinguish them from silk, and as the surviving wall paintings show, the limbs of the wearer could be discerned easily through the fine fabric. Fine and expensive garments were to remain one of the essential attributes of success.

In Mesopotamia, Postgate (1994, 235–40) says that textiles had developed to industrial proportions by the end of the third millennium and was an important industry relying on huge flocks of sheep for

Fig. 20.4 Statuette of the time of Gudea "Patesi" of Lagaš, after 2,600 B.C. Louvre, Paris.

wool and flax for linen. Most of the evidence comes from institutional (temple and palace) records and one text records some 600 tonnes of wool from the capital city in the Ur III period (2150–2000 B.C.). Postgate says that this is entirely consistent with the contemporary Lagaš archives that mentions a factory in Guabba of over 6,000 workers, the great majority being women and children. The state employees worked in these factories for specified periods and outside these periods they were free to work on their own account. They appear not to have been slaves. Payments were made in kind and were mostly of grain, wool or cloth, and oil, but other commodities are sometimes mentioned. The fine quality of clothing made at this time is illustrated in Fig. 20.4, which shows a statuette of the time of Gudea "Patesi" of Lagaš, in 2,600 B.C. or later. The dress is typical of Sumer and has inset embroidered or woven bands.

The following labour account from Postgate (1992, text 12.6) records the level of organisation and specialization fostered in the Ur III period. The types and grades of cloth were specified with great precision, and the quantities produced were recorded along with the work-days required.

From the labour accounts of a textile workshop at Larsa
1 robe, fine, *laharītum*, 6 minas (3 kg):

4 months	plaiting
2 days	cleaning
2 days	trimming
1 month	sewing up
2 days	beating
2 months	teasing
4 days	stripping and—ing
1 month	sewing, teasing and finishing
4 days	stretching and pressing
1 month	releasing
1 month 10 days	blowing

384 textile-workers (days), quota for one fine *laharītum* robe of 6 minas (3 kg).

[There follows similar entries, then a sub-total:]

134 textile workers (days), quota for a second quality robe of 6 minas; 536 textile-workers (days), quota for 4 second quality robes of 6 minas each; 3,390 (daily) quotas (and) 150 holidays = 3,546 textile workers (days), quota and holidays, deducted giving a remainder of 1,134 textile-workers (days) from Ipqu-Adad.

Such expensive garments are also recorded in Anatolia, in the Hittite Law Code. Gurney (1990, 70) records the controlled prices of garments as:

Fine garment	30 shekels
Blue woollen garment	20 shekels
Head band	1 shekel
Fine shirt	3 shekels
Large linen cloth	5 shekels

For comparison an ox for ploughing is valued at 15 shekels and a sheep at 1 shekel.

Greece and Rome

As illustrated in Chapter Three (Fig. 3.6), manual work was no disgrace in Homeric Greece and even the wife of Odysseus worked at the loom although she was well supplied with slaves. Of course she had an ulterior motive because she had promised her suitors that she would choose a husband when she had completed her weaving, but to avoid the decision she wove all day and unpicked every night, but nevertheless her preoccupation with the loom was not considered unreasonable for a person in her position as the wife of a great landowner and leader. Similarly Arete, Queen of the Phaeacians, span and wove in her palace, "For in the fine cloak and tunic she saw him [Odysseus] wearing she recognised some clothes that she herself had made with her women's help." And so it remained throughout most of history that spinning and weaving were equally acceptable for slave or mistress and remained mostly a domestic activity despite the early development of the factory system in Mesopotamia.

In the palace of Alcinous, where Arete was Queen (Odyssey, VII), we are told that 50 maids were employed. "Some grind the apple-golden corn in a hand mill, some weave at the loom, or sit and twist the yarn, their hands fluttering like the tall poplar's leaves, while the soft olive oil drips from the close woven fabrics they have finished. For the Phaeacians' extraordinary skill in handling ships at sea is rivalled by the dexterity of their womenfolk at the loom, so expert has Athene made them in the finer crafts, and so intelligent." The mention of olive oil dripping from the woven fabrics is interesting. Oil is often used on raw wool, after it has been washed and cleaned, to enable it to be combed and prepared for spinning more easily.

Glotz (1965. 181) notes that at Athens in the Archaic period the textile industry broke away from its domestic roots while allowing them to continue. It concentrated on the provision of luxury items but did not show any great volume of production. Meanwhile, more common stuffs were produced in the home. In the fourth century B.C. specialisation had made considerable progress and Glotz says that Athens specialised in the manufacture of men's clothing, Megara in the worker's *exomis*, Corinth in blankets and *kalasireis* of fine wool, and also made linen cloth. Pellene made cloaks, Patrae was filled with women working in its weaving workshops, Cos produced silk goods, Chios, Miletos, and Cyprus exported hangings, embroidered garments and carpets, Tara produced linen stuffs, and Syracuse transformed Sicilian wool into fine textiles of many colours. Textiles were produced in almost every home and in every town. The textile industry, even after it had become specialised, nevertheless produced only small quantities.

Xenophon (*Memorabilia* 2.7.5–6, 12, 14) records a conversation between Aristarchus and Socates that sheds light on cloth manufacture at Athens: "What of men's and women's cloaks, little tunics, capes, and sleeveless tunics?" Socrates said. "Indeed yes, all these are useful." Aristarchus replied. "Then," said Socrates, "don't the women in your household know how to make these things?" "I think they can make them all." "Really?" Don't you know that—Demeas of Collytus by making capes; Menon by making cloaks; and most of the Megarians keep well nourished by making sleeveless tunics?" "Yes, by Zeus," Aristarchus replied, "for these men have purchased barbarians [slaves] that they can force to make what is useful. But I have freeborn people and relatives"; see Humphrey et al. (1999, 396). The implication here is that Aristarchus would be unable to compete in the clothing trade because his freeborn relatives either do not work so hard as slaves, because they cannot be driven, or because they cost more in wages and keep. It is also clear that good profits are to be had in the cloth trade and eventually Aristarchus is convinced by Socrates to put his female relatives to productive work.

The main elements of a simple vertical, warp-weighted loom are illustrated in the Boeotian black-figured skyphos shown in Fig. 20.5. This is of the fifth century B.C. and illustrates the story of Odysseus and Circe, but the loom is of more interest. It consists of two vertical beams that rested, probably, against a wall. The cloth was wound onto a horizontal crossbeam at the top and the warp hung down and was tensioned by a row of small weights at the bottom. Alternate strands of the warp were divided into two groups, called sheddings, and held separate by means of a horizontal bar, the heddle, which held one shedding in loops of string. One shedding is at the rear and the other at the front, and the space between them is the shed. A shuttle with its spun thread is shown in Fig.20.5 and this is passed by hand through the individual warp threads leaving the horizontal weft behind. Necessarily the weft is laid some distance from the cloth and must be pushed up into position. The horizontal bar called the reed, on which the shuttle seems to rest, is used for this purpose. When it is raised the reeds, which are inserted between the warps, gently compress the weft into the woven cloth. It may be that the shuttle was woven in and out of the warp by hand but this is a tedious business and at some point it was discovered that a horizontal bar, the heddle, could be used to move the rear shed to the front or returned to the rear, thus creating the shed through which the shuttle could pass without having to weave

Fig. 20.5 A vertical, warp-weighted loom of the fifth century B.C. Boeotian black-figured skyphos from Thebes.

in and out of individual warp threads. The sequence of operations is thus:

1. Raising and lowering alternately the shedding of warp threads.
2. Passing the shuttle.
3. Driving up each weft-thread after the shuttle is passed using the reed.
4. Winding the cloth on the cloth roller.

Pollux (*Lexicon*, 10.124–5) has provided a list of the necessary equipment that should be in the women's quarters for spinning and weaving. Two items he mentions are unfamiliar, namely the *onos* and the *epinetron*. These were ceramic covers fitted and shaped to provide a stable base on the upper leg for twisting the rove before hand spinning a thread. Pollux writes: "The most essential equipment for use on a daily basis has already been discussed. In a few words one can pull together the rest of those items that pertain to women's quarters: woven baskets and baskets with narrow bottoms and the smaller types of both, the *onus* upon which they spin and the *epinetron* and the spindle and the circular whorls, the skein of yarn, the weaver's shuttle and the comb of the loom; and the upright loom as well as the side beams of the loom; and the weaver's rod [to attach the alternate threads of the warp] and the beam along the vertical beams

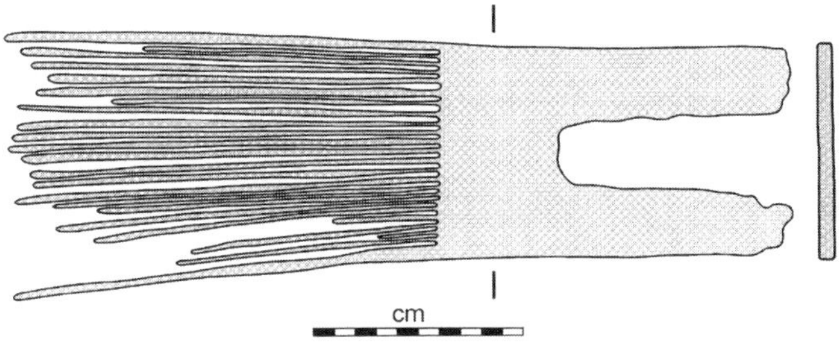

Fig. 20.6 Iron wool comb from Caistor-by-Norwich. After Liversidge (1973, 167).

of the loom [between which the web hangs down] and the long beams of the loom [for the warp threads] and the loom weights, and the flat blade [to strike the threads home]"; Humphrey et al (1999, 364).

This type of vertical loom required the weaver to stand at his work, since the weaving was done at the top and as the cloth is made it is wound on to the top bar to maintain a constant working height. During the Roman Empire a vertical double-beamed loom was developed and this allowed the weavers to sit as they worked. Servius (*On the Aeneid* 7.14) makes this clear when he writes: "—Our ancestors used to weave standing, just like today when we see the linen weavers." Similarly Festus writes: "The manly garments named *rectae* [upright tunics] are those which are woven upwards by standing weavers", Humphrey et al (1999, 364).

Before the yarn reached the weaver it had to go through numerous processes. After wool was shorn or plucked from the sheep or goat it was first cleaned to remove burrs and dirt and it was probably washed and dried and perhaps beaten to loosen the fibres. Wool was then either combed or carded. There is evidence, Fig. 20.6, that in Britain during the Roman period it was combed. As mentioned above, the wool could be sprayed with water and soaked in oil prior to combing and the comb would be heated and past through the wool until it adhered to the comb. The wool was then removed and was ready for spinning.

The preparation of flax and hemp was different from that of wool. The seed capsules were removed and the stalk fibres separated from the woody tissue by laying the stalks in warm water until they loosened from the bark (retting). The loosened fibres were laid in the sun and the woody tissue broken. After breaking, the tissues were laid across the end of a bench and beaten by striking the suspended fibres with the edge of a board. The flax was then drawn through the teeth of a comb to remove the last bark and produce the final strands suitable for spinning. The waste fibres from flax were often used to make sackcloth.

Spinning is the same for flax, wool and cotton and the purpose is to twist the short fibres together to grip each other tightly and form a continuous thread or yarn. Until the thirteenth century this was done with distaff and spindle. The distaff was a rod on which the prepared fibres were arranged. The spindle was a separate rod with a round weight, a whorl, attached to it. The spinner teased out and twisted some fibres from the distaff and attached them to the spindle.

Then the spinner set the spindle spinning and pulled out the fibres slowly. The spinning spindle twisted the pulled fibres into a single strong thread. When the thread was long enough for the spindle to reach the floor spinning stopped and the spun thread was wound onto the spindle and the process repeated until all the wool was spun. Fig. 20.7 shows some Roman spinning equipment, namely a bone handle probably from a distaff, a pottery spindle whorl found at Great Chesterford, and two bone needles.

Medieval Period

Throughout the Dark Ages in Europe cloth making continued in the home but little or no development occurred, and the techniques used in the ninth and tenth centuries A.D. would have been familiar to those living a millennium or more earlier. The vertical, warp weighted loom and the distaff and spindle continued in use and wool continued to be fulled by trampling underfoot. Nevertheless cloth making was an important activity. Amongst the Vikings who settled in Iceland in the late ninth century, cloth was commonly used as money, for silver was not particularly useful and their tough, homespun cloth became a more practical medium of exchange. Certainly, it was their chief export. The rate was initially 6 ells per ounce of silver but later it declined to 48 ells per ounce before recovering to 36 ells per ounce in about 1200; Foote and Wilson (1979, 55).

In England, wool production from huge herds of sheep became

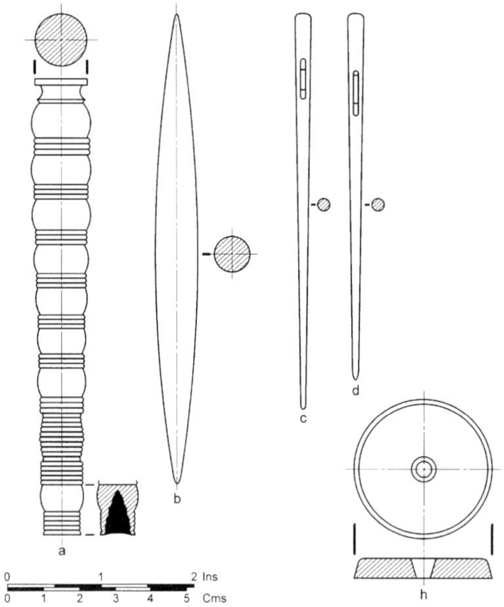

Fig. 20.7 a: bone handle, probably from a distaff, Great Chesteford. b, c, d: bone spindle and needles and pottery spindle whorl, Great Chesterford. After Liversidge (1973, 167).

a speciality of the monasteries, particularly the northern Cistercian monasteries such as Fountains, Rievaulx, Kirkstead, Revesby, and Spalding. A list of 180 wool producing monasteries, mainly Cistercian, was compiled by Pegolotti, a member of the Florentine merchant house of Bardi, in about 1297; Hockey (1970, 57). It quotes the prices for three grades of wool: best, medium, and lockwool (short clippings), as well as the number of sacks available annually. These figures are difficult to interpret and use as the numbers in each grade are not given, the monasteries may also have sold to, or bought from, other merchants, and what they sold was, in any case, the surplus after their own requirements had been met. The average Cistercian yield was 19½ sacks per monastery, while that of the Yorkshire houses was 38½ sacks and that of the eight southern Cistercian houses was 13½ sacks. One sack is reckoned to be about 364 lb (165 kg), so the Yorkshire houses were each producing about 3.2 tonnes of wool per year. The wool exported from the whole of England increased to about 5,000 tonnes per year by 1350.

English wool was considered to be, at that time, the finest available

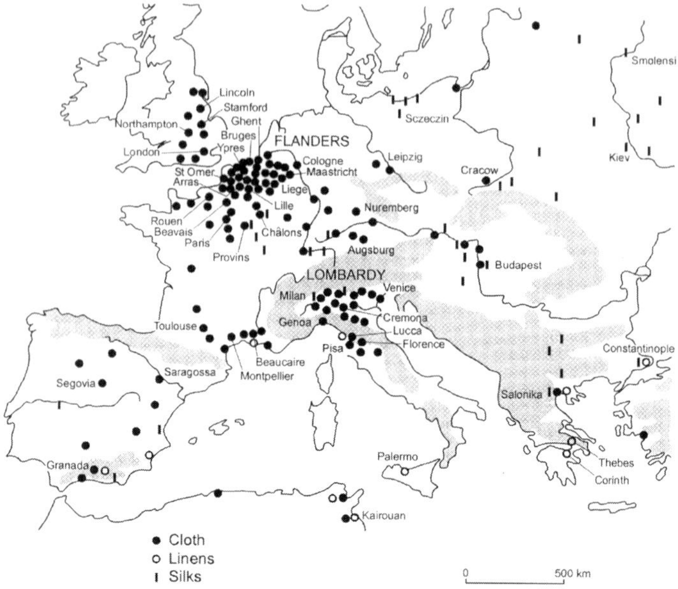

Fig. 20.8 The textile cities of Europe. After Gimpell (1993, 101).

and was eagerly sought after. It was unnecessary for English wool merchants to travel in Europe to sell their product but rather the customers came to England to buy up the supplies in advance of production. Thus the monastic houses sold, and were paid, for their wool before it was produced and they were often contracted to do so for several years in advance. Much trade took place at the weekly, monthly or annual fairs and markets that were springing up everywhere. But although English wool was in great demand it was made into fine cloth on the continent, for England at this time did not have the necessary industrial skill. The great textile centres of Europe grew up in the Low Countries and in northern Italy, as the map, Fig. 20.8, illustrates. Flemish towns such as Ypres, Ghent, and Bruges, French towns such as Arras, Saint-Omer, and Douai, and Italian towns such as Florence, Lucca, Venice and Cremona, were the chief centres. It was in these towns and cities that a capitalist economy developed in textiles and employed many thousands of wage earners. The Low Countries led the way and some important technical improvements occurred.

In the twelfth and thirteenth centuries remarkable progress was made in textile manufacture. First, in the twelfth century, the ancient vertical, warp weighted loom was largely replaced by the horizontal treadle loom, Fig. 20.9 (bottom). Secondly, at the end of the century, water powered fulling mills were introduced, and finally, in the thirteenth century, the hand spinning wheel first appeared, Fig. 20.9 (top). In none of these cases do we know much about their origins. It is, perhaps, important to realise that two types of spinning wheel developed. The earliest involved separate processes for spinning the thread and then winding the thread onto the spindle, as was done with a distaff and spindle. The second spinning wheel combined drawing out, spinning, and re-winding into a single process, but was not invented until about 1480.

The first device had two wheels of different diameters, a large wheel and a small one connected to a spindle. The slowly rotating large wheel drove the smaller wheel at a much greater rotational speed by means of a rope drive. Prepared fibres of wool, flax, or cotton were mounted on the spindle and the spinner used the right hand to turn the large wheel while using the left hand to draw out the fibres. These were drawn out at an angle of 45 degrees to the axis of the spindle. As the large wheel turned, the small wheel rotated the spindle and the fibres were drawn off the end and twisted immediately by the rotating spindle. Thus the spinner drew out the thread while turning the wheel until an arm's length was spun. The thread was then held at 90 degrees to the spindle and the wheel was rotated backwards to wind the thread onto the spindle. Spinning with this wheel continued to be a two-step process, spinning and then winding on, as it had been with the distaff and spindle.

The combed or carded material was usually processed twice on the spinning wheel. In the first stage the combed or carded material was applied to the spindle and while one hand turned the wheel the other drew out the combed or carded material, which, receiving a slight twist from the spindle, were made into thick threads called rovings. These were then wound back onto the spindle to form cops. In the second stage the rovings were spun into yarn using much the same process, but the tread was drawn out finer and given more twist. The advantages of the spinning wheel were great but even so many textile guilds in the old textile cities at first prohibited the use of spinning wheels although they quickly dropped their opposition and the wheel became the main method of spinning fibres. However, spinning by distaff and spindle continued even after the invention of the spinning wheel.

The old vertical, warp-weighted loom was replaced in the twelfth century by a horizontal, foot-operated, counter-balanced treadle loom, Fig. 20.9 (bottom). This loom consisted of two horizontal beams. The cloth beam was nearest the weaver and the warp beam was at the opposite end. Warping this beam had to done with care to ensure an even tension and to help in preventing breakages the thread was treated with size to strengthen it. In simple weaving the alternate warp threads were separated into two groups (sheddings). The free ends of one shedding were threaded through the eyes of the heddles before being attached to the cloth beam and the tension adjusted. A heddle is a device made of two strips of wood with strings, like the teeth of a comb, stretched between them. Each string was actually a string loop with one section tied off to produce an eye. One of the sheddings was chosen to pass through these eyes, usually one warp per eye. When the weaver pressed the pedals of the loom the string attached to the pedal would lower a heddle thus pulling down the warp threads that passed through its eyes. This heddle was also attached to another heddle by means of an overhead pulley. The other shedding was passed through the eyes of this second heddle. When the first heddle was pulled down the second heddle ascended raising its threads. This opened a horizontal space in the warp threads, called the shed, through which the shuttle was passed from hand to hand trailing the weft thread behind it. The weft thread was laid some distance from the cloth so it was gently beaten into the cloth using a tool called a reed. This also ensured the weft remained straight. The reed was a wooden frame, rather like a heddle but fitted with teeth made of reeds rather than string, and hinged to the frame above the loom near the pulley system. The warp threads were passed through the reed during warping. To beat the cloth down the weaver pulled back on the reed, pressing the weft threads into place with the teeth. As it was produced the cloth was wound onto the cloth beam and the warp wound out from the beam at the other end of the loom.

Different patterns could be produced by variations in the number of warp threads passing through the eyes in the heddles and, of course, by changing the colour of thread on the shuttle. A loom with two heddles produced plain or tabby woven cloth. Looms with four heddles produced diamond, twill and herringbone patterns. With the passage of time more and more heddles were added to enable more complex patterns to be woven. The reed, sometimes called a

Fig. 20.9 Top: wheel for spinning and bobbin winding. Bottom: pedal loom with two heddles, twelfth century.

sword or comb, and used to beat down the cloth, was replaced by a batten, supported in a heavy wooden frame from the main frame, and its weight and free swinging motion improved its beating action.

Another technical development of the late twelfth century was the introduction of mechanised fulling by waterpower in place of the traditional manual method of treading under foot. As mentioned in Chapter Six, the Domesday Book gives no indication that water-power was used in fulling and there is no mention of water-powered

Fig. 20.10 Wooden hammers driven from a watermill through a set of cams.

fulling mills in England until the end of the twelfth century. After weaving, the wool (or cotton) needed to be fulled or scoured. Fulling is a trampling and washing process used to clean and thicken the cloth and close the gaps in the weave. The cloth was soaked in one of several alkaline detergents made from plant ashes and stale urine or fuller's earth (fine hydrated aluminium silicate). The cloth was originally trampled under foot in a vat and was removed periodically and rearranged for evenness. After a certain time it was removed, washed in a stream and then beaten with sticks. The cloth could shrink as much as 20% to 50% in this process.

Fulling was mechanised to mimic the action of trampling underfoot by arranging large wooden hammers to rise and fall back on the cloth in a vat filled with the appropriate solution. The action was very different from that of the corn mill, which was, hitherto, what the watermill had driven. Instead of a purely rotary motion, a reciprocating motion was demanded and this required the use of a camshaft. The action of the camshaft and the hammers of a fulling mill is illustrated in Fig. 6.11 of Chapter Six and Fig. 20.10 shows a set of water powered hammers still working in France, but used for the preparation of paper pulp rather than for fulling cloth.

Carus-Wilson (1941) and Pelham (1944) believe that the introduction of mechanised fulling created a revolution in the English textile industry of the thirteenth and fourteenth centuries. Pelham notes that we are accustomed to associate industrialisation with the growth in urban areas at the expense of the countryside, but in the thirteenth century the reverse process occurred. The introduction of water powered fulling mills took place on the upper courses of streams in order to take advantage of the waterpower. The map produced by Pelham and shown in Fig. 6.12 of Chapter Six, confirms Carus-Wilson's historical arguments. It is apparent that there is:

1. a concentration in the western half of the country, especially in the Lake District, the West Riding of Yorkshire, and the southern Marches of Wales.
2. A concentration on high ground and avoidance of the lower, navigable sections of the main rivers.
3. An avoidance of the older centres of manufacture, many of which were situated in the towns and cities on these lower sections.
4. A high proportion of early mills in the West Riding, the mid-Severn valley, the Cotswolds, the Kennet Valley, and Devonshire.

The concentration in the west of the country is simply because that is where the high ground with fast flowing streams are found. In steep valleys only a short leat from the main stream is necessary to create a significant head of water, which could be used by the more efficient overshot mill. On the slow moving rivers of the plain, on which many of the older textile towns were situated, the main power source was the undershot mill (current mill) which, turned very slowly (about half the speed of the current) and although fit for grinding corn, if sufficiently wide, it developed insufficient torque and power to drive a set of fulling hammers. Consequently there was a shift of fulling from the older textile centres to the country manors, and to save transport costs there was likely to be a corresponding shift in spinning and weaving. This shift was of some additional advantage because it freed manufacturers from the restrictions of the craft guilds who were strong in the towns. In about 1540, when Leland made his journey through the English counties he mentions several fulling mills of particular local importance, and all situated in the west of England. At Fenny Bridge, Devon, he found the river was divided into four channels to serve the grist and fulling mills; the last abbot of Cirencester spent £500 building two fulling mills that were of exceptional importance to the town; and Leland also comments on fulling mills at Dursley, Frome, and at Holyfield, Herts.

Lighter fabrics were scoured rather than fulled. The cloth was

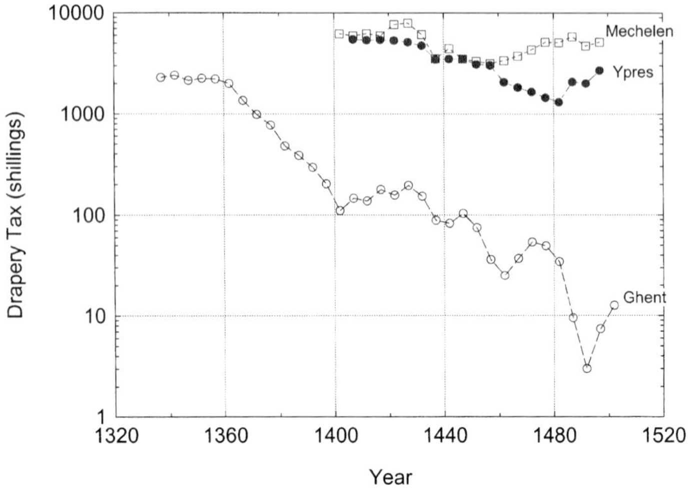

Fig. 20.11 Decline in the Flanders textile industry in the fifteenth century. Carus-Wilson and Coleman (1963) and Bridbury (1982).

washed for about two hours in hot water and worked with soap or a mixture of urine and lye after which it was rinsed in cold water and wrung out before being placed in a pit of fuller's earth and boiling water for half an hour. It would then be rinsed again and hung out to dry. It was then stretched on a frame or tenter (two horizontal rails on vertical posts) and would be brushed with teasles or hedgehog skins, and cropped. After this the cloth was bleached, pressed and perhaps dyed.

The pre-eminent position of the Low Countries slipped away during the fourteenth and fifteenth centuries to a rising textile trade in England and in Italy, Fig. 20.11. Ghent virtually ceased production and the trade at both Mechelen and Ypres declined. As early as 1239 Klemm (1959, 86) notes that there was a remarkable development of a capitalist economy for textile work, an example of which was the establishment in Florence of the woollen cloth industry. This received a considerable boost in 1239 when Benedictine Monks brought their knowledge of the superior processes of Flemish textile production from the north to Florence. Here it developed in the two following centuries into a combination of large centralised workshops with wage earning workers, and of widely scattered home workers. As a result there soon arose a far-reaching division of labour. In

1388 there were in Florence more than 200 textile workshops, where 30,000 persons lived by the production of cloth, and such great industries, although exceptional, indicate the beginning of the breakdown of the medieval economy. Raw material for this industry was provided from England by the Florentine and Italian bankers such as the Bardi, the Frescobaldi, and the Perruzzi, who were in England to collect Papal taxes. They had the capital to buy up the production from the monastic and other suppliers for several years in advance. In consequence the Low Countries found it more difficult to find supplies.

Gimpel (1993, 99) notes that unrest and disruption of the textile trade became a regular feature in Europe during the late thirteenth century. The first recorded was in Douai in 1245, and in 1274 the weavers and fullers of Ghent left the city and migrated to the Duchy of Brabent. Revolt was particularly strong in 1280 when there were armed eruptions followed by severe repression in Ypres, Bruges, and again at Douai. Textile workers migrated to other places, including England and Italy, taking with them their knowledge and experience. A fresco in Manchester Town Hall, painted in the Victorian era, celebrates the arrival of Flemish weavers in Manchester in 1363.

Similarly, during the period of England's Hundred Years war with France (1338–1453) remarkable changes were happening in the English cloth trade, Fig. 20.12. Wool that had formerly been exported to the continent for conversion into cloth was increasingly processed at home. In the two centuries between 1350 and 1550 the export of wool declined by an order of magnitude, from about 30,000 sacks (5,000 tonnes) to about 3,000 sacks (500 tonnes). At the same time exports of broadcloth increased by two orders of magnitude, from about 2,000 broadcloths to about 100,000. The main period of this change was in the years immediately following the Black Death and the Peasants revolt, from 1350 to 1400. This is a remarkable period of sustained growth, especially when it is remembered that the Black Death reduced the population of England, as elsewhere, by about a third, and that many manors stood idle for want of labour. It surely indicates that the English peasantry was at last throwing off the yoke of serfdom and seeking employment, or additional employment, elsewhere.

English wool was produced in great quantities and was, until the Spanish Merino wools were developed in the mid sixteenth century, the finest available. The Flemish and Italian clothiers became dependent

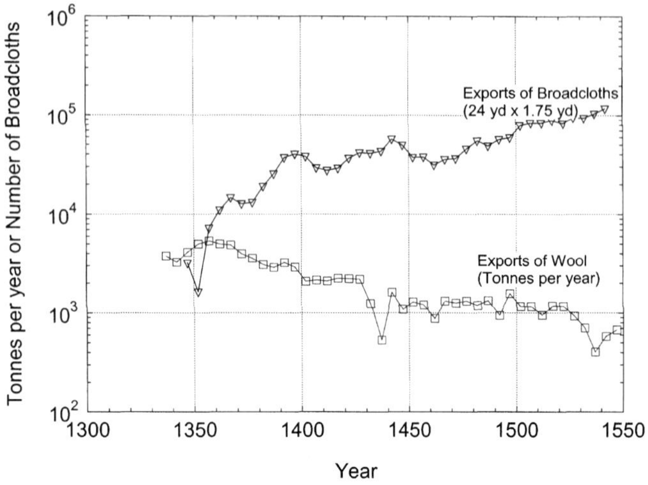

Fig. 20.12 Changes in English wool and cloth exports, 1340–1540. Carus-Wilson and Coleman (1963) and Bridbury (1982).

of them. Edward I or his advisors perceived, instinctively, that foreign demand for English wool was relatively inelastic, especially for the finer, luxury end of the market, and that prices could rise considerably without reducing demand. He had a virtual monopoly of good quality wool. In 1275 Parliament allowed Edward I to levy the first export duty on wool amounting to one mark (67p) per woolsack (165 kg); perhaps about 7% of the average price. In 1336, on the eve of the Hundred Years War, and at a time when the Flemish industry was undergoing a period of change, Edward III imposed an additional subsidy of £1 per sack, which he increased to £1.67 per sack in 1338, and to £2 per sack in 1341, giving a total tax bill of £2.33 per sack (£2.50 per sack for aliens). These rates were regularly confirmed by Parliament until 1362 when it was reduced by a half. The price of wool had become an instrument of government and a source of income for continental wars. The following year the capture of Calais enabled it to become the compulsory wool staple (monopoly) for all wool exports from England to Northern Europe, and it was designed to pass the tax onto foreign buyers rather than domestic growers. Because of deflation the price of wool had fallen so the tax burden now amounted to about 25% or more. English clothiers bore no such burden; finished cloth was taxed at

14d (5.6 p) per broadcloth, and they were free to export finished cloth at much lower prices than their continental rivals who had to pay the heavy export taxes on raw wool. Thus, inadvertently perhaps, taxation policy to pay for continental wars also stimulated English cloth production and export and diminished raw wool exports. Another factor that stimulated English cloth production was that, as a relative newcomer, it was freer of restrictive practises that were imposed by the Guild system to defend their members.

The Renaissance and Early Modern Period

At the end of the fifteenth century the old process of spinning yarn, in which spinning and winding were done as separate operations, was replaced by a new spinning wheel in which these processes were done at the same time, thanks to the invention of the spindle flyer. Such an invention considerably increased the productivity of spinners. Fig. 20.13 shows a drawing of a spinning wheel from the *Medieval Housebook* of about 1480. This is now thought to be the earliest representation of the new spinning wheel. It is not shown with the treadle drive that came to be almost universal, but it has the essential element, namely the spindle flyer. The flyer shown is a quite advanced design, suggesting that it originated much earlier than 1480. Leonardo da Vinci also sketched a spindle flier, Fig. 2.1 of Chapter Two, which illustrates the action rather more clearly than Fig. 20.13. The flyer consists of several parts. A hollow spindle was mounted horizontally between two arms so it could spin freely. A spool attached to the spindle also moved independently. Spindle and spool had separate rope drives from the large wheel. The rope from the main wheel looped twice, once round the spool drive and once round the spindle drive.

The prepared wool was fastened to a distaff attached to the main frame. The fibres were initially pulled out by hand and twisted and passed through a hole into the centre of the hollow spindle. It then emerged radially through another hole, part way down the spindle. From here it passed over the flyer, held in place by hooks at the end of the flyer. It was then secured to the spool. When the wheel was turned it would spin the spindle and the thread passed over the flyer and twisted. Then, because the spool was spinning faster than the spindle, the spool would draw out the thread and wind it on.

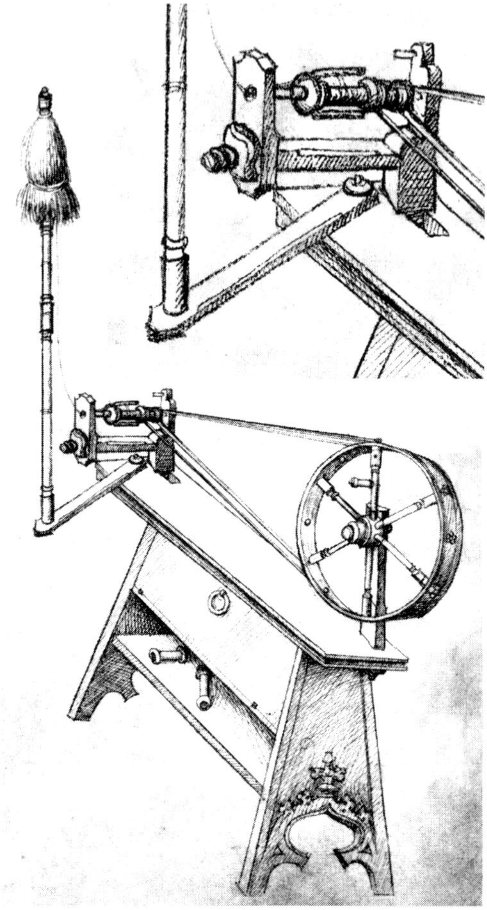

Fig. 20.13 The spinning wheel, *c.* 1480.

As the spool filled, its radius increased and it began to wind even faster, causing the thread to have less twist in it. To alleviate this the thread was held out by the operator or moved to a different hook on the flyer. The flyer made the spinning process continuous because the spinning and winding-on occurred simultaneously. The manufacture of yarn was still done in two processes; the first creating a thick, slightly twisted rove, and the second creating the finished thread from the rove.

The simple horizontal loom remained unchanged for some centuries after its introduction and it was adequate for plain or relatively simple patterns, but as the demand for more complex patterns developed so did the loom's complexity. For weaving very large and complex patterns the drawloom was introduced but, as with most of the early textile machinery, its origins are unknown. Probably it was introduced into the silk weaving centres of Italy from the East where it had long been used in silk weaving. The drawloom had two devices for separating the sheddings of warp to clear a space for the shuttle. In addition to the usual heddles that the weaver operated by treadles, sets of cords were also used to raise the warp threads. These were gathered into groups as required by the pattern. The cords were not lifted by the weaver but by a draw-boy who was seated on top of the loom.

The situation was improved in Italy and France in the early seventeenth century when a mechanical draw-boy was devised, which allowed the assistant to stand at the side of the loom and still control the cords. But the system was still prone to human error and this stimulated the invention of wholly automatic draw-boys. These developments occurred mainly in France, where the weavers of figured silks were very advanced, and eventually lead to the development of the Jacquard loom. The Jacquard loom is, in fact, a conventional loom fitted with a Jacquard attachment to enable it to weave complex patterns and even pictures. It was really the inventions of three men over a period of about 75 years that produced the famous Jacquard loom.

The first invention came in 1725 when Basile Bouchon added a mechanical draw-boy to a conventional loom. This was a mechanism that selected the cords to be lifted from a perforated pattern in a roll of paper. The holed paper passed round a four-sided "cylinder", which was pushed towards a selecting box and met with needles carrying the warp-controlling cards. The needle that met un-perforated paper slid away while the others passed through the holes. The weaver operating his foot treadle then drew down the selected cords. Thus a different set of cords could be drawn for every passage of the shuttle. This, of course, eliminated the possibility of errors by the operator but it was still necessary to have a human operator to move the selected cords, so although quality improved the costs remained the same. Falcon improved this mechanism three years later by increasing the number of needles and using a rectangular perforated card for each shedding motion. The cards were strung together to form a continuous chain that would generate repeat patterns.

The next development came in 1745 when Jacque de Vaucanson made an improved loom. He mounted a selecting box above the loom, where it acted directly on hooks that controlled the warp. These hooks passed through needles and were raised by a metal bar. The needles were selected by a perforated card passing around a sliding cylinder, and were operated without the help of an assistant. However, it is not known if this device ever worked successfully, but it was in attempting to improve it that Jacquard invented his famous attachment.

Jacquard was commissioned to overhaul Vaucanson's loom, and had to do so without the operating instructions, which were missing. He succeeded, and in 1801, at the Paris Industrial Exhibition, he was able to demonstrate his improvements, and in 1804–5 he introduced an improved version. This was an automatic, selective, shedding device, mounted on top of the loom, and operated by a treadle controlled by the weaver. Like the drawloom, every warp runs through a loop in a controlling cord and is held taut by a weight. Each cord is suspended from a wire (a hook) that is bent at the bottom to hold the cord, and bent at the top to hook it around the lifting mechanism. To allow only those warp threads that are necessary to be raised, some of the hooks must be dislodged from the lifting mechanism (a rising horizontal bar). Horizontal needles connected to the hooks are used to do this. As the perforated pattern card moves into place on the cylinder (actually a rectangular block) the needles pass through the holes in the card, and the warps are raised. Where there are no holes, the needles are pushed back against a spring on the opposite end of each, thus pulling the hooks away from the raising bar and consequently these warps are not raised.

Cotton

In about 1538 when John Leland journeyed through Lancashire he found the manufacture of cotton cloth already well established. He writes: "Bolton on the Moor has a market whose main commodities are cottons and course yarn. Several villages on the moors around Bolton manufacture cotton", see Chandler (1993, 270). Leland does not mention cotton manufacture at Manchester, which he visited, but Camden, in 1582, mentions that it is famous for its manufacture of stuffs called "Manchester cottons", although these are considered to be a kind of coarse woollen cloth. Wadsworth and Lacy Mann (1931, 4) trace the beginnings of Lancashire cotton to the indigenous woollen and linen industries carried on as the usual secondary occupations of the smallholder and his family, using home produced wool or flax. Cotton first definitely appears shortly after 1600. A petition to Parliament in 1621 says "about twenty yeeres past diverse people in this Kingdome, but chiefly in the Countie of Lancaster, have found out the trade of making of other Fustians, made of a kind of Bombast or Downe, being the fruit of the earth growing upon little shrubs or bushes, brought into this Kingdome by the Turkie Merchants, from Smyrna, Cyprus, Acra, and Sydon, but commonly called cotton wooll—There is at least 40 thousand peeces of Fustian of this kind yeerely made in England—and thousands of poore people are set on working of these Fustians." Fustian is cloth having a linen warp and a cotton weft.

It is rather surprising that the manufacture of this exotic material should thrive in south Lancashire for it was quite remote from London and other markets and transport was, initially, not easy. The area around Manchester and the north of the county was filled with mosses that were impossible to cross without sinking, even on foot, and the condition of the roads in the county left much to be desired. Defoe (1986, 544), writing in about 1725 says "—on the road to Manchester we passed the great bog or waste called Chatmos, the first of that kind that we see in England, from any of the south parts hither.—we found many of them in this country—it will bear neither a horse nor man except in an exceeding dry season, and then not so as to be passable." The Pennines effectively cut off access from the east, and the Cumbrian fells cut it off from the north, but from the west the port of Liverpool, which grew with enormous speed between 1680 and 1720, rivalling and eventually outstripping Bristol, proved to be an excellent gateway for the import of raw cotton and the export of finished goods. It is usually said that cotton was attracted to Lancashire largely because of the damp, humid climate and the availability of fast streams of very soft water running off the Pennine slopes, which gave a plentiful supply of water power for driving the necessary mills and for processing the cloth. Of course, cotton became established in Lancashire before waterpower was used for spinning or weaving, but it was used in fulling.

By 1700 as much as 1000 tonnes of raw cotton per year was

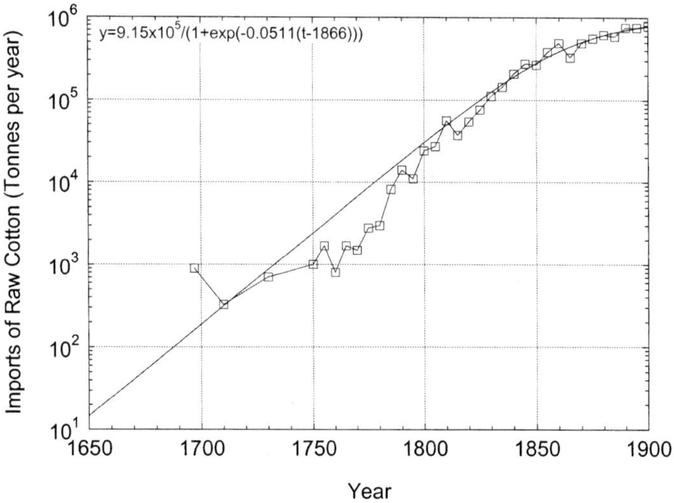

Fig. 20.14 Imports of raw cotton into the UK, 1697–1879. Vincent (1881, 213), Mitchell (1978, 251).

being imported into the United Kingdom and this increased rapidly after about 1750, and approached a million tonnes per year by the end of the nineteenth century, see Fig. 20.14. The average growth rate was about 6% per year from 1750 to about 1860 and the "heroic age" is considered to be 1780–1820. As mentioned above, exports of raw wool in the fourteenth century amounted to about 5,000 tonnes per year, and were the staple of the medieval economy. Cotton imports reached this figure in about 1780 and grew to be 100 times larger in the following century.

By the end of the early modern period, and the beginning of the industrial period, cotton manufacture in the United Kingdom, had become the most important industrial activity and was easily the most important export commodity, see Fig. 20.15. Cotton on its own out-performed all other sectors of the economy and textiles as a whole, namely cotton, wool, linen, and silk manufacture were dominant. The exports of coal, iron and steel, or machinery, did not compare with the export of cotton goods until the end of the nineteenth century. The relatively inexpensive supplies of raw cotton made it possible for cotton to dominate. In the eighteenth century, these came in the main from the West Indies, The Levant (Turkey, Italy and Venice), with small amounts from Europe, Africa, and the

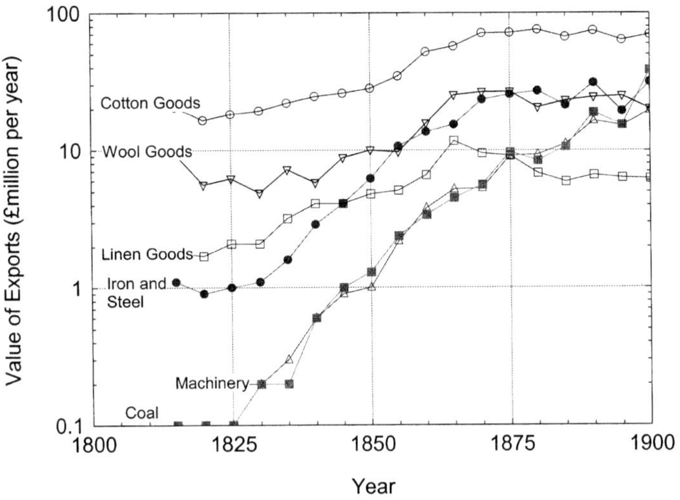

Fig. 20.15 Value at current prices of the principal domestic exports, Great Britain 1814–29, UK 1826–1900. Mitchell and Deans (1962, 302).

East Indies. In the nineteenth century imports from India, and later from the United States dominated. But the most important factor was a series of inventions that improved the productivity of spinning, weaving and the preparation and finishing processes.

The industry was also aided by legislation. During the seventeenth century many calicoes and muslins were imported from India through the East India Company and this reduced the market for wool. In 1678 Parliament, under pressure from the wool manufacturers, decreed that in all burials the shroud should be of woollen cloth and that cotton, silk, or other materials were banned. It was not a popular measure except in the wool districts. From 1700 attempts were made to restrict this trade further and to encourage home production of woollen cloth by having an import tax levied. For example, in 1700 the wearing of cotton goods was forbidden in England and in 1721 an Act of Parliament imposed a fine of £5 on the wearer and £20 on the vendor of cotton goods. The woollen manufacturers, who feared the popularity of the imported cottons from India, were largely responsible for promoting this legislation. But the demand for cotton was not to be denied and this legislation also hurt home production of cotton goods, consequently a compromise was reached and in 1730 an Act was passed allowing goods with a linen warp and a cotton weft to be sold on paying a moderate tax of 6d per square yard. Of course such legislation prevented the import of Indian muslins and calicos without prohibiting English cotton manufacture, and it was withdrawn in 1774, when English cotton production became less expensive and was, in a few years, to overtake the production of wool. At the same time the tax was reduced to 3d per square yard.

John Kay, a broadcloth weaver from Bury, Lancashire, made the first of the inventions that eventually led to cotton's dominance when in 1733 he devised his flying shuttle. Although a native of Bury his father had started a woollen business at Colchester, which the son eventually inherited. In 1730, when he invented his flying shuttle, Kay was working in Bury as a reedmaker. He had already made improvements by making the vertical reeds of thin polished brass wire, a distinct enhancement on the cane that was then commonly used. This made the reeds more durable and enabled fabrics of finer and more even texture to be produced. Before Kay's invention of the flying shuttle it had been thrown through the shed by hand, and for broadcloth this required two assistants; one to throw the shuttle from the right to the centre, the other, reaching between the warps to send it on its way to the left and then, after shedding and beating, he would return it to the centre where the first assistant would pick it up and continue the process. Kay devised a simple mechanical device that the weaver controlled with a cord and not only eliminated the need for two assistants but also more than doubled the speed of work and improved the quality of the cloth.

Fig. 20.16 shows Kay's flying shuttle attached to the reed, which is used for beating up the weft after it has been laid. The warp, it will be remembered, passes through the vertical spaces between the reeds, and the heddles separate the two sheds to form the space through which the shuttle must be passed. The problem was to get the shuttle through the shed as quickly as possible and without the aid of assistants. To do this Kay built a ledge, sometimes called the shuttle race, on the bottom of the reed along which the heddles would lay one of the sheddings of the warp and over which the shuttle could move. The shuttle is held in one of the boxes, E, at either side of the loom and attached to the reed. These boxes are fitted with a block of wood, called a picker that slides along a wire in the box and is connected by a cord to a handle in the centre of the loom. When the weaver jerks this handle the picker accelerates

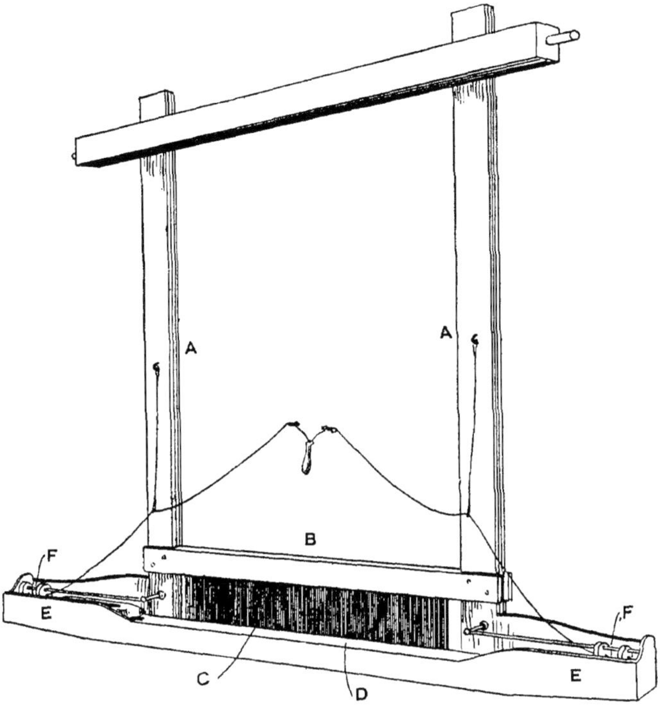

Fig. 20.16 Kay's flying shuttle attached to the hinged reed of a loom. A = swords, B = hand tree, C = reed, D = shuttle race, E = pickers. After Murphy (1910, IV, 119).

the shuttle and sends it flying over the warps on the shuttle race and into the opposite box, leaving the weft behind. Shuttle speeds of 5 to 10 m/s are attained. The reed is then used to compress the weft into the cloth in the normal way and the heddles are operated to form a new shed. A sharp jerk in the opposite direction returns the shuttle to the original box. To withstand such treatment the shuttle itself had to be adapted, although these were later developments and did not form a part of Kay's original patent. In particular, the shuttle was fitted with small wheels so it could glide over the shuttle race without damaging the warps. For its first fifty years or so it was usually called the wheel shuttle rather than the fly or flying shuttle. He also made the shuttle longer and heavier to avoid sending it accidentally against the warps and damaging the web. Finally, the method of dispensing the weft from the spool in the shuttle was improved to avoid breaking the more delicate weft. In ordinary shuttles the weft spool was loosely fitted to each end and left to revolve, dispensing the weft as the shuttle was moved and the spool turned. Kay fixed the spool to one end of the shuttle so the weft came off at the free end and emerged from an eye he had cut in the side of the shuttle. This allowed the spool to be longer and hence to carry more thread and with less risk of it breaking. Wadsworth and Mann (1931, 452) say that the improved shuttle was exhibited in Rochdale in 1735, where it was immediately copied and was the origin of considerable litigation that lasted for many years.

By these means one weaver could weave more cloth than two or three weavers had managed under the old system. The animosity this raised amongst handloom weavers forced Kay to flee to Leeds where he continued his work as a reedmaker. His device was quickly taken up in the woollen industry and copied without permission or fee and he was impoverished by lawsuits. By 1745 Kay was back in Bury working with Joseph Stell of Keighley on a smallware loom operated by mechanical power rather than by manual power. Although this loom does not seem to have been successful, Kay's inventions were feared and in 1753 a mob broke into Kay's house in Bury, destroying everything and forcing him to flee, apparently to Paris, where he continued in business, with considerable encouragement from the French Government, with machines smuggled out of England. He is thought to have died in France in poverty and obscurity. In about 1760 Kay's son, Robert, perhaps working with his father, invented the drop shuttle box, an imaginative device that brought different shuttles into line with the shed so that wefts of different colour were more easily laid. Wadsworth and Mann (1931, 405) remark on the reversal of Kay's treatment in England and in France. "In England he met with jealousy and dislike, but the weavers, after a good deal of preliminary grumbling, pirated his invention. In France he was patronised by the Government and given every facility for popularising the fly shuttle, but its general use died out within a generation."

Although the flying shuttle increased the productivity of weaving it created difficulties for the spinners. It is usually reckoned that between 4 and 10 spinners were required to supply one weaver, depending on the fineness of the yarn, so increasing the weaver's output could not, in itself, increase the production of cloth. Kay's flying shuttle created a demand for spinning and drew more and

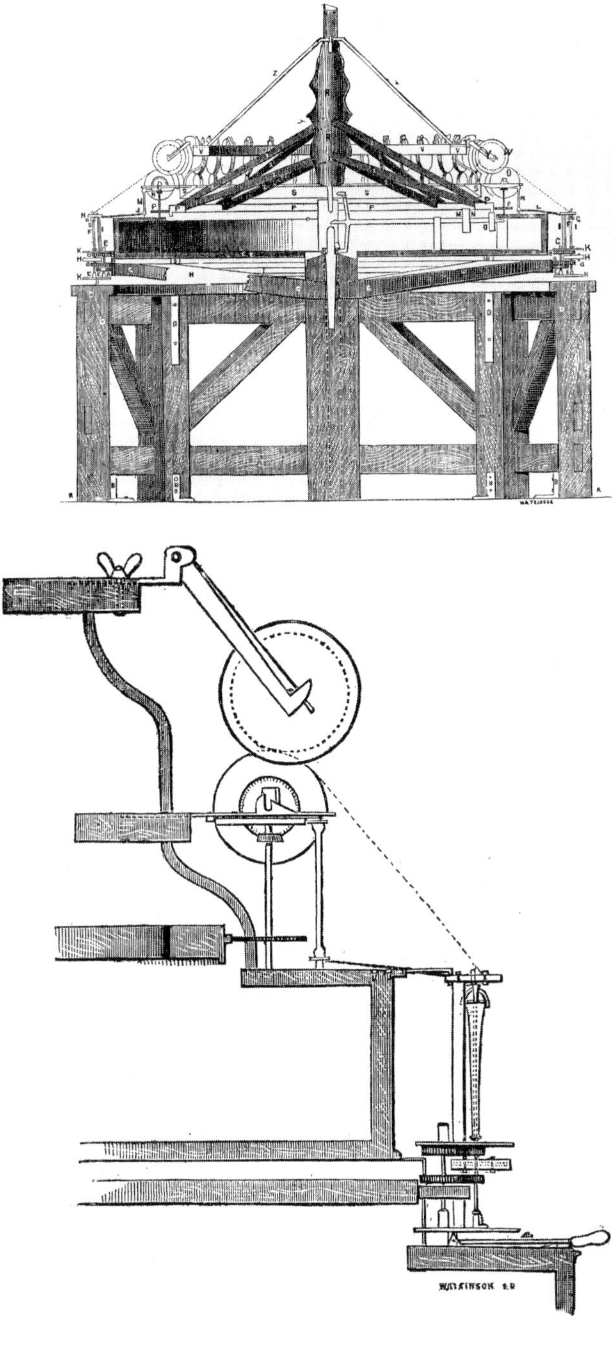

Fig. 20.17 Paul's second spinning machine, 1758.

more inhabitants into the industry; it reduced the demand for weavers, although the ever increasing demand for cotton ensured that very few went out of business; and it eased the weaver's task and improved quality. However, the most important asset of Kay's invention was that it was perfectly adaptable for automatic weaving by power.

The need for improved spinning wheels was obvious and the textile districts seem to have been a hotbed of invention at this time. When Kay was working on his flying shuttle Lewis Paul and John Wyatt were devising the first spinning machine to use rollers to draw out the thread in the modern manner, see Fig. 20.17. Paul was the son of a physician and described himself as a gentleman of Birmingham, and claims to be the inventor, whereas Wyatt was a skilled mechanic and a man of some means, who certainly contributed much of his skill and finance to its development. Paul's patent for roller spinning was taken out in 1738. Roller spinning uses two pairs of rollers of small diameter between which the rove passes. The second pair of rollers rotates faster than the first pair and thus pulls the thread faster than the first pair can deliver it, consequently the thread is stretched and its thickness reduced. Two spinning mills using these machines were set up, the first in Birmingham and the other in Northampton. "—asses walking round its axis" were used to drive the machinery. Neither mill was successful. Roller spinning did not become firmly established for another thirty years when Richard Arkwright patented his invention in 1769 and built the first successful spinning mill at Cromford in the early 1770s. However, the factories using Paul's spinning machinery quickly demonstrated that they were unable to maintain a sufficient supply of carded wool or cotton and this turned Paul's mind to the invention of a carding machine, for which he obtained a patent in 1748. The invention has the first suggestion of a circular or continuous carding engine with a comb for stripping off the carded wool or cotton, Fig. 20.18; it was introduced into Lancashire in about 1760, shortly after Paul's death.

Other inventors were equally busy but only partially successful. Laurence Earnshaw of Mottram, a man well known locally for his mechanical genius, invented a machine to spin and reel cotton in one operation in about 1753, but not much is known of his invention. He showed it to his neighbours and then destroyed it so that it might not take the bread from the mouths of the poor; see Wadsworth and Mann (1931, 472). Two years later a clockmaker,

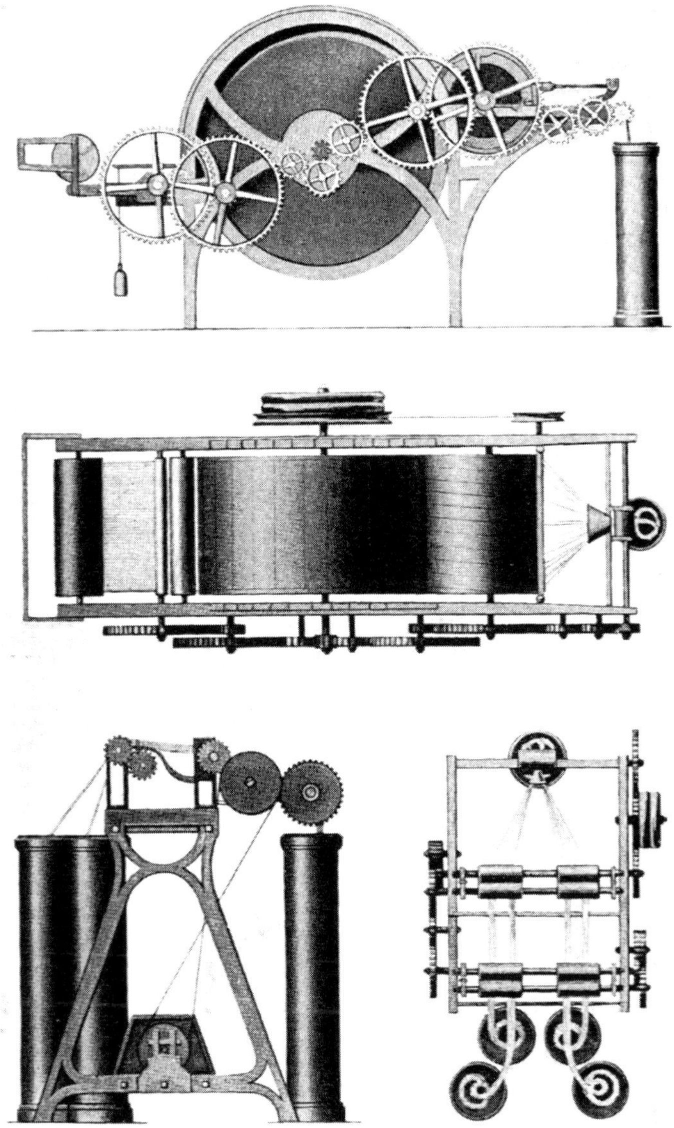

Fig. 20.18 Top: a carding machine. Bottom: a drawing frame. Baines (1835).

James Taylor of Ashton-under-Lyne, patented an automatic spinning machine, similar in many respects to that invented later by Hargreaves. His patent describes it as "an engine to be worked whether by men, horses, wind or water for spinning Cotton Wool into Yarn which will spin more and better Cotton Yarn than ever before made use of and make a truer evener and softer Yarn than any common hand can Spin and lays the Harle or Staple more straight and close than any Yarn yet produced." The cotton roving was wound round spindles standing "upright horizontally in a parallel line", which is not too clear but seems to suggest several vertical spindles in a horizontal line. These were connected to a central drum "that works a slip and rope string which turns the spindles", which suggests that the spindles were driven by a rope from a single drive wheel. Apparently the machine simulated the action of the medieval distaff and spindle in that the roving was drawn out a few feet, twisted as required, and then wound. However, he was forced to give up his machine by the ill treatment he received from spinners who, by their poverty and dependence on spinning, were prejudiced against any improvement.

Nevertheless, the shortage of spun yarn was great and in 1764 James Hargreaves invented the spinning jenny, for which a patent was taken out in 1770. Hargreaves was born probably in Blackburn in about 1740, and in 1760 Robert Peel, grandfather of the famous statesman, employed him to improve the carding machine that had just been introduced into Lancashire. He is said to have thought of his improved spinning wheel after observing that a conventional machine continued to rotate and spin after it had fallen over so the wheel and spindle rotated about a vertical axis. It was an accident that must have occurred many times but it took some insight and ingenuity to realise that in this configuration it was possible to place a number of spindles side by side so that several yarns might be spun at the same time. Thus his spinning jenny as he called it, was built on the same principle as existing spinning wheels but it rotated about a vertical axis and it used more spindles, see Fig. 20.19.

Essentially, Hargreaves made a frame containing a row of eight rovings and a row of eight spindles. The rovings extended to the spindles, passing between two horizontal beams that could be opened or shut like a parallel ruler. When shut they clamped the rovings and they could be drawn along the frame from the spindles by the spinner's left hand, by which they were stretched. At the same time

Fig. 20.19 Hargreaves's spinning jenny. Baines (1835, 157–8).

the spinner's right hand rotated a large wheel that caused the spindles to rotate rapidly, and thus the rovings were spun into thread. By returning the clamp to the original position, and letting down a presser wire, the thread was wound upon the spindles.

At first Hargreaves used eight spindles and from 1764 to 1768 he used his spinning jenny in his own home and also supplied it to neighbours; an act of generosity that was to cost him dearly. But in 1768 spinners using the traditional wheel became anxious and a mob from Blackburn and elsewhere attacked his home and destroyed his jenny and loom. He and his family moved to Nottingham where he went into partnership with a Mr James and, in 1770, he belatedly took out a patent on his invention. However, the secret was out and spinning jennies became common in Lancashire and Yorkshire. Hargreaves sued for infringement of his patent but the case was abandoned when it was learned that he had already sold them in Blackburn. A few years after his death in 1778 there were 20,000 hand jennies of 80 spindles each at work in England. Jennies having up to 120 spindles were made and as it was hand operated it could be installed in cottages at relatively small cost. It remained popular for spinning woollens but it was soon replaced for spinning cottons.

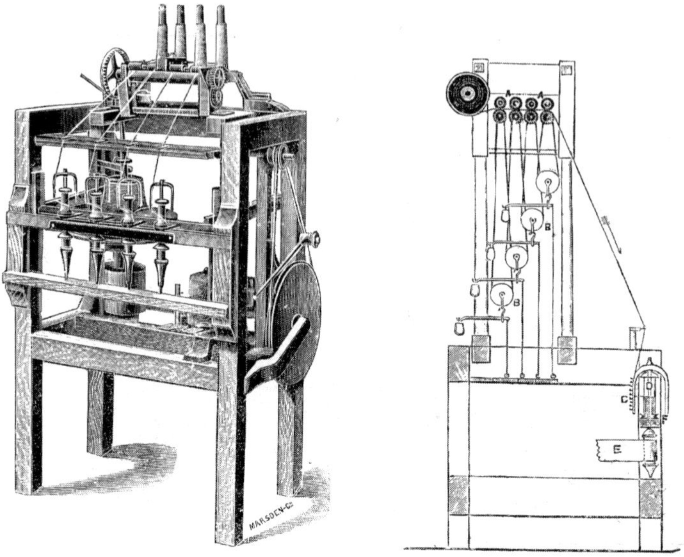

Fig. 20.20 Arkwright's improved spinning frame, or water frame *c.* 1775.

At much the same time Richard Arkwright evolved his machine for spinning cotton by rollers. As mentioned above, roller spinning was a fundamentally different process from the traditional spinning wheel or from Hargreaves's spinning jenny. Arkwright was born in Preston in 1732 and by 1760 he was working as a barber and wig-maker in Bolton. The decline in the fashion for wigs is supposed to be the reason for Arkwright taking an interest in mechanical inventions related to spinning and weaving which, at that time, were a constant source of interest and conversion in Lancashire as elsewhere. Kay's flying shuttle had created a need for improvements in spinning that was apparent to all. In 1767 Arkwright asked a clockmaker of Warrington, also called Kay, to make some articles in brass, and shortly afterwards he gave up his business, moved back to Preston, and with the assistance of Kay and his friend Mr Smalley, he devoted himself to his new invention. When his machine was perfected Arkwright, Smalley and Kay the watchmaker moved to Nottingham, the main centre of the stocking trade, and set up a small spinning mill. A patent was taken out in 1769.

The spinning frame, Fig. 20.20, consisted of four pairs of rollers acting by tooth and pinion. The top roller was covered with leather hearing it discussed by some Manchester businessmen. He was born at Marnham, Nottingham in 1743, educated at Oxford, and was presented to a living at Doncaster, Yorkshire. He was admired in his lifetime at first as a poet but in midlife he took an ever-increasing interest in mechanical devices. He happened to be at Matlock in 1784 when he fell into conversation with some Manchester businessmen and was prompted to build a power driven loom. By the following year he had a machine that required the strength of two powerful men to work it even at a slow rate and the springs that threw the shuttle "were strong enough to have thrown Congreve's rocket". Nevertheless he took out a patent in 1785, after which, he says, he condescended to see how other people wove and was astonished by their easy mode of operation. Armed with this experience and knowledge he then improved his own loom; taking out another patent in 1787, see Fig. 20.22. The main improvement he made concerned:

1. controlling the cloth to ensure its forward edge was always in the same position when the batten swung to beat up the weft. This implies control of the letting-off and taking-up motions,
2. a mechanism to stop the loom automatically when the warp threads broke,
3. a mechanism to cease winding up the cloth if the weft failed to be delivered by the shuttle,
4. strengthening the warp by sizing it on its passage to the heddles,
5. regulating the supply of warp and the winding up of woven fabric, and
6. adjusting the temple stretcher to keep the cloth to an even width.

The machine finally constructed and used by Cartwright was thus very complex and imperfect in many of its details but it showed the way to others who were able, eventually, to perfect it. Cartwright himself invested some £30,000 on his power loom alone, building a power-loom factory at Doncaster, driven by horses originally and later, in 1789, by a small steam engine. A steam-powered mill, using 400 of Cartwright's power looms, was introduced into Manchester in 1791 and it was said that the wages bill for weaving was reduced by a half. However, Knott Mill, the factory in which they were installed, burnt down in the following year, probably by saboteurs who feared for their livelihood. After this disaster his Doncaster factory was short of capital and was abandoned in 1793 leaving Cartwright deeply in debt after his investment in his power looms.

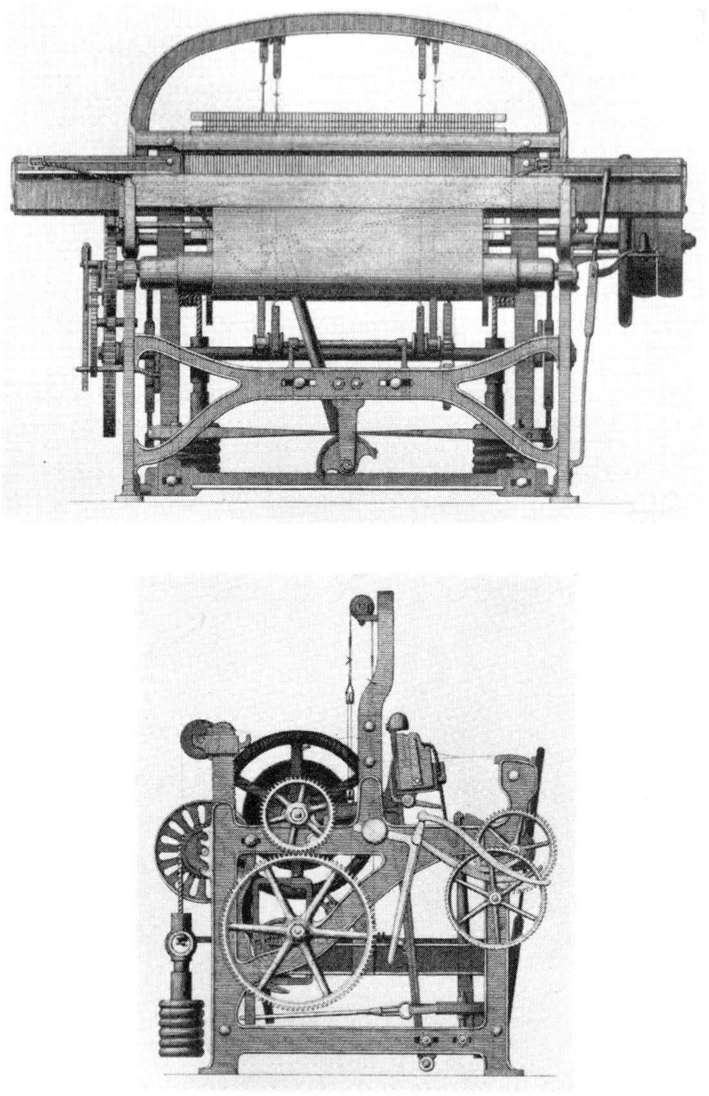

Fig. 20.22 Cartwright's power loom. Baines (1835).

Other inventors were busy improving the power loom. At the time Cartwright was working on his second loom, Jefferey, a physician of Paisley, independently invented a power loom similar to Cartwright's. It too suffered by having too powerful a spring to propel the shuttle and the batten with consequent breakage of the warp or weft, but Jefferey devised a method of checking the shuttle, which saved the weft. In 1793 Andrew Kinloch erected several looms in Glasgow and was employed to erect 40 more for Robert Miller. Three years later Miller applied eccentric wheels, now called wipers or tappets, to drive the shuttle, which reduced the damage to the warp. In 1800 some of Kinloch's looms were erected at Stalybridge, in the Manchester cotton district, 15 of which were later transferred to West Haughton. Manufacturers, somehow, were improving Cartwright's power loom and putting them to work. William Horrocks, according to Murphy (1910, IV, 170), did most to bring the machine to working order. In 1804 Cartwright's patent ran out and by 1806 he discovered that his invention was now a source of considerable profit to the Lancashire weavers. After a letter of complaint to a friend in Manchester a group of 50 Manchester firms petitioned Parliament to bestow a significant recognition on Cartwright for the services he had given to his country by his invention of the power loom. Cartwright too petitioned Parliament and in 1809 he was awarded a grant of £10,000, which again restored his independence.

Thus the period from 1720 to 1820 saw the application of new inventions to improve the productivity of all textiles and is rightly recognised as one of the roots of industrialisation. But these improvements, impressive though they are, were to be swamped by the developments of the next hundred years.

WEAPONS TECHNOLOGY

CHAPTER TWENTY-ONE

WARFARE

Introduction

"It is scarcely an exaggeration to say that *mechanical* invention until the thirteenth century A.D. owed a greater debt to warfare than to the arts of peace"; so thought Mumford (1967, 227). To support this statement he mentions the Bronze Age chariots that preceded the use of wagons, the chariot-scythe that was used to mow down men before it was used to mow crops, and the mechanical and optical devices invented by Archimedes to defend Syracuse before they were put to more constructive industrial use. Weapons were the first to benefit when bronze replaced copper and when iron replaced bronze. The debt, of course, extends beyond the thirteenth century and through to the present. Gunpowder was used in guns and cannon long before it was used in mines; rockets were used as incendiaries before being used to launch satellites, submarines were used to attack shipping before being used to explore the oceans, and jet engines were used in fighter aircraft before being used in passenger transport. Warfare is still a driving force of mechanical invention.

All engineers were originally military engineers who, when they were not engaged in building defensive works or making weapons or military vehicles, were able to undertake civil projects. It was comparatively late that a distinction between military engineers and civil engineers emerged, and even later that the distinction between civil engineers and mechanical engineers became necessary. Thus military matters, and warfare in particular, have always been important in engineering, and military necessity has often provided the funds and the labour essential for technological innovation. However, warfare and engineering have not always been mutually advantageous. Total war, in which a large part of a society's wealth is devoted to victory, is usually beneficial to engineering innovation but other forms of conflict may actually retard engineering developments, for example, by banning or restricting the development and use of new weapons and technology. The chivalrous knight considered that certain weapons (the bow, the crossbow and firearms) were dishonourable and would prefer death to dishonour. War has always had rules. There is, in fact, a complete spectrum of aggression from total warfare through various types of warfare limited by concepts of chivalry and fair play, to ritual warfare, and sporting contests. All of which have been useful ways of satisfying disputes, honour, ambition or aggression without putting the very survival of the society at risk. These are quintessentially male activities and perhaps for this reason engineering has been mainly a male profession.

In this chapter the root of human aggression in the instinctive behaviour of animals is examined. Man has retained his animal instincts, but the all important inhibition instinct, that prevents animals from killing their own kind, has been suppressed in man by the use of weapons that place a psychological and a physical distance between killer and victim. From the very earliest of times man has been a maker of weapons and the new research, attesting that in prehistory warfare was ubiquitous, is presented. To prevent the worst effects of warfare it has often been ritualised by society and examples of such wars in primitive and in historical societies are given. Finally, the serious global wars that account for 80% of all war fatalities are examined, albeit briefly.

Animal and Human Aggression

A vicious person we say behaves like an animal and a brutal environment we describe as a jungle. We base such views on our knowledge of hunting by animals, but it has often been noted that animals, unlike man, kill only to eat and very rarely kill their own species, although murder and cannibalism sometimes occur. Man not only kills to eat but also frequently kills his own kind in wars, in anger (murder), in human sacrifice to gods, or in cannibalism. Clarke (1971, 197) notes that, "Nature is no redder in tooth and claw than is a slaughter-house". In fighting and hunting, animals use claws, hooves, teeth, beaks, horns and any other weapons provided by evolution. Man, on the other hand, is simply not equipped in this fashion and would find it very difficult to skin and dissect a dead animal without the use of tools; and men certainly could not hunt and bring down prey without weapons or traps. Animals, of course, do not use

Fig. 21.1 Black-backed jackals fighting; partly controlled by instinct. Kgalagadi Transfrontier Park, South Africa.

weapons although monkeys sometimes throw stones and apes sometimes use clubs. Nevertheless, the hunting specialisations of animals provide examples which man has copied. We think of the power and strength of the lion and tiger, the speed of the antelope and the cheetah, the size of the elephant, the protective armour of the tortoise and the turtle, the camouflage of most animals, and the relative safety to be found in herds, shoals or flocks. In strategy too animals have much to teach us, as exemplified by the hunting strategies of a pride of lions.

Storr (1968, 31) says that, "Aggression between members of the same species is necessary if the strongest are to be selected and to flourish. Yet aggression must not get out-of-hand, or else the species would destroy itself." Lethal conflict between animals of the same species compromises their survival and seldom occurs. Presumably, species that did engage in lethal conflict have, with the notable exception of man, died out. Perhaps the most important work on aggression in animals is that of Lorenz (1996, but first published in 1963), for which he was awarded a Nobel prize. Lorenz pointed out that a form of warfare does exist amongst animals of the same species but it is ritualised and seldom results in death. Such conflict usually occurs between males for the possession of females. Stags for example, grow their antlers and horns specifically for the purpose of fighting for access to the females. The fight between two males is a trial of strength and the winner is the one who persists the longest. Although the fight looks fierce, death or even injury rarely occurs. The antlers of stags are never used in serious fights against other species and if attacked they defend themselves with their forelegs and hooves. When the rutting season is over the antlers are discarded. The most fearsome weapons possessed by animals are mainly for the purpose of frightening rivals into submission, although they may sometimes be used for hunting or defence against other species. Fig. 21.1 shows an apparently savage ritual fight between two jackals.

Many animals, whether on land, in water or in the air, are territorial, and whether living singly, in a family group, or in a larger group, they mark and defend their territorial boundaries. The purpose of such territories is to establish an area from which sustenance can be drawn and families raised without too much competition and conflict. Usually these territories are marked out at the start of a breeding season. They are marked by scent, by birdsong, and in many other ways, but the signals are recognised by intruders of the same species and if they wish to claim the territory they must engage in a ritual fight. An animal on its own territory will fight vehemently but when crossing its boundaries into a neighbouring territory it normally becomes less aggressive, unless it is a young animal looking to establish a new territory or take over an existing one.

The arousal of aggression in an animal, including man, is signalled in a number of ways and this arousal serves to prepare an animal for physical struggle. There is often an increase in pulse rate and of blood pressure with an increase in the peripheral circulation and a rise in blood glucose. Many animals may attempt to exaggerate their size. A cat arches its back; some animals, perhaps even man, cause their hair to stand on end; and others stand upright or turn their largest profile to the intruder. Sometimes the animal changes colour, or screeches or beats its chest. Much strutting and staring may occur while the opponents apparently consider their prospects. The subsequent fight, if it proceeds so far, may use formidable looking weapons but death is unusual, serious wounds are uncommon and the loser, after signalling his defeat, withdraws. The evolutionary advantage of such behaviour is that neither animal is debilitated and the loser is free to challenge for another territory, which may be held by a weaker male, and so only the strongest males gain and hold territory and have access to females.

Lorenz (1996, 30) writes "This territorial aggression—gives an ideal solution to the problem of the distribution of animals of any one species over the available area in such a way that it is favourable to the species as a whole." He notes, however, that the same result is often achieved without aggression. It may suffice if animals of the same species are offended by the other's smell, as in the scents deposited at their territorial boundaries by cats. Another method is sexual selection by the female. "Wherever we find exaggerated development of colourful feathers, bizarre forms etc in the male, we may suspect that the males no longer fight but that the last word in the choice of a mate is spoken by the female, and that the male has no means of contesting this decision". Birds of paradise, pheasants, and many other animals adopt this strategy.

Territoriality is instinctive in animals and when the size of territory is reduced, through population growth or artificially as in many zoos, they become more aggressive and fights to the death then become frequent. Clarke (1971, 214) quotes Zuckerman's observation of the baboons at London Zoo in 1929–30. When females were introduced into a hitherto peaceful colony of male baboons, violence and conflict erupted with the result that 30 of 33 females and 8 of the 61 males died by violence. At the time this was thought to be normal behaviour for baboons but fieldwork has since shown it to be vastly irregular. Monkeys and apes in the wild are very peaceful and violent death is exceptional. Both animals and men, it seems, require their personal space and an aggressive reaction is generated if this space is violated. Sitting next to a stranger when other, more remote, seats are available is considered offensive, and standing within 50 cm of a stranger in an otherwise empty room is equally offensive and may lead to aggressive exchanges. It is an act of aggression if the room is empty, but it would not cause offence if the room was crowded and there was no alternative; indeed, it is an essential condition for a good party. In addition to these psychological effects, crowding may also introduce hormonal imbalances that cause fighting, stress, and disease. High population density, shortage of food or of habitat causes lethal conflicts, but in less stressful conditions ritual conflict is the norm.

An important element in ritual fights is the generation, in members of the same species, of an inhibition against further aggression.

Lorenz (1996, 206) notes that "A raven can peck out the eye of another with one thrust of its beak, a wolf can rip the jugular vein of another with a single bite. There would be no more ravens and no more wolves if reliable inhibitions did not prevent such actions". These behaviour patterns are called submissive or appeasement gestures. Lorenz prefers the latter. Appeasement gestures have evolved in many animals and are essential if the species is to survive. In appeasement, the losing animal does everything to avoid stimulating further aggression in its opponent. Usually this is in two stages. Firstly, the animal switches off its own aggressive signals. It takes a considerable time for most animals to loose their aggression, once roused, so to ensure aggression ends quickly and without harm the second stage of appeasement is to offer up to the victor the most vulnerable part of the body. This has immediate effect and the victor may simulate an attack on the most vulnerable part without actually doing harm. Thus wolves use their teeth to attack an opponent's throat but when beaten the loser will turn its head, exposing its jugular vein to the victor. The victor bites the air near the jugular and shakes his head, symbolising his victory and assuaging his aggression without doing harm. Similarly a jackdaw exposes to the aggressor's beak the unprotected base of his skull (the place these birds attack when they mean to kill). Almost everyone is familiar with the appeasement gesture of dogs, which is to turn over on their back exposing their vulnerable underbelly to the victor, and even to humans it is disarming. This behaviour in animals is instinctive and the speed with which aggression is reduced (inhibited, is the technical term) is equally instinctive. Animals with sharp and murderous teeth and claws backed by powerful muscles need such inhibitions if the strongest and fiercest are to fight for the right to breed without destroying their species in the process.

Man is now definitely known to be quite closely related to the primates namely the chimpanzees, the gorillas, the orang-utans, the gibbons, and the monkeys. Washburn (1991, 61–69) has recently reviewed the anatomical and biochemical evidence and concludes that whereas the anatomical data is ambiguous the DNA evidence is convincing and offers quantitative proof independent of the opinions of the tester. In the hybridisation test, strands of DNA from two species are mixed, allowing them to match up and form a double helix. The residual genes that do not match up are washed out and measured using a radioactive marking technique. The amount

Table 21.1 Genetic Distances from Humans by DNA Hybridisation. Sibley and Ahlquist (1984).

	Chimpanzee	*Gorilla*	*Orangutan*	*Gibbon*	*Monkey*
Millions of Years	4	5	10	12	20

of residue is a measure of the genetic distance between the species. The relation between man and the apes is shown in Table 21.1 where the estimates of millions of years are the shortest probable estimates from a variety of workers.

As man is indeed related to the primates, it seems obvious to most people that man too must possess some of the aggression and inhibition instincts of animals. Certainly, we engage in ritual warfare and defend territory. We have "achieved the difficult task of avoiding killing without destroying the important functions performed by fighting in the interest of the species. All culturally evolved forms of fair fighting from primitive chivalry to the Geneva Convention, are functionally analogous to—ritualised combat in animals", Lorenz (1996, 241). He goes on to discuss that most ritualised form of warfare, namely, sport, the main function of which he considers to lie in its discharge of man's aggressive urge.

It would also be useful if we could have an animal's inhibitions against killing when presented with suitable appeasement signals so that it would be impossible for us to kill our fellow men except under extreme provocation. Almost the only way we can kill without the aid of weapons is by strangulation, and very few of us, even when fighting for our life, could look into the pleading eyes of a defeated victim and continue to choke him until he expired, so perhaps we are sensitive to appeasement signals. Mankind's problem is that he is not equipped with adequate teeth and claws, which makes him a poor natural killer, and he must, therefore use weapons. The invention of weapons, according to Lorenz (1996, 208), threw mankind's social norms out of gear so that willingness to kill and emotional impunity to killing have increased. Mankind finds it easy to kill his own species because the use of weapons introduces a psychological and a physical distance between the victor and the loser so the victor is unable to see or to respond to the loser's appeasement signals. With no appeasement signals the victor kills without remorse.

Thrown sticks and stones, the sling, the spear, and the bow introduced a greater physical distance, which made it impossible to see appeasement signals, and so we have lost, or are losing, that particular inhibition. Modern weapons such as bomber aircraft and guided bombs and missiles increase the distance even more, so the military are able to kill innocent civilians and children without shame. We even invent phrases such as *collateral damage* or *friendly fire* to hide the horror of what we are doing. Ritual warfare, which limits the amount of force and the conditions under which it is used, sometimes saves us from the worst excesses.

Lorenz's work seemed to many to imply that man was naturally warlike, which contravened the traditional opinion that man was inherently peaceful and his aggression was learned as a result of environmental pressures. In particular it was held that warfare was not inevitable. This controversy produced the great *nature* versus *nurture* debate. A group of scientists meeting in Seville issued a set of five propositions in May 1986 (the Seville statement) intended to discourage the belief that war was inevitable. The first proposition was that "It is scientifically incorrect to say we have inherited a tendency to make war from our animal ancestors", and the other four were in similar vein. In a society threatened by nuclear weapons such statements were comforting. They were adopted by UNESCO and published by scientific journals and societies although they were recognised as being themselves scientifically incorrect. The first statement, for example, implies that animals make war; they do not and no one has suggested otherwise. Lorenz had contended only that animals make a form of non-lethal warfare and that we had inherited aggressive instincts from our animal ancestors. It is now accepted by most scientists that man, like other animals, has an aggressive drive and Fox (1991, 84), for example, who worked for a peace organisation for 12 years, says of the great debate that "It is hard to see what all the fuss was about." Man's aggressive drive is similar to his sex drive, and like the sex drive it is aroused by appropriate external stimuli (erotic in the case of the sex drive) or by internal (hormonal) conditions, or by a combination of the two. Equally important is the fact that these drives can be inhibited by sublimation or they can be ritualised.

Whatever theories are developed regarding man's aggression, it remains a fact that man has been pre-occupied throughout his history with the manufacture and improvement of weapons, and man's earliest relics are of stone tools, most of which appear to be weapons. According to Boswell in his *Life of Samuel Johnson*, Benjamin Franklin defined man as a tool-making animal, and although Johnson objected that many a man had never made a tool, it did not reduce its validity and in the following century anthropologists and archaeologists, anxious to distinguish early man from the apes, took up his idea. Darwin and the earliest anthropologists thought the fundamental event was that apes emerged from the forest into the savanna to seek food, and to survive in the new environment they developed an upright stance, bipedal walking, and used tools. Ardrey (1965, 205) noted that all the stone tools used by early man and illustrated in countless textbooks were, in fact, weapons, see Fig. 21.2, and he thought that "man's cultural efforts is not truly that of the tool; it is that of the weapon". Ardrey would have defined early man as a maker of weapons but thought it too controversial. *Man the toolmaker* creates a positive and comfortable image but *man the weapon maker* suggests an aggressive and less comfortable beast, but it may be more accurate. Since making weapons and tools is the province of the engineer we could say that it is his engineering that defines mankind, but perhaps it would be better to say that the root of engineering, ingenuity, is what differentiates man from other beasts. Fortunately we do not have to define man in terms of a single characteristic. Potts (1991, 41–60) lists the major features, which distinguish modern humans from animals, and he classified them under four major headings, namely:

Brains: Large brain, extreme dependence on tool use, complex symbolic language, development of cognitive skills via complex transmission of information across generations.
Diet: Protein from large animals, high amounts of starch, delayed consumption of food, food sharing sometimes extensive.
Locomotion: Bipedal walking, open vegetation habitats.
Social: Strong economic and sexual bonding, long period of subadult status, home base focus of social life.

The traditional view is that the fundamental shift in human evolution occurred when hominids left the forest for the open savanna and the main adaptations to this new environment was the use of tools and bipedal walking. The use of tools (actually weapons) was related to the dangerous predators faced by early hominids in the open savanna, and bipedal walking developed because height increased

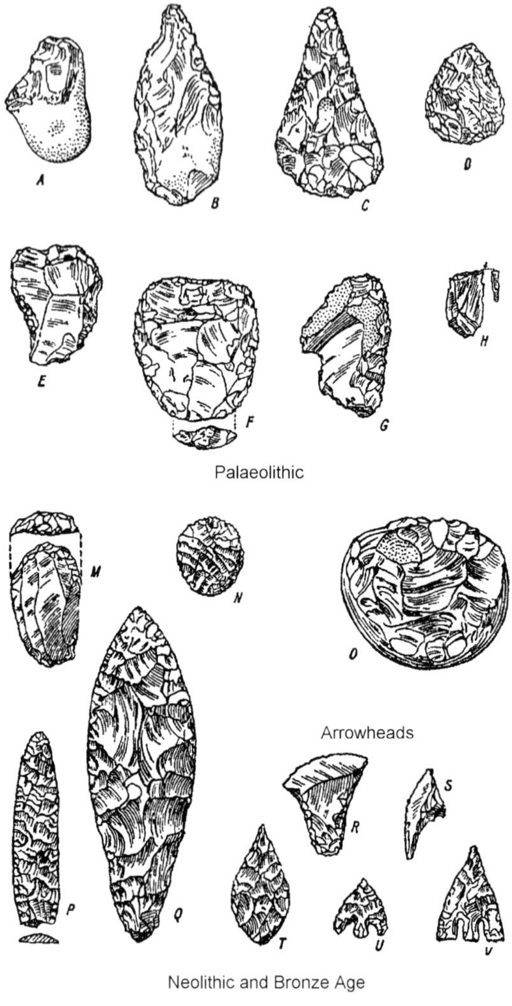

A. Early pebble tool
B. Abbevillian axe
C. Middle Achueulian axe
D. ovate with s shaped edge
E. Clactonian flake tool
F. Levalloisian flake tool
G. side scraper
H. Burin

M. Neolithic scraper
N. Bronze Age scraper
O. Knife with polished edge
P. slug knife
Q. flint dagger
R, S, T, U, V. Arrowheads.

Fig. 21.2 Examples of Palaeolithic and Neolithic tools (weapons). After Wood (1967, 260–61).

the distance that could be seen and freed the hands to carry weapons. These two factors are traditionally thought to be the prime movers in human evolution, and they generated: changes in diet from vegetarian to meat-eating, reduction in size of canine teeth, which were replaced by the use of tools, and increase in brain size, which led to language and to culture.

The earliest hominids existed at least 4 MYA (million years ago), and bipedal walking and prolonged infancy are their earliest characteristics. Living at a home base, large brains, hunting, eating meat, and the regular manufacture of tools are ascribed to a later phase, usually coincident with the origin of Homo about 2.5 MYA. Potts (1991, 41–60) questions this traditional view, and it is useful to summarise his criticisms. Fossil footprints from Laetoli and the anatomy of several hundred fossil finds from Ethiopia and Tanzania demonstrate conclusively that apes began to walk and to lose their canine teeth about 4 MYA. However, in areas where hominid fossils are found, there is evidence (fossil pollen, micro fauna, and macro fauna) that only about 2.4 MYA did major expanses of savanna grassland replace forest. The forest was becoming confined to large patches separated by savanna and bipedal walking may have developed to facilitate movement between these forests. Thus upright walking developed to conserve the forest way of life. Potts believes that the original development of habitual bipedalism, regular habitation of open savannas, and the evolution of anatomically modern bipedalism were significantly separated from one another in hominid evolution.

The earliest evidence for the manufacture of stone tools or weapons is 2.5 MYA but tools made of materials such as wood are unlikely to be preserved and it is uncertain whether the earliest bipeds used perishable tools. Stone-technology 2 MYA was simple, largely involving the production of sharp-edged flakes and cores by stone on stone percussion. Hominid brain size began to expand from 2 MYA and continued to the present time. The earliest stone tools and the origin of a large brain approximately coincided and this has suggested an intimate connection between stone technology, brain size, and the overall mental ability of hominids. However, brain size in the earliest specimens of *H. erectus*, for example, is approximately 25% smaller than brain size 1.2 MY later; yet during this period, stone tools remained stagnant and were dominated by hand axes and by small cores and flakes. Consequently the assumption that stone technology and brain size are related may have to be reconsidered.

The discovery of early archaeological sites has suggested that by 2 MYA hominids possessed the basic characteristics of the hunter-gathers, especially the transport of food to a home base where food was processed and which also served as the focus of a social life. Potts notes that this interpretation of early archaeological sites, has met extreme criticism from a number of studies because evidence of bone damage indicates that both hominids and large carnivores were attracted to the same specific places. This overlap in hominid and carnivore occupation implies that these early sites did not serve as home bases. Only later in time did an ecological separation between hominids and carnivores occur which allowed hominids to create safe home bases. Potts suggests that such an ecological separation occurred about 700 TYA (thousand years ago) in eastern Africa. A second issue concerns whether early hominid hunter-gatherers were, in fact, hunters. The presence of tool cut marks on animal bones from 2 MYA is clear evidence that hominids had access to meat but reliance on hunting, as opposed to scavenging, for animal protein now appears to have arisen after 1.5 MYA. The earliest clear evidence for systematic hunting of a single species does not occur before about 100 TYA in southern Africa, and not until after 30 TYA in Europe.

The existence of stone tools has also caused some anthropologists to claim that language also must have existed at least 2–2.5 MYA. However, Potts notes that innovations in stone and bone technology are associated with the first symbolic representations, such as cave paintings and mobile sculptures, most examples of which date to about 20 TYA. Prior to 100 TYA the lack of diversity in tools and the nature of archaeological sites reflect the lack of the innovation that characterizes culture in the modern sense. This contrasts with the long-held assumption that once tools appear in the prehistoric record, culture with the basic elements, such as language, also appears.

The cave paintings of the Upper Palaeolithic give very limited information about the tools and weapons used at that time. Most cave paintings, in fact, are of animals. Fig. 21.3, from Les Trois Frères, shows hybrid animals followed by a man apparently disguised as a bison. The reindeer nearest the man has a bison's head, turned towards the man, and a swollen vulva. The man wears a bison's head and fur pelt, and operates some kind of bow. At Niaux in the Pyrenees is a bison that incorporates a natural hole to represent the

Fig. 21.3 Top: hybrid animals and man in a bison skin holding a bow? Bottom: wounded bison with arrows or spears.

eye and other holes to represent wounds from which engraved barbed darts (arrows or spears) project. These are amongst the earliest pictures of weapons.

Later cave paintings from Mesolithic Spain, date from 12000–7000 B.C. and are rich in hunting scenes with warriors in animated and lively movement. Fig. 21.4 (top), despite its diagrammatic construction, very clearly shows a man with a bow and arrow shooting at a reindeer; it is the first definite appearance of the bow and arrow in painting. Lively hunters carrying bows and arrows and wearing clothes are also depicted in Fig. 21.4 (bottom).

Warfare in Prehistory

In the earliest times, the life of the hunter-gatherer was short, which implies that the birth rate must be high, and thus the need to bear

Fig. 21.4 Left: the bow and arrow appears, Cogul, Lerida, Spain. Right: hunters with bow and arrows, Valltorta and Cueva de los Cabalos, Spain.

and rear children throughout a woman's adult life was essential. In order to ensure the survival of the group, women were, for most of their adult life, either pregnant or nursing, or perhaps both. In addition, women were and are on average somewhat weaker and smaller than are men. For these reasons, they were less able to undertake the more vigorous activities such as hunting wild animals or defending territory. They assumed the homely tasks of child rearing, food gathering and food preparation. The men were the hunters, made the weapons for hunting, travelled away from the home, made contact with other groups, engaged in trade, quarrelled, fought, built defences, engaged in raiding, and made weapons for war. Women would have made better ambassadors. Bearing children was not seen as risking a life for the sake of the group, although it was cherished through the worship of fertility gods, but hunting fierce animals and defending the settlement were considered acts of valour meriting great prestige. The reward for such valour, at least amongst some present-day primitive societies, was sexual access to women. This ensured that it was the strongest and bravest who were allowed to reproduce, and it also encouraged others to follow their example.

It is sometimes assumed that warfare and violence were rare in the distant past when man was first a forager (hunter-gatherer). With a low population density, pressure on food and other resources must have been slight and there was, according to present day idealists, no need for conflict. Mumford (1967, 216), for example, says "In the earlier stages of Neolithic culture, there is not even a hint of armed combat between villages: possibly even the massive walls around ancient towns like Jericho—performed a magico-religious function before they were found to furnish a decided military advantage." He goes on to compare the skill and courage of the surviving Palaeolithic hunters with the plodding gardeners and farmers of the Neolithic. Mumford thought that warfare emerged only when the priests demanded victims for human sacrifice. It is more comfortable to believe that man emerged from his peaceful idyll (if it ever existed) only when populations grew to the point where groups were forced to compete for the available water, food, shelter and other resources. If this was so then it happened quite early because it is now reasonably clear that violence, conflict, and warfare has certainly existed since the end of the last ice age and perhaps much longer. The main evidence for this is from extensive archaeological and osteological data; see Martin and Frayer (1997). The osteological evidence of trauma in human skeletal remains includes embedded arrow points, depressed cranial fracture, knife cuts on bone etc, all of which may be found in a surprisingly large proportion of surviving human skeletal remains, and are taken to indicate violent, as opposed to natural, death. The archaeological evidence of warfare is mainly related to defensive works and fortifications in and around human settlements and the way these, and the location of the settlements, have changed over time. Mumford was quite wrong. Neolithic culture does show extensive evidence of warfare and the "plodding gardeners and farmers" of the Neolithic were sufficiently fierce to take over and hold the land already occupied by Mesolithic hunters.

Lambert (1997, 77–103) presented evidence for lethal projectile wounds and sub-lethal, healed cranial fractures in human bone from around the Santa Barbara Channel of California. Her material extended from about 6000 B.C. to about 1800 A.D. and she found that physical violence was highest during periods of ecological stress brought on by greater population density and periods of drought. Lambert took the mortuary remains from 30 prehistoric and historic sites and examined them for evidence of depressed cranial vault fractures. Most modern accidents leading to head trauma cause linear fractures in the cranium. Depressed cranial fractures correspond to a more or less circular depression of about 13 mm diameter and 1.4 mm deep in the skull and are caused by violent contact with weapons such as rocks, clubs and spears. Such fractures range from non-penetrating indentations, which are rarely lethal, to complex

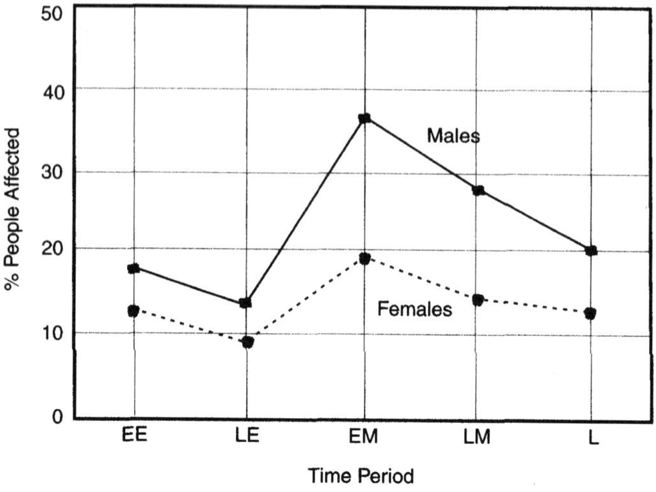

EE = 6630–4050 B.C.; LE = 4050–1490 B.C.; EM = 1490 B.C.–580 A.D.; LM = 580–1380 A.D.; L = 1380–1904 A.D.

Fig. 21.5 Sex differences and temporal variation in the frequency of healed cranial vault fractures. After Lambert (1997, 87).

penetrating fractures where bone fragments are driven into the brain. Of 753 individuals examined, Lambert found 128 (17%) with healed cranial vault fractures, and 90% of these were circular or ellipsoidal in shape, implying the use of some type of rounded club. Only 10 crania had penetrating fractures, signifying that most blows injured but did not kill the victim.

The proportion of cranial vault fractures was found to vary with age and with gender. Children were rarely affected (3%) whereas 9% of adolescents, 17% of young adults, and 24% of mature adults under 40 exhibited healed cranial vault injuries. Surprisingly, only 16% of adults over 40 years of age had such injuries. This is understandable if those sustaining head injuries died younger than those not sustaining such wounds.

Fig. 21.5 shows the variation in the frequency of healed cranial vault fractures with gender and with time. The sample is of 608 individuals of known sex, 319 male and 289 female. Cranial injury was more common in males than females and the difference is statistically significant. Sub-lethal cranial vault fractures are present from all periods and indicate that physical violence occurred throughout

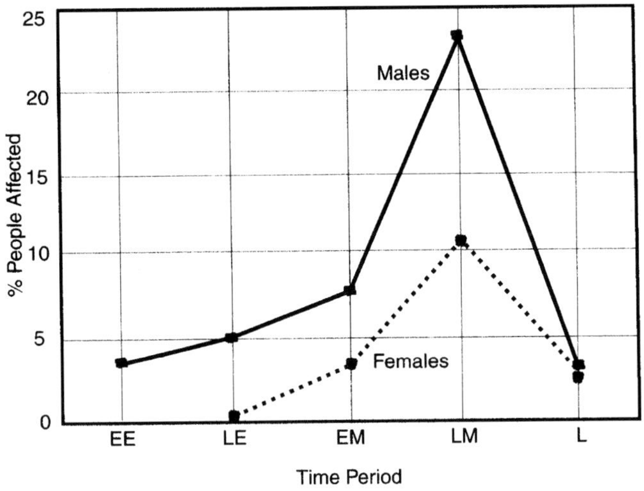

EE = 6630–4050 B.C.; LE = 4050–1490 B.C.; EM = 1490 B.C.–580 A.D.; LM = 580–1380 A.D.; L = 1380–1904 A.D.

Fig. 21.6 Sex difference and temporal variation in the frequency of projectile injuries.

prehistory. Violence, however, appears to have peaked during the period from 1490 B.C. to 580 A.D. and to have been very high during the period from 580 to 1380 A.D. These periods are known to be a time of rapid population growth, increasing sedentism, and increasing dependence on marine resources.

Differences in the location of the injury were found. In males, about 65% of injuries were to the front, 25% to the side, and only 10% at the rear. In females, about 35% were to the front, 40% to the side, and 25% to the rear. The frontal injuries are taken to represent face-to-face conflict. The side injuries occurred with greater frequency on the left side, suggesting a blow when facing a right-handed aggressor. Injury on the right side could occur from a left-handed aggressor or when fleeing from a right-handed aggressor.

Lambert also investigated projectile injuries caused by stone, wood, cane, or metal tipped projectiles. The frequency of these injuries is such that the probability of accident must be ruled out in the majority of cases. Because of their lethal intent, such injuries are a better indicator of lethal violence than cranial head injury. In modern tribal groups, projectile injury is most frequently observed in feuding, raiding, or open warfare. Such injuries reveal themselves when stone points and tips of fractured points are found embedded in bone. Scars on bone surfaces caused by projectiles that did not embed are also found. Frequently a projectile must have struck a vital organ and caused death without leaving evidence on the skeleton. Where one or more such a projectiles remain within the body they are often recovered during excavation from beneath the skeleton and provide additional evidence of projectile injury. The temporal and gender frequencies of projectile injuries, shown in Fig. 21.6, are similar to those for cranial vault fractures. The distribution of such wounds shows that most were sustained in the thorax (33%), the pelvis/abdomen (27%), and the head/neck (25%). The lower limbs sustained about 9%, and the upper limbs about 6% of projectile wounds. About 70% of these injuries would cause death.

Lambert notes that the period of maximum violence ended about 150 years before European explorers first arrived in the Santa Barbara area and European contact, therefore, had no role to play in the highest levels of violence in the area. The violence observed in the prehistoric and historic record, she notes, correlate well with the resource stress model of violence, although she was unable to test other models that emphasise reproductive resources (women) and kin selection. Unfavourable climatic conditions leading to drought and skeletal evidence of ill health were taken to be the indicators of resource stress.

Frayer (1997, 181–213) presented further skeletal evidence of prehistoric violence, from about 5000 B.C. He examined the late Mesolithic site at Ofnet, Bavaria, in which a large number of men, women, and children were slaughtered and selected parts (head and neck vertebrae) were buried in two mass graves. It is one of the earliest and most convincing sites providing evidence of prehistoric violence. Frayer considers there were 31 individuals in one pit and a further 6 in a second pit. The peri-mortem nature of the cranial wounds, their constant size and shape, and evidence that they were placed in the grave at the same time, leads him to conclude that the site is of one, or possibly two, prehistoric massacres. The demographic profile of the main pit was not that of a natural population, so it appears that either the massacre occurred when the majority of the adult males were absent or that they were buried elsewhere. Frayer found no evidence of cannibalism and he is not hopeful that an adequate account of the events leading to the massacre will ever emerge. What is evident, however, is that these Mesolithic men, women, and children were victims of a massacre and ceremonial burial and that violence between groups of European hunter-gatherers was probably usual.

Keeley (1997, 303–19) has discussed the archaeological evidence of war and conflict between hunter-gatherers and agricultural invaders (the Linienbandkeramik, or LBK, people). LBK settlers colonised a large area of land on the North European Plain stretching from Ukraine in the east to Northern France in the west. Late Mesolithic hunter-gatherers already occupied this region but their way of life soon crumpled and the change to LBK farming took no more than a few hundred years. Contrary to the traditional views, Keeley finds that this encounter was not peaceful. Evidence that the Early Neolithic frontier was violent include:

i. finds of homicide victims in both Late Mesolithic and LBK burials,
ii. fortified LBK border settlements,
iii. no-man's land between LBK and final Mesolithic settlements, and
iv. war weapons in the LBK material culture (they were not hunters).

Keeley is not so reticent as Frayer in interpreting the gruesome finds at Ofnet and takes the remains to be the result of head hunting by Late Mesolithic foragers.

Mesolithic skeletons, although relatively few in number frequently show evidence of violent death, so much so that in Keeley's opinion the Mesolithic is becoming recognised as the period in which true warfare first appears in Europe. One of the most dramatic finds was a mass grave at Talheim, Germany, which contained 34 men, women and children, all of whom had been killed by one or more blows to the head, delivered by the distinctive LBK adzes and axes. He notes that hunting of wild game by the early LBK settlers was minimal but arrow points occur with high frequency in their settlements, which suggests that they were not for hunting but for weapons of war.

Defences, especially in the western part of their distribution, enclose many LBK settlements. Such "enclosures" are V-shaped ditches, two or three metres deep, backed by palisades, and having "baffled" gates, and sometimes a gate tower. The LBK distribution map in Fig. 21.7 (left) shows that these fortifications were clustered along the limits of their settlement zone, and clearly, the threat to the LBK settlements came from the Mesolithic population beyond the Neolithic

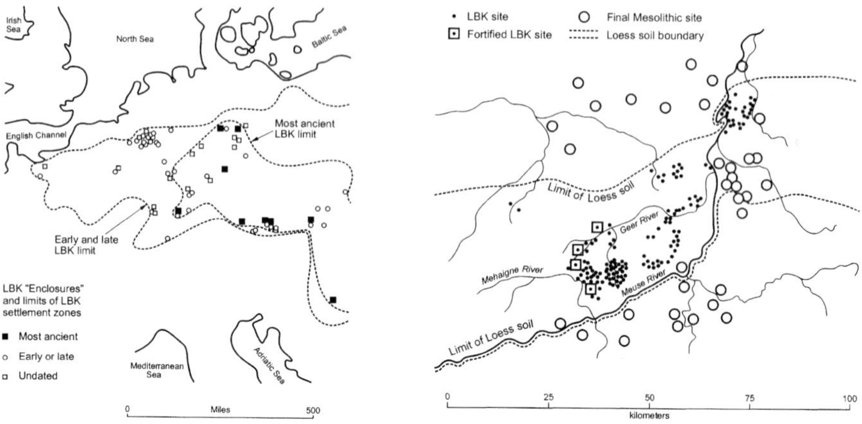

Fig. 21.7 Left: map showing the association between LBK fortifications and the limits of LBK settlements. Right: Map of final Mesolithic and Early Neolithic (LBK) sites. After Keeley (1997, 309–10).

frontier, and *vice versa*. Fig. 21.7 (right) shows the separation between the nearest Mesolithic and Neolithic settlements; such settlements are close only where the "trench" of the Meuse/Maas intervenes.

The exact form and motivation of prehistoric warfare is difficult to define but at least some visual evidence of warfare in the Mesolithic is available. For example the rock painting in the rock-shelter of Las Dogues, Castellón de la Plaña, eastern Spain, shows two groups of archers in combat, see Fig. 21.8 (top). The group of ten or eleven archers on the left is being attacked vigorously by a group of eighteen or nineteen archers from the right. The artist manages to convey by their body posture that the small group is withdrawing while the large group is pressing forward. Fig. 21.8 (bottom) shows two groups of archers attacking each other. Their warlike intention is unmistakable. Clark (1963, 81) notes that warlike images such as these are confirmed by finds of one or more arrowheads embedded in human skeletons. When more than one arrowhead is found it seemingly rules out hunting accidents, although, of course, execution is not excluded. Executions, in fact, did occur in the Mesolithic period. Fig. 21.9 shows two such a scenes. In Fig. 21.9 (left) a line of ten bowmen, the "firing squad", are shooting at a hunter lying on the ground. The hunter props himself up on his arms but his legs beat the air and arrows fly towards him. Fig. 21.9 (right) shows

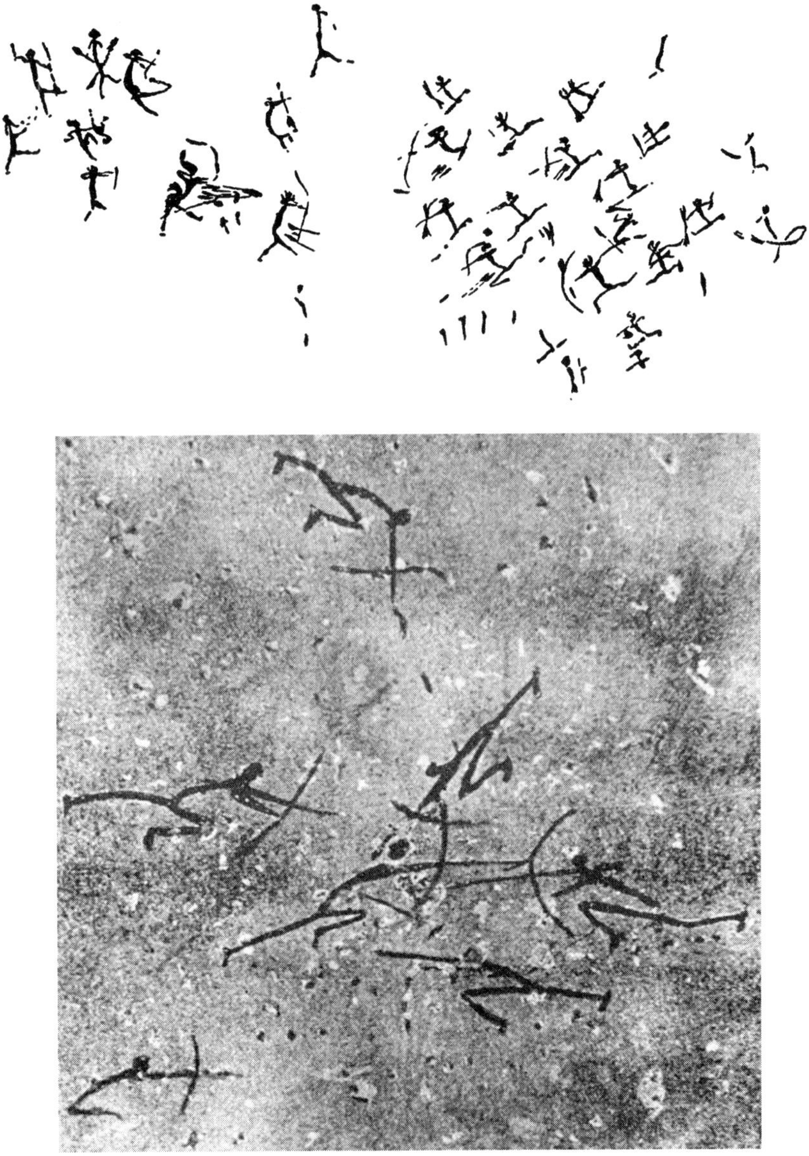

Fig. 21.8 Top: rock painting showing archers in combat. Bottom: archers in battle. Castellón de la Plaña, eastern Spain.

Fig. 21.9 Left and Right: two Mesolithic scenes showing executions. Gasulla Ravine, Spain.

a man lying on his back, pierced by missiles, while two women, with arrows in their hands look on, and in the background there appears to be a group of men, possibly the firing squad. Similar scenes occur elsewhere and Kühn (1956, 73) suggests that the victim may have contravened some law of the chase.

Warfare in Primitive Societies

All the above is hard evidence of a violent prehistory. The ethnographic record, that is, the record of primitive societies of the present or recent past, confirms that warfare and violence are ubiquitous. Almost no society is totally peaceful although the Eskimo of the Arctic, the Mission Indians of North America, the Arunta of Australia, and the Andaman Islanders of the South Pacific, come close to this ideal. In their analysis of warfare, Ember and Ember (1997, 1–19) divided primitive societies into five types of subsistence regime: forager, horticulture, intensive agriculture, mixed farming, and pastoral. Each of these was classified as either pacified (forced to stop fighting by external settlers) or non-pacified. They found that in 57% of non-pacified societies warfare was constant and only in 9% of such societies was it absent or rare, see Table 21.2.

Table 21.2 Frequency of Warfare in Primitive Societies. Ember and Ember (1997, 5)

	Absent or Rare	*Every 3–10 years*	*Every two years or so*	*Every year in a season*	*Constant or at any time*
All Societies (134)	28%	13%	9%	12%	38%
Non-pacified Societies (90)	9%	9%	9%	17%	57%

Ember and Ember also determined war frequency in five subsistence regimes, Table 21.3. The results indicate that hunter-gatherers experience war less frequently than the other subsistence groups, although even for hunter-gatherers it occurred every two years or so. It seems likely that the hunter-gatherers, being mobile, could simply move away from possible conflicts, especially if they were unlikely to win, whereas the other groups, having fixed assets in the form of growing crops and stored surpluses, were less able to avoid conflict. Ember and Ember, however, tested a large number of theories about what might predict more war or less war but most theories failed.

Table 21.3 Frequency of warfare in Non-Pacified Societies by type of subsistence. Ember and Ember (1997, 7)

	Absent or Rare	*Every 3–10 years*	*Every two years or so*	*Every year in a season*	*Constant or at any time*
Hunter-Gatherer		X	M	X	
Horticultural			X M	M X	
Intensive Agriculture					M
Mixed					M
Pastoral			X	M	X

M = Median, X = upper or lower limit

Nevertheless, they found that two possible predictors of more war were:

i. threat of natural disasters (weather or pests), and
ii. socialisation for mistrust.

They found that in their sample hunter-gatherers were fortunate in having fewer unpredictable natural disasters and they found that hunter-gatherers had less socialisation for mistrust (were more trusting), and therefore they experience war less frequently.

Conflict and warfare in primitive societies certainly existed, and still exists, but it was carefully graded and was seldom the all-out warfare of modern times in which a nation's entire productive resources are demanded. Warfare was often ritualistic and designed to solve problems in a honourable way without endangering the group's survival. Anthropologists recognise that the probability of violent conflict increases with the social distance between the offended parties. If the social distance is small (the parties are related) then lethal fights are usually avoided and warfare is best described as feuding. There is an obligation to settle the feud relatively quickly and by peaceful means. If the social distance is greater then there is no such obligation; warfare may be ritualised and disputes may last a long time. However, particular fights may be brought to an end for any number of reasons without loss of face and the dispute may be taken up again at a later time. Such fights may be an important social activity, a form of bonding, as the myths relating the lives of a group's heroes make plain. But killing or even serious wounding is not common. Serious wars of conquest against socially and physically distant neighbours are unlikely unless, for example, population pressure is great and there is a need for more land.

The present day Yanomamö of Brazil and Venezuela, for example, are trained from youth to be fierce and to exhibit violence to women. Female infanticide is practised to increase the number of fierce males. Amongst their own villages they demonstrate an escalating degree of violence, see Chagnon (1993). At the lowest level, chest-pounding and side-slapping duels usually take place at inter-village feasts and arise between members of different villages over insults, trade, food, or women. After taking hallucinogenic drugs to foster violence, the man steps forward to beat his chest. A representative of the other village takes up the challenge and strikes his opponent's chest several times. To demonstrate toughness the man does not respond immediately but eventually demands the right to strike back. After exchanging blows for some time one or both are disabled or winded and after this ritual fight both sides swear their eternal friendship and the problem is forgotten.

The next level of violence is the club fight, which is usually related to adultery. The plaintiff carries a long pole, sharpened at the end if it is serious, to the village centre and shouts insults to the offender. If the challenge is accepted then the plaintiff and the offender take turns to receive a blow on the head, starting with the plaintiff. After this, others join in, picking sides and wielding clubs. To prevent a death the village headmen intervenes with bow and arrow and threatens to shoot anyone who continues the fight, but even so a villager may die and the attacker must escape. Such a death inevitably leads to raiding between villages. In one case, Chagnon reports that the raiders chased the inhabitants of a large village out of their homes under a hail of spears. The inhabitants of the large village recovered, chased the raiders, and a second spear fight took place. Several men were wounded and one died later, but the intention had not been murderous.

After this, further raids occurred in which the intention was to kill someone (anyone) from the opposing village. The raiders hid outside the opposing village until a suitable victim appeared, killed him or her, and then fled. This, of course, called for reprisal raids. Chagnon considers these raids more warlike than the spear fights, where killing was not the intention. The ultimate act in this escalation of violence was the treacherous feast in which one of the warring villages persuades a third village to invite its enemy to a village feast where they are surprised and as many as possible are killed. The widows are distributed among the victors.

Such warring is not particularly valiant and it was not undertaken to obtain territory or in response to shortages of water, food or other essentials. Chagnon believed it was to establish a village's reputation so that its women, who were always in short supply since female infanticide was practised, could not be acquired on favourable terms. The displays of fierceness were intended to deter seducers, wife-stealers, and raiders. In wars with non-Yanomamö neighbours however, their behaviour was quite different; it was genuinely fierce and undertaken to obtain additional territory and resources.

There are many primitive peoples who engage in ritual war and Fig. 21.10 shows one such group in South Africa, but perhaps it is

Fig. 21.10 Group of native boys and men with new shields of hard bullock hide, assegais, or spears. South Africa, 1880.

only necessary to refer to one other example. The Dani of western New Guinea consist of about 50,000 people split into a dozen alliances, Clarke (1971, 209–10), and were studied by a team of anthropologists in 1961. War, more correctly ritual war, was a constant preoccupation. Apparently, they engaged in ritual war to promote the success and well being of their social order. Their health, welfare and happiness depended on the pursuit of aggression against traditional enemies; something Thomas Jefferson did not think about when considering the American Constitution. Formal battles lasted for a whole day if the weather was satisfactory and the decision to fight was taken early in the morning and shouted across the no-man's land separating the two alliances. The morning was spent in preparation and by noon the two sides face each other at a distance of about 500 m. Small groups of 50 to 200 warriors advanced to within 50 m of the enemy, discharged their arrows for 10 to 15 minutes and then retired. Most energy was spent in dodging arrows. After 10 or 20 such raids some warriors began to return home and arrows were replaced by insults. Eventually all returned to their homes. In any year perhaps 10 or 20 deaths occurred in such fights, but this was not considered excessive. The normal death rate amongst the Dani from all causes is probably in excess of 1,000 per year so deaths due to warfare amount to less than 2% of the total and is comparable to that in Europe during the last few hundred years.

Ritual Warfare in History

Ritual warfare, as practiced in virtually all primitive societies, was carried over into the historical period and through to the present. In antiquity, the proportion of the male population that engaged in warfare was normally quite small. This was undoubtedly related to productivity; a large agricultural and industrial surplus is required to sustain a standing army or navy. A militia made up of trained amateurs was cheaper but was less effective and was less able to stay under arms for long periods. Consequently, the military, within early societies, formed a social-elite who were usually very close to the centre of political power. Slaves were not admitted to any level of military organisation except in times of dire necessity. Greek triremes, for example, were normally rowed by Greek citizens and not by their slaves. Total warfare, in the modern sense, did occur, but if the social distance was not great, warfare was restricted so the participant societies were not permanently damaged. Such ritual wars took place according to strict rules designed to limit the objectives and the extent of the conflict. It is useful to examine typical examples.

Perhaps the most famous ritual conflict is that which took place about 1000 B.C. between David and Goliath and described in the Bible (1 Samuel 17). The Philistines were an immigrant people who came from Crete and settled along the coastal strip of Canaan in the 13th and 12th centuries B.C. They were the neighbours and enemy of the Israelites, and culturally superior despite the impression given in the Bible. During one conflict, which seems to have been for the acquisition of slaves, the Philistines drew up their forces on a hill and the Israelites, under King Saul, camped on an opposite hill; but neither side attacked the other. It seems that they had decided to settle their dispute by the single combat of a selected champion from each side. The Philistine champion was Goliath of Gath, a man said to be almost three metres tall. He wore a bronze helmet, and bronze plate armour, and on his legs were bronze greaves. One of his weapons was a bronze dagger and the other was a heavy spear with an iron head. A shield bearer marched ahead of him for protection. For forty days he came out equipped before the Israelites shouting, "I am the Philistine champion; choose your man to meet me. If he can kill me in a fair fight, we will become your slaves; but if I prove too strong for him and kill him, you shall be our slaves and serve us." David, a shepherd boy visiting his brothers in Saul's army, volunteered to fight Goliath and convinced Saul that he was worthy by recounting the times he had killed bears and lions attacking his father's herds. Presumably, he had killed these creatures by stunning them with a slingshot and then dispatching them with a knife. The plan was agreed and David, after rejecting Saul's offer of armour, took to the field against Goliath armed only with a sling and five, no doubt substantial, pebbles. David, of course, had the advantage of both speed and range and stunned Goliath with a direct hit on the forehead, he then used Goliath's own bronze dagger to cut off his head. Seeing this the Israelites raised the war cry, pursued, and slaughtered the Philistines all the way to their home city of Gath, and on their return, they plundered the Philistine camp. Despite the slaughter referred to in the Bible it is likely that the majority of the Philistines found safety behind their city walls, and that the Israelites were more intent on plundering the Philistine camp. The agreement for single combat, made no doubt to prevent a general slaughter, seems to have been partially successful.

Another example is provided by the conflict between the Argives and the Spartans over the possession of Thyreae, an area on their frontier. Herodotus (I, 82) tells us that in the middle of the sixth century B.C., the Argives and the Spartans agreed that 300 picked men on each side should fight it out and that the remainder of their armies should return home least they be tempted to intervene. When darkness put an end to the battle, only three of the 600 were left alive, two Argives and one Spartan. The two Argives claimed victory and returned to Argos but the Spartan remained under arms, stripped the bodies, and carried their equipment to his camp. Both armies claimed the victory, the Argives because they had more men left in the field and the Spartans because they were left in possession of the field. Consequently a fresh battle began in which, after severe losses on both sides, the Spartans were victorious. In this case, the attempt to limit the battle was unsuccessful; but the attempt was made.

Herodotus (4.1) also records an early conflict between the Paeonians and the Perinthians. It seems that an oracle had told the Paeonians to march on the Perinthians and once they faced each other they were to wait until the Perinthians called on them by name before attacking, otherwise they were to hold back. The Perinthians took up their positions on the outskirts of their town, a challenge was issued and three champions from each side—a man, a horse, and a dog—engaged respectively in single combat. Seeing that two of their three champions were getting the better of the tussle, the Perinthians in great delight shouted out 'Io Paeon', the cry of victory, whereupon the Paeonians supposed that this was the cry mentioned by the oracle. They attacked immediately, won a resounding victory, and left few of their enemies alive. Again the attempt to limit the battle had been made but had failed.

It is likely that limited war was usually more successful than in the two cases quoted above. Strabo (X, 1, 12) describes a dispute that took place in the late eighth century between the cities of Chalcis and Eritria over of possession of the Lelantine plain. "In general these cities were in accord with one another, and when differences arose concerning the Lelantine plain they did not so completely break off relations as to wage war in all respects according to the will of each, but they came to an agreement as to the conditions under which they were to conduct the fight. This is shown by a certain inscribed pillar which forbids the use of long-range missiles." Presumably, the rules carved into the pillar were obeyed.

Long-term ritual warfare with limited objectives was often in the interests of the attackers and was tolerated by the defenders. Herodotus (1:17) recounts that: "Alyattes carried on the war that he had taken over from his father, against the Milesians. His custom each year was to invade Milesian territory when the crops were ripe—. The Lydians refrained from demolishing houses in order that the Milesians, having somewhere to live, might continue to work the land and sow their seed, with the result that they themselves would have something to plunder each time they invaded their country."

On another occasion, according to Herodotus (9:26), after the Tegeans and others had occupied the Peloponnese, the Heracleidae, whom they had displaced, attempted to force their way back. The Tegeans and others formed a common army and met the invaders at the Isthmus. Hyllus, the leader of the invaders, proclaimed that there was no need for the two armies to risk their lives in a general engagement; it would suffice if a champion chosen from the Peloponnesian army met him in single combat upon agreed conditions. The Tegeans and their allies accepted this proposal and engaged upon oath that, if Hyllus were victorious, the Heracleidae should be allowed to resume the ancient rights of their family, and that if he were vanquished, they should withdraw their army and make no further attempt upon the Peloponnese for a hundred years. The

Tegean king, Echemus, was chosen to fight and he engaged with and killed Hyllus in single combat. In recognition of this, the Tegeans were granted certain privileges, such as fighting in the place of honour on the right wing, whenever joint operations of their armies occurred. They claimed this honour, but lost it to the Athenians, at the battle of Plataea, in 479 B.C., when a confederation of Greeks defeated the invading Persians under Xerxes and Mardonius.

At Plataea too the possibility of ritual or limited war was offered, by the invader, but was declined. According to Herodotus (9:48) the Persians sent a herald to the Spartans saying "why should we not fight with equal numbers on both sides, with you (who are supposed to be the most valiant) as the champions of Greece, and ourselves as the champions of Asia? Then, if it seems a good thing that the rest should fight too, they can do so after we have finished; otherwise let us settle it between us, and let the victor be considered to have won the battle for the whole army." The herald was ignored. Limited war to settle serious disputes with other Greeks was often acceptable but limited war against an Asiatic invader seeking to conquer the whole of Greece was inappropriate; the social distance was too great. In effect the invaders were treated as a territorial animal would treat a hostile invader of a different species; they would fight to the death.

The Etruscans also attempted to limit technology in their wars. Iron weapons were particularly effective against bronze and Pliny (XXXIV 139) notes: "In a treaty granted by Porsenna to the Roman people after the expulsion of the Kings, we find it specifically stated that iron shall be used only for agriculture." The Etruscan, Lars Porsenna, seems to have been master of Rome towards the end of the sixth century B.C., and it appears that by restricting Roman access to iron weapons he hoped to curtail their ambition. It was a century later that Rome, after beating off attacks from Gaul, embarked on her period of great expansion and Etruria was soon overrun.

Garlan (1976, 15), discussing warfare in the ancient world notes that Athens, between 490 B.C. and 338 B.C., was at war more than two out of every three years and never enjoyed peace for as long as ten consecutive years. But such warfare in early societies was often contrived and limited in scope so that the dispute could be settled without endangering the survival of the groups involved. These attempts were not always successful, but neither did they always fail. Such an approach to warfare was not conducive to the invention of new weapons and did not encourage technological development. New weapons, like the 'long range missiles' described by Strabo, or the iron weapons feared by Porsenna, were liable to be forbidden in ritual wars. As the population increased, and states developed, a single war was unlikely to threaten extinction and therefore ritual wars became less common. Rome, however, was ever warlike, mainly because they were interested in territorial acquisitions. Garlan says that: "During the whole of Roman history down to the Empire the temple of Janus which 'signified when open that Rome had taken up arms and when closed that peace reigned all about her' was according to Livy (I,19,2–3) closed on only two occasions from the time of Numa Pompilious (beginning of the seventh century) to that of Augustus (end of the first century B.C.)". Thus, for the best part of six centuries Rome was at war and busily expanding its borders.

The Romans were more concerned in building an empire than in settling lesser disputes, and did not engage much in ritual warfare. However, this was not the case with many of their enemies and both Posidonius and Diodorus describe duels between warriors when fighting the Celts. Diodorus (5.29) writes of the Celts: "It is also their custom when they are formed for battle, to step out of the line and to challenge the most valiant men from among their opponents to single combat, brandishing their weapons in front of them to terrify their adversaries. And when any man accepts the challenge, they break forth into song in praise of the valiant deeds of their ancestors and to boast of their own high achievements." Pleiner (1993, 28) notes that Q. Fabius Ambustus won an equestrian duel with a Gallic chief in a skirmish near Clusium in 387 B.C. and before the battle of Anienum (*c.* 360 B.C.) a Gallic warrior provoked the Roman forces until Titus Manlius accepted his challenge to a duel and, with the permission of his commander, Manlius killed the Gaul and took his necklace (torque) as a trophy. After Manlius' victory the Gauls retired from the field. Pleiner quotes three or four other such instances and concludes that the ancient institution of single combat prevailed in the distant past among the Romans, but they had nearly forgotten it. The Celts, it seems, kept the tradition alive until the second century B.C.

Ritual warfare, however, survived in the chivalry of the medieval period and in the Geneva Convention of modern warfare. The feudal system developed during Charlemagne's time and was important in providing knights (cavalry) to his army. After his death the Frankish kingdom declined and much of Europe was reduced to a state of anarchy in which the owner of every manor became a minor king. Manor houses were fortified and protected with moats, and petty raids against neighbours became common. On a larger scale Europe was attacked by Vikings, Saracens, Magyars and Slavs and Christianity was threatened. To defend themselves against such tyranny, and in the absence of a central power, the larger landowners associated to make their property secure and work for a common good.

By about 1030 A.D. the external threat was much reduced and the nobility, trained in the arts of war, needed other distractions if they were not to turn against each other. They were civilised by the Church who recruited them to fight on its behalf. In 1095 Pope Urban II proclaimed a Crusade to free Jerusalem from the Heathens. For the next 200 years various crusades kept European knighthood busy, and mostly out of Europe. Urban also declared that every person of noble birth on attaining the age of twelve should swear before a bishop that he would defend widows and orphans, and that women of noble birth should be his special care. The protection of noble women was to become a favourite cause of knighthood; in addition to a wife most knights also had their lady whom they swore to protect and honour with deeds of great valour, either on the battlefield or at jousting tournaments. The Church became less keen on this idea, with its implications of unfaithfulness, and tried to discourage it. John of Salisbury, see Oakeshott (1960, 185), thought that knighthood was the armed hand of the state, its members must be selected carefully to serve the king but never to the exclusion of their duty to serve the church. Vincent of Beauvais wrote: "the use of an organised knighthood lies in protecting the Church, attacking disloyalty, reverencing the priesthood, avenging the wrongs of the poor and keeping the country in a state of quiet." Urban's plea to protect women and orphans was quietly dropped. The knights bound themselves to these duties by a solemn vow taken at a religious ceremony after an all-night vigil, but most, no doubt, preferred to be knighted on the battlefield. Thus the orders of knighthood and the ideals of medieval chivalry were created and moulded.

The earliest orders of knighthood were formed in the sixth and seventh centuries, and many were formed at a later date to support the crusades. In England a priest conferred knighthood, after confession and consecration of his sword, but the Norman conquerors did not approve of this and it was suppressed in a synod held at

Westminster in 1100 A.D., and thereafter only the monarch conferred it. According to semi-historical records, Hereward the Wake, as an act of rebellion against the Normans, was knighted in the old style by the church, see Swanton (1998, 37). He behaved in the manner expected of knights, even when they were outlaws. He adopted a suitable lady to defend; when besieged he appeared before the castle walls every morning to engage in single combat with chosen champions from the besieging army; although he was happy to fight against more than one knight at a time he considered it unfair and would call-off an attack by his own knights on a single enemy knight. Always he preferred to act honourably even if it was against his own best interest. The orders of knighthood proved very popular for over 245 were created in many different countries, although most of these were founded after 1500 when feudalism was mostly dead. Ultimately Cervantes mocked their ideals in *Don Quixote*, but these ideals were carried over into serious wars, sometimes with advantage and sometimes disastrously.

Examples of misguided chivalry, some would say of gallantry, abound in medieval warfare. Oakeshott (1960, 198), for example, notes that at the battle of Bouvines in 1214 the Flemish knights refused to charge a body of infantry because they were not gentlemen, in consequence of which they lost the battle. Chivalry and the sense of fighting fairly also influenced their attitude to new weapons, such as guns and cannon. Greener (1910, 55) translates Muntluc (*Michaud et Poujoulat*) who writes, "I—have seen brave and valiant men killed with it in such sad numbers, and it generally happened that they were struck down to the ground by those abominable bullets, which had been discharged by cowardly and base knaves, who would never have dared to have met true soldiers face to face and hand to hand. All this is very clearly one of those artifices which the devil employs to induce human beings to kill one another." Fear of guns was very real and Shakespeare's mocking verses in Henry IV (i.iii), "and, but for those vile guns, He would himself have been a soldier", seem very apt. The English had adopted much the same attitude towards crossbows when they were first introduced.

It was, in fact, this preoccupation with nobility and dislike of new weapons that made knighthood ridiculous. Medieval knights were driven by an instinctive lust for warfare that satisfied the same basic needs, as ritual warfare in primitive societies, and as sport does today. The words written above in relation to the Dani tribe, seem equally appropriate to medieval knighthood or to modern sportsmen: "they engaged in ritual war to promote the success and well being of their social order. Their health, welfare and happiness depended on the pursuit of aggression against traditional enemies". Medieval knighthood had the excuse that it acted on behalf of the State, the Church, or women and orphans of noble birth. Very few died in ritual warfare and the remarks attributed to James I, that the defensive armour of his knights was a double protection, preventing the bearer from being injured or from injuring others, emphasized the ritual nature of knightly combat before its eventual demise. The decline of chivalry was noted in 1475 by William of Worcester, see Douglas (1969, 1134), who wrote "But now of late, more the pity, many who are descended from noble blood and are born to arms—set themselves—to learn the practice of law or customs of land, or of civil matter, and so do waste greatly their time in such needless business as to hold courts—also there to embrace and rule among your poor and simple commons of bestial countenance that wish to live in rest. And he who can be a ruler and put himself forward in such matter is, as the world goes now, more esteemed among all estates than he who has spent 30 or 40 years of his days in great jeopardy in your ancestors' conquests and wars". His solution to this state of affairs was "that the sons of princes, lords, and most of those that are of noble blood—while they are green of age, be exercised in—using jousts, to learn to run with the spear, and handle an axe, sword, dagger, and all other defensible weapons, wrestling, leaping, and running, to make them hardy, free, and well-bred." His plea fell on deaf ears, for the times had moved on.

Major Wars

In modern times warfare has been scarcely less frequent than in earlier ages. Wright (1942) and Richardson (1960), and numerous researchers since, have collected statistics on wars. Richardson classified them according to magnitude, which he takes to be the number of fatalities, and defines the magnitude of a war as the logarithm to base ten of that number. Thus a war of magnitude 3 has $10^3 = 1000$ fatalities, and a war of magnitude 7 has $10^7 = 10$ million fatalities. A "war" of magnitude 0 has one fatality (murder). The number of

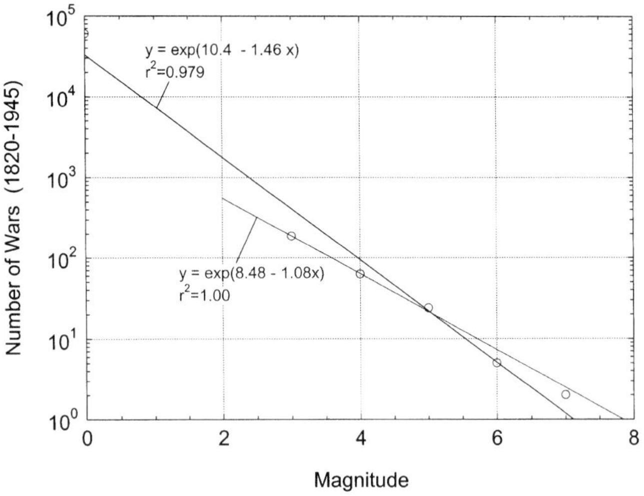

Fig. 21.11 Frequency and magnitude of wars in the period 1820–1945. Richardson (1960, 128).

wars in Richardson's list occurring in the period 1820–1945 is shown in Table 21.4 and is plotted in Fig. 21.11.

Table 21.4 Frequency and magnitude of wars, 1820–1945. Richardson (1960, 128).

Magnitude	7±1/2	6±1/2	5±1/2	4±1/2	3±1/2	0±1/2
Number of Wars	2	5	24	63	188	60000

Richardson could find data on wars of magnitude 3 to 7 and he could find data on murder, magnitude 0, but data for magnitudes 1 and 2 (3 to 316 fatalities) proved difficult to find because such small conflicts are not often of interest to historians and there are, in any case, a large number of such conflicts. The total number of deaths caused in wars of a given magnitude during the period 1820–1945 is illustrated in Fig. 21.12. As expected the largest wars, World War I and II, accounted for most deaths and wars of magnitude 2 and 3 produced the smallest number of deaths, but what is surprising is that conflicts of magnitudes 0 and 1 also produce a

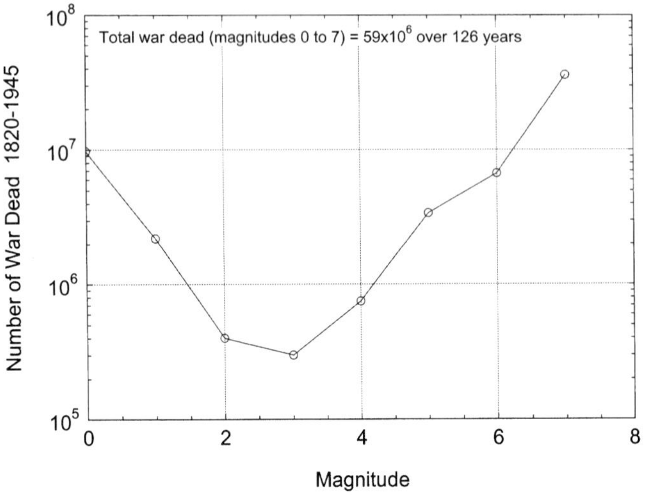

Fig. 21.12 Total number of deaths in wars of various magnitudes in the period 1920–1945. Richardson (1960, 150).

large number of deaths. Magnitude 0 (murder), in fact, is the most deadly magnitude after the two World Wars. The total number of deaths in all wars, including murder, during this period is estimated to be 59 million. The average world population is estimated to be 1.5 billion, so assuming a death rate of 20 per thousand per year the total number of deaths between 1820 and 1945 is about 3.75 billion and so the proportion dying as a result of war is about 1.6%. This is significant rather than overwhelming, but it does mean that death from war is comparable to the death rate from some diseases.

The frequency with which individual states engage in war may also be determined from Richardson's list, Table 21.5. It appears that Britain, during the period 1820–1945 was the most belligerent state, being engaged in a total of 28 out of 94 wars. France was engaged in 21 out of 94 wars and only Sweden among the larger states was able to avoid war altogether. On this basis Britain could be considered the most aggressive state but of course most Britons would claim that they are very rarely the aggressor and are simply the most frequently attacked state. The nation most involved in war has changed frequently, as Wright's list of wars for the period 1480–1941 shows. The most belligerent nations in consecutive periods of about fifty-years were: Austria (1480–1550), France (1550–1600),

Table 21.5 List of belligerents in wars 1820–1945. States with population<one million are omitted. Richardson (1960, 173).

	Magnitude 6.5 to 7.5	*Magnitude 3.5 to 7.5*
Britain	2	28
France	2	21
Russia	2	18
Turkey	1	15
China	1	14
Spain	0	11
Germany or Prussia	2	10
Italy or Piedmont	2	10
Austria (various)	2	9
Japan	2	9
USA	2	9
Egypt	0	7
Greece	2	6
Sweden	0	0
Total Wars	2	94

Spain (1600–1650), Britain (1650–1700), Britain (1700–1750), Russia (1750–1800), Britain (1800–1850), Britain (1850–1900), Britain, Italy, Russia and Japan (1900–41).

One of the reasons given by Richardson for Britain's participation in so many wars is that, having a large empire, Britain had a large number of frontiers to defend. Wars tend to occur between states sharing a common frontier (the sea may be a common frontier) and only rarely do armies have to pass through an independent state to attack an enemy not sharing a border. Modern missiles and aircraft have enabled states not sharing a border to engage in war directly but formerly this was more difficult. Richardson thought there should be a correlation between the number of wars a country has fought and its average number of frontiers. For this purpose the sea counted as a single frontier, and if two countries touched at more than one point (either at home or in its colonies) this also counted as a single frontier. Countries change their frontiers from time to time and to allow for this Richardson calculated the average number of frontiers over a specified time; thus fractional frontiers appear. The results are shown in Fig. 21.13. In general the frequency of wars rises in proportion to the number of frontiers but there is considerable scatter in the data because, of course, other factors are also important. The British Empire, having an average of 22.5 frontiers and 26 wars in 126 years, is a little above the average, as is France, and Italy, but Russia, the USA, Greece and Japan all engage in war more frequently than their frontiers suggest is reasonable. Portugal is a particularly peaceful nation, having only one war whilst defending 9 frontiers. Six countries did not engage in war at all during this period. These countries are Sweden, with only 2.3 frontiers, Ecuador with 3.5, Venezuela with 4.0, Switzerland with 5.0, and Persia with 5.6.

Richardson also collected data on the ostensible causes of the wars, the participants, the outcome, its duration, and any other facts he felt relevant. During the 133 years from 1820 to 1952 Richardson found there were 209 wars, about 1.57 new wars per year. Wright (1942, 129) found 299 wars in the 432 years between 1500 A.D. and 1931, which is about 0.7 new wars per year, and suggests, erroneously according to Richardson, that warfare is becoming more frequent. The discrepancy between the lists is that the definition of war is different, and conflicts that are included in one list may be excluded from the other.

However, wars, like other infrequent events, do not usually occur at evenly spaced intervals. It may be suspected that the frequency of wars follows a normal distribution and has a mean value and standard deviation. However, events that are not too frequent usually follow a Poisson distribution, that is, the number of years, n, in which the event occurs x times per year (x = 0, 1, 2, etc.), in a total period of N years, is related to the mean frequency, λ, by

$$n = N\lambda^x \exp(-\lambda) \,/\, x! \qquad 21.1$$

This distribution has the considerable advantage that it is specified by only one parameter, the mean frequency, and does not need a separate standard deviation. The standard deviation, in fact, is equal to the mean. The probability of an event not occurring in a given period (x = 0) is exp(-λ), to occur once in a given period it is λexp(-λ), to occur twice in a given period it is λ^2 exp(-λ), and to occur x times (= 3, 4 etc) one simply multiplies the probability of the previous period by λ/(x-1).

Richardson applied this equation to both his own and to Wright's list of wars and found there was quite good agreement. Of wars having a magnitude 3.5 to 4.5 (3,162 to 31,620 fatalities) Richardson

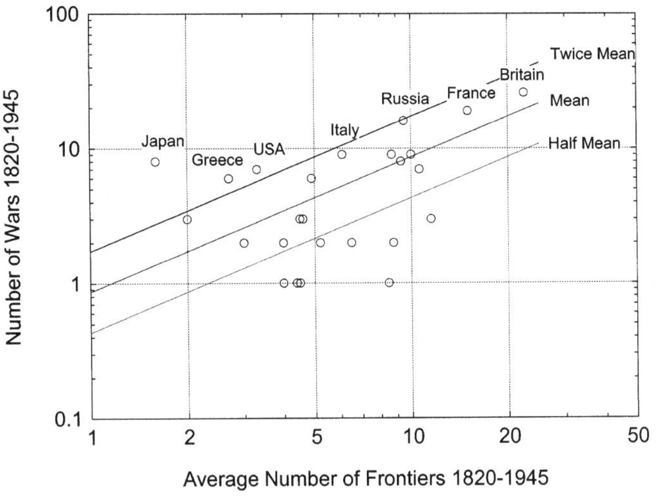

Fig. 21.13 Correlation between the average number of frontiers and war frequency.

Table 21.6 Frequency of wars of magnitude 3.5 to 4.5, 1820–1929 from Richardson's list. λ = 59 wars in 110 years = 0.536 wars per year.

Wars per year	*0*	*1*	*2*	*3*	*4*	*Total*
Number of such years	65	35	6	4	0	110
Poisson Distribution	64	35	9	2	0	110

found there were 59 wars in the 110 years from 1820 to 1929, about 0.54 new wars per year, and the number of years in which there was 0, 1, 2, 3, or 4 new wars is specified in Table 21.6. Wright found 299 wars in 432 years and the number of years in which there was 0, 1, 2, 3 or 4 new wars per year is specified in Table 21.7. The agreement in both cases is quite good, thus the Poisson distribution tells us how often wars of a certain magnitude will occur but, of course, it does not specify exactly when or where they will occur.

Global wars, having a much greater magnitude than normal, are very infrequent, about one per century, and it might be expected that they also have a Poisson distribution, but this is not so. It may be that there have been an insufficient number of such wars to form a reasonable sample, but it is more likely that such wars have a fundamentally different frequency distribution. To understand how a

Table 21.7 Frequency of wars of all magnitudes, 1500–1931 from Wright's list. $\lambda = 299$ wars in 432 years $= 0.692$ wars per year.

Wars per year	*0*	*1*	*2*	*3*	*4*	*Total*
Number of such years	223	142	48	15	4	432
Poisson Distribution	216	150	52	12	2	432

frequency distribution can change it is useful to consider the frequency of vehicles passing a fixed point on a road. When the frequency is very low (less than one vehicle per minute) there is a large distance (1–2 km) between vehicles and they do not interfere with each other's speed very greatly. Under these conditions the frequency distribution corresponds to Poisson's relation. As the number of vehicles using the road increases the vehicles come closer together and a vehicle's speed is then controlled to some extent by the vehicle ahead of it and the frequency distribution begins to change. When they are dangerously close, the distance between vehicles becomes fixed at about 50 m, and the frequency of vehicles passing a given point is then constant (it is the vehicle speed divided by the distance between vehicles, about 0.5 vehicles/s). Thus on a crowded road vehicles do not pass a fixed point at random intervals but at fixed intervals of about 2 s.

The same phenomenon arises with wars. Wars that occur at random are unrelated to the wars occurring immediately before or after them. Such wars are likely to be small or medium sized wars between neighbouring states, and the previous or subsequent war may be between different states and for different causes. Large scale, global wars are different. According to the power cycle theory, the leading states have followed a path, indexed for example by their gross national product or military spending, in which this function, relative to their rivals, increases exponentially, reaches a maximum, where it may remain for some time, and then declines relative to their rivals. It is during this phase, when a leading power is in relative decline and rivals are increasing in power and wealth that major wars are likely to occur. They tend to occur between the same economically dominant combatants, usually to gain or to preserve economic leadership, and they undergo a characteristic cycle having a frequency of about one global war per century during the modern period (1500–2000 A.D.).

Global wars are also those having the largest magnitude, that is, the largest number of fatalities or other casualties relative to other wars. Like all wars, their magnitude has increased with each succeeding war. The last global war (World War I and II) was of magnitude 7 and as the world population is about magnitude 8.7 it is clear that any further escalation in global wars may be the last. Modelski and Morgan (1985, 473–502) have estimated the number of fatalities in all wars between 1494 and 1945 and conclude that the five major conflicts account for almost 80% of the total.

Many states have passed through the characteristic cycle of relative growth, power, and relative decline, but not all have been involved in global wars during their period of decline. Italy peaked in the fifteenth century, Portugese and Spanish power in the sixteenth century, Sweden, The Netherlands, and France in the seventeenth century, Britain in the eighteenth and nineteenth centuries, Germany at the end of the nineteenth century, and the USA in the twentieth century. A useful summary of some of these cycles may be found in Modelski and Thompson (1989, 23–54). They note that more than a dozen attempts have been made to investigate cycles of wars since 1937 and the perceived cycles range in length from 20 years to 200 years depending on the choice of data. If the data conforms to a Poisson distribution then a regular cycle is not possible because the events will be random. Global wars, however, do seem to be regular, although no one is sufficiently pessimistic to suggest that they are inevitable. The next one is due in about 2020 A.D.

Toynbee (1954, 234–87), impressed by the research carried out by economists on cycles of growth and decline in national wealth (Kondratieff cycles), looked for similar cycles of war and peace and he eventually recognised five cycles leading to major wars in modern history. The cycle he adopted consists of:

i. premonitory wars (which may not always occur),
ii. a general war,
iii. a breathing space,
iv. supplementary wars,
v. a general peace.

Toynbee thought that in a system of two or more states their relative strengths need regulation from time to time as some states become stronger and others weaker. Improving states may attempt to exploit their newfound strength and it is the difficult task of the other states to discourage their expansion. This is done through diplomacy and, if this fails, through a major war in which some support the challenger and others oppose, so a general war on a large scale breaks out. Occasionally there is a premonitory war immediately before the general war. Such global wars are exhausting and the subsequent peace is frequently an unsatisfactory compromise, designed more to end the fighting than to ensure long-term peace. Consequently, after a breathing space, a supplementary war occurs. The supplementary war is usually less destructive than the general war, but it is nevertheless a global war, and as a result a clear victor emerges and a more genuine peace prevails. This general peace lasts until the leading nation loses vigour and one or other of the most prominent nations feels sufficiently strong to challenge for leadership. Economic leadership, of course, has advantages because the leading state is also the strongest state militarily and is able to arrange world trade to suit itself. It should be noted, however, that no state has ever risen to a position in which it has had a completely free hand. The wishes of other nations have always to be considered, even if they are ultimately ignored. According to Toynbee this cycle of events has been repeated five times in modern history, see Table 21.8, although only the middle three cycles are reasonably similar. The periods of war have been getting shorter, but more intense, whereas the periods of peace have been getting longer. This, Toynbee thought, was more disruptive to society, and he clearly preferred the relatively mild, though continuous wars of 500 years ago to the short but very intense and damaging wars of the twentieth century.

The basis on which Toynbee selected his major wars was intuitive but Modelski and Thompson (1989, 36) arrived at the same conclusions by a more scientific method. They identified world power (militarily and economically) with sea power by using a 500 year series of the inventory of the capital ships of the navies of the global powers, Fig. 21.14. They found four complete cycles in which the proportion of capital ships of the leading powers increased during a major war, reached a maximum just after that war, and then declined. During the next major war their proportion of the world's capital warships might continue to decline, if they lost their position as the leading power, or might increase again if they were victorious. This cycle turns out to be reasonably regular and the wars associated with the rapid rise in the capital warships were taken to be the major global wars. As mentioned above, they were the same wars that

Table 21.8 Toynbee's Cycles of War and Peace in Modern Western History.

	Overture 1494–1568	*1st Cycle 1568–1672*	*2nd Cycle 1672–1792*	*3rd Cycle 1792–1914*	*4th Cycle 1914–*
Premonitory Wars			1667–68		1911–12
General war	1494–1525	1567–1609	1672–1713	1792–1815	1914–18
Breathing Space	1525–36	1609–18	1713–33	1815–48	1918–39
Supplementary Wars	1536–59	1618–48	1733–63	1848–71	1939–45
General Peace	1559–68	1648–72	1763–92	1871–1914	1945–
	Valois-Habsburg War.	Spanish-Dutch War. Thirty Years War.	Anglo-Dutch War. War of Spanish Succession.	Revolution and Napoleonic War. More Revolutions	World War I and World War II

Toynbee identified. A small difference is that they add the Portuguese war in the Indian Ocean to the first cycle, because it was a result of this that Portugal emerged as the first global power of modern times although they did not dominate Europe. They owned about 60% of the world's capital warships and were the first to exploit the newly discovered lands. In 1494, under the Treaty of Tordesillas, Spain and Portugal agreed to divide between themselves the unexplored part of the world; a treaty that was later confirmed by the Pope. The Portuguese discovered and claimed Brazil (1500), Sumatra (1509), the "Spice Islands" i.e. Indonesia (1511). They discovered the sea route to India round the Cape of Good Hope, where they established a trading port at Goa. In 1542 they reached Japan and in 1557 Macao became a permanent Portuguese port in China.

A criticism of Modelski and Thompson's method is that it seems to rule out the possibility that a landlocked country, or one without a strong navy, can ever become the world's leading power. However, their list of winning coalitions and challengers in the five cycles of global war, see Table 21.9, seems to confirm that sea power is essential. This, perhaps, is reasonable if control of world trade is a major objective of global wars, but the emergence of air power may change this in the future. The list does not reflect random events. England (or Britain), for example, has always been a member of the winning coalition, and the same may almost be said of Portugal, the Netherlands and the USA, whereas France has been the losing challenger three

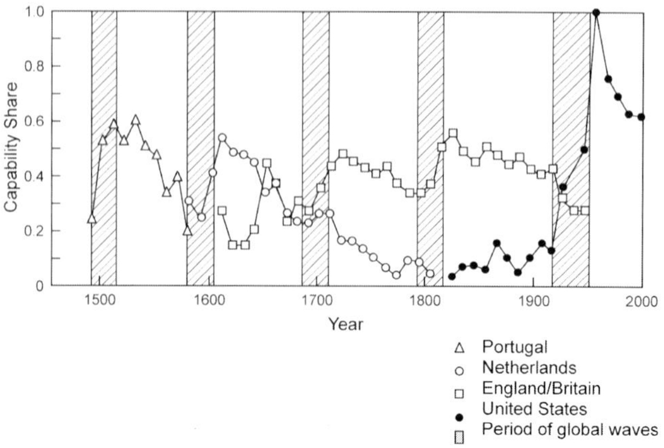

Fig. 21.14 Proportion of total capital warships held by various world powers in the period 1494–1974. Data determined at ten-yearly intervals. After Modalski and Thompson (1989, 36).

Table 21.9 Winners and losers of global wars (Modelski and Thompson, 1989, 46).

Global War	*Date and Name*	*Winning Coalition (Summary)*	*Challenger*
1	1494 Wars of Italy and Indian Ocean	Portugal*, England, Burgundy, Spain	France
2	1580 Dutch-Spanish Wars	U.P. Netherlands*, England, France	Spain
3	1688 Wars of the Grand Alliance	England*, U.P. Netherlands	France
4	1792 Wars of French Revolution and Napoleon	Britain*, Russia	France
5	1914, 1939 World Wars I and II	United States*, Russia/USSR, Britain	Germany

* The emerging world power.

times and a member of the winning coalition once. Modelski and Thompson remark that it is not far fetched to say that the same states (the oceanic coalition) have won all five global wars whereas the continental challenger has always lost.

CHAPTER TWENTY-TWO

ANCIENT AND CLASSICAL WARFARE AND WEAPONS

Introduction

The three major components of war are strategy, logistics and tactics, and at least an elementary knowledge of each is necessary to a proper understanding of the value of the various weapons systems that have proved useful from time to time. The introduction given here is based on Jones (1988). Strategy is the art of war, that is, the art of moving and disposing troops or ships so as to impose on the enemy the place, time, and conditions for fighting favourable to oneself. It concerns the manoeuvring that takes place before a battle is joined and unless both commanders are satisfied that they can win, one or other will retreat, if that is possible, to find more favourable conditions. Bringing the enemy to battle is no easy task, for a commander will not accept battle unless he believes he can win, or he has no other choice. Logistics is equally important, and may be considered an element of strategy, but it is often ignored outside military circles. It concerns the supply of water, food, fodder, and all the other essentials needed to maintain an army in the field. Denying an invader access to food and water, usually by destroying crops and poisoning wells, is a particularly efficient way to gain the ultimate victory, especially if the enemy is unlikely to be defeated in battle. A better-known logistics strategy is the siege, in which an attacker denies supplies to the defender, but he has, of course, to maintain his own supplies and this if not always easy. Tactics, on the other hand, concerns events after a battle is joined and relates to the movement of troops on the battlefield, to enable stronger elements on one side to operate against weaker elements of their enemy. It also includes skill in handling weapons systems.

If fighting is at close range (hand to hand) it is then usually described as shock fighting. Typical weapons used in shock fighting are clubs, axes, daggers, swords, hand-held spears, pikes and pistols. Fighting at longer range requires missiles and typical weapons are the thrown javelin or spear, slingshot, bows and arrows, ballista,

Table 22.1 General characteristic of army units.

	Shock (short range)	*Missile (long range)*	*Mobility (speed)*
Heavy Infantry	Very Good	Medium	Poor
Light Infantry	Poor	Very Good	Medium
Heavy Cavalry	Good	Poor	Good
Light Cavalry	Medium	Medium	Very Good

muskets, cannon, and rockets. These weapons may always be fired by troops on the ground (infantry) but mounted troops (cavalry) are limited to those weapons that do not frighten the horses or unseat the rider. The accuracy of mounted cavalry is also usually less than that of the infantry when using the same weapons. Modern cavalry, who go to war in armoured vehicles, are not so restricted. Cavalry, of course, adds speed and manoeuvrability to an army and both infantry and cavalry may be used either in shock action or in missile action. The modern view is that cavalry can capture territory, but to hold territory you need infantry. The degree of protection (armour) offered these troops varies from light to heavy and this clearly affects their speed of movement and the speed and accuracy with which they can fire their weapons. Thus the weapon systems available to a commander are traditionally divided into four types, although their actual arms and armour have varied greatly throughout history, they are: light infantry, heavy infantry, light cavalry, and heavy cavalry. Table 22.1 summarises their main characteristics when fighting with pre-industrial weapon systems.

In shock fighting, protective armour is an advantage and mobility is not required thus heavily armoured infantry and cavalry are usually the most effective arms and lightly armoured infantry the least effective. In long-range battles armour is less important but speed and the ability to fire weapons accurately are essential. Consequently, the accuracy and speed of firing of lightly armoured infantry is advantageous. Lightly armoured cavalry have the advantage of speed but the accuracy and frequency of fire from horseback is not so good as from the ground, but of course, the cavalry can make good this deficiency by dismounting.

Another important consideration is the length of a campaign. In an emergency there is little doubt that all able-bodied citizens would rally to the defence of their own country and property but for long foreign expeditions they are usually inappropriate. Militias are amateur armies, usually semi-trained, and the troops have other, perhaps more important duties, such as sowing or harvesting crops, and they are, therefore, unable to remain under arms for long periods. If agriculture and civilian life is sufficiently productive then professional armies can be maintained out of the surplus. Professional armies, although expensive, can stay in the field for longer periods and are better able to undertake extended wars in neighbouring countries. If the time available for a campaign is short then the commander is restricted to a raiding strategy, but if he has a professional army capable of staying under arms for an extensive period then he can use a persisting strategy, assuming that he has the wherewithal to keep it in the field.

In addition to set-piece battles the commander has to be proficient in logistics to ensure adequate supplies to his own troops and to deny them to his enemy and he must be proficient in siege warfare, that is, with the defence of his troops behind barricades or city walls and with the numerous methods used to raise a siege. If a commander is restricted to a raiding strategy then he will have difficulty in maintaining a siege for any length of time. On the other hand, an army restricted to a raiding strategy can often carry its own supplies for the duration of the conflict and logistic support is then relatively unimportant. Logistics is particularly important for professional armies using a persisting strategy.

At home, or when passing through friendly territory, armies usually pay for the goods and services they require, but in enemy territory they live off the land, foraging and taking what they require. If an area can sustain a certain size of population for a year then the arrival of an army of comparable or greater size puts a severe strain on food and water resources, and the length of time that the increased population can be sustained is reduced. The army must send troops ever-greater distances to obtain supplies. A "scorched earth" policy, of burning growing crops and poisoning or destroying water supplies to deny them to an invader, is a very effective logistics strategy. An army, whether on the move or stationary, must continually be on the lookout for supplies for both men and horses and these matters often determine the course of a campaign. Throughout the pre-industrial period water transport was cheaper and faster than land transport so armies, whenever possible, relied on water transport for supplies. Thus they tended to move along the coast, so they could be supplied by ships, or along the navigable river valleys.

Prehistory

Of tactics, strategy, and warfare little is known from this period but defensive works are known and sieges must have occurred. Warfare and weapons were important in all the earliest societies of which we have any knowledge and as we have seen, fights with bow and arrows were depicted on rock walls in the Mesolithic period, and there is ample evidence of violent death from prehistoric skeletal remains. Ardrey (1965, 205) considered that all Palaeolithic tools were, in fact, weapons but regardless of the truth or otherwise of this it is certainly true that axes, daggers, spears, spear-throwers, and bows and arrows were invented and used, not only for hunting but also in war.

Men's first weapons were probably the club and the flint hand-axe. Clubs, being of wood, are unlikely to survive in great numbers from prehistory but the stone hand-axe is quite common and examples are shown in chapter 16, Fig. 16.1. Hand-axe production, whether as a tool for clearing forests or as a weapon for hunting or warfare, was prolific, as the extensive remains of flint mines throughout Europe confirm. An example of such a flint mine, Grimes Graves in Norfolk, is discussed in chapter 16. Particularly fine polished hand-axes were produced, so evidently the time spent polishing paid for itself in a more efficient chopping tool. Weapons, however, were almost always beautifully made.

Knives, daggers and spearheads developed from the hand-axe, and they too were often highly decorated. In Predynastic Egypt, around 3300 B.C., flint-working techniques reached their zenith and metals began to be used. Fig. 22.1 shows an obsidian spearhead and a flint knife with a handle covered in gold leaf. The spearhead is forked and has neatly serrated edges except at the round end where it joined the shaft. Usually Egyptian stone tools were made of local flint but occasionally imported material such as obsidian was used. The knife too is particularly good, having a gold foil handle fastened to a plaster base with rivets. The handle is decorated with three dancers, depicted in Fig. 22.1, and the reverse shows a multi-oared boat.

The earliest spears are from the Lower Palaeolithic. A stout wooden

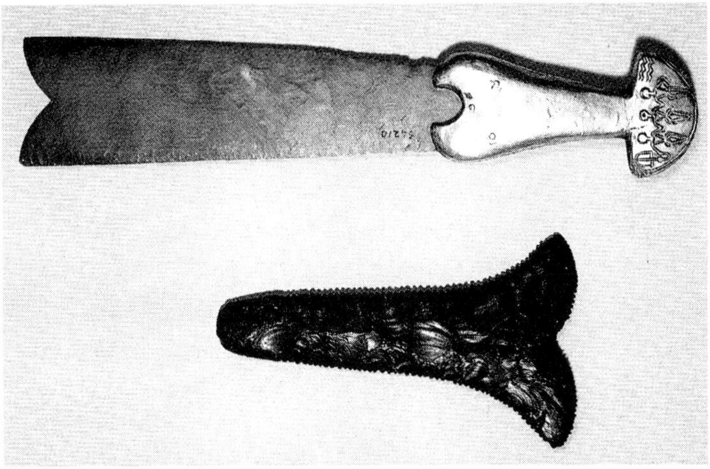

Fig. 22.1 Egyptian lithic implements. Top: a knife with gold handle. Bottom: an obsidian spearhead. Predynastic period, 3370–3240 B.C. Cairo Museum.

spear with a point has been found at Clacton, Essex and at Verden, Lower Saxony, Clark (1963, 60). They were made of yew and the Verden spear was found between the ribs of an extinct type of elephant. Related to the spear is the fishing harpoon, with barbed hooks. This was a late Palaeolithic invention. Spear-throwers were commonly used to increase the mechanical advantage of the arm and to impart a greater velocity to a thrown spear, both to increase its range and its striking velocity. Usually spear-throwers were about 30 cm long, made of antler bone, and were often highly decorated, even from the earliest times. There is a famous scene in the Lascaux shaft, Fig. 22.2 (left), which appears to be a representation of a bird-headed spear-thrower and a single barbed spear. The painting is probably pre-13000 B.C. and suggests that the spear and spear-thrower were in use before the barbed harpoon. In other examples the ends of spear-throwers are decorated with ibex, or with other animals. Fig. 22.2 (right) is an illustration of a Mesolithic hunting scene in which four relaxed hunters are shooting at a herd of deer. The bows are not particularly powerful for they do not seem to penetrate very far and the deer continue running even after several hits. Probably the hunters had to chase the injured deer until it weakened sufficiently to be caught and killed.

Arrows, and in particular, arrow heads, have survived in much greater numbers than bows and for this reason archaeological research

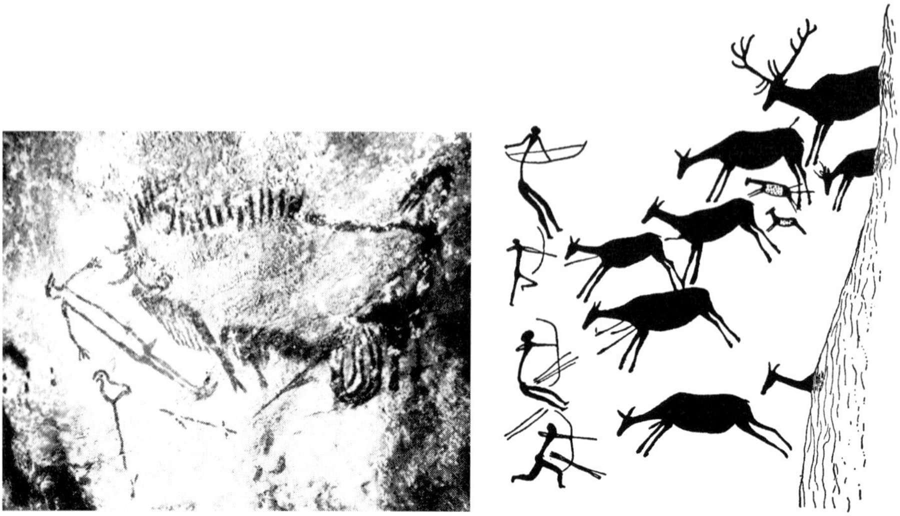

Fig. 22.2 Left: scene from Lascaux showing a bird-headed spear-thrower. Right: Mesolithic hunting scene, Valltorta Ravine, Spain.

has concentrated in this area. Collins (1975, 116) says that arrow points dating from about 12000 B.C. have been found but the earliest satisfactory evidence for arrows is a hoard of over 100 from the waterlogged sediments at Stellmoor near Hamburg, dating from 9000–8200 B.C. They were made of pine and at one end there was a notch for the bowstring. In many cases these arrows were made in two parts with a forked upper part of 15 to 20 cm length into which the wedge shaped tips of the lower part fitted. At the other end was a slot for the arrow point. Although undoubtedly arrows, and one was found in the breast of a reindeer, no actual bow was found. In the Mesolithic, small triangular flints known as microliths were sometimes used to tip and barb arrows. They were fitted into slits at the end of the arrow, bound tightly using animal sinew or gut, and sometimes held in resin, probably birch-pitch. Fig. 22.3 illustrates two early examples from Clark (1963, 70). In Fig. 22.3 (right) the action of penetration with a point, and cutting with an edge, are done by separate microliths, but in Fig. 22.3 (left) a single microlith is designed for both purposes. Later arrowheads had chisel-shaped points and in the Neolithic, arrowheads having projecting tangs or of a lozenge shape, became common.

Flint arrowheads remained in use well into the Bronze Age and the earliest tang and barb forms made in metal were copies of flint

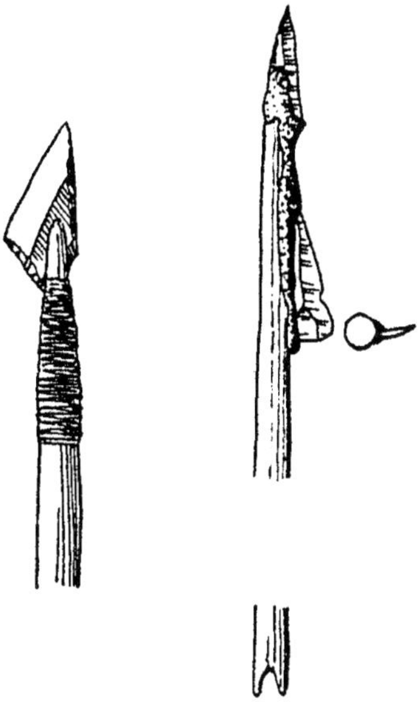

Fig. 22.3 Left: oblique arrowhead of Danish Early Coastal culture, showing method of mounting. Right: Maglemosian arrow set with microliths, Loshult, Scania, south Sweden. Clark (1963, 70).

prototypes. The reluctance to use metal arrowheads was probably related to the cost and to the considerable risk of them being lost in use. Clark notes that modern experiments have shown that obsidian arrowheads fired into a target of animal tissue at a distance of about 10 m penetrated 76 cm as opposed to the 56 cm achieved by a steel arrowhead, but the conditions of the trial are not sufficiently well defined. Properly conducted trials would use gelatin as a tissue simulant rather than animal tissue, which is variable in its properties. The results are only comparable if the ratio of arrow-mass to cross-sectional area, and the striking velocity are the same. The material of the arrowhead, so long as it is sufficiently hard and sharp, should be unimportant.

Prehistoric arrows seem to have been longer than modern arrows. The latter are typically about 72 cm ± 3 cm but prehistoric arrows vary from 75 cm to 102 cm. The shaft diameter was usually in the range 0.7–1.0 cm. From the earliest times arrows were fletched with feathers to stabilise their flight. Although no feathered arrows have survived from this period they are depicted in Palaeolithic art.

The earliest bow is from Holmegaard, Denmark and is of the Mesolithic period from about 6500 B.C.; see Fig. 22.4. It was made

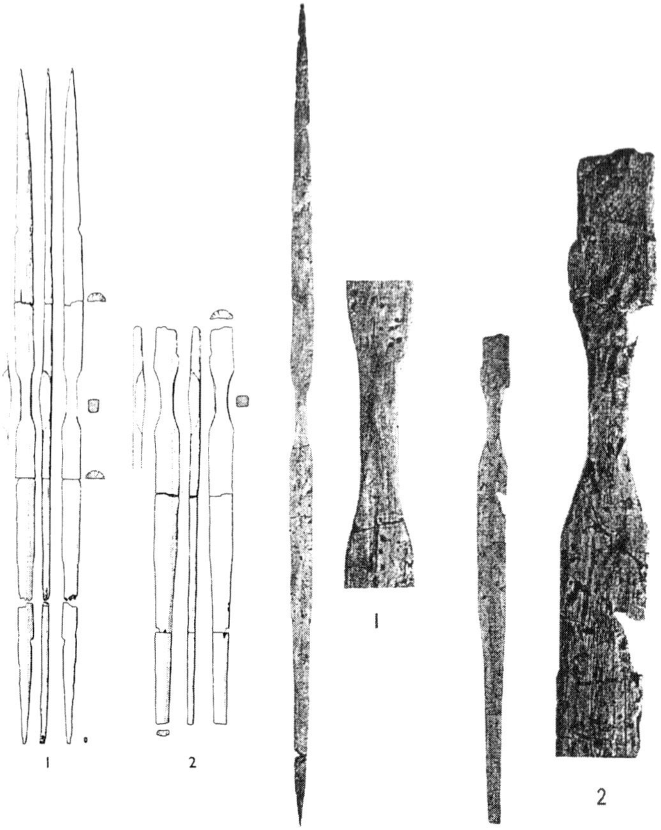

Fig. 22.4 Mesolithic elm bow from Holmegaard, Denmark. Clark (1963, pl. VI).

of elm, 154 cm long, and although complete with a well-defined grip, it was broken. The limbs were tapered. Width was 4.4 cm (limb), 2.5 cm (grip), and thickness was 2.0 cm (limb) and 2.9 cm (grip). A similar, but half complete bow was also found. The bow and arrow was widely used by Mesolithic hunters but they should have been less common amongst the early Neolithic farming communities. However, Clark notes that bows are much more numerous from Neolithic contexts even though they have survived only on bog or lake sites which inhibit the decay of wood. Presumably Neolithic farmers used bows as weapons of war as well as for limited hunting.

The earliest bows are assumed to be self-bows, that is, bows made from single pieces of wood as in Fig. 22.4. However, Clark (1963,

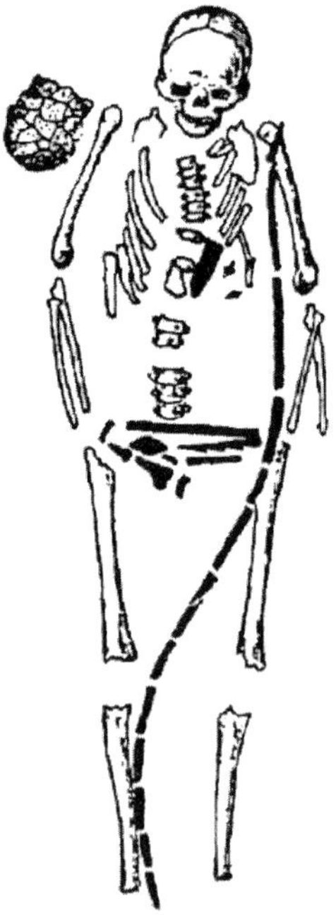

Fig. 22.5 Traces of composite bow, marked by broken antler stiffeners, found with the burial of a stone-age hunter of the Serovo stage in the Lena Valley, Siberia. After Okladnikov in Clark (1963, 51).

50–51) notes that the composite bow developed in environments in which timber of an adequate toughness and resiliency was unavailable. Composite bows occur as early as the third millennium B.C., as shown, for example, by a bow found in Siberia, Fig. 22.5, which was stiffened with antler splinters, and presumably given resilience with a sinew backing. The self-bow developed into the famous English medieval longbow, whereas the composite bow developed into the equally famous Turkish bow. The merits of the various types of bow have been hotly disputed and will be examined later.

Clark notes that prehistoric bows were almost always D-sectioned, although it is not clear if the convex side faced the bowman, as is the modern practice. The neutral axis passes through the centroid of the section so to avoid excessive tensile stress the convex side should be in compression and facing the bowman. The wear patterns suggest that some prehistoric bows adopted the modern arrangement, but presumably others did not. Prehistoric self-bows, like modern bows, were cut from yew, elm, or other suitable wood but there is no indication of the modern practice of retaining the sapwood on the back of the bow away from the archer and the heartwood facing the archer.

Other items associated with archery are: arrow shaft smoothers, arrow straighteners, and wrist guards. Arrow shaft smoothers are a pair of stones having a semi-circular depression of the same diameter as the arrow shaft, and it is rubbed along the shaft to give a smooth, straight shaft. Arrow straighteners are usually of bone with a drilled hole, which is passed over the shaft and used to take out any bends. Wrist guards are usually smooth bone plates that may be strapped to the wrist as protection against the twang of the bowstring.

Ancient Civilisations

The earliest records of the third millennium B.C. show that the weapons systems were light infantry and mounted infantry (chariots) and they were used in shock action, missile action, and in sieges. Egypt during the Old Kingdom is considered to be relatively peaceful but it owed this peace to its geographical position, being mostly surrounded by inhospitable country, but before it was unified within these borders it too was a place of conflict. The Narmer palette, Fig. 22.6, marks the beginning of official hieroglyphic writing and commemorates Narmer, identified with King Menes, who unified Lower and Upper Egypt about 3000 B.C. On one side he is shown wearing the crown of Lower Egypt, preceded by flag bearers, and inspecting bound and decapitated enemies. At the foot of the palette he is shown as a bull, breaking down the walls of a city and trampling over his enemy. On the obverse the king wears the crown of Upper Egypt and brandishes a mace to smite a kneeling prisoner, probably of Lower Egypt. This was to become the tradition pose of warrior pharaohs. The submission of the Delta is depicted by a Horus falcon (the king and god of the capital of Upper Egypt), which holds a prisoner by a rope through the nose and stands on plants of the Delta. The Narmer palette is the earliest symbolic record we have of a military campaign. Although details are lacking, it is clear that Narmer, king of Upper Egypt, attacked and captured the walled cities of Lower Egypt, seizing and decapitating many prisoners, and unifying Upper and Lower Egypt.

Fig. 22.6 The Narmer palette commemores Narmer's unification of Upper and Lower Egypt, about 3000 B.C. Cairo Museum.

Fig. 22.7 The location of important early urban sites. After Collins (1975, 172).

The process of urbanisation in Egypt is not well defined but according to Whitehouse (1975, 171) it occurred slightly later than in Mesopotamia. Fig. 22.7 shows the location of the more important early urban sites, most, but not all of which are situated in the valleys of the Nile, Tigris, Euphrates, and Indus. It seems that only the larger alluvial valleys were sufficiently productive to sustain urban life.

Cities began in Mesopotamia when a flourishing, but otherwise unexceptional agricultural society began to use its surplus of produce and people to create specialist occupations and to build walled towns. The most significant developments, according to Postgate (1992, 24) were in the Uruk period, from 4000 B.C. to 3200 B.C. The earliest city walls are those of Kabira, an Uruk period settlement on the Euphrates in Syria. Uruk itself was a large city of some 400 ha and the walls, which are of the Early Dynastic period, about 3200–2350 B.C., are 9 km in length. Surviving texts refer to some 35 cities at this period, most of which were walled, although many were much smaller than Uruk. Postgate (1992, 241) is of the opinion that it is naive to suppose that the earliest cities of Mesopotamia did not maintain some kind of professional army. All armies show considerable variety in equipment, expertise, and social class, and those of Mesopotamia were no different. The sling, spear and the bow and arrow were already developed as sophisticated weapons, and no doubt there was a permanent class of professional soldiers close to the king to advise on strategy and tactics and to train the citizens. Unlike Egypt, which was unified at an early date, Mesopotamia remained a collection of city-states with many frontiers and many wars.

Weapons made of metal became increasingly important after about 3000 B.C. Gurney (1990, 129) gives details of a Hittite ceremony that was enacted at the unidentified city of Gursamassa in honour of the god Yarris. A surviving text says: "The young men are divided into two groups and receive names; the one group they call 'men of Hatti', the other group 'men of Masa'; and the men of Hatti carry weapons of copper, but the men of Masa carry weapons of reed. They fight each other and the 'men of Hatti' win; and they seize a captive and devote him to the god." The latter phrase sounds rather like human sacrifice. In Gurney's opinion this ceremony may commemorate one of the earliest battles using metal (copper) weapons and fought against a people with less sophisticated weaponry. Copper weapons from the earliest periods are not often found because metal was valuable and when artefacts were worn out they were often sent for scrap, re-melted and re-used.

The Stele of Vultures (about 2450 B.C.) in the Iraq Museum, Baghdad, illustrates three classes of infantry armed with shoulder axes, adzes (which could be agricultural tools) and spears. The spearmen were protected with shields, probably of leather studded with metal. Postgate believes the spearmen were the more effective arm for they survived to the Ur III period (2150–2000 B.C.). The bow was, of course, a well-established weapon in hunting but military bows appear to be rare. Naram-Sin holds one on his victory stele and there is a good illustration of what appears to be a quite advanced design of reflex bow that comes from Mari in the Early Dynastic III period (2600–2350 B.C.), Fig. 22.8. The image shows an archer and a spearman sheltering behind a wicker shield. The bow is definitely of reflex shape, but whether it is made of the same materials or has the same power as later reflex bows is not known. The archer is unable to carry a shield and operate a bow at the same time but in this case a spearman and a wicker shield protect him.

The Stele of the Vultures and the Standard of Ur (Fig. 3.5, Chapter Three) both illustrate third millennium chariots, or battle carts. Those on the Standard of Ur have four-wheels, are drawn by four animals (possibly onagers), and carry two soldiers. One was the driver and the other was equipped with copper-tipped javelins kept in a quiver at the front of the cab. The entrance was at the rear and the front was built-up to give some protection to the occupants. It was used in shock action and is shown being driven over the bodies of the defeated foe. These vehicles must have been unwieldy, with their solid wooden wheels and primitive construction. There seems to be no steering mechanism to allow the vehicle to turn without skidding. Nevertheless they added speed to the army and no doubt terrorised their opponents. Postgate (1992, 246) says they lost their paramount role under the Akkad Dynasty (2350–2150 B.C.) and by the Old Babylonian period (2000–1600 B.C.), heavy battle carts seem confined to ceremonial or ritual use in the service of the gods. Very light, two-wheeled chariots drawn by the newly introduced horse replaced them. Cavalry, in the sense of a body of men riding on horseback, does not appear till late in the second millennium, probably because the necessary skills, to control a horse and shoot a weapon with any degree of accuracy while riding bareback, had not developed.

Siege warfare must have been the most common. An early Mesopotamian text of about 2350 B.C. mentions battering rams and other correspondence is full of references to siege towers. An example is an early Old Babylonian letter from Esnunna (2000–1800 B.C.). "Say to my lord: The troops are well. The city is safe. The garrison of my lord is strong. Even if the Amorites should make war for ten years and bring ten battering rams, ten siege towers, and twenty ladders, I will remain strong in my city. My lord should not worry";

Fig. 22.8 Archer, with a reflex bow, and a spearman holding a protective wicker shield. Incised shell plaque from Mari, Early Dynastic III. 2600–2350 B.C. After J.-Cl. Margueron in Postgate (1992, 249).

Postgate (1992, Text 13.8). Clearly, siege warfare was well developed in both Egypt and Mesopotamia at the start of the historical period. This is further confirmed by a text, discussed in chapter 1, Fig. 1.30, which describes a school mathematical exercise in which the volume of earth required to build an inclined plane against a city wall is calculated. The inclined plane is explicitly stated to be for the purpose of capturing an enemy city "hostile to Marduk". From the beginning, it seems, religion was used to justify war.

In Egypt the army did not progress so fast and battle carts remained unknown. As the relatively peaceful Old Kingdom broke up and was replaced by the turmoil of the First Intermediate Period and the Middle Kingdom, mural decoration of tombs began to show soldiers besieging cities, and independent nome chiefs began to recruit and train local armies. Examples of such troops are shown in Fig. 22.9 (top).

Fig. 22.9 Top: Nubian archers. Bottom: Egyptian spearmen. Painted wood, Tomb of Mesheti, Middle Kingdom, 11th dynasty, about 2000 B.C. Cairo Museum.

The group of forty Nubian light infantry are un-armoured and equipped only with bow and arrow as they march in step and in neat rows and files. They hold a bow in one hand, arrows in the other, and do not have a quiver. The bow required two hands to operate so archers cannot also carry a shield and they are not, therefore, well protected for shock action. A group of forty Egyptian shock troops is illustrated in Fig. 22.9 (bottom). A wooden shield protects each of them, but they are otherwise unprotected, and they carry a short stabbing spear. Relative to the Nubians they qualify as heavy infantry but they would not fare well against invaders armed to Mesopotamian standards.

At the start of the second millennium new weapons were developed in Asia. Bronze weapons appear, and about 1700 B.C. came the light horse drawn chariot. Chariots were to become the principal weapon of the Hittites who established themselves in Anatolia. Egypt however, remained militarily backward and it was not until the invasion of the Hyksos that the horse, the wheel, the chariot, bronze weapons, and the reflex bow, were introduced into Egypt. The Hyksos, "rulers from foreign lands" but about whom not much else is known, captured most of the Delta and Lower Egypt by about 1650 B.C. and it was not until about 1550 B.C. that Ahmose I drove them out and took the throne, founding the New Kingdom. The Egyptians, after this, became more warlike and the sons of Pharaoh, who in the Old Kingdom held administrative offices, were now depicted as generals at the head of an army. The chariot shown in Fig. 22.10 (left) is that of Amenhotep III (1386–1349). The king stands on his horse-drawn chariot and carries a bow, a whip, and the reins in his hands. He has a quiver of arrows on his back and a larger one attached to the body of the chariot. The horse carries four Nubian captives with bound hands and there is a fifth captive astride the chariot shaft and a sixth protrudes from the chassis. The horse is elaborately harnessed and the characteristic high neck harness is apparent and clearly chokes the horse as it pulls. Such choking harnesses continued in use throughout antiquity and Fig. 22.10 (right) shows a Roman example, although, of course, they did not use chariots when fighting but only in sporting or ceremonial events.

Speed was introduced into warfare when the light horse-drawn chariot with spoked wheels replaced the old heavy battlewagons. Its place of invention is not known but it makes its appearance around 1700 B.C. in Anatolia, Babylonia, Mittani and a little later in Egypt.

Fig. 22.10 Left: Chariot on Stele of Amenophis III (detail), Cairo Museum. Right: Roman chariot on a stone fragment, Arles Museum.

It revolutionised warfare. Archers shooting at long range from speeding chariots, usually against other speeding chariots, dominated the battlefield and the infantry found it hard to attack and resorted to armour to defend themselves. Infantry, of course remained essential for the chariots could not fight in rough or mountainous terrain where there was little room for manoeuvre. Also the infantry were essential for siege work. Gurney (1990, 86) is of the opinion, backed up by Hittite texts on the training of horses, that an Aryan clan from Mittani brought knowledge of horses and horse training into Anatolia and that this explains the sudden development of light horse drawn chariots. The Hittite and Egyptian chariots were similar to each other. Both had wheels with six spokes but the Egyptian chariots carried a driver and a warrior whereas the Hittite chariots carry a driver, a warrior, and a shield bearer. Pharaoh is not usually shown with a driver. The warriors fight with the lance and the bow. The larger crew of the Hittite chariot was advantageous in close fighting, and the separation of attack and defence was also advantageous, but the extra man produced a heavier, slower chariot.

The first battle of which we have a reasonable account is the Battle of Megiddo in 1479 B.C., between the Egyptians and a group of Asiatic allies. This account of the battle is condensed from Breasted (1912, 285–90) and illustrates the most mobile form of warfare, chariotry, and the most static, the siege. Tuthmosis III invaded Asia with 25,000 to 30,000 men. Marching at the rate of about 28 km per day he passed through Tharu, Gaza, and 140 km further on he reached the southwest slopes of the Carmel Ridge. The Asiatics under the king of Kadesh had occupied the fortress of Megiddo in northern Israel on the Plain of Esdraelon to the northeast of the Carmel Ridge. Tuthmosis was determined to attack and took the direct route through the pass rather than either of the longer routes around the Ridge. His officers opposed this saying: "Will not horse come behind horse and man behind man likewise? Shall our advance-guard be fighting while our rear-guard is yet standing in Aruna [southeast of the pass]?" This was sound advice but Tuthmosis ignored it and personally lead the way. Emerging from the pass they were spotted by the Asiatics, but they had expected Tuthmosis to march round the ridge and consequently their main force was to the south and had to be brought back to the fortress of Megiddo. Thus Tuthmose was able to clear the pass and gather his army; he camped to the west of Megiddo. The following day the enemy chariots emerged to defend the city and drew up opposite the Egyptian chariots. Tuthmosis attacked and the enemy almost immediately gave way: "they fled headlong to Megiddo in fear, abandoning their horses and their chariots of gold and silver—; the people of this city having closed it against them, lowered clothing to pull them up into this city. Now if only the army of his majesty had not given their heart to plundering the things of the enemy they would have captured Megiddo at this moment". "Their champions lay stretched out like fishes on the ground. The victorious army of his majesty went round counting the spoils,—They brought the whole booty which they had taken, consisting of hands [severed from the slain], living prisoners, of horses, chariots, gold and silver." Evidently the camp of the king of Kadesh had fallen to the Egyptians.

Tuthmosis besieged the city by building a wooden wall around it: "they measured this city, surrounding it with an enclosure, walled about with green timber of all their pleasant trees. His majesty himself was upon the fortification east of the city, inspecting what was done." However, before the wall was complete the king of Kadesh escaped northward and the remainder surrendered after seven months. Amongst the spoils were 924 chariots, including those of the kings of Kadesh and Megiddo, 2,238 horses (2.4 per chariot), 200 suits of armour, the armour and tents of the king of Kadesh, 2,000 large cattle, 22,500 small cattle, household furniture, gold, silver, lapis lazuli etc. They also harvested the fields, taking 113,000 bushels, sufficient to feed 14,000 people for a year (8 bushels per person per year).

The discovery of bronze, and later of iron, made remarkable changes in weapons technology. Bronze, unlike copper, was able to take and retain a sharp edge and iron is even better in this regard. The new, harder metals lead to the invention, in the second millennium, of the sword and of metal-armour. Armour probably developed to protect the charioteers, and perhaps the infantry, against the attacks of the enemy chariots and their archers. Being expensive, armour may well have been the preserve of the wealthier warrior who usually fought in the chariots. Armour was mentioned amongst the spoils captured at Megiddo but the relatively small number of suits (200) suggests that it was either expensive or impractical. An early example of Mycenaean body armour, of the fifteenth century B.C., is illustrated in Fig. 22.11 (bottom left). The cuirass is roughly formed from cylinders of bronze and must have been stiff and cumbersome to wear and for this reason Drews (1993, 174–5) thought it more likely to have been worn by a charioteer than by the infantry. The helmet is much lighter, being made of boars' tusks. A similar helmet appears on an ivory statue depicting the head of a Mycenaean warrior of the thirteenth century B.C., now in the Athens Archaeological Museum. Some evidence that charioteers rather than infantry used armour is contained in a linear B clay tablet from Knossos, Fig. 22.11 (top). These are usually lists of stores or equipment and some of the signs are ideograms depicting the objects listed. In Fig. 22.11, the list shows ideograms of a chariot, a bronze cuirass, and a horse, which suggests that these items belonged together. Metal armour at this time was crudely made although as early as the third millennium B.C. the Egyptians had considerable skill in making life-size metal statues by beating copper sheets over a wooden core to which it was nailed, Fig. 22.11 (bottom right). Apparently they had the skill to make armour plate but not, evidently, the need.

A serious objection to the use of metal armour is that most metals, when exposed to intense sunlight, become objectionably hot to the touch, and are capable of producing pain or even skin burns. In hot countries it would be very uncomfortable or even dangerous. The medieval crusaders, when fighting in Palestine, very sensibly wore a white surcoat to reflect the sun's rays but there is no evidence that this solution was adopted at earlier times. Nevertheless body armour continued in use as illustrated, for example, by the

Fig. 22.11 Top: Linear B clay tablet from Knossos showing ideograms of a chariot, a bronze cuirass, and a horse. After Hood (1974, 113). Bottom left: Mycenaean armour found in the grave of Dendra. Fifteenth century B.C. The Archaeological Museum, Nauplia. Bottom right: Life size copper statue of Pepi I, *c.* 2281–2241 B.C. Cairo Museum.

famous Mycenaean "Warrior Krater" of the late thirteenth century B.C., Fig. 22.12 (left). This shows six warriors each armed with a small shield and a spear and wearing a chiton, short breastplate, helmet, and knee length leggings. It is the first appearance of what might be described as heavy infantry. A female figure in black, standing on one side, raises an arm in farewell and one imagines they are leaving to fight the invading Dorians that were to crush their civilisation. The breastplate appears to be bell shaped, the helmet

Fig. 22.12 Left: Mycenaean "Warrior Krater". Thirteenth Century B.C. Athens National Archaeological Museum. Right: bronze breastplate and helmet. Seventh century B.C. Argos Museum.

has side cheeks and a tall crest and the whole are comparable to the late Geometric bronze armour shown in Fig. 22.12 (right).

In the eastern Mediterranean the infantry carried shields of various shapes. A figure eight shape enveloped warriors on three sides from neck to ankles but provided room for arm movements at the waist. An alternative was a half cylinder shield giving protection from neck to shins, but less room for the manoeuvre of weapons. The Hittite shield was rectangular with side cut-outs. All these shields are suitable for defence against missiles but are not particularly useful in hand-to-hand fighting. In the thirteenth century B.C. the round shield appears and proved to be the most appropriate for shock action. It was smaller than most shields, being from 60 to 100 cm diameter, and had a conical surface receding from the central boss. It was made of wood, covered with leather and often had a metal boss and rim. They covered a relatively small proportion of the body but were lighter, more manoeuvrable, and gave ample space for fighting with the right hand. Drews (1993, 179) believes that such shields were introduced to the eastern Mediterranean by barbarian skirmishers from the west although their ultimate provenance in unknown.

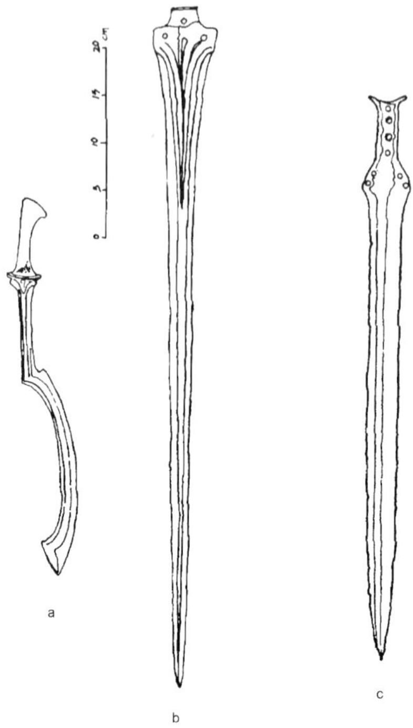

Fig. 22.13 a. Bronze sickle sword from tomb of Tutankhamun. b. Bronze Anatolian rapier, *c.* 1400 B.C. c. Bronze Naue Type II sword from Mycenae. After Drews (1993, 197).

Metal greaves came into use about 1200 B.C. but were used only in the Greek world and after about 1150 B.C. they vanish, only to reappear in the eighth century B.C. Their demise is related to the poverty of the Greek Dark Age that followed the catastrophic events at the end of the Bronze Age and led, amongst other things, to the loss of literacy. The protection offered by greaves, however, is very slight. The earliest examples are of 2 mm thick bronze and Drews (1993, 177) says that modern experiments show that even a thickness of 3 mm can be entirely cut through by a slashing sword.

Short knives and daggers had long been known and some very fine examples of worked flint knives with ornate gold handles have survived and may be seen, for example, in Fig. 22.1. Swords for slashing or thrusting are an invention of the second millennium B.C., but they remained a rarity until iron-working methods had developed. Iron-working seems to have been developed first by the Hittites in Anatolia. As was mentioned in chapter 16, iron was treated initially as a rare metal and the Hittite king Hattusilis III, in the thirteenth century B.C. was able to send a single knife blade of iron as a fitting present to foreign royalty.

In fighting with a thrusting sword it is desirable that the point can be moved from place to place with great speed so that weak spots may be attacked before movement of the shield or body protects them. Such a sword should be light, pointed, straight, and have a centre of gravity close to the hilt where torque and thrust are applied. A slashing sword, on the other hand, relies on its momentum for effect and it is desirable to have the centre of gravity a long way from the hilt so that its action is more like that of a battleaxe. It should be relatively heavy, curved, and sharpened along the edges. An infinite range of possibilities exists between these extremes and most swords may be used for both thrusting and slashing. As a general rule eastern swords tend to be slashing swords whereas western swords have relied more on thrust.

Stabbing weapons having a blade length less than about 35 cm are usually considered to be knives or daggers; between 35 and 50 cm it is a dirk, between 50 and 70 cm it is a short sword, and blades longer than 70 cm are long swords. The earliest near Eastern sword was the "sickle sword" which was usually less than 30 cm blade length and used for slashing. It was well distributed and one was found in Tutankhamun's tomb, Fig. 22.13 (left). Rapiers, designed essentially for thrusting were inventions of the middle Bronze Age and Fig. 22.13 (centre) illustrates a typical Anatolian example of about 1400 B.C. It is 79 cm long and tapers from 7.5 cm to 3 cm at a quarter length and 2 cm at half length. Charioteers carried such swords, some of over a metre in length, for defence if they were forced to abandon their chariot. About 1200 B.C. the famous Naue Type II sword, Fig. 22.13 (right), was introduced from central and northern Europe; it quickly became the standard and its use extended from the near east to Britain and from Scandinavia to Italy and the Balkans. Its great advantage was that it could be used both for thrusting and for slashing. Hittite infantry soldiers equipped for shock fighting are illustrated in Fig. 22.14. On the left he wears a belted kilt and a helmet, and carries a sword and a battle-axe. The helmet has an extension hanging behind the nape of the neck. The sword was used for slashing. Egyptian carvings show Hittite infantry armed with a long spear. The lack of armour, except for the helmet, and the lack of a shield seems remarkable. Even the king, Fig. 22.14 (right) is protected only with a long padded dress and helmet. He carries a spear, a sword, and a sceptre.

However, charioteers rather than infantry were relied on in battle and the object was always to catch their enemy in the open where chariots may be used effectively. As ever it was difficult to bring an enemy to battle and Gurney (1990, 90) quotes a Hittite text that records a successful strategy adopted by Mursilis II (1321–1295 B.C.) against a certain Pitaggatallis. "—the army I ordered to advance in battle order. And because (the enemy) had outposts, if I tried to surround Pitaggatallis, the outposts would have seen me, and so he would not have waited for me and would have slipped away before me. So I turned my face in the opposite direction towards Pittaparas. But when night fell, I turned and advanced against Pitaggatallis. I marched the whole night through, and daybreak found me on the outskirts of Sapidduwa. And as soon as the sun rose I advanced to battle against him; and those 9,000 men whom Pitaggatallis had brought with him joined battle with me, and I fought with them—and I destroyed my enemy."

The most famous Hittite battle took place against the Egyptians at Kadesh in 1275 B.C. It is described in great detail in an Egyptian text. Briefly, the Hittite army camped at Kadesh on the river Orontes and concealed its position and strength from the Egyptian scouts. Deceived, the Egyptian commander Ramesses II moved quickly to the west of Kadesh leaving two of his divisions straggling a long way behind. A strong detachment of Hittite chariotry passed unnoticed behind the city, crossed the Orontes, and attacked the centre of the Egyptian column as it moved into camp. The Egyptians would have been annihilated, had not a detachment arrived, most fortuitously, and caught the Hittites unprepared as they pillaged the camp. Ramesses had a lucky escape and withdrew has army back to Egypt where he presented the battle as a great victory.

A good impression of warfare and weapons at this period is given by a reconstruction painting from a relief of Ramesses II at the siege of Dharpur carved on the inner south wall of the Hyperstyle Hall in the Ramesseum, Fig. 22.15. The king is shown on an enlarged scale shooting arrows from his charging chariot. He is alone in the chariot and controls the horses with reins tied around his waist. The horses trample over the Hittite enemy and the Hittite chariotry is in confusion. Other chariots carry an archer and a driver/shield bearer. The Egyptian infantry use clubs and shields, short swords

Fig. 22.14 Left: Hittite soldier armed with axe and sword. Right: Hittite king with spear, sceptre and sword. Breasted (1912, 383).

and shields, or bows and one Hittite is shown riding bareback on a horse and indicates the increasing ability to control a horse and fight. The city is attacked with scaling ladders and is defended by bowmen and men with spears. Body armour is absent and the Egyptians are naked except for a loincloth whereas the Hittites wear a long robe reaching almost to the ankles.

The end of the Bronze Age in the eastern Mediterranean was marked by a series of catastrophes in which more than forty important sites in Greece, Crete, Anatolia, Cyprus, Syria, and Southern Levant, were destroyed in or about 1200 B.C. Drews (1993) discusses the various theories put forward by historians, such as earthquakes, migrations, drought, the emergence of iron technology or the collapse of agricultural or other important systems. He rejects these ideas for sound archaeological reasons and he concludes that the destruction was the result of military invasions by barbarians. The light, horse drawn chariot, had dominated Bronze Age battles for 500 years but had become vulnerable to attack by infantry who were now armed with javelin and sword. A javelin, heavier than an arrow and shorter than a spear, could stop horses, which could not be stopped by the bow. Infantry became more important but the chariot did not decline immediately although it became, in the first millennium

Fig. 22.15 The capture of Dharpur by Ramesses II. Clayton (1994, 150).

Fig. 22.16 Assyrian warfare. Bas-relief from walls of Ashur-Nazir-Pal III, king of Assyria, 884–860 B.C. British Museum. Found at Nimrud.

B.C., heavier and carried one or two shield-bearers to give additional protection to the driver and the archer. The horses, however, remained vulnerable. Nevertheless, chariots continued in use, in Western Europe, until the first century A.D. when, for example, the Britons under Queen Boudicca used them unsuccessfully against the Roman heavy infantry and cavalry.

Egyptian warfare and siege techniques seem elementary especially when compared with the methods used 400 years later by the Assyrians. A scene comparable to Fig. 22.15 is shown in Fig. 22.16.

Fig. 22.17 Top: a mounted Assyrian soldier. Bottom: an Assyrian war chariot. From the time of Ashur-Bani-Pal (668–624 B.C.). Rostovtzeff (1945, 112).

This is a bas-relief from the walls of Ashur-nazir-pal III, who was king of Assyria 884–860 B.C. It shows the king, enlarged in scale, protected by a shield bearer and firing arrows at a besieged city. The king wears a cap, is clothed from head to toe in a thick padded robe, and carries a sword in addition to his bow. His shield bearer wears a helmet, a padded knee length robe, some form of greaves, and is armed with a sword. Archers, similarly attired and equipped, fire on the city. In front of the king is a moving tower with two archers and it is positioned to attack the city and to defend the soldiers who operate a battering ram suspended from a pair of heavy chains. On the left two more soldiers are using crowbars to break down the city walls and from the city walls archers fire on the Assyrian attackers and are encouraged by women. Many defenders lie dead or are shown falling from the walls.

Changes also occurred in the mobile arms, namely the emergence of cavalry, the decline of the light horse drawn chariot, and its replacement by a heavier chariot. Light cavalry and heavy cavalry had emerged. The Assyrian light cavalry is illustrated in Fig. 22.17 (top). An archer sits astride a horse. He has no saddle or stirrups, so his seat is by no means firm, but he does sit on a cloth and the horse seems to be protected with thick padding. The rider is protected with helmet and a cuirass, probably of padding or of horn or metal pieces sewn onto a cloth. His legs are also protected. He fires a bow and has a quiver of arrows and a short sword in his belt.

The heavy Assyrian chariot, Fig. 22.17 (bottom), replaced the light chariot of the Egyptians and Hittites. It carried a crew of four, namely: a driver, an archer, and two shield bearers. All four wear helmets with earflaps, and a cuirass made of either padded material or of horn or metal pieces sewn onto cloth. The wheels are large, heavy and clumsily made and the weight is carried well forward of the axle, which gives minimum resistance when travelling over sand but not when travelling over a hard surface.

To give some idea of the proportions of the various arms, the first effective opposition encountered by the Assyrians was that of the kings of Hamath and Damascus, who defeated Shalmaneser in 853 B.C. with a force of 63,000 infantry, 2,000 light cavalry, 4,000 chariots, and 1,000 camels. The battle that ensued caused heavy losses but resulted in the abandonment of the Assyrian campaign.

Greek and Roman Warfare

Clark (1963, 82) has described the decline of the self-bow in northwest Europe. After a long period of dominance, from the ninth millennium to the middle of the second millennium, the bow entered on a period of decline during which it was only one of a number of effective weapons, including swords and metal-headed lances, that developed with the advent of bronze, and later of iron metallurgy. The bow continued in use for hunting and its decline, therefore, seems to be related to the introduction of effective metal armour in warfare. Arrows shot from relatively weak bows were ineffective against armour but swords or spears, thrusting into narrow gaps in the armour, were more successful. Metal armour, however, was expensive and its adoption must have been slow. The picture is further confused because after the Dorian invasion of Mycenaean Greece, writing and even, perhaps, metallurgy was lost and did not recover until the late Geometric period. The bow continued to play an important though diminished role and did not become dominant again until it was made much stronger so it could pierce armour plate, as was possible with the medieval English longbow and with the Turkish bow. The bow remained useful against lightly armoured troops, but it was only one option and it was not central in Greek and Roman armies.

In the full Bronze Age the lack of surviving bows, and the extreme rarity of flint or bronze arrowheads, is remarkable. Clark (1963, 83) notes that not a single arrowhead was found at many important British Bronze Age sites and at many others only one or two occur. For example, at Mildenhall, Suffolk there was only two barbed and tanged arrowheads from a rich assemblage of worked flints. Elsewhere in northwest Europe the situation was the same and arrowheads, which were once common, disappeared from the record. In Greece, the earliest area of European bronze production, the beginning of the Mycenaean Age had already relegated the bow to hunting and Clark observes that works of art showing the bow usually depicted it for hunting stag or lion, whereas in scenes of combat the chariot borne warrior was armed with a powerful thrusting lance and a sword. Later in Mycenaean culture the slashing swords came into use, together with defensive armour, including metal helmets, corselets and greaves of the type found in the middle Danube area. In

Fig. 22.18 Re-curved, composite bows illustrated on a calyx krator of *c.* 450 B.C. showing Artemis and Apollo shooting the children of Niobid. Orvieto.

the Mycenaean culture the spear seems to have been used for thrusting, but in Proto-Geometric and Geometric times a pair of casting spears or javelins (weapons of the 'Iliad' and 'Odyssey') was preferred.

In late Geometric times archers regained some importance in Greek warfare. Vase paintings, Fig. 22.18, suggest that bows were the composite Asiatic type, combining elements of wood horn and sinew, and bronze arrowheads were the three-winged variety introduced to Asia Minor by the Cimmerians. Such arrowheads have been found as far west as France and Flanders. In comparison to the sword and the spear, the bow remained of little importance in northwest Europe during the Pre-Roman Iron Age, both in the central areas of the Hallstatt and La Tène cultures and in the outlying Celtic provinces. The site of La Tène itself yielded only 12 examples of arrowheads against 166 swords and 264 spearheads.

The Homeric warrior of the twelfth century B.C., as imagined sometime in the ninth to the seventh centuries B.C. when the stories were written down, was armed with spear and sword, and protected by a helmet, cuirass, greaves, and an ox-hide shield strengthened at the edge with bronze. Light chariots were no longer used, although they continued to appear on vase paintings in the company of gods and heroes, and the method of their use appears to have been forgotten. The Greek Homeric hero was imagined to ride out to battle in a chariot, and as bows and arrows or lances were not used he dismounted before fighting an opposing champion. He used his spear either as a thrown javelin or in thrusting. The infantry at Troy were lightly armed and seem often to have been spectators to fights of single combat that were essentially ritual in nature. Ritual wars, as we have seen, remained common in Greece well into the historic period. Warfare, however, changed with the Spartan's use of heavy infantry (hoplites). Large armies, mostly of heavily armoured infantry, were drawn up in opposing lines, several rows deep, and charged each other trying to break or to outflank the enemy. The hoplites carried a large shield on the left arm, and were armed with helmet, cuirass, greaves and spear. Their right side was relatively unprotected but gained some shelter from their neighbour's shield. Consequently there was a tendency for an attacking line to move to its right. This produced a simple overlap on the right wing of each line of hoplites and gave them a local numerical advantage when this overlap attacked the enemy flank, so they were usually victorious on the right wing. The right wing became the place of honour.

The Spartans adopted a tactic of marching a proportion of their right wing along the enemy's flank immediately before battle, which invariably gave them an immediate advantage. Epaminondas (*c.* 420–362 B.C.) commanded the Theban armies at their victories at Leuctra and at Mantinae and raised Thebes to be the most powerful city in Greece for a time. He was successful because he improved tactics by thinning his lines of hoplites to build up a column 50 rows deep on one flank, with which he attacked while holding back the rest of his line. This column, out-numbering the 12 rows of the opposition's line, was invariably successful and was then in a position to attack the enemy's flank when the main battle started. The phalanx of heavily armed infantry was always vulnerable on its flanks and needed to be protected.

By the early part of the fourth century B.C. mercenary troops known as peltasts were increasingly used. These professional troops were armed with a javelin and light leather shield, which gave them their name. They were more mobile than the hoplites, whom they could attack with javelins and then retreat faster than the hoplites could charge. Peltasts and hoplites are illustrated in Fig. 22.19. In the Corinthian war, the Athenian Iphicrates successfully used a force of peltasts, with improved weapons, against the Spartans.

Philip of Macedon used a phalanx of 16 ranks armed with spears 2 m to 3 m long. Although his phalanx resembled and had the

a *b*

Fig. 22.19 Left: Hoplite. Right: Peltast.

defensive characteristics of a hedgehog, it remained vulnerable to flank attacks and required protection. Philip's son, Alexander the Great, developed cavalry and light infantry to provided this protection. He also appreciated the value of reserves by retaining his best cavalry until the critical point in a battle when he personally led them to deliver the final blow. Later, Alexander changed the phalanx by substituting from the fourth to the 15th row an equal number of ranks of lightly armed troops, carrying bows and javelins.

Cavalry was expensive and was recruited from the wealthiest strata in Greek society. The hoplites formed the bulk of the army and were drawn from the middle classes while the poorest citizens served as light infantry or as rowers in the navy. The numbers of troops involved in Greek wars were fewer than those used earlier in Asia. In the Peloponnesian War (431 B.C.) Athens had an army of 13,000 hoplites formed into 10 regiments of 1,300, and each regiment was divided into four battalions. The cavalry was only 1,000 in number. There was a territorial army of 16,000 men. At Mantinae in 418 B.C. the Spartan army consisted of 7 regiments of 512 hoplites, subdivided into 16 platoons of 32. Each platoon had its own commanding officer and achieved great flexibility of movement on the battlefield but nevertheless they were defeated by Thebes.

Essentially the responsibility for the supply of arms lay with individual soldiers who were expected to buy or make their own weapons. In Hellenica (III, 4,17) Xenophon tells us that at Ephesus, in 395 B.C.

when troops were preparing for war under the Spartan King Agesilaus: "The market was full of all sorts of horses and weapons, offered for sale, and the copper-workers, carpenters, smiths, leather-cutters and painters were all engaged in making martial weapons, so that one might have thought that the city was really a workshop of war." At the sea battle of Artemissium, in 480 B.C., Herodotus (8:17) says that, of the Greeks, "the most conspicuous were the Athenians—and in particular Cleinias, the son of Alcibiades, who was serving in his own ship manned by two hundred men, all at his own expense". When the state did supply arms they deducted the cost from the soldier's pay. Probably a soldier would take greater care of his weapons if they were his own. However, the provision of a navy was a great expense, for not every citizen was as wealthy as Cleinias. The navy in which he served had, in the main, been bought by the state. "The Athenians had amassed a large sum of money from the produce of the mines at Laurium, which they proposed to share out amongst themselves at the rate of ten drachmas a man; Themistocles, however, persuaded them to give up this idea and, instead of distributing the money, to spend it on the construction of two hundred warships for use in the war with Aegina. The outbreak of this war at that moment saved Greece by forcing Athens to become a maritime power. In point of fact the two hundred ships were not employed for the purpose for which they were built, but were available to Greece in her hour of need" (Herodotus 7:144). The Greek hour of need was, of course, the Persian invasion by land and sea. The silver mines at Laurium, south of Athens, were let by the State for an annual percentage of their revenues, and the discovery of a rich vein of silver, in 483/2 B.C. was a great windfall that was used wisely.

Dionysius the Elder, tyrant of Syracuse, was more vigorous in the provision of arms. In 399 B.C., when he was at war with Carthage, Diodorus (XIV, 41, 3–42, 2) says that:

> He gathered skilled workmen, commandeering them from the cities under his control and attracted them by high wages from Italy and Greece as well as Carthaginian territory. For his purpose was to make weapons in great numbers, and every kind of missile, and also quadriremes and quinquiremes, no ship of the latter size having yet been built at that time. After collecting many skilled workmen he divided them up into groups in accordance with their skills, and appointed over them the most conspicuous citizens, offering great bounties to any one who created a supply of arms.

> And since the Syracusans enthusiastically supported the policies of Dionysius it came to pass that rivalry rose high to manufacture arms. For not only was every space, such as the porticoes and backrooms of the temples as well as the gymnasia and colonnades of the market place crowded with workers, but the making of arms went on apart from such public places in the most distinguished homes. In fact, the catapult was invented at this time in Syracuse, since the ablest skilled workers had been gathered from everywhere into one place. The high wages as well as the numerous prizes offered to the workmen who were judged to be the best stimulated their zeal. And over and above these factors, Dionysius conversed with them in kindly fashion, and rewarded the most zealous amongst them with gifts and invited them to his table. Consequently, the workmen brought unsurpassable devotion to the devising of many missiles and engines of war that were strange and capable of rendering great service.

Dionysius was, in fact, a man of low birth, although not a slave. A political agitator and popular leader, he made himself sole commander against the Carthaginians. He was, apparently, an indifferent general, but nevertheless ruled Syracuse from 405 to 367 B.C., governed half of Sicily, and extended his conquests on the Italian mainland. He won the literary prize at Lenaea with a play called the "Ransom of Hector" but, alas, he died from a well-known literary failing, namely, over-drinking when celebrating his success. Being of lowly birth he was, no doubt, free of the revulsion that middle and upper class Greeks felt, at this time, towards the practical arts and, because of his genuine interest, the artisans responded with new or better quality weapons; this seems to have compensated for any lack in military skill. The new weapons included ships having four and five banks of rowers (quadiremes and quinquiremes) and the first artillery, namely the catapult.

That the bow had ceased to be a significant weapon is confirmed by Pausanius, writing in the second century A.D. (*Guide to Greece 1.23.4*) he says "no Greek region uses them except for Crete. We know that the Opuntian Lokrians were already regimental spearmen in the Persian wars, although Homer has them marching on Troy with bows and slings. Even among the Malians archery has not survived". Probably, the powerful re-curved, compound bows demanded much strength and extended training that resulted in bodily deformation, as did the English longbow at a later period, and this would offend Greek ideas of physical beauty. Bodily deformation was one reason why the Greeks despised the mechanical trades. Powerful bows, how-

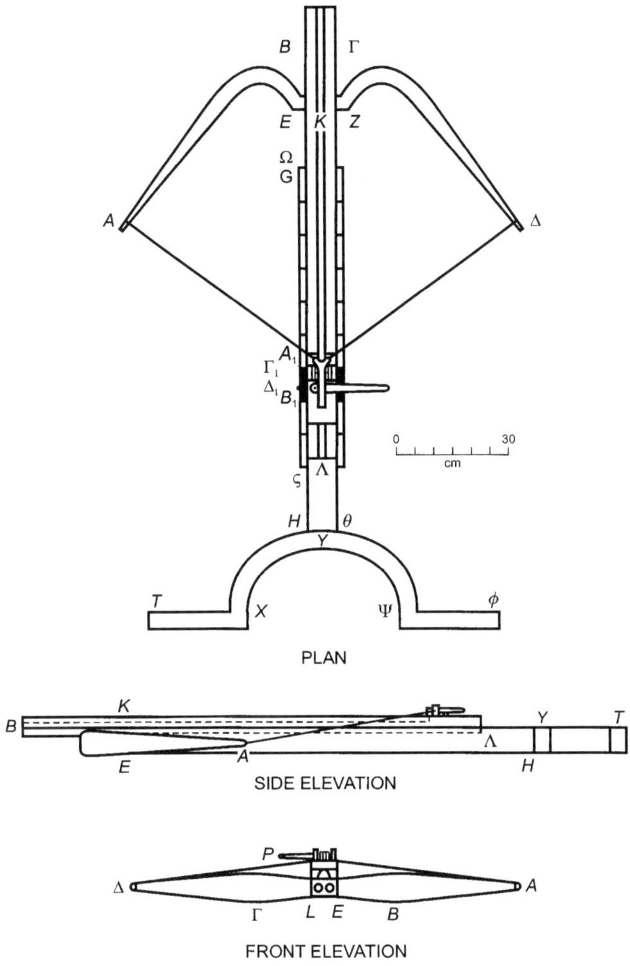

Fig. 22.20 Gastrophetes or belly bow with composite arms. After Marsden (1969, 6).

ever, came into use to throw spears and stones and were described in great detail in technical manuals. Five such treatises have survived and give detailed instruction on the manufacture of Greek and Roman artillery. The earliest is by Biton, about 240 B.C., followed by Philon of Byzantium about 200 B.C., by Vitruvius in about 25 B.C., and by Heron of Alexandria who wrote two manuals and flourished about 75 A.D. Heron's work appears to be based on a lost work of Ctesibius of Alexandria who lived in the third century B.C. All these works are available in translation and are discussed in detail by Marsden (1969, 1971). Marsden believes that it was not the catapult but the *gastrophetes* (belly bow) that was invented at Syracuse in 399 B.C., see Fig. 22.20. This is a bow with a very strong stave that required more than the strength of a man's arm to pull back. The technique was to use two arms and to support the stock against the belly. Increasing the width and thickness of the stave of either a conventional self-bow or a composite bow could have generated the increased draw back force. Marsden believes that the composite bow was preferred.

In the composite bow, the stave has a thin wooden centre and is reinforced with horn and sinew. Both these materials can sustain a greater stress and strain than wood without fracture and are, therefore, more suitable. The bows shown in Fig. 22.21 are comparatively modern Asian reflex bows, described by Payne-Gallwey (1903), and are shown in the unstrung and the strung positions. Payne-Gallwey says that such bows, although quite short, require a pull of 150–160 lbs (667–712 N) and demanded greater strength than he possessed. Stringing such a bow presents special problems and Payne-Gallwey illustrates the method adopted in Turkey and the identical method illustrated on a Greek vase, Fig. 22.22. Such bows, being very powerful, can achieve remarkable flight distances, particularly if light arrows are used. Payne-Gallwey (1903, 19–22) reprints a letter of one of his ancestors who witnessed, in 1795, an arrow shot to a distance of 482 yards (441 m) using a short Turkish composite bow. None of the experienced western bowmen present could draw the Turkish bow to its full extent and the extreme range was attributed to the strength of the bow and bowman and the lightness of the arrow. Payne-Gallwey claims to have shot very light arrows over 415 yards (380 m) and he does not discount Turkish claims, recorded on memorial stones, that arrows have been shot up to 838 yards (766 m). Heavier arrows are required, however, to penetrate armour and so for military and hunting use, heavier arrows having a shorter range are preferred.

Ancient composite bows may not, of course, have been as difficult to string and draw as those reported by Payne-Gallwey, but as Marsden (1969, 10) points out, some supporting evidence that this may have been the case is contained in the "Odyssey" where a competition was arranged for the hand of Odysseus's wife Penelope. The winner was to be the suitor who was able to string Odysseus's composite, re-curved bow. None succeeded until Odysseus himself, dis-

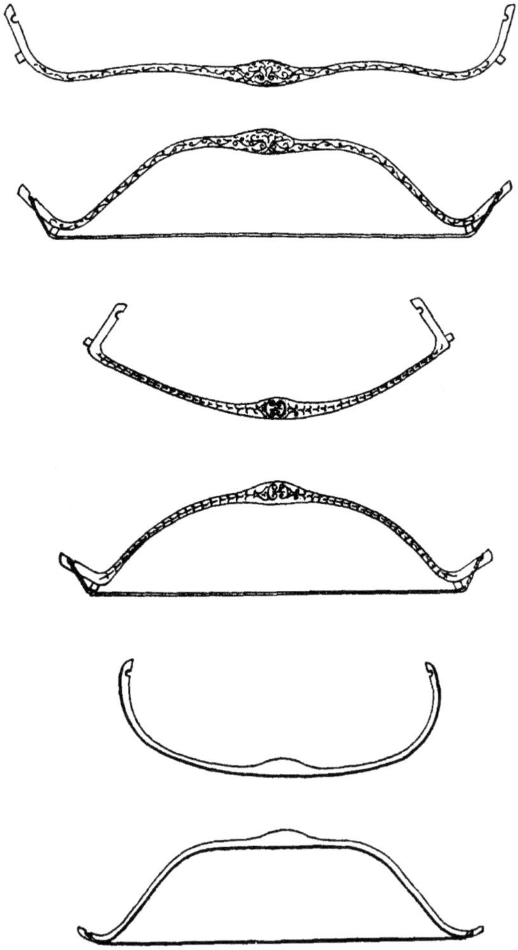

Fig. 22.21 Asian reflex bow showing the unstrung and strung positions. The composite construction of the stave is of horn, wood, and sinew. Payne-Gallwey (1903, 17).

guised as a beggar, made the attempt. After inspecting the horn to ensure it was free of wormholes he strung the bow with a single, skilful movement. The passage makes sense only if a powerful, composite, re-curved bow is to be re-strung.

If the *gastrophetes* was stronger than existing composite bows then it is very likely that it too had a composite structure. Two further developments were the introduction of a winch at the back of the stock, to enable the draw back force to be increased even further, and the introduction of a base on which it could be held easily and

Fig. 22.22 Stringing Turkish and Greek bows. Payne-Gallwey (1903, 10).

pointed in any desired direction. The latter required a universal joint between the base and the *gastrophetes*. It was possibly at this stage that they were referred to as catapults rather than *gastrophetes*. Such weapons were built having a curved stave-length of about 3 m and they could now throw very long and heavy arrows. They were adapted also to throw stones. Marsden estimated they could have thrown stones of over 2 kg in mass a distance of about 275 m, thus exceeding the range of conventional hand-bows by over 100 m.

The new stone-throwers could be used from a relatively safe distance to batter the walls of a town or city during siege operations. Battering rams were normally used for this purpose but they had the disadvantage that they must be operated immediately under the walls and they were, therefore, vulnerable to attack from the city's defenders. However, for a catapult to be effective in this role, the size of stone and its velocity needed to be increased. Heron put the matter succinctly: "They sought to make the arms of the bow more powerful, but they could not realise their intention by the use of composite bows." The resilience of the compound bow was replaced by a pair of torsion springs made from stretched sinew. Exactly when and where the torsion spring was invented is not known but torsion, in ropes, in screw-presses, and in wringing-cloths, was quite familiar and may have prompted the invention. The torsion spring underwent much development before an optimum arrangement was found. Fig. 22.23 shows the general arrangement and the front elevation gives a good view of the torsion spring. It consisted of strands of hair or sinew stretched over iron bars and held by a wooden frame. The strands passes through holes in the frame and at the mid-point rigid arms are inserted and a bowstring is attached to the tips. Rotating the strands exerts a large torque on the arms making them difficult to draw back without the aid of a winch at the opposite end of the stock. An arrow or stone is positioned, a catch is undone, and the weapon releases its missile.

The diameter of the torsion spring, more accurately, the diameter of the hole in which the torsion spring operated, was the fundamental dimension. The Greek design manuals for these ballistae laid down a simple formula for this dimension and from it all other dimensions were scaled, Marsden (1969, 1971). For example, the height of the spring, L, was taken to be 6.5 diameters and the overall length of the weapon was 30 diameters. For a simple arrow-thrower the spring diameter, D, was taken to be proportional to the length of the arrow, l

$$D = l/9 \qquad 22.1$$

The equation works in any units but dactyls were usual (1 dactyl = 19.3 mm). For the stone thrower the spring diameter was related to the weight of the stone, m, by the equation

$$D = 1.1(100m)^{1/3} \qquad 22.2$$

The weight of the stone is in Attic minas (1 mina = 0.96 lb = 0.4366 kg) and the spring diameter is in dactyls.

Artillery was also used to great effect by the Romans, and Josephus had much to say about their throwing engines in his account of "The Jewish War", written in the first century A.D. In particular he tells us that at the siege of Jerusalem the defenders mounted on the walls some artillery pieces they had captured earlier from the Romans, but they had no real idea how they operated. "These, however, most of the men were unable to use owing to inexperience, but following the instruction from deserters a few managed to use their engines after a fashion—". On the other hand the Roman artillery, in the hands of experienced engineers, was fearful. Josephus says: "The engines of all the legions were masterpieces of construction, but none were equal to those of the Tenth; their spear-throwers were more powerful and their stone-throwers bigger, so they could repulse not only the sorties [of the Jews from the city] but also the fighters on the wall. The stone missiles weighed half a hundredweight [25 kg]

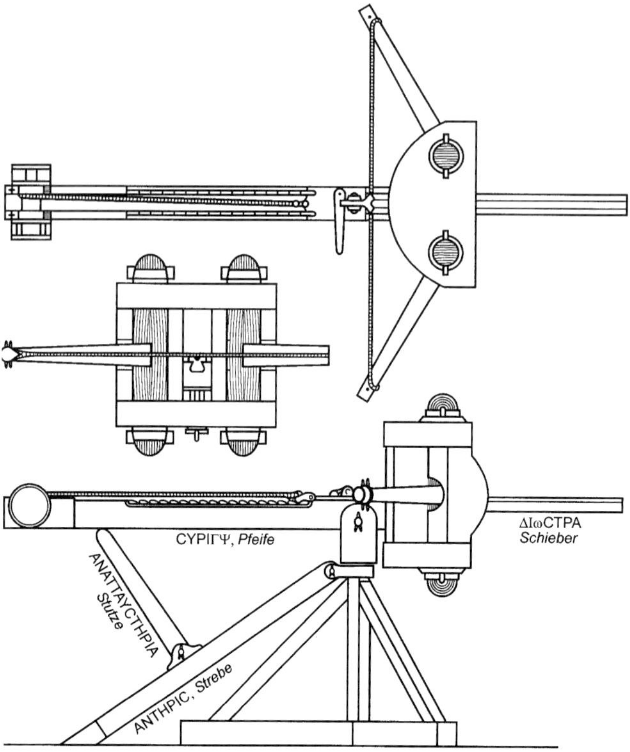

Fig. 22.23 Torsion catapult, after Usher (1929).

and travelled four hundred yards [366 m] or more; no one who got in their way, whether in the front line or far behind, remained standing." At the earlier siege of Jotapata, where Josephus was in command, he tells us that Vespasian had set up an artillery ring of 160 engines to attack the walls. "In a synchronised barrage the catapults shot lances into the air, and stones weighing nearly a hundredweight [50 kg] were discharged from stone-throwers, together with firebrands and a dense shower of arrows, driving the Jews not only from the wall but also from the area inside traversed by the missiles; for a host of Arab bowmen with all the javelin men and slingers let fly at the same time as the artillery." Later the Romans "looked over the walls and saw the Jews collecting in one spot to receive their water-ration; so they made this spot a target for their spear-throwers, causing very heavy losses." Evidently, Roman artillery were not only numerous but they were used to shoot volleys in conjunction with other arms, and they were remarkably accurate, being able to pick out particular targets, such as the queue for water.

The writings of Josephus and of Philon and others, see Marsden (1969, 99–115), describe the many ways in which artillery was used in sieges. They provided covering fire to support sallies by the besieged, and the attackers used them to repulse such sallies and to attack personnel. But their greatest use was against battlements and walls and in support of scaling towers and rams, to which the defenders responded by directing their own artillery against such scaling towers and against the attacking artillery.

The Roman army, that created and sustained the Empire over a period of several centuries, was a formidable force relying chiefly on superb organisation and tactical genius. They were not particularly inventive but took existing weapon systems and developed them to their limits. Josephus describes the Roman army in its usual marching order. "The light-armed auxiliaries and bowmen formed the vanguard, with orders to repel sudden enemy rushes and reconnoitre woods suspected of concealing ambushes. Next came a body of heavy-armed Roman troops, mounted and unmounted. These were followed by ten men from every century carrying, beside their own kit, the instruments for marking out the campsite. After them came road makers to straighten out the bends in the highway, level rough surfaces, and cut down obstructive woods, so the army would not be exhausted by laborious marching. In the rear of these the personal baggage of the commander and his senior officers was concentrated under the protection of a strong cavalry force, behind which rode Vespasian himself with the cream of his horse and foot and a body of spearmen. Next came the legionary cavalry; for each legion has its own troops of 120 horses. Mules carrying the battering rams and other artillery followed these. After them came the generals, the prefects of the cohorts, and the tribunes, with a bodyguard of picked troops, and behind them the standards surrounding the eagle, which is at the head of every legion, as the king of birds and most fearless of all: this they regard as the symbol of empire and portent of victory, no matter who opposes them. The sacred emblems are followed by the trumpeters, and in their wake came the main body, shoulder to shoulder, six men abreast, accompanied as always by a centurion to maintain the formation. The servants of every legion marched in a body behind the infantry, looking after the soldiers' baggage by the mules and other beasts. In the rear of all the legions

Fig. 22.24 Roman torsion catapult depicted on Trajan's column.

marched the bulk of the mercenaries [actually, auxiliaries], followed by a protective rearguard of light and heavy infantry with a strong body of cavalry."

Clearly, this was no raiding party but was an army designed to capture and occupy territory. A Roman army on the march is depicted on Trajan's column and Fig. 22.24, taken from the column, shows a Roman torsion catapult drawn by two mules. The two torsion springs are particularly prominent. By the fourth century A.D. Vegetius says that every century of every legion had a mobile catapult with a crew of eleven men, and for siege warfare every cohort was equipped with an onager, so called because the machine had a violent kick during recoil.

Josephus also writes that the Romans "—never give the enemy a chance to catch them off their guard; for whenever they invade hostile territory they rigidly refuse battle till they have fortified their camp. This they do not construct haphazardly or unevenly, nor do they tackle the job with all the manpower or without organized squads; if the ground is uneven it is thoroughly levelled, then the site is marked out as a rectangle. To this end the army is followed by a large number of engineers with all the tools needed for building. The inside is divided up ready for the huts. From the outside the perimeter looks like a wall and is equipped with towers evenly spaced. In the gaps between the towers they mount spear-throwers, catapults, stone-throwers, and every type of ordnance, all ready to be discharged. Four gates are constructed, one in each length of

Fig. 22.25 A panel from the Gundestrup cauldron, *c.* 200 B.C., showing Celtic cavalry and infantry.

wall, practicable for the entry of baggage-animals and wide enough for armed sorties, if called for.—The erection of the outer wall and the buildings inside is accomplished faster than thought, thanks to the number and skill of the workers. If necessary a ditch is dug all round, six feet [2 m] deep and the same width."

The tactical skill and engineering efficiency of the Roman army is apparent to anyone who reads of Caesar's exploits, but the "barbarians" whom the Romans fought, then came to rely on for their defence, and who ultimately defeated them, were also fearsome warriors. The Danish Gundestrup cauldron, Fig. 22.25, dates from about 200 B.C., long before the Roman expansion into northern Europe, and illustrates a scene of Celtic infantry and cavalry. If the Greek historian Polybius is relied on, we would believe the Celts were naked when they fought, but here we see that they were clothed, armoured, and shod. The infantry wear breeches, a tunic, and a cap and carry a shield and spear. The cavalry wear crested helmets, carry a spear, and their horses are partly covered by a saddlecloth and neck cloth. At this time human sacrifice had not been replaced, as it had in less barbarous societies, by animal sacrifice and the scene also shows, on the left, a victim being ceremonially drowned by being put head downwards into a vessel of water. Human sacrifice persisted until the barbarians were conquered by Rome. Caesar describes how some tribes in Gaul made large images of wickerwork into which they placed living victims, usually criminals of some description, before setting them alight.

Great improvements in the technique of manufacturing steel swords occurred in the first millennium A.D. The Roman *Gladius* was a short stabbing sword carried as the common weapon of every Roman soldier, but the "barbarians" preferred a longer, stronger and more flexible weapon. Metallurgical examination has revealed a complex, patterned structure on Celtic swords blades dating from as early as 500 B.C. This structure is the result of iron-welding, by heating two or more bars to a high temperature and then hammering them to form a single bar. According to Pausanius (*Guide to Greece X.16.1*) iron-welding was invented by Glaukos of Chios. Glaukos made a famous iron stand holding a silver bowl and dedicated it to the Gods at Delphi. The iron stand was famous because its elements were not held together by pins or nails but by the welded joint itself. Despite Greek claims to have invented iron-welding, Celtic iron swords made by welding together several thin bars of iron, each bar having a slightly different carbon content, have been found from as early as 500 B.C., see Pleiner (1993, 165–9).

As discussed in the chapter on manufacture, steel could not easily be made before the early-modern period. Smelting furnaces could not reach the melting temperature of iron (1530°C for pure iron and 1150°C for iron with 3.5% carbon) but they could reach about 800°C. At this temperature the iron ore, mainly iron oxide, reacts with the surrounding carbon monoxide, one of the products of burning charcoal with an inadequate air supply, so the oxygen is removed from the ore leaving a solid bloom of nearly pure iron. This is taken from the furnace and may be hammered to any desired shape while it is hot, usually to remove impurities of slag still remaining from the ore. It is known as wrought iron, although harder than copper and bronze it is not sufficiently hard for edge weapons. Steel, formed from pure iron by the addition of up to 1% carbon, is very much harder and was formed by bringing hot wrought iron into contact with carbon (charcoal) for extended periods of time and then hammering to redistribute the carbon that diffuses through the surface. Small quantities of steel could be made in this way and was welded to softer iron to give swords a steel edge and a soft iron core.

Celtic smiths did not always use their steel to best effect but Pleiner (1993, 136) found that steel was used to edge almost 60% of swords having sections covering the whole of the cross-section, and about 40% of those where the section covers one cutting edge. The blade could be formed from a single bar of metal, which was then drawn

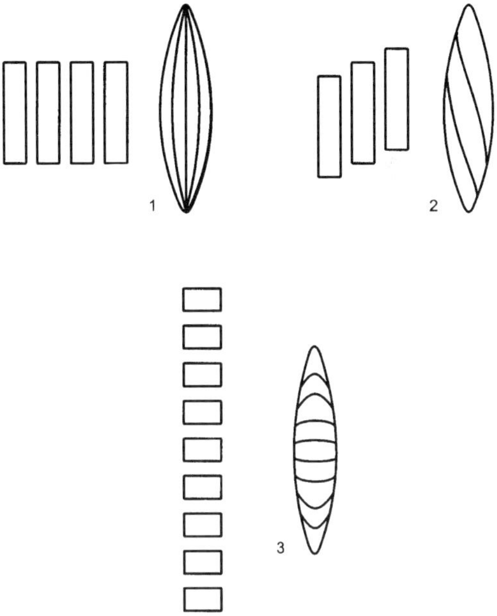

Fig. 22.26 Three principal schemes of building up Celtic sword blades. After Pleiner (1993, 136).

down, or by welding two or more flat bars together. Fig. 22.26 illustrates the principal methods found by Pleiner to build up Celtic sword blades from separate bars of iron or steel. They may be arranged:

i. broad-face horizontal, so that the weld seams run from cutting edge to cutting edge,
ii. with some lateral displacement so that the edges do not line up exactly before forging,
iii. by butt-welding two or more flat components with the broad faces vertical so the weld seams run parallel to the cutting edges.

The edges were usually made of harder steel and the core of softer steel, but this was not always the case. Microscopic examination does not provide evidence of quench hardening in water or oil but the hardened edges were probably obtained by rapid, perhaps accidental, cooling in air.

Initially the Celtic cutting sword, and its method of use, was greatly feared by Greeks and Romans but ultimately, once they had devised tactics to defeat it with their short, stabbing sword, they considered

Fig. 22.27 Sword showing three bands of pattern welding.

it inferior. In the first century A.D. Tacitus (*Agricola* 36) says: "—Agricola exhorted four battalions—to bring things to the point of the sword and to hand-to-hand fighting; a manoeuvre familiar to them from long service and embarrassing to the enemy whose shields were small and whose swords too long; for the British swords, without points, were of little use against locked lines and in fighting at close quarters." And a century later Polyaenus (*Strategemata* 8.7.2) wrote "—as protection against Celtic swords which slashed down and cut open heads, [Camillus] had helmets constructed entirely of iron so that the swords would either glance off or be broken; he had shields sheathed with copper, as wood alone was no protection against the blows—For the iron of the Celts was badly wrought; it bent easily and their swords buckled and were no use for fighting". Certainly it was true, as Pleiner found, that about 50% of Celtic swords were of wrought iron with little of no steel, consequently the claim that their swords buckled was probably correct. It is supported by, for example, by the Viking myths in which battling heroes occasionally stop to straighten their sword.

True pattern welding, Fig. 22.27, became prominent from the third century A.D. and reached its peak in the sixth and seventh centuries. By the tenth century pattern welded swords were rare in Europe but oddly enough pattern welded gun barrels were retained for sporting guns well into the twentieth century. The pattern shown in Fig. 22.27 was obtained by a twisted rod technique and was characteristic of Turkish blades until the nineteenth century.

Analysis

A Note on the Ancient Formulae for Catapults

As the weight of a stone is proportional to the cube of its diameter it is clear that the equation for the stone-thrower is derived from that used for the arrow-thrower. If this is so then we may equate Equ. 22.1 and Equ. 22.2. Taking the mass of an arrow to be $\rho\pi d^2 l/4$. where ρ is the density of the arrow material in mina/dactyl^3, and d the diameter of the arrow-shaft gives the relation between length and diameter as

$$\frac{l}{d} = \sqrt{76217\rho} \qquad 22.3$$

If the density of a wooden arrow shaft is 0.011 mina/dactyl^3 (680 kg/m^3) then the length to diameter ratio is 29:1. Thus a two-metre arrow would have a shaft diameter of 58 mm. This seems reasonable and confirms that the two design formulae are related.

The maximum stress and the strain energy stored in a torsion spring is given by

$$\sigma = \frac{G\theta}{2}\frac{D}{L}$$
$$U = \frac{\pi}{8}\sigma\theta D^3 \qquad 22.4$$

σ is the maximum stress, G the modulus of rigidity, θ the angle of rotation (twist), D the spring diameter, and L the spring height. Thus if the springs for different weapons have the same angle of twist, modulus of rigidity, and the same diameter to height ratio, then they all have the same maximum stress, regardless of size. If the stress and angle of twist is constant then the energy stored, U, is proportional to the cube of the diameter.

Some proportion η of the strain energy of the spring is converted to the kinetic energy of the projectile mass, m, and of the parasitic mass, m_p, then

$$\eta U = (m + m_p)\frac{v^2}{2} = \frac{\pi}{8}\eta\sigma\theta D^3 \qquad 22.5$$

v is the velocity of the projectile. Substituting from Equ. 22.2 gives the velocity as

$$v = 2\sqrt{\frac{133\pi}{8}\frac{\eta\sigma\theta}{1 + m_p / m}} \qquad 22.6$$

For geometrically similar catapults all the terms on the right hand side of Equ. 22.6 are independent of size and thus all the catapults designed to the spring formulae have the same projectile velocity and the same range. The larger catapults simply throw heavier projectiles. Of course, modern artillery pieces are not designed on this principle and there is no logical reason why ancient pieces should be, except that it guarantees a satisfactory level of performance. Such formulae may define the optimum conditions, given the materials available, but they also retard development and experimentation. Increasing the size of the spring undoubtedly increases the mass or the range of the projectile but it also places increased stresses on the component parts, so these would also need to be made stronger.

CHAPTER TWENTY-THREE

MEDIEVAL AND EARLY MODERN WARFARE AND WEAPONS

Medieval Weapons and Warfare

The first significant developments in warfare, after the fall of the Roman Empire in the west in 456 A.D., were related to improvements in cavalry. In the eastern empire a heavy cavalry, clad in mail shirt and using stirrups, evolved and led, eventually, to the dominance of the shock tactics of the mounted knight on medieval battlefields. Horseshoes appear to have been invented earlier, for the Celts used them as early as 50 B.C. and numerous horseshoes and nails have been found in Roman Britain and in Gaul. Horn grows quickly in hot dry regions, and on the horse's native grassland the wear rate of hooves is not excessive, but on hard or rocky surfaces the wear rate increases and in damp, cold climates, the growth rate declines so the hoof must be protected, usually with an iron hoop, to avoid lameness. This may account for the early and more frequent appearance of horseshoes on the cold, wet, lands north of the Alps. Horseshoes, however, need to be replaced regularly and this contributed to the great expense of keeping a horse.

Fig. 23.1 shows a seventh-century horseman riding bareback, apart from a saddlecloth, and armed with lance, sword, and round shield. Although the horse was, probably, shod and was controlled by means of a bit and reins, the horseman remains insecure for want of a proper saddle and stirrup. These were equally unknown to the ancient civilisations as to the Greeks and Romans, all of whom rode their horses bareback or with a simple saddlecloth. The stirrup provides a base for the rider's feet against which he can push to maintain his position on the horse. This is particularly useful in warfare when a rider wishes to strike an enemy with sword, or spear, or when he wishes to deflect a blow with his shield. It enables a charging knight to transfer some of the horse's momentum to the spear point. It also provides an easy way to mount. The stirrup seems to have been invented in China in the fifth century A.D. and to have spread

Fig. 23.1 The horseman's stone of Hornhausen, seventh century. Landesmuseum für Vorgeschichte, Halle.

throughout the far and near east before its arrival in Carolingian France during the eighth century. Its arrival in France corresponded with the rise of the Frankish cavalry and with the rise of the feudal system. This system enabled the king to retain armed knights for his wars and reward them with lands and financial independence. It has been argued by some historians that the feudal system developed as a direct result of the stirrup, which made a retinue of expensively equipped knights an essential element in warfare. Certainly Charlemagne, in an edict of 805, linked the owning of land to the obligation to own a mail shirt. The modern view, however, does not support such an easy explanation of feudalism, and warriors, in any case, followed a particular leader for many different reasons. The Vikings in the ninth and tenth centuries were attracted to their leader by family ties or by the prospect of plunder and adventure. In tenth century Poland knights were paid directly by their leaders, but the prospects of conquest and ownership of new lands was important to the Norman knights who invaded England with Duke William. In all cases there was pressure on the successful leader to reward his followers with grants of land, the revenue of which could be used, in part, to equip them for knightly service in time of war. Knights in protective armour and mounted firmly on a horse bred to take this weight, and armed with lance, mace, or sword, came to dominate the battlefield.

The Bayeux tapestry, Fig. 23.2, illustrates eleventh century warfare and weapons very well. In Fig. 23.2 (top) mail armour is carried on poles and is being loaded onto the sailing ships prior to invasion. The porters are also carrying helmets, held by their nasal bars and swords slung over their shoulders or carried by the hilt. The middle figure depicts the Norman cavalry (knights) and archers attacking the Saxons. The archers (light infantry) are shooting bows and are equipped with plenty of arrows held in the left hand or in a quiver but they do not appear to have any secondary weapons. Their only protection appears to be a padded helmet and padded or leather clothing, although one archer is shown in chain mail. The bows seem to vary in length relative to the archer. Some seem to be longbows but others are quite short. The knights are shown securely seated in a saddle with proper stirrups. They are clad in chain mail, some of which extends down the leg to the foot, and they wear a helmet with a nose bar and carry a kite-shaped shield. They are armed with spears, used in a variety of positions, and swords. Some carry maces, which may have been quite effective against chain mail. The horses are unprotected. Fig. 23.2 (bottom) shows the Saxon knights, dismounted and fighting shoulder to shoulder. They are equipped in much the same way as the Norman knights. Arrows are shown protruding from the kite shaped shields and in the lower border knights lie dead.

The shield wall of the dismounted knights at Hastings, to judge from the Bayeux tapestry, was a most inadequate affair. Nevertheless, the most successful defence against charging knights (heavy cavalry), once they had passed through the hail of arrows from the light infantry, was a line of pikes provided by a heavy infantry. The heavy cavalry was effective only in shock action at close quarters and if prevented from getting amongst the enemy they were ineffective. Pikemen could pack together in a line more closely than the attacking knights and thus each knight had to face two or more heavy infantry with pikes of two or three metres length from which their horses shied or bucked.

By the eleventh century composite crossbows made of wood and horn were in use. Famously, a crossbow was used to kill William II

Fig. 23.2 Top: mail armour, helmets and swords being loaded into ships for transport to England. Middle: Norman knights and archers advancing to battle. Bottom: Norman knights attacking the shield wall of the dismounted Saxon knights. Bayeux Tapestry, eleventh century.

while hunting in the New Forest. It was, evidently, a new and terrible weapon and in 1139 its use against Christians was banned by order of the Pope; the ban was ignored. During the century and a half following the Norman invasion four crusades against the infidels in Palestine were organised and these did much to bring new ideas and techniques into Europe. In particular, siege techniques improved; throwing engines and battering rams made a re-appearance after being almost forgotten during the Dark Ages, although they were used by the Vikings when they besieged Paris. At the Siege of Acre in 1191, Richard of Devizes and Geoffrey de Vinsauf, Hassall (1973, 102) report that the French king constructed a petrariaed, a stone throwing engine, which he called the "Bad Neighbour" and used it to attack the walls both night and day, but the Turks also built one, called the "Bad Kinsman" which often fired at and broke the "Bad Neighbour". The French king re-built it and eventually succeeded in shaking down the principal wall. The Duke of Burgundy, the Templars, and the Hospitallers, also built petrariae at Acre and one was built at the common expense at which a priest constantly preached to collect money for its maintenance and to hire people to fetch stones for casting. At length these machines knocked down about two poles length of the tower Maledictum. King Richard of England took over a large machine constructed by the Count of Flanders and also built four machines of his own. One of these, called "Berefred", had "steps to mount it, fitting most tightly to it; covered with raw hides and ropes, and having layers of most solid wood, not to be destroyed by any blows, nor open to any injury from the pouring thereon of Greek fire, or any other material. He also prepared two mangonels, one of which was of such violence and rapidity that what it hurled reached the inner rows of the city market place. These engines were plied day and night, and it was well known that a stone sent from one of them killed twelve men with its blow; the stone was afterwards carried to Saladin for inspection; King Richard had brought it from Messina, which city he had taken. Such stones and flinty rock, of the smoothest kind, nothing could withstand; they either shattered in pieces the object they struck, or ground it to powder." A modern reconstruction of a medieval ballista is shown in Fig. 23.3 (top) and a reconstructed battering ram with a protective roof is shown in Fig. 23.3 (bottom). A range in about 300 m has been obtained with such reconstructed machines.

Fig. 23.3 Reconstructed medieval siege weapons at the Chateaux des Baux de Provence. Top: a ballista. Bottom: a battering ram.

"Greek fire", mentioned above, was a variable mixture of naphtha, quicklime and various other compounds that was used from about 500 B.C. to spread fire. It was ignited and thrown from fire pots and was a particularly useful defence against siege towers and wooden artillery. It was also used as an artillery projectile and was used in naval warfare to set fire to enemy ships, Fig. 23.4. Josephus, in Williams and Smallwood (1981, *III, 278*), tells us that Vespasian used three siege towers fifty feet [15 m] high and encased in iron

Fig. 23.4 "Greek fire" in naval warfare, fourteenth century.

on all four sides "so that they should be too heavy to overturn and almost proof against fire". In the seventh century Greek fire played an important part in the defence of Byzantium.

From these throwing engines developed a new and more terrible engine, namely the trebuchet; see Fig. 23.5. This was a late medieval invention. Torsion springs and bow staves were insufficiently strong to cast very heavy stones against the enormously strong walls of medieval towns and castles and in the search for greater kinetic energy the throwing arm was propelled by the fall of a very heavy weight in a large container. The weight amounted to several tonnes and was raised, for example, by means of a suitably geared treadmill or a system of pulleys having a large reduction ratio, which pulled down the throwing arm and raised the heavy weight. In Fig. 23.5 (top) the trebuchet is about to release its projectile. The man at the rear has a rope that operates the release mechanism and the twelve men holding the four ropes at the front are, evidently, intended to pull downwards to increase the momentum to the descending weights in the container, although as Payne-Gallwey (1995, 253) points out, they would not have much practical effect. The projectile is not placed directly on the end of the throwing arm but is attached to it through a sling and it slides along a central plank between the treadmills. This plank is grooved to ensure the projectile does not move out of line when it accelerates. A suitable release mechanism of ropes is arranged to liberate the projectile at the desired angle.

Leonardo da Vinci illustrated a trebuchet, Fig. 23.5 (middle), which used a dead horse as a projectile. It was intended that the decaying

Fig. 23.5 Top: a trebuchet with its arms pulled down by means of two treadmills. Middle: a trebuchet illustrated by Leonardo da Vinci and used for casting a dead (diseased) horse into a besieged town. Payne-Gallwey (1995, 271, 253). Bottom: modern reconstruction of a medieval trebuchet at the Chateaux des Baux de Provence.

Table 23.1 Performance of French Medieval Throwing Engines, Beffeyte (2000, 9–19).

Name	*Period of Use*	*Range m*	*Projectile Mass kg*	*Rate of Fire Shots/h*	*Number of operators*
Pierrière	11th–15th C	40–60	3–12	60	8–16
Bricole	12th–15th C	80	10–30	60	16
Mangonneau á roues de carrier	12th–15th C	150	100	2	12+
Trébuchet	12th–16th C	220	125	1–2	60+

carcass should spread disease in the besieged town and it was an early use of biological warfare. The trebuchet was the direct competitor of the cannon in siege operations and was, for a long time, preferred because of its accuracy. Variations in the burning rate of gunpowder produced large variations in the range of early cannon.

There was in fact, a range of stone-throwing engines varying in the size of stone, range thrown, rate of fire, and the number of men needed to service it. The smallest were quite portable but the largest were very heavy engines and were constructed on-site, consequently they were used only in sieges. Table 23.1 summarises the performance of some reconstructed machines reported by Beffeyte (2000, 9,19). The smallest machines, pierrière and bricole, rely on manpower, rather than weight, to propel the stones, and projectiles of 1 kg have attained a velocity of about 40 m/s. Such engines were commonly mounted on the ramparts to defend a town or castle against attackers. A larger engine, using counter weights to throw the projectiles, was the mangonneau. In this engine men in a treadmill raised the counter weights. Beffeyte also describes a large trebuchet with a counter weight of 6 tonnes which could throw a 136 kg projectile a distance of 200 m with a dispersion not exceeding 12 m longitudinal and 4 m lateral. A modern reconstruction of a medieval trebuchet is shown in Fig. 23.5.

Long-range hand weapons consisted, at this time, of longbows and crossbows. The crossbow was used quite early in China, Fig. 23.6. This bronze crossbow mechanism is in the Burrell Collection, Glasgow, and the inscription, inlaid in gold and silver, dates it to 147 A.D. This is many centuries before they appeared in Europe although Payne-Gallwey (1903, 43) says there is good evidence of hand held

Fig. 23.6 Chinese crossbow mechanism, 17.1 cm × 13.9 cm, 147 A.D. Burrell Collection, Glasgow.

crossbows in Vegetius, about 385 A.D., whom he translates as "with them were the javelin men, who from their bow in hand or crossbows directed their arrows". This, he says, is supported by two bas reliefs of the fourth century at the museum at Puy, in France. However, there are no other references, either literary or pictorial, to crossbows in Europe until the tenth century, when, according to Payne-Gallwey, the city of Senlis was successfully defended by crossbowmen when attacked by the Belgium king Lewis in 947. Lothair also used them at Verdun in 985.

Only one crossbow was depicted on the Bayeux tapestry and it appears that although they were used, it was not a common weapon. Probably, however, they were used mainly for hunting, and as mentioned above, a bolt from a crossbow accidentally killed William II when he was hunting in the New Forest in 1100. Certainly their low rate of fire, in comparison to other bows, was a disadvantage in battle. Crossbows were banned by the Pope in 1139 but were reintroduced by Richard I in his wars in France and Palestine. Richard I also died by the crossbow. Payne-Gallwey quotes Richard Le Breton "Thus perished by the crossbow, which the English count dishonourable, King Richard, who first introduced the crossbow into

France." From this time, about 1200, until about 1470, the crossbow remained a popular weapon in continental armies, and was in fact given an honourable place in most battles before it was superseded by handguns. The best crossbowmen were considered to be Genoese who were often hired for service both on land and at sea. Some 20,000 Genoese crossbowmen fought on French ships at the battle of Sluys in 1340 and, according to Froissart, 15,000 Genoese crossbowmen formed the front rank of the French army at Crécy in 1346, but they failed to shoot because it is said the strings of their bows were wet. In view of the fact that they did advance against the English is more likely that their crossbows were simply out of range when the English archers started shooting. The range of the crossbow was probably about 180 m whereas the longbow was perhaps 230 m. This was an important difference, especially when added to the longbow's increased rate of shooting. The subsequent charges by the French heavy cavalry, under a hail of English arrows, lost many horses and men, despite their armour, and they were unable to attack effectively. Despite being outnumbered 3:1 Edward won a resounding victory. The longer range and faster rate of shooting of the longbow proved effective.

Fig. 23.7 (left) shows crossbows in use on the fore and after castles of a rowed galley and Fig. 23.7 (right) shows a somewhat later development, namely, the windlass crossbow, which proved to be necessary when wood or wood and horn bow staves gave way to steel staves in about 1370. Although this windlass bow could outshoot the long bow it was very slow to operate, as Fig. 23.7 illustrates, and a longbow could, according to sources quoted by Payne-Gallwey, outshoot a crossbow by ten or twelve arrows to one, although in experiments he found that crossbows could shoot one arrow per minute whereas the longbow could shoot six arrows per minute. The older method of tensioning the crossbow was to use a belt claw. At the bowman's waist was a belt claw. To tension the crossbow he places the bow stave under his foot, bends down to hook the belt claw under the string and then stands up, tensioning the bow until the string drops into a catch and the bow can be loaded with a quarrel. During this time he is, of course, vulnerable.

The men of Gwent, according to Gerald of Wales (1978, 1:4), writing in the twelfth century, were particularly skilled in the bow and arrow. He records that at the capture of Abergavenny Castle an arrow penetrated the oak doorway of the tower, as thick as a man's

Fig. 23.7 Left: crossbowmen and harquebusiers prepare for battle on a rowed galley, Valturius, 1472. Right: windlass crossbows illustrated in Froissart's Chronicles. The windlass lies on the ground when shooting. Payne-Gallwey (1903, 4, 8).

palm, about 10 cm. On another occasion he says that a Welsh arrow penetrated a man's thigh, his iron cuishes, the leather skirt of his tunic, his saddle, and finally penetrated so deep into the horse that it killed the animal. This weapon relied on the strength of the archer and the long hours of daily training he was willing to undertake to acquire the necessary skill and strength. The crossbow, in comparison, was more expensive to make but it required relatively little training or manual strength. The longbow first came to prominence at the battle of Falkirk in 1298 when Edward I defeated William Wallace. The Scottish heavy infantry armed with 4 m pikes adopted a four-deep, all-round defensive formation in which the front row knelt holding their pikes at an angle from the ground and the other rows held their pikes horizontal. The cavalry and archers were held in reserve until Edward's attack collapsed, as it had done previously. Attack by the English heavy cavalry again failed to penetrate the thicket of pikes so the king brought up his longbows to concentrate their arrows on particular sections. Serious casualties created gaps in the Scottish lines and the subsequent attack by the English heavy cavalry was successful.

Later successes at Crécy, 1346, and Agincourt, 1415, against superior numbers, confirmed the superiority of the longbow, at least in the hands of trained bowmen. French attempts to adopt the longbow failed because, apparently, they were unwilling to undertake the extensive training that must have resulted in archers having disproportionate arm and chest muscles. Despite improvements in the crossbow and in handguns the longbow remained an English weapon, even into the sixteenth century. A large bundle of longbows and arrows, for example, were brought up from the wreck of the *Mary Rose* that sank in 1545. Many needed a pull of over 100 lb [445 N] to shoot them. To encourage the use of longbows laws were passed forbidding the employment of crossbows in England, even for hunting. Most of these acts were passed in the sixteenth century when it was becoming clear that handguns were at least equal, and perhaps superior, to both crossbows and longbows.

Armour, of course, was the main defence against both long-range and shock weapons and a careful balance had to be maintained between the weight of armour and the penetration that offensive weapons could achieve. Too much armour was heavy and reduced the manoeuvrability of the knight it defended. Chain mail continued to be used until the end of the thirteenth century, as Fig. 23.8 demonstrates. This is one of the famous scenes from the French Maciejowski Bible of about 1275 and illustrates the storming of a castle and the shock action of mounted nights. The defenders wearing kettle hats throw down rocks or wield axes. The attackers climb scaling ladders, or mine beneath the walls with a pick. The miner is defended by a crossbowman about to release an arrow by pressing an elongated trigger. The mounted knights wear chain mail with a surcoat and a variety of helmets from a simple chain mail coif, to the *spangenhelm* and the great helm. A *spangenhelm* is shown being cleaved by a sword. There is little protection for the horses.

Fig. 23.8 also shows a two-handed slashing sword wielded by a mounted night and slicing through the chain mail of his mounted opponent so it appears that chain mail no longer gave sufficient protection. Similarly the lower figure shows a lance piercing chain mail and propelled by a knight firmly seated on his horse in a high saddle. At the beginning of the fourteenth century a transition from chain mail to plate armour began. Plate armour is heavier but more effective against arrows and the thrust or slash of swords and axes. At first it was used to cover the most vulnerable areas, as illustrated

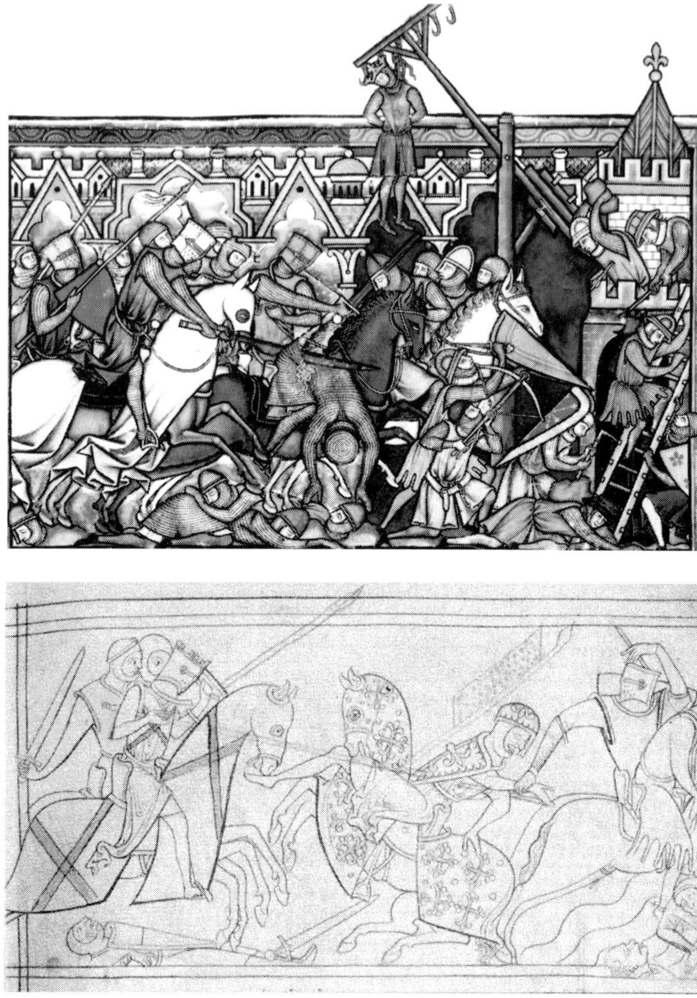

Fig. 23.8 Top: storming a castle and shock fighting by mounted knights wearing chain mail. Some skulls are protected by a mail coif, some by a "great helm" and some by a conical "spangenhelm". Pierpont Morgan Library, New York. Bottom: The horses are caparisoned to provide some slight protection against either arms or weather. The high cantles and pommels of their saddles give the knights a firmer seat. *The Lives of the Two Offas*, c. 1250, The British Library, London.

by the memorial brass of Sir William Fitzralph in about 1323, Fig. 23.9 (left). Here, chain mail is worn all over and is given additional protection by strapped solid plates to cover the arms, shins, and feet. The head is covered with chain mail but would be further protected with a great helm. The great helm had the disadvantage

Fig. 23.9 Transitional armour. Left: memorial brass of Sir William Fitzralph, *c.* 1325. Victoria and Albert Museum, London. Right: tomb effigy of Walther von Hohenklingen, 1386. Schweizerisches Landesmuseum, Zurich.

of having to be removed quite often, to speak or to see, so they were often made with an open face and fitted with a hinged visor that could be raised more conveniently. The action of a knight in raising his arm to lift his visor when approaching a colleague is said to be the origin of military saluting. When the shape of helmet changed to a more conical form with a visor it was called a basinet and is, perhaps, the most familiar type of knight's helmet.

Fig. 23.9 (right) shows the effigy from the tomb of Walther von Hohenklingen who was killed at the battle of Sempach in 1386, and it illustrates a mixture of padding and plate armour that must have been more comfortable than chain mail and plate. The basinet, breastplate, gauntlets and leg defences are all solid plate. Under the breastplate is a padded jacket with puffed sleeves. His neck protection seems to be leather or some other padded material and may be reinforced with mail.

The thickness of armour varied considerably. Jones, in Hardy (1992, 233–4), gives the following data:

German basinet, 1380,	1.52 mm to 3.81 mm
German basinet, 1370–80,	1.27 mm to 2.44 mm
Italian basinet, 1370–80,	1.52 mm to 3.05 mm
Italian basinet, 1370–80,	2.54 mm to 4.57 mm
Pair of Italian cuisses, 1390,	1.27–1.78 mm
Composite half armour, Italian, 1470,	2.03 mm to 2.79 mm

He concludes that armour on the top of the head was thickest and varies from 2.5 to 4.6 mm, followed by the chest where it varies from 2.0 to 2.8 mm, and the legs where it varies from 1.27 to 1.77 mm. The steels used also varied quite markedly. Jones, in Hardy (1992, 232–3) found steels having 0.1% to 0.2% carbon and tensile strength from 310 MPa to 1086 MPa. Surface hardening by hammering and carburisation was known but, because of non-metallic impurities in the steel, toughness and ductility would be lower than modern steels. It is a myth, of course, that armoured knights were so heavy they had to be lifted onto their horses by a winch, or that once they were unseated they were unable to rise from the ground. Armour typically weighed about 20 to 40 kg, less than what a modern soldier might carry, and a knight would be able to mount and dismount reasonably and would be able to move relatively freely, if slowly. Muddy or marshy ground, of course, was a problem.

By the end of the fifteenth century full plate armour was in use and armour protection had been extended to the horse, as Fig. 23.10 (left) illustrates. The horse armour includes a *chanfron* for the face and head, a *peytral* for the chest, and a *crupper* for the flanks. Such armour not only protected the horse against arrows shot from long range but also protected them against the pikes as they tried to penetrate and break the lines of heavy infantry. Armour, like most other things, was subject to fashion, as a comparison of the two suits shown in Fig. 23.10 makes clear. The German Gothic armour contrasts with the late Elizabethan armour of Lord Buckhurst with its pronounced doublet and point at the waist. The plain areas were originally blued, by heating to about 350°C, and the steel plates were etched and mercury gilded. Despite all this decoration the armour was intended for use in war and it is thought that Lord Buckhurst ordered it in anticipation of the arrival of the Spanish Armada in 1588.

Pikemen remained the main defence against attacking knights but they developed an offensive role in the fourteenth century. The Swiss adopted a formation of 50 ranks and fifty files so that 2,500 men occupied a front of only 50 m and a depth of 25 m. Usually they arranged themselves in three rectangular blocks and they drilled together so they were able to move without losing formation. When attacked they stood still, faced outwards in all four directions, and planted their halberds to form an impenetrable thicket. Being an all-round defence they could not be outflanked, and casualties in the front lines were transferred quickly to the interior and replaced by fresh soldiers. Thus the Swiss square was able to charge and attack an enemy while remaining relatively invulnerable. Eventually they were armoured with steel helmet and breastplate and given a pike having a long steel point. After defeating Charles the Rash, Duke of Burgundy, on three occasions in 1476 the Swiss square was considered invincible. Numerous armies demanded their services as mercenaries and the German Landknechts copied their methods. The halberd was adopted as the Swiss national weapon.

The invention of gunpowder has been ascribed to the Chinese, who seem to have used it for fireworks and for fire-lances. In Europe, Berthold Schwarz [Black], a Franciscan monk of Freiburg, in southern Germany, was once thought to have invented both gunpowder and guns in 1313, and a statue in his honour graces the main square in Freiberg. In many countries gunpowder is still known as black-powder. However, it seems that nothing he has written has survived. In England Roger Bacon is recognised as the first to record gunpowder's existence when he wrote down its composition in 1249 as an example of the marvels of science that are greater than those of magic. He wrote that saltpetre and sulphur when mixed with charcoal will produce a loud explosion and bright flash when touched by fire, and he recognised its usefulness in frightening armies. As Bacon's letter was dedicated to the Bishop of Paris, who died in 1249 it is clear that gunpowder was invented before this date, although probably not by Bacon. Saltpetre is formed naturally by the action of soil bacteria on proteins exposed to air and it grew best in tombs and in dung heaps, for example near stables. Saltpetre was gunpowder's

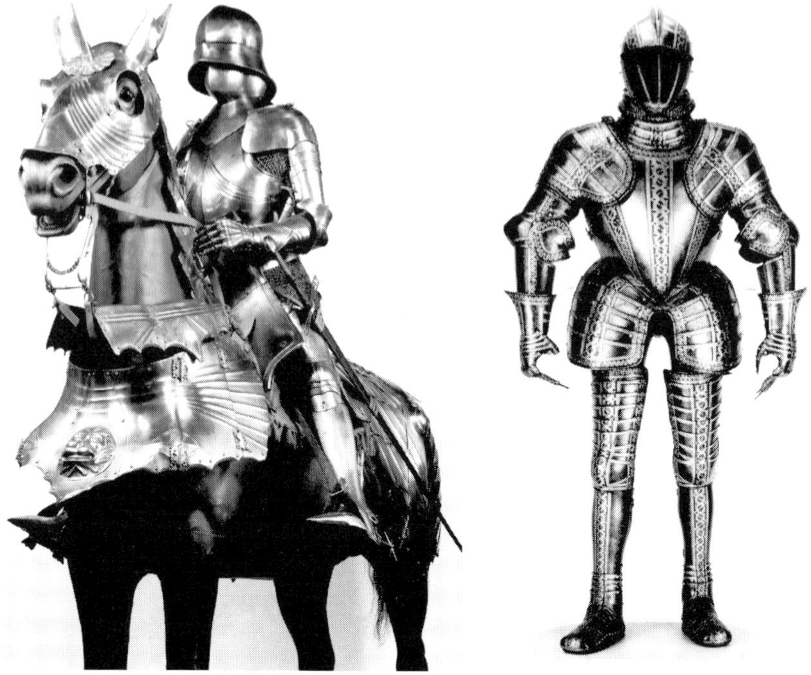

Fig. 23.10 Left: German equestrian armour in the Gothic style. Probably made for Waldemar VI of Anhalt-Zerbst (1450–1508). The Royal Armouries, London. Right: Full-plate armour of Lord Buckhurst, about 1587, made in the Royal Workshops, Greenwich. Weight 32 kg. Wallace Collection, London.

main ingredient and contains a significant amount of molecular oxygen that is essential for rapid combustion.

As Table 23.2 shows, the composition of gunpowder was by no means constant and changed significantly with time and place. Biringuccio (1540, 413) makes different recommendations for heavy guns, medium guns, and light guns such as harquebuses and pistols. By the eighteenth century it had become stabilised at the composition still in use today, namely 75% potassium nitrate, 15% carbon, 10% sulphur. Gunpowder is a simple mechanical mixture of these ingredients and ideally the particles of each constituent should be reduced to a molecular scale before mixing, but in practice the constituents are reduced to a very fine powder. Interstices exist between the particles and the grains and during combustion gases emerge from the surfaces and surge through the interstices, spreading the

Table 23.2 The changing composition of gunpowder.

Date	*Authority*	*Saltpetre*	*Charcoal*	*Sulphur*
1249	Roger Bacon	41.2	29.4	29.4
1350	Arderne	66.6	22.2	11.1
1540	Biringuccio	50.0* 66.7** 83.3***	33.3* 20.0** 8.3***	16.7* 13.3** 8.3***
1647	Master Gunner Nye	66.6	16.6	16.6
1670	Sir J. Turner	71.4	14.3	14.3
Modern		75% (Potassium nitrate)	15%	10%

* Heavy guns. ** Medium guns. *** Harquebuses and pistols.

combustion quickly through the whole mixture. Gunpowder is reasonably stable and does not detonate but it can be ignited to burn very quickly by impact or by increasing its temperature above about 300°C. Ignition is retarded when the gunpowder is damp. Confinement, as in a gun barrel, raises the pressure and increases the rate of combustion to explosive rates.

Biringuccio (1540, 413) tells us that the ingredients "are mixed in by crushing or milling them very fine and then they are granulated and dried—But since these powders ignite very easily when they are being made, they would not be without danger to the one who makes them if he did not prevent this by moistening. Therefore take care not to crush them when dry, both to avoid this danger and also because they crush better. Moisten with ordinary water to a certain degree of moistness so that it sticks together when squeezed in the hand." He describes various methods used to reduce the constituents to a fine powder. In Fig. 23.11 three stamps are in use, one for each constituent, after which the powders are mixed.

After the loss of Normandy in the early thirteenth century the English kings attempted to regain their possessions, during what is now known as the Hundred Years War. Ultimately they failed, but their campaigns in France produced notable victories of considerable technical interest. In particular: Edward III's first use of guns and gunpowder at Crécy in 1346, the devastating power of the English longbow against fully armoured knights at Crécy and again

Fig. 23.11 Gunpowder stamps for reducing the three constituents of gunpowder to a fine powder prior to mixing, 1470.

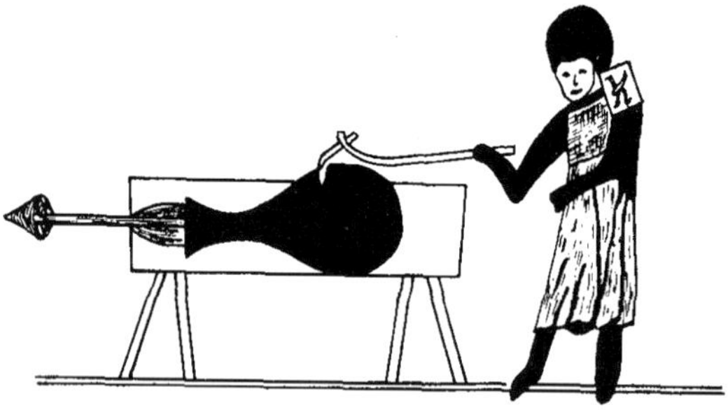

Fig. 23.12 The first illustration of a cannon, 1327, from Christ Church, Oxford MS 92.

at Agincourt in 1415, and earlier in the same year, Henry V's capture of Harfleur which was, according to Hogg (1963, 205), the first investment in history during which artillery proved its worth.

It is not known when or by whom guns were invented but it is clear that guns and gunpowder first appeared in Western Europe about 1300, some 50 years after Bacon first described the composition of gunpowder. Needham (1986, 10) believes that the true gun first appeared in China in about 1280. This was some three and a half centuries after the Chinese discovery of gunpowder, during which period they used it in the fire-lance. The first European illustration of a cannon occurs in 1327, see Fig. 23.12, which is re-drawn from an illuminated manuscript of Walter de Millemete, Christ Church Library, Oxford. It appears to be propelling an arrow from a horizontal pot lying on a table. Twenty years later, as mentioned above, the English used cannon at the battle of Crécy. Although he did not mention guns in the first edition of his *Chronicles*, Froissart later compiled a shorter version in which he writes that at Crécy "The English had with them two of the bombards and they made two or three discharges on the Genoese who fell into a state of disorder when they heard the roar". Evidently, as Bacon predicted, the roar of the guns was as effective as their projectiles in scattering soldiers, in this case the Genoese archers that lead the French attack, but most of the credit at Crécy belongs to the English and Welsh archers and their longbows, which effectively destroyed the French heavy cavalry (knights) before they were able to bring their shock weapons into action. Fig. 23.13 from Froissart's *Chronicles* illustrates the battle and shows a bombard in the margin.

Other writers also mention the use of guns at Crécy, and Hogg (1963) has discovered in the Wardrobe Accounts of the Tower of London that numerous items of munitions were purchased and shipped for Edward's Normandy campaign. For example, on the 4th March 1346 "10 guns with mountings, six pieces of lead, 5 barrels of powder, 100 large leaden shot for the same guns" were purchased. As the king did not sail to France until July it is evident that these and other pieces were purchased for the Normandy campaign. Acquisitions continued through into November when, for example, there was "2021 lb saltpetre and 446 lb sulphur purchased". On the 1st September, after the battle, Edward ordered his artillery in the Tower to be shipped to Calais where he appears to have had at least 20

Fig. 23.13 Battle of Crécy, 1346. A bombard is shown in the margin.

guns for his siege of that port. Evidently his cannons were not considered effective in battering down the town walls for he chose to conserve his new weapons and to starve Calais into surrender, which they did after holding out for a year.

Before the end of the Hundred Years War cannon had been improved to the point where they made effective battering weapons against walled cities. Hogg (1963, 202) notes that in 1384 the castle and town of Cherbourg, in English hands, had "ten guns to throw stones, two of which are of iron and eight of other metal; seven of the ten guns casting large stones 24 inches in circumference, and the three remaining casting large stones 15 inches in circumference". They also had "200 lb of powder, 26 lb of saltpetre and 24 lb of pure sulphur". In 1405, Henry IV laid siege to Berwick Castle with his artillery and it is reported that the first shot overthrew part of one of the towers, whereupon the defenders immediately surrendered.

A letter from Henry V, announcing the capture of Harfleur, gives a very chivalrous account of the siege. He says they arrived before the town on the 17th August 1415 and because of "the strength and position of our cannon, and our other ordnance, the people who were within the town made great urgency to have diverse parleys with us", see Moriarty (1989, 153). After many discussions the king resolved to attack on the 18th September, but those within the town proposed that they would offer hostages and agreed to deliver the town at 1.00 am on the following Sunday if the Dauphin had not come to their rescue. Henry was so keen to entice the Dauphin into battle that he allowed the town's leaders safe conduct through his lines to seek his assistance. However, "Sieur de Hankeville and the others of his company, returned on the [appointed] day at eight of the clock in the forenoon—without any rescue being offered by our said adversary, his son, or any other party; and the keys of the town were then fully delivered and rendered into our hand—".

It all sounds very courteous and agreeable, and one might suspect that an element of ritual warfare was present, but an army chaplain, quoted by Hogg (1963, 204), gives a more believable account saying, "The king gave himself no rest by day of by night until he had fitted and fixed his engines and guns under the walls, and planted them within shot of the enemy and had covered them against the shot and offensive weapons of the enemy—". The siege continued with sallies from the French, mining operations under the walls, and all the usual stratagems of siege craft. "In the meantime our king, with his guns and engines, so battered the said bulwark and the walls and towers on every side (especially where the enemy had arranged their guns and cross-bows towards us) that within a few days by the impetuosity and fury of the stones, the same bulwark was in great part broken down;—and although our guns had dismantled the bulwark, walls and towers during the day, the besieged by night, with logs, faggots and other barricades refortified the streets; and even thus stopped up the lanes with clay, earth and mud in great thickness; "so that the stones from our guns, coming down upon them, might be swallowed up in them—". Hogg concludes that Harfleur was a gunners' victory and that without cannon the siege would probably have failed. After the siege the guns were sent to Calais and consequently they played no part in the subsequent battle at Agincourt.

The first siege cannon needed to compete with the trebuchet and thus they quickly developed to a very large size. The earliest seem to have been made mostly from wrought iron and later of cast bronze. Fig. 23.14 shows the construction of *Mons Meg*, a wrought iron cannon of 1449 now on display in Edinburgh Castle. It was made at Mons and Hogg (1963, 11) writes that Treasury accounts of 24th June 1497 mention that it was fitted with a new "cradill", probably constructed for the siege of Norham Castle in that year. Its constructional details were revealed by an X-ray examination

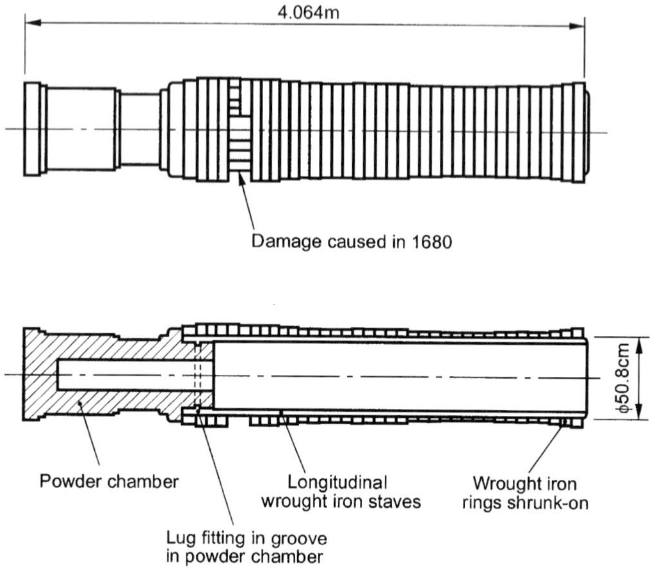

Fig. 23.14 Constructional details of *Mons Meg* (1449). After Crossland (1986, 151).

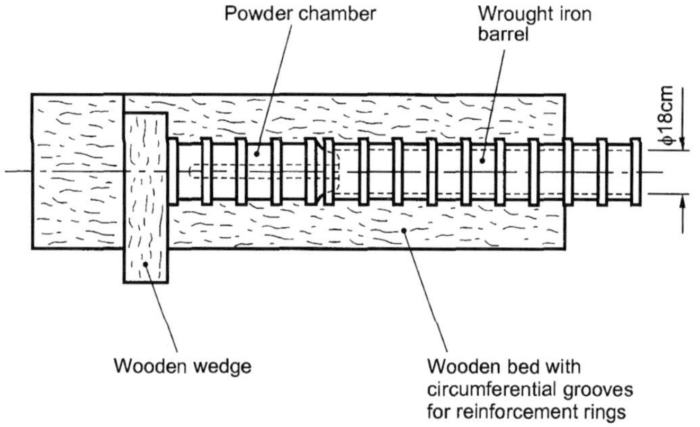

Fig. 23.15 Breech-loading wrought iron naval gun, early sixteenth century. After Crossland (1986, 153).

undertaken by Mr R. Smith of the Royal Armouries and reported in Crossland (1986, 151). The bore is 20 inches (50.8 cm) and it fired a granite ball of 330 lb (150 kg). Crossland believes the powder chamber was probably made by hammer welding small pieces of wrought iron together over a mandrel until the required size and shape was attained. The barrel is made of longitudinal wrought iron staves, about 10 cm wide and 2.5 cm thick. A lug on the staves fits into a groove in the powder chamber around which they are arranged and the staves are held in place by 37 wrought iron hoops 3.5 inches (8.9 cm) thick. The hoops were probably shrunk on to bind the structure together.

Evidently the art and craft of the cooper had been adapted to gun barrel manufacture, and may be the reason we call them *barrels*. Throughout most of the pre-industrial period the clearance between the projectile and the bore of most cannon was about 0.5 inch (1.3 cm), to ensure that the ball loaded easily, consequently the bore need not be made with great precision. Crossland estimates that the bursting pressure of the powder chamber was about 320 MPa, and the yield pressure about 99 MPa. For the barrel, where it meets the chamber, he estimates the bursting pressure was 91 MPa and the yield pressure 48 MPa. At the muzzle the corresponding pressures are 55 and 29 MPa. To give a 150 kg stone a muzzle velocity of 100 m/s, enough to cast it 1,000 m at the optimum inclination, requires a mean pressure of about 1.4 MPa and a maximum pressure of about 3 MPa in this gun so it had an adequate factor of safety.

Because of their method of construction on a mandrel, most early wrought iron cannons were open at each end of the barrel and breech loading was usually necessary, Fig. 23.15. Such breech-loaders were not fitted with seals and the high-pressure gas eroded the surfaces as it escaped. This problem was, no doubt, responsible for the development of cannons cast in bronze, and later in iron, and for the development of muzzle-loaders. As Table 23.3 confirms, the earliest cannon were made of wrought iron but after about 1450 cast bronze cannons became the norm, although Filarete mentions a cast iron bombard in the Castello of Milan in the form of a lion in 1463.

Renaissance and Early Modern Warfare

Greener (1910, 55) translates a renaissance French author who writes of gunpowder: "No use has yet been made in France, in 1547, of that terrible weapon against men. The French used it with good effect against some castles in 1338, but they would blush to employ it against their fellow creatures." He goes on to say that the English had no such scruples. Despite this reluctance to use firearms against

Table 23.3 Some Early Cannon, Crossland (1986, 151).

Date	*Name*	*Material*	*Weight tonne*	*Calibre cm*	*Weight of stone ball kg*
1377	Duke of Burgundy	Wrought Iron?	–	–	91
1377	Duke of Burgundy	Wrought Iron	–	56	204
1382	Great Bombard of Ghent	Wrought Iron	13.2	63.5	317
before 1424	Mont St Michel	Wrought Iron	–	48	–
			–	38	–
1449	Mons Meg	Wrought Iron	5.1	50.8	150
1464	Mahomet II	Bronze	17	63.5	310
–	Mahomet II (1430–81)	Bronze	40.7	73.4	434
1548	Great Gun of Beejapore	Bronze	40.7	73.4	434
1586	Great Gun of Moscow	Bronze	38.8	91.5	1016

men, as opposed to the walls of castles or towns, hand weapons, which could only be effective against men, were developed from the earliest times. Moreover, the gun carriage (block trail carriage) had replaced the gun cart by the end of the fifteenth century, making it a more manoeuvrable weapon that could be used on the battlefield against infantry.

Hand held firearms dating from about 1400, such as one found in South Schleswig and now in the Copenhagen Museum, developed from small cannon. They consisted of a small barrel attached to a wooden stick, were usually cast of bronze or iron, loaded at the muzzle, and ignited by hand using a slow match or hot wire. It sounds, and no doubt was, a dangerous procedure but from these developed the hand-ignited harquebus and the harquebus with a matchlock mechanism for firing. After about 1567 these heavy muskets in the Spanish style were fired throughout Europe from a simple forked rest and they were very successful in penetrating armour. They effectively chased the heavily armoured knight from the battlefield.

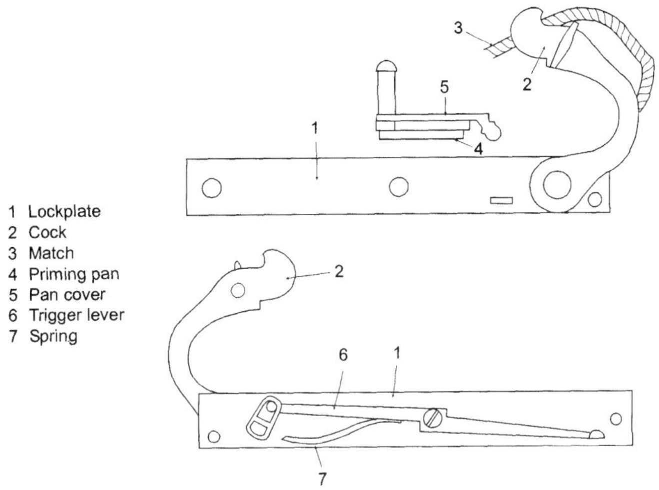

Fig. 23.16 Matchlock mechanism.

The hand cannon was ignited by a slow match, that is, by a length of cord soaked in saltpetre and allowed to dry. It replaced the hot wire and burned slowly with a glowing end and was used to ignite a small amount of gunpowder in a flash-pan. The flash-pan communicated with the main charge in the barrel through a small hole. Gunners were greatly encumbered with a heavy gun (harquebus), fork support, powder horn, lead bullets in a pouch, cloth or paper packing, rammer, slow match, and the means of lighting or re-lighting his match. The matchlock, Fig. 23.16, mechanised this process to some extent. The mechanism seems to have been invented in Germany about 1460–80. A glowing slow match was held in a serpentine fastened to the weapon. A covered pan, usually on the side of the barrel, was filled with priming powder, the powder charge poured into the muzzle with a lead or stone bullet and paper or cloth packing, and the charge was rammed. The slow match must be ignited, placed in the serpentine, and kept alight and glowing until needed. The cover on the priming pan was raised and on the trigger being pressed the glowing tip of the slow match was plunged into the priming powder in the pan, igniting it and, through a connecting hole, the gunpowder in the chamber. Although an improvement it was a very inconvenient system and rates of fire were not high and misfires frequent. Often the weapons would be discharged only once or twice during the course of a battle lasting several hours

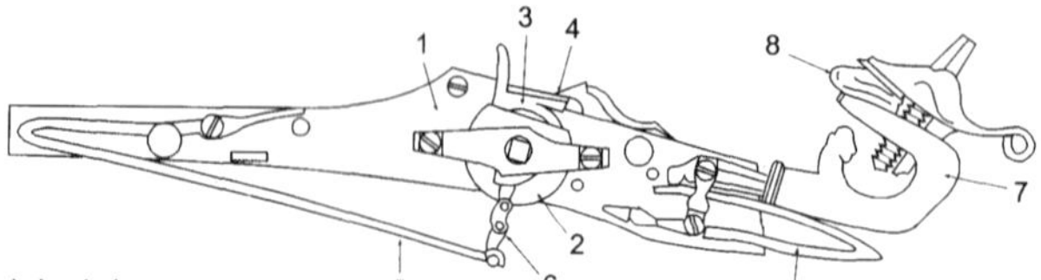

Fig. 23.17 The wheel lock mechanism.

although, of course, they could be fired faster than this. The matchlock attained parity with the crossbow in Spanish armies by 1500, Guilmartin (1974, 147), and replaced the crossbow on Venetian war galleys in 1518. Crossbows were unusual after about 1540. It was superior, not because it was technically better than the crossbow (they had a marginally better range, penetrating power, and rate of fire) but the harquebus was simpler and more rugged. It was used in Europe until about 1700 and in India and Japan until the middle of the nineteenth century.

The great inconvenience of the matchlock was that the slow match had to be maintained during the whole period of a battle. In wet weather this was difficult and at night it gave away the firer's position. Leonardo da Vinci drew a wheel lock mechanism in the *Codex Atlanticus* and a somewhat later drawing of 1505 is found in a Nuremberg manuscript of one Martin Löffelholtz. In 1518 wheel lock guns were banned throughout the Holy Roman Empire but by 1530 they were in common use all over Europe. The wheel lock, Fig. 23.17, eliminated the glowing match. It was a complicated and relatively expensive mechanism, not unlike a modern cigarette lighter, and called for the skills of both locksmith and clockmaker. A spring was wound up, and when released by the trigger it rotated a steel wheel against which a flint or pyrites, held by a serpentine, was pressed by a spring, thus generating a stream of sparks. At the same time the cover from the priming pan was removed by the mechanism and the sparks ignited the primer, which ignited the main charge in the usually way. Wheel lock mechanisms needed to be rewound between firing, for which a key was provided. Losing the key rendered the gun useless and was something that was guarded against most carefully. The mechanism increased the cost of a gun, apparently by two or three times, and for this reason it never completely replaced the matchlock mechanism but it was definitely more convenient and reliable. Various devices were introduced to ensure that it could not fire accidentally and even winding mechanisms were introduced to eliminate the need for a winding key. Another improvement was a waterproof priming pan to keep the priming powder dry even in bad weather. It was much used in sporting guns but was not generally taken up by the infantry.

Unlike the matchlock with its smouldering fuse the wheel lock pistol had the advantage that once wound up it could be retained in its holster until required. It was found ideal for firing by the cavalry from horseback and it soon came to replace the lance. A cavalryman needed one hand to control his horse and had only one hand free for a weapon. He preferred to be armed with two or more primed pistols in his belt and a sabre to draw when these were fired, rather than with a sabre and lance. Thus the heavy cavalry (knights) having been forced to abandon their heavy armour by the penetrating power of the harquebus re-invented themselves when wheel lock pistols became available. The German cavalry (Reiters) developed a new tactic in which a deep formation was used. The front ranks fired their pistols and then rode to the rear to perform the tedious work of reloading while the second and third ranks fired in their turn. This tactic, called a caracole, required considerable practice but it was often, if not invariably, successful. They were vulnerable while re-loading. There is something slightly comical about two bodies of cavalry performing a caracole by facing each other on horseback, shooting, and then queuing to take and receive another shot. What made this possible, beside their gallantry, was that they had retained some armour, see Fig. 23.18, which gave adequate protection against pistols but was inadequate against the harquebus. Such armour often bears the proof mark where the maker fired a pistol to dent the armour without actual penetration. Jones (1988, 196) points out that at the battles of Moorkerhyde in 1574 and Ivrey in 1590 the winning side abandoned the caracole in favour of a sabre charge. Nevertheless the cavalry could now attack heavy infantry and a defensive wall of pikemen with their pistols, keeping their distance until they had cleared sufficient gaps in the line for a successful sabre charge. In attacking light infantry armed with more powerful harquebus and musket, the speed of the horses could carry them through the danger zone, and once in amongst the enemy they usually had the advantage.

Fig. 23.18 Armour showing dents made by proofing bullets. Left: German cavalry armour, 1630, weight 20.25 kg, (Wallace Collection). Right: English troopers or harquebus armour, *c.* 1650, The Royal Armouries.

The snaphance and flintlock mechanisms, Fig. 23.19, are rather similar to each other and they became very popular because they were reliable and less expensive than the wheel lock mechanism. The earliest written evidence for the snaphance lock comes from Italy and Sweden in 1547, Dolinek and Durdik (1993, 191), and they spread quickly throughout Europe although they were not perfected until about 1610. The main spring was usually on the outside of German, Scandinavian, Russian, and Spanish locks but on the inside of English, Scottish and Florentine locks. The steel against which the flint must strike was sometimes separate but more often it was attached to the cover of the priming pan. On pulling the trigger, the cock swings forward to scrape the flint down the face of the steel, pushing it away. If the steel is attached to the pan cover then it also lifts the cover exposing the priming powder to the stream of sparks. If the steel is not connected to the pan cover then the cover must be

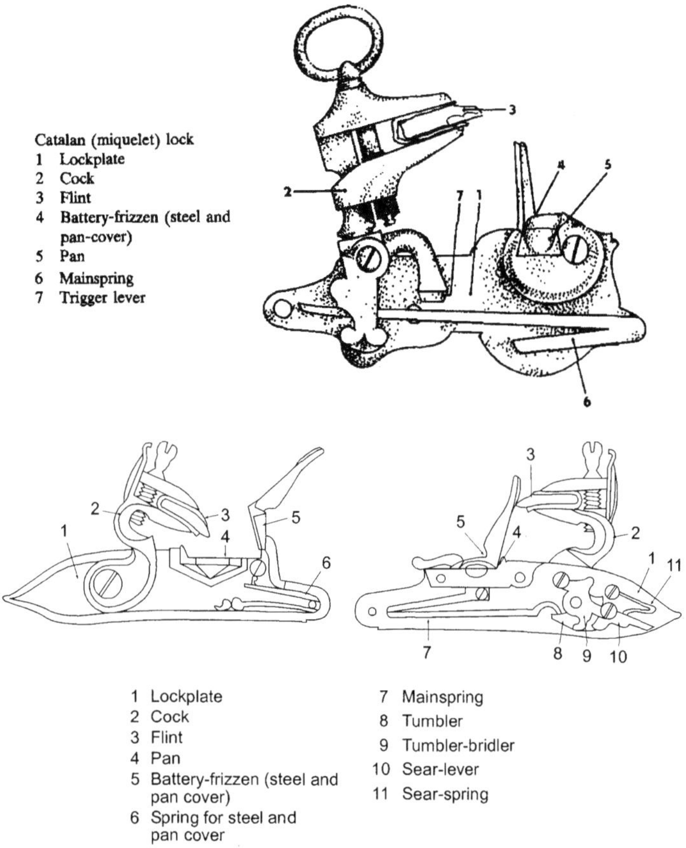

Fig. 23.19 Top: the snaphance. Bottom: the flintlock.

opened by some other mechanism. As with the wheel lock, many devices were used to prevent the mechanism firing accidentally. Two-thirds of Cromwell's New Model Army, created in 1644–45, was equipped with muskets, mainly matchlocks but also many flintlocks or snaphances, and the remaining third were equipped with pikes. Although the cavalry wore breastplate, back plate and helmet, the pike-men were usually without armour.

The armies of the 1690's were provided with large numbers of flintlock muskets and new tactics emerged. The old plug bayonet had blocked the muzzle so it could be used either as a musket or as a pike, but the new socket bayonet, which consisted of a blade attached to the side of the muzzle allowed the weapon to fire with the bayonet attached and by 1700 the socket bayonet was universal.

All infantrymen could function as either heavy or light infantry, and pike-men, the main defence for centuries against heavy cavalry, became obsolete. The pistol and sabre, as we have seen, allowed the horseman to operate either as heavy or light cavalry, making it, for a time, supreme against infantry on open ground, but the socket bayonet restored the balance. Heavy infantry muskets outclassed cavalry pistols in terms of range and penetration, and a mounted sabre attack against a wall of bayonets was as unlikely to succeed as it had been against a wall of pikes; the horses simply turned away from the cold steel and the cavalry were unable to penetrate the line of "pike-men".

The flintlock mechanism proved more reliable than the matchlock. The latter's success rate, according to Jones (1988, 269), was only about 50% but the flintlock ignited the primer and fired the bullet on 60–70% of attempts and later this increased to about 85%. The flintlock also had a greater rate of fire. A musketeer firing a matchlock first filled his pan from a powder horn, opened a small wooden cartridge and emptied the powder into the barrel, took a ball from its pouch and put it in the barrel along with a piece of cloth, took his ramrod and rammed the cloth and ball down on the powder, took a burning match, ensured it was lit, fastened it to the serpentine, took aim, and fired.

The flintlock musketeer had a paper cartridge, Fig. 23.20, containing the ball and the requisite amount of powder. To fire, he bit off the end, retaining the ball in his mouth, used some of the powder to fill the priming pan, poured the rest into the barrel, took the ball from his mouth and put it in the barrel with the paper cartridge, rammed the paper and ball onto the powder, and was then ready to fire. In this way he was able to fire two or three times per minute. Stronger barrels allowed an increase in the charge mass and a 28 gram ball had a muzzle velocity of about 300 m/s, giving it the ability to penetrate 7.5 mm of armour. Although they had a shorter barrel length and bores of less than 19 mm they had a range and power comparable to the older Spanish musket, which required a forked rest. But the flintlock could fire three times faster and was more reliable in its firing. Being a smoothbore it was sufficiently accurate to hit a man at about 100 m but aimed-fire at greater distances was too erratic. Loading a flintlock required less space and thus the line of musketeers could be closer than a line of harquebusiers (0.6 m rather than 1 m) and the faster rate of fire meant

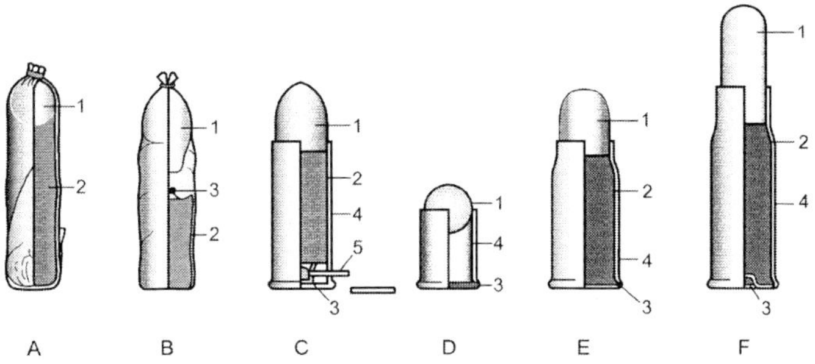

Principal types of cartridges:
A Paper cartridge for muzzle-loaders
B Paper cartridge for Dreyse needle-guns
C Lefaucheux pin-fire cartridge
D Flobert cartridge
E Rim-fire cartridge
F Centre-fire cartridge

1 Projectile
2 Powder charge
3 Priming cap (fulminate compound)
4 Cartridge case
5 Firing pin

Fig. 23.20 Cartridge ammunition.

that, when firing on command one rank at a time, the number ranks could be reduced from 6 to 3. Volley fire, from one or more ranks, and free fire when the enemy was at close range (less than 100 m) were normal.

Increased reliance on guns implied a greater need for the manufacture of gunpowder. Biringuccio (1540, 415) notes that this is a dangerous operation because friction heats the powder so that it easily ignites. He says: "In order to compound it, it was once customary to weigh each of the materials separately and then to proceed to crush them together. Then it occurred to some others to crush each thing by itself [Fig. 23.11], sift it, and then to compound them. Nowadays as the best and quickest method, take the quantity of saltpetre that you wish to put in the work and put it in a kettle with as much water as you think will suffice to dissolve it when it is put on the fire to heat. After it is dissolved, remove and place it in a firm place on the ground, and put a suitable quantity of charcoal, in pieces just as you have it for less trouble, and by stirring mix them with the said dissolved saltpetre. Then have your sulphur finely ground and well sifted and proceed to throw it on top while you stir the charcoal and saltpetre with a stick, as is done with cheese on macaroni. By stirring continually as well as you can, proceed to mix it. Then put the composition thus made to dry out somewhat. To make it fine and to incorporate it you have only to crush the charcoal, and, at the end, having made it fine and impalpable if

Fig. 23.21 Making gunpowder from Thomas Malthus's Practique de la Guerre, Paris, 1650. The British Library, London.

possible, diligently dry it of every trace of moisture that it may contain, and then sift it very well with a fine sieve. Then moistening it again with ordinary water or with some vinegar, force it through a riddle or sieve to granulate it [this is known as corning]. Dry it out again so you can use it for your needs and, when dried, put it in dry wooden vessels to keep it".

The method is illustrated in Fig. 23.21. In the background the constituents are delivered to a set of stamps driven by a water wheel. In the tubs at *A* the saltpetre is mixed with water and is filtered and drained. At *B* the tubs are heated over a fire to dissolve the saltpetre. The man at *C* mixes in the charcoal and sulphur and the man at *D* works a sieve with rollers to force the mixture through the holes to create damp gunpowder grains of a suitable size. The method of using the sieve was described in Biringuccio (1540, 413), who says that "The manner of corning [granulating] all sorts of powder is with a sieve made, with a thick skin of parchment, full of little round holes, into which sieve the powder must be put, while it is dank, and also a little ball, that when you sift, may roll up and down upon the clots of powder, to break them, that it may corn, and run through the holes of the sieve."

Gunpowder manufacture was an important national asset, crucial for the defence of the realm, and it was essential that the ingredients necessary for its manufacture were secure. Saltpetre was the critical ingredient and occurred naturally in the rotting residues of dove houses, stables and so on where animal fluids collect. It formed faster in hot climates and saltpetre was initially imported from the Near East, until demand increased the price to the point where local supplies were sought. England followed France in setting up saltpetre commissioners having the right to enter premises such as stables, cellars, cattle sheds, and sheep pens and to remove all the salt they could find. In 1558 Elizabeth I granted this monopoly to Richard Hills and George and John Evelyn. George Evelyn was the grandfather of the famous seventeenth century diarist. By the end of Elizabeth's reign the monopolist's were abusing their position and they were very unpopular. The monopoly was abolished in 1624 but to ensure an adequate supply an act of Charles I, (17 Car. I. c. 21, 1625), proclaimed that the floors of such buildings likely to yield saltpetre should not be paved or boarded or covered in sand or gravel but were to "lie open with good mellow earth" to encourage the growth of saltpetre. More saltpetre-men were appointed with powers to enter any premises and take away appropriate deposits and constables and other officers were instructed to provide the necessary carts and horses for its transport. Finally, "None shall bribe any saltpetre-man for the sparing of any ground fit to be wrought for saltpetre." The saltpetre-man remained a very unpopular visitor and a cause of much discontent and complaint. Several other schemes for collecting saltpetre were tried but none were successful. For example, an act ordered every householder to save and store all human and as much animal waste fluid as possible, but it proved unworkable and the supply of saltpetre for gunpowder remained a problem, both for Britain and for other countries, until the discovery and importation of rough nitre from the East Indies and from the vast nitre beds of Chile.

Progress in the scientific understanding of ballistics was slow. One of the earliest writers on artillery was Nicolo Tartaglia who in 1537 published his "Nova Scientia" and laid down the important axiom that the maximum range of any piece was attained when it was fired at an elevation of 45 degrees; an observation based no doubt on practical experience. He also invented the gunner's quadrant, Fig. 23.22, which held the field as the ranging instrument for guns until it was superseded by the tangent scale in the eighteenth century. Biringuccio in 1540 also describes the use of the gunner's quadrant but does not mention the angle for maximum range. Maximilian I is also said to have invented the gunner's quadrant. Fig. 23.22 not only illustrates the gunner's quadrant but also shows the difference between direct and indirect fire and demonstrates a practical awareness that the angle of descent is steeper than the angle of ascent, although this effect is greatly exaggerated in the drawings. It also illustrates that increasing the height of the cannon above the target increases its range.

William Eldred in his book "The Gunner's Glasse" 1646, records thirteen words of command to load and fire a cannon and the average rate of fire was 8 shots an hour. Eldred says "One may well make 10 shots in an hour if the pieces be well fortified and strong; but if they be ordinary pieces, then 8 is enough, always provided that after 40 shots you refresh and cool the piece, and let it rest for an hour, for fear lest 80 shots shall break the piece, being not able to endure the force and the heat." This last restriction was important and applies still to modern artillery. After firing about 80 rounds the barrel becomes very hot, and there is a danger that gunpowder may ignite after contact with the hot barrel for a few minutes. Such a pre-ignition would be a disaster if the crew was still loading at the muzzle or if the gun was not trained on its proper target.

Fig. 23.23 illustrates the general proportions and shape of seventeenth century cannon and Table 23.4, based on the recommendations of Thomas Smith in "The Complete Souldier" (1628), gives the principal dimensions and performance of the various types from the large cannon of 8 inches [200 mm] bore to the smallest falconet of 2.25 inches [57 mm] bore; see Hogg (1963, 27). Thomas Smith also notes that: "A bullet of iron to the like bullet of marble stone is in the proportion as 15 to 34. A bullet of lead to the like bullet

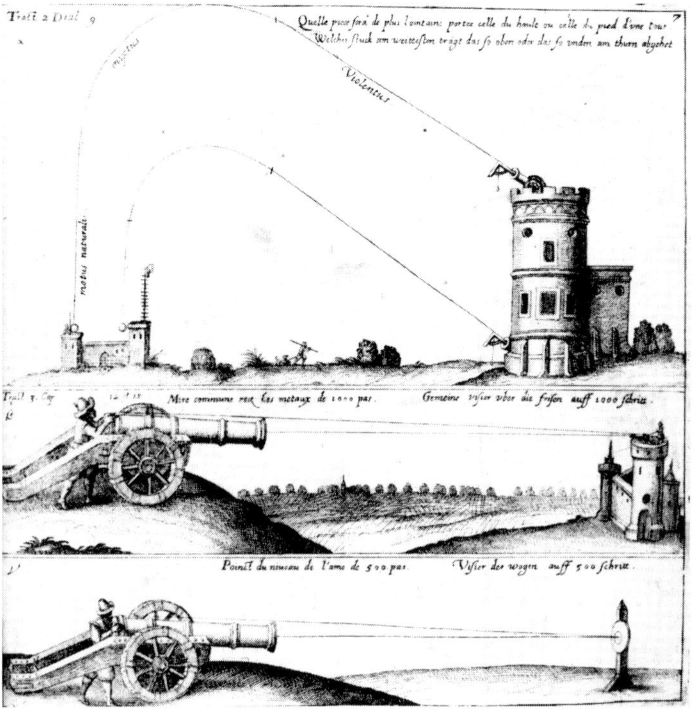

Fig. 23.22 Laying guns, Norton, R. "The Gunner", 1628, British Museum. Top: indirect fire mode using a gunner's quadrant. Bottom: direct fire mode.

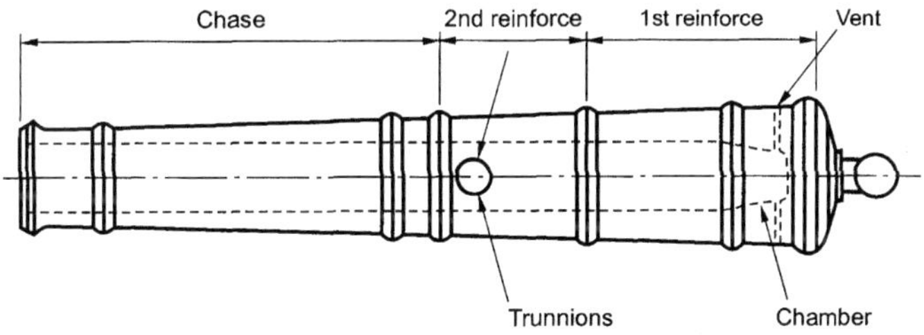

Fig. 23.23 A typical cannon. After Crossland (1986, 153).

of marble stone is in the proportion as 4 to 1. A bullet of lead to the like bullet of iron is in the proportion as 28 to 19. Powder is to Serpentine in the proportion of two to three." He recommends that guns should be proved with a weight of powder equal to that of the iron bullet; then with 25% more powder and finally with 50% more powder.

Table 23.4 Dimensions of Great Ordinance. Based on Thomas Smith, "The Complete Souldier" (1628).

	Cannon	*Cannon Serpentine*	*Demi-Cannon Eldest*	*Demi-Cannon Ord.*	*Demi-Cannon*	*Culverin*	*Demi-Culverin*	*Saker*	*Falcon*	*Falconet*
Calibre Inches (cm)	8 (20)	7.5 (19)	6.75 (17)	6.5 (16.5)	6 (15.2)	5.5 (14)	4.5 (11.4)	3.75 (9.5)	2.75 (7)	2.25 (5.7)
Shot Diameter Inches	7.5 (19)	7.25 (18.4)	6.5 (16.5)	6.25 (15.9)	5.75 (14.6)	5.25 (13.3)	4.25 (10.8)	3.5 (8.9)	2.5 (6.4)	2 (5.1)
Shot Weight lbs (kg)	64 (29)	52 (23.6)	36.625 (16.6)	32 (14.5)	24.5 (11.1)	19 (8.6)	11.75 (5.3)	6 (2.7)	2.333 (1.1)	1.125 (0.51)
Charge Weight lbs (kg)*	32 (14.5)	26 (11.8)	20 (9.1)	18 (8.2)	16 (7.3)	15 (6.8)	9 (4.1)	5 (2.3)	2.25 (1.0)	1.5 (0.68)
Weight of Piece lbs (kg)	8000 (3628)	7000 (3175)	6000 (2721)	5600 (2540)	5000 (2267)	4600 (2086)	3000 (1360)	1900 (862)	750 (340)	400 (181)
Length of Piece feet (m)	12 (3.7)	11.5 (3.5)	11.25 (3.4)	10.5 (3.2)	11 (3.3)	13.25 (4.0)	11 (3.3)	9.5 (2.9)	7 (2.1)	6 (1.8)
Metal Thickness at Touchhole inches (cm)	8 (20.3)	7.5 (19.1)	6.75 (17.1)	6.5 (16.5)	6 (15.2)	5.5 (14.0)	4.5 (11.4)	3.75 (9.5)	2.75 (7)	2.25 (5.7)
Metal Thickness at Trunnions inches (cm)	6.75 (17.1)	5.667 (14.4)	5.25 (13.3)	5 (12.7)	4.75 (12.1)	4.5 (11.4)	4 (10.2)	3.25 (8.3)	2.333 (5.9)	2 (5.1)

Length of Carriage Planks feet (m)	16.75 (5.1)	16.25 (5.0)	15.5 (4.7)	15.25 (4.7)	16 (4.9)	18.25 (5.6)	16.125 (4.9)	14 (4.3)	10.25 (3.1)	8.667 (2.6)
Men to Draw Each Piece	90	80	65	60	65	50	36	24	16	10
Horses to Draw Each Piece	16	14	11	10	9	8	7	5	3	2
Yokes of Oxen to Draw Each Piece	17	15	12	11	10	9	8	6	3	2
Thickness of Metal at Neck End inches (cm)	4 (10.2)	3.75 (9.5)	3.375 (8.6)	3.125 (7.9)	3.1 (7.9)	3 (7.6)	2.625 (6.7)	2.5 (6.4)	2 (5.1)	1.75 (4.4)
Point Blank Range Yards (m)	300 (274)	340 (311)	370 (338)	350 (320)	340 (311)	420 (384)	380 (347)	300 (274)	260 (238)	220 (201)
Maximum Range Yards (m)	1500 (1372)	1600 (1463)	1800 (1646)	1700 (1554)	1600 (1463)	2100 (1920)	1800 (1646)	1500 (1372)	1200 (1097)	1000 (914)

* Corne powder

Using the theory set out in Chapter Twenty four, the performance of the guns specified by Thomas Smith in Table 23.4 may be estimated. The results are shown in Table 23.5. The muzzle velocity was estimated from the maximum range, ball diameter, and ball weight specified by Thomas Smith and by assuming a drag coefficient of 0.3. The penetration of wrought iron armour and oak were calculated from the muzzle velocity, diameter and mass of the balls and are valid, therefore, only for short range, direct fire. Some interesting results emerge. In particular, the weapon with the greatest range, muzzle velocity, and penetration appears to be the culverin. This is attained by the use of a long barrel, about 30 calibres long compared to about 15 calibres of a cannon, and is all the more remarkable because compared to the cannon it fires a smaller diameter ball, uses half the gunpowder, is half the weight of a cannon, and requires fewer men, horses or oxen to pull it.

The gunner's quadrant had to be calibrated for the particular gun and gunner to be of use. The gun was usually elevated with levers beneath the breech but in 1574 John Skinner an English master gunner introduced an elevation screw to give finer control. Aiming was by looking along the length of the barrel although Biringuccio in 1540 recommended a piece of wax attached to the muzzle as an aiming mark.

The possibility of a high and continuous rate of fire was one that attracted inventors from an early time, particularly in relation to small arms. They usually took the form of several pre-loaded barrels that were revolved manually to bring each one into line with the firing mechanism. Such weapons were cumbersome and often dangerous to the user. They first appear in the sixteenth century and are known as the pepperbox design. They were originally fired by a flintlock mechanism but revolvers were not really successful until percussion caps were adopted.

An early example is by James Puckle who took out his English patent for a revolving gun in 1718. He described it as a portable gun called a 'Defence', for bridges, breaches, lines, and passes, ships boats, houses, and other places. He says "it discharges soe often and soe many bullets, and can be soe quickly loaden as renders it next to impossible to carry any ship by boarding". It was mounted on a tripod but was a simple revolver with a pre-loaded cylinder containing nine rounds. The cylinder had to be unlocked and turned manually to bring up the next round. In trials he managed to demon-

Table 23.5 Estimated performance of Thomas Smith's Great Ordinance (Table 23.4).

	Muzzle Velocity (m/s)	*Penetration of Wrought Iron Armour (cm)*	*Penetration of Oak (cm)*
Cannon	128	18.5	35.3
Cannon Serpentine	135	17.9	33.6
Demi-Cannon Eldest	149	18.2	34.8
Demi-Cannon Ord	145	16.6	31.2
Demi-Cannon	141	14.5	27.0
Culverin	175	18.6	36.0
Demi-Culverin	158	14.4	28.5
Saker	148	10.0	19.2
Falcon	133	6.6	12.6
Falconet	127	4.6	8.5

strate a rate of fire of 63 rounds (7 sets of chambers) in 7 minutes, which was a very creditable performance in 1718. He seems to have designed his gun to fire round bullets against Christians but square bullets against the Turks! His own description of its parts (Fig. 23.24) is: *1* the barrel of the gun, *2* the set of chambers charged put on ready for firing, *3* the screw upon which every set of chambers play off and on, *4* a set of chambers ready charged to be slopped on when the first set are pulled off to be recharged, *5–12* refer to the tripod, *13* a charge of 20 square bullets, *14* a single bullet, *15* the front of the chambers of a gun for a boat, *16* the plate of the chambers for a gun for a ship shooting square bullets against Turks, *17* for round bullets against Christians, *18* a single square chamber, *19* a single round chamber, *20* a single bullet for a boat, *21* the mould for casting single bullets.

Puckle tried to form a company to develop his invention during the South Sea bubble period of 1720, but satirist hinted that the gun would only wound its shareholders. Puckle died in 1724 and is now remembered mainly as the author of "The Club", a dialogue between father and son. Real progress with revolvers had to wait until the nineteenth century. In 1818 the American inventors Elisha Collier and Artemus Wheeler developed a much improved hand rotated flintlock revolver and in 1836 Samuel Colt patented his percussion revolver.

The science of gunnery improved greatly when Benjamin Robins,

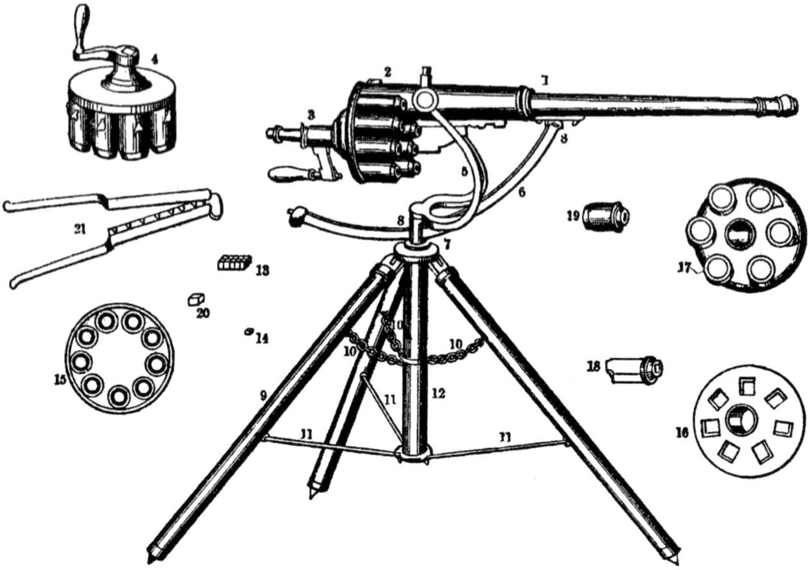

Fig. 23.24 Puckle's tripod mounted revolver of 1718. Knight (1884, 850).

a mathematician and engineer published his "New Principles of Gunnery" in 1742 (re-printed in 1972). His three main innovations were: the ballistic pendulum, Fig. 23.25 (left) by which he was able to measure a bullet's velocity; a theoretical method of calculating its velocity at the muzzle; and measurements showing how the bullet is retarded by atmospheric drag.

His ballistic pendulum was a block of wood backed by a heavy metal plate and suspended from a pivot which, when hit by a pistol ball swung through an arc (measured by the plate pulling a piece of cloth through a sprung elbow) from which the velocity, momentum, and kinetic energy of the ball may be calculated. The ballistic pendulum remained in use throughout the nineteenth century and even into the twentieth century. His method of calculating the striking velocity of the bullet can be expressed as

$$v = \frac{ML}{m}\sqrt{\frac{g}{R}} \qquad 23.1$$

M is the mass of the pendulum, m the mass of the bullet, L is the length of ribbon pulled through the spring, g is local gravity, and R is the radius of the pendulum from the pivot.

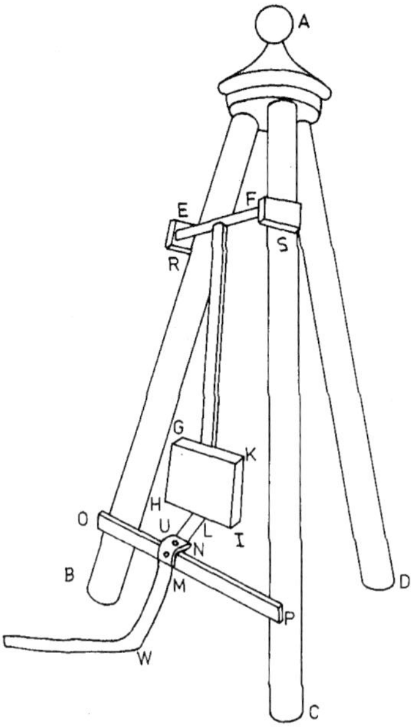

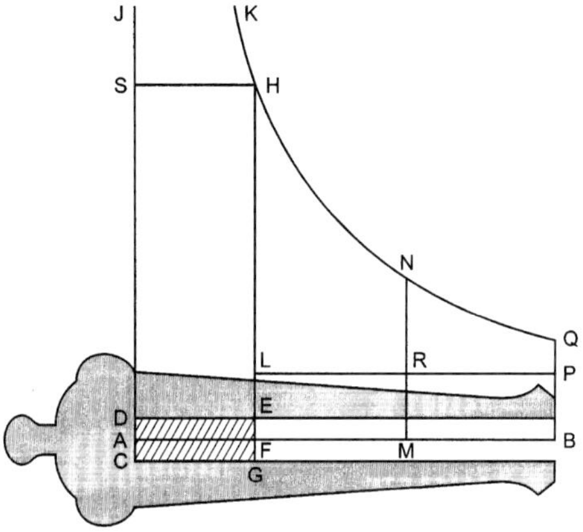

Fig. 23.25 Left: Benjamin Robins ballistic pendulum, 1742, by which he was able to measure the bullet velocity. Right: Robin's assumed pressure distribution in a gun barrel from which he calculated the bullet velocity. Robins (1972, 25, 17).

Robins also had a good understanding of how the gas pressure and the force on the bullet varied as it moved down the barrel, Fig. 23.25 (right). He measured the pressure rise of various amounts of gunpowder when fired into a large evacuated vessel and estimated that when fired in a gun the pressure rise would be 1,000 times atmospheric pressure (101 MPa). He assumes that all the gunpowder is burnt before the bullet moves and this gives the ordinate *FH* in the figure. He also assumed the pressure was inversely proportional to distance from the breech, which is equivalent to assuming the expansion process is isothermal, and thus he was able to determine the force accelerating the bullet down the barrel and by integration he determined the work done and the muzzle velocity. His method is equivalent to the equation

$$v = \sqrt{\frac{2p_1V_1}{m}\ln\left(\frac{V_2}{V_1}\right)} \qquad 23.2$$

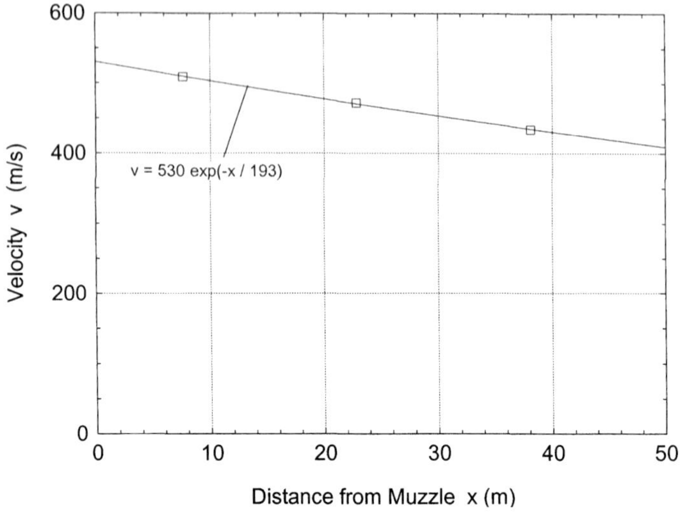

Fig. 23.26 Loss of velocity with distance for a musket bullet. Measured by Robins using his ballistic pendulum.

v is the muzzle velocity, p_1 is the initial pressure (101 MPa), V_1 is the initial volume between the rear of the bullet and the breech, V_2 is the volume between the muzzle and the breech, and m is the mass of the bullet. In his trials he used a 0.75 inch (19 mm) bore musket, the length *AF* was 2.625 inch (66.7 mm) and the length *AB* was 45 inch (1143 mm). His spherical lead bullet was 41.67 grams and the charge mass was half the bullet mass, 20.8 grams. From Equ. 24.2 the muzzle velocity is 507.8 m/s (1666 ft/s), which compares favourably with Robins' own calculation of 1668 ft/s. The velocity he measured when he fired this musket into the ballistic pendulum was 1642 ft/s. He was the first to have any real idea of the pressure within a gun or the velocity of its bullet.

Until this time the idea that a gas as insubstantial as air could retard a heavy lead ball in its flight was thought impossible. Robins did not believe this and using his pendulum he determined how the velocity decreased as range increased, Fig. 23.26, and derived formulae for the motion of a projectile taking account of air resistance and gravity; all of which he presented in papers to the Royal Society. For direct fire at short-range the velocity falls exponentially with distance, see Equ. 38 in chapter 24, and thus Robins' measurements are equivalent to a muzzle velocity of 530 m/s and a drag coefficient of 1.27. Wind tunnel measurements of the drag coefficient of spheres at this velocity (a Mach number of about 1.4) show that the drag coefficient should be 0.97. Robins' measurements, therefore, are quite respectable considering the equipment he was using.

Robins also investigated rockets and wrote a paper to the Royal Society in May 1749 on "Rockets and the Heights to which they Ascend", and the following year further work was presented describing some experiments he had made with others on the flight of rockets. He died in 1751 while working on military fortification projects in India.

Rockets are usually considered as modern weapons but they are surprisingly old. The Chinese, as early as the tenth century onwards used rocket-propelled fire-arrows, Needham et al (1986), and further reports come from 1232 when the Chinese employed rockets against Tartars. By 1250 the Arabs had introduced them into Europe and Timur, the Turkish conqueror, appears to have used rockets against Delhi in 1399. Rocket-launched explosive grenades existed by the thirteenth century and rockets were used at Chioggia in 1380 during a war between Genoa and Venice. Biringuccio (1540, 433) illustrates large fire-tubes, Fig. 23.27 (top) and smaller fire-tubes attached to the end of lances, Fig. 23.27 (bottom), and describes their manufacture. Biringuccio says they are made in the form of rockets or squibs, 30 cm or more long, and are filled with coarse gunpowder, pieces of Grecian pitch, sulphur, grains of common salt, iron filings, crushed glass, crystalline arsenic, and the like. They are fitted with a fuse, and fastened firmly to the end of a pike or lance with the fire-emitting end facing the enemy. When ignited they emitted "hot tongues of fire about two braccia [120 cm] long, full of explosions and horror". They were used on land by the infantry to frighten the horses of attacking cavalry, and they were used at sea to ignite enemy ships and rigging.

From fire tubes it was but a short step to rocket spears. Probably the oldest surviving self-propelled missile is the rocket spear (fire arrow) in the Royal Army Museum at Stockholm, Fig. 23.28. It was made in 1564. Renström (1992, 1) describes the device in some detail and in addition to X-ray studies and chemical analysis he has made theoretical studies to determine the weapon's performance. He also made a modern reproduction that was successfully demonstrated at the 1992 International Ballistics Symposium in Stockholm, where it

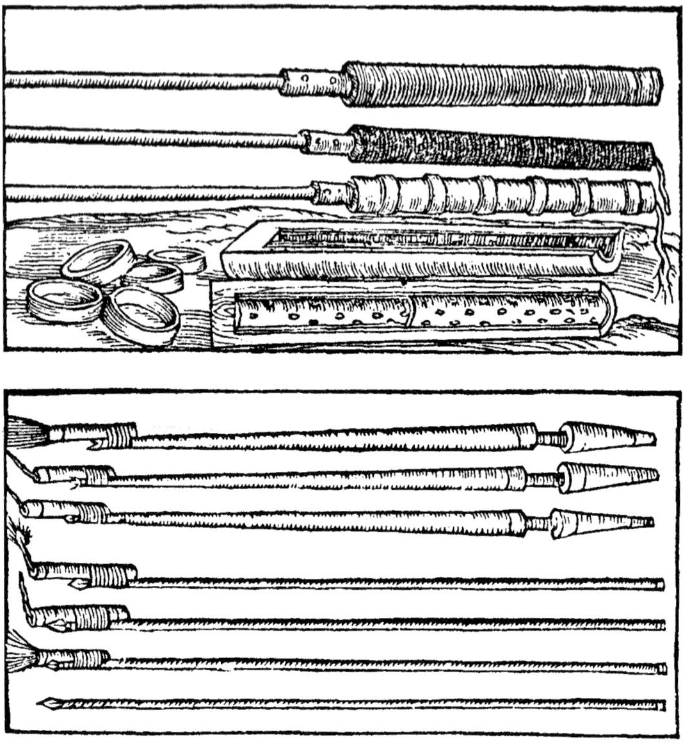

Fig. 23.27 Top: large fire-tubes reinforced with wire and bands, and a lined wooden tube serving as a light gun. Bottom: lances and pikes with small fire tubes attached. Biringuccio (1540, 427, 433).

attained the predicted range and performance. It was probably the first gun-launched, rocket-assisted projectile. Weighing 7 kg, it could travel up to 2,100 m carrying an incendiary warhead of 3.5 kg.

Something is known of the designer, despite the secrecy that surrounded the weapon. In 1544 Gustav I employed Gilius Packet of Danzig, as a specialist in casting cannons. Twenty years later, in 1564, he was contracted to improve projectiles and it is almost certain that Gilius Packet, or one of his close colleagues, made the fire arrow now in the Stockholm Army Museum. The fire arrow was certainly fired in battle. It appears that Erik XIV, eldest son of Gustav I, used them during his siege of Varberg in southern Sweden in 1565 when some 24 incendiary arrows were fired into the town and a further 9 were fired against the castle. Probably, they were

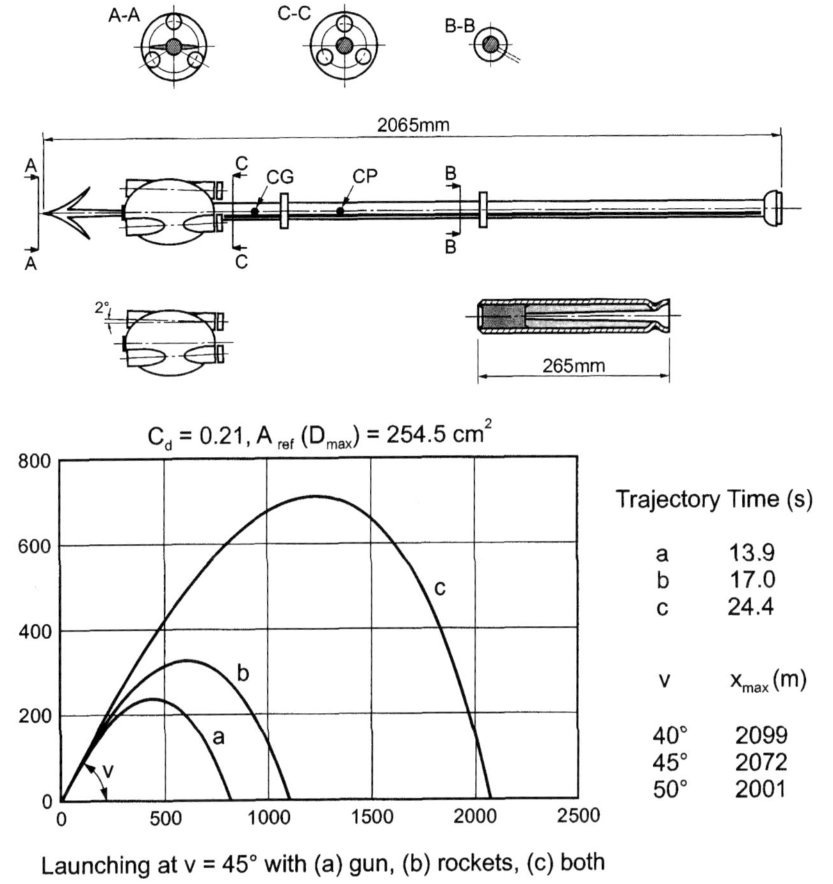

Fig. 23.28 Top: Swedish fire arrow, 1564, with shoulders for a sabot, three rocket motors having converging-diverging nozzles, and an incendiary warhead. Bottom: Flight characteristics computed by Renström (1992).

launched from a cannon by means of a sabot (two half cylinders) that wrapped around the shaft between the shoulders, and fell away during flight. The three rocket motors near the point then ignited on a fuse to prolong the spear's flight. The rocket motors were made of compressed gunpowder and, surprisingly, X-ray photographs have revealed that the gunpowder was moulded to the shape of convergent-divergent nozzles, exactly like modern rockets. The motors were also pointed slightly inwards towards shaft to improve stability. Careful measurement on the original shows that the centre of pressure is well behind the centre of gravity, Fig. 23.28, so the device was stable in flight. It appears, however, that the secret of such weapons died with its designer.

In 1650 Kasimir Simineowicz, a Polish artillery officer, proposed a rocket that was to have three powder filled tubes, but unlike the Swedish fire arrow where the three motors fired at the same time, those of Simineowicz were arranged in stages so that the first tube started the flight, then as this burned out it ignited the second and so on. He also suggested stabilising the flight by means of fins instead of a stick. There is, however, no record of their use.

Rockets appear frequently to have been used in celebrations and firework displays and at the an exhibition of fireworks in Green Park, probably to celebrate the treaty of Aachen in 1748, Benjamin Robins, as mentioned above, determined that three of the rockets ascended to a height of 550 yards [503 m], and one to a height of 600 yards [548 m]. He calculated that on level ground they might be seen from a distance of 50 miles [80.5 km] and could be useful for military signalling, Ellicott (1750, 578). Robins had more rockets made at the Royal Laboratory, Woolwich, and other gentlemen had rockets made to their own design. About twelve rockets were fired in September 1749. Most did not rise above 400 yards, one to about 500 yards and one to about 600 yards. They were seen by lookouts at distances of 38 miles, and 40 miles. Further tests were ordered using larger rockets. Mr Samuel Da Costa, of Devonshire Square, "a gentleman of an extraordinary Genius in Mechanics", Mr Banks, "a Gentleman who had for many Years practis'd making Rockets", and two other persons, agreed to make the rockets varying in diameter from 1.5 inch to 4.5 inches. Some 28 rockets were fired. A 1.5 inch rocket reached 743 yards, a 2.5 inch rose to 880 yards and another to 1071 yards. The 3.5 inch rockets ascended to 1109 yards and 915 yards and Da Costa's 4 inch rocket ascended to 1190 yards. Another 4 inch rocket rose to 700 yards, turned, and nearly reached the ground before it went out. The experimenters concluded that the larger rockets were too expensive and uncertain, and that the best size was 2.5 to 3.5 inches diameter. The work, however, was not taken further because Da Costa died shortly before his rockets were fired and Robins, the principal experimenter, died in India the following year.

Rockets were well known in India as incendiaries. Hogg (1992, 106–7) says that in 1657 a rocket falling near the magazine destroyed

Fig. 23.29 Royal Horse Artillery training with Congreve's rocket in 1835.

the fort of Bitar in India, but the British did not realise their military potential until 1792 when besieged at Seringapatam. It was reported that a single rocket killed three men and wounded four others. These rockets weighed 4.5 kg, were attached to a 3 m stick and had an effective range of about 1600 m. A second engagement at Seringapatam, in 1799, where more rockets were fired, prompted the British Army to ask the Royal Laboratory at Woolwich for rockets, but none were available. William Congreve, the son of the Comptroller at Woolwich, and employed there in his own right, set about developing one. The rocket's main attraction was that it was very light and exerted no recoil force on its launcher, so it could be used from small boats or by individuals to discharge projectiles to the same range as artillery, and without the problems of hauling heavy ordnance. Congreve's rocket consisted of a tubular steel casing about 60 cm long, filled with gunpowder, on the end of a 3 m stick, Fig. 23.29. It was launched from a light tube supported on two legs and it had a range as great as current field guns. It was eventually made in ten sizes and Congreve also developed shrapnel and incendiary warheads, to improve its effectiveness.

Hogg (1992, 107) says that in October 1806, during the Napoleonic Wars, 18 British ships fired 2,000 rockets into Boulogne, setting fire to the town, and in 1807 the same was done at Copenhagen and at Walcheren. At Leipzig, in 1813 the Rocket Brigade sent over from Britain helped turn the tide of battle in the Allies' favour (Prussia, Poland, Austria, and Sweden). The rocket's chief effect, however, was that it terrorised the enemy without inflicting very much actual damage. Eventually Congreve's rocket was superseded. Hale, an American engineer developed an improved rocket that was stabilised by fins in the rocket jet that caused it to spin in flight, thus eliminating the need for a long stick. Hales, rocket was first utilized by the Americans in the Mexican war (1846–8), but was then dropped in favour of conventional artillery. The British employed it for a short time on their Abyssinian campaign in 1867, but it was little used subsequently. After 1870 rapid improvements in artillery soon outstripped the performance of rockets and it was not until the Second World War that rockets were developed to outperform artillery.

Developments in firearms in the eighteenth century were modest. Variations of the "Brown Bess" musket served the British Army from about 1720 to about 1840. The Short Land Pattern was 106.7 cm barrel length and was adopted by the infantry in 1768 and the cheaper India Pattern musket of 0.75 inch (19 mm) bore replaced this from 1793. New ideas began to emerge towards the end of the century and through the nineteenth century and these included the introduction of percussion caps, rifled barrels, the re-introduction of breech loading, and automatic fire.

Percussion, by striking ordinary gunpowder between two hard surfaces, ignites it but the rate of burning and susceptibility to misfire is the same as when ignited by sparks. Fulminates, on the other hand, detonate when struck but the rate of burning, several kilometres per second, is far too fast for combustion in the main gun chamber. They were, however, suitable for use in small quantities as priming powders. Fulminates were discovered in the eighteenth century and many accidents occurred in their preparation, most notably in 1788 to the French chemist Berthollet, who gave them up. In England, Howard manufactured fulminate of mercury mixed with saltpetre to produce a very sensitive powder, known as Howard's powder, but their first practical application to guns was in 1807 when the Reverend Alexander Forsyth, a Scottish clergyman, took out a patent that used fulminates as a detonator to ignite gunpowder in firearms, Fig. 23.30 (top). He arranged a "scent bottle" containing fulminate as a part of the lock so that when rotated it deposited the correct amount into the vent and rotating back returned it to where the fulminate could be struck by a hammer and firing pin. He also arranged that the priming pan was not open but closed so that none of the gases could escape except by the vent hole to

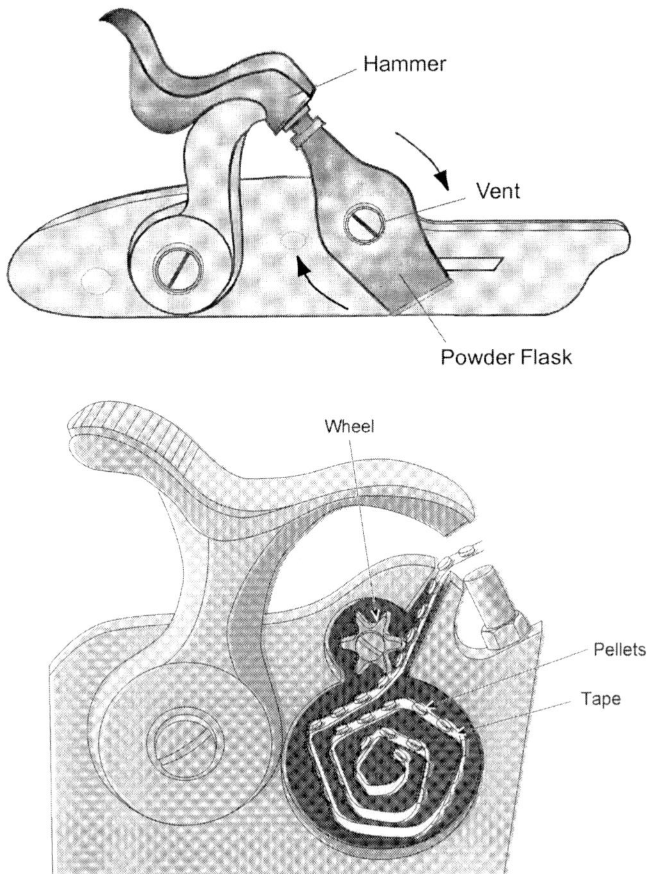

Fig. 23.30 Top: Forsyth's "scent bottle" percussion lock, 1807. Bottom: Maynard's taped percussion cap, *c.* 1850.

the main chamber. His design was not particularly successful and was not taken up, but others soon improved on it. These used deposits of fulminate fitted into various metal or paper tubes, discs, or pills, later made up as priming-tape on a moveable strip of paper. Maynard's primer tape, Fig. 23.30 (bottom) appeared in about 1850. However, a copper percussion cap, fitted over the end of a nipple, proved the most advantageous system. Compared to a flintlock it was much simpler, having no jaws, flint, steel or pan. Several English gun-makers claimed its invention and it spread to the Continent, remaining in use for about 50 years until the end of the muzzle-loading era. Conversion of flintlocks to percussion was relatively easy and percussion ignition and the percussion cap quickly replaced the flintlock mechanism. The British Pattern percussion musket of 1839 was a transitional type, made up of parts intended for flintlocks. Most armies changed to percussion ignition around 1840.

The improved stability and accuracy of projectiles when spinning in flight was well known to archers and the designers of cross-bows frequently arranged for the bolt to spin, sometimes by means of a spiral groove that rotated the bolt as it left the bow. Rifled guns were first used in the sixteenth century in Germany but did not prove successful, probably because the rifled grooves fouled quickly and made it difficult to load and ram home a bullet from the muzzle. Greener (1910, 623) says that the Landgraf of Hesse had a troop of riflemen in 1631, and ten years later Maximilian of Bavaria had several troops armed with rifled harquebus. Greater accuracy and range were the advantages of rifled muskets but their great disadvantage was that they fouled easily which made them difficult and slow to load. They were preferred as sporting or hunting guns rather than as weapons of war where rate of fire rather than accuracy was important. Nevertheless, interest remained and in 1763, the Danish army introduced a 27 mm rifled musket. In the American War of Independence 1775–83, the hunters and frontiersmen who rallied to the cause naturally used their rifled hunting guns and quickly gained a reputation for marksmanship and for their small bore, 11.2 mm Pennsylvania or Kentucky rifle. In France, Louis XIII had armed his bodyguard with rifles and later two men from each light cavalry regiment was similarly equipped and were afterwards formed into a regiment of carabineers. Napoleon, who knew a thing or two about guns and war, withdrew them.

Rifled barrels, however, were too good to be discarded. The "Saturday Review" of 4th July 1885, carried an interesting and informative article summarising their history. Rifle Brigades, it said, were the crack shots of the army and a few riflemen were attached to most regiments. In 1690 a troop of Life Guards was supplied with eight rifled carbines, in 1800 the 60th Rifles were armed with the seven-grooved Baker rifle of 15.5 mm bore. This rifle could make it dangerous for a "head and shoulders" at 250 m, whereas the smooth bore was so erratic it was little use trying to hit a single man at more than 100 m. However, the smooth bore loaded easily whereas the seven-grooved Baker rifle was troublesome and after a few rounds it was very difficult, having become partially blocked.

Greener (1910, 629) says that with the Baker, mallets were issued to hammer the bullets into the muzzles. By 1836 it was replaced by the larger 17.9 mm two-grooved Brunswick rifle, firing a belted bullet, which loaded easier and shot a little better than the Baker, but was not a great improvement. It fouled easily and the fumbling in cold weather to get the belt of the bullet into the groove at the muzzle was a great drawback. Neither of these could attain long-range rifle fire (over 700 m) and the whole of the combat was normally within easy view of the commanders of both sides and consequently they were never taken by surprise fire.

The nineteenth century was a period of great invention and development in firearms and the era of the muzzleloader and the spherical bullet was coming to an end. The problem of loading a rifled barrel was partially solved in 1826 when Delvigne devised a rifle with a powder chamber of smaller diameter. The ball was loaded easily down the rifled barrel and came to rest against the chamber. A few sharp taps with the ramrod deformed the lead bullet, expanding it into the rifle grooves. However, the deformed bullet had an erratic flight and much of the accuracy of the rifling was lost. In 1835 W. Greener designed an oval bullet, Fig. 23.31 (left and centre), one and a half diameters in length, with a flat end and a hole into which fitted a tapered plug. The bullet's diameter was small enough to allow easy loading and it could be loaded either way. Upon firing the pressure forced the tapered plug into the hole, expanding the bullet so that it filled up the rifled grooves in the barrel. In trials it surpassed expectations. The bullets were fired into sand and on recovery were found to have perfect rifling marks, but the design was rejected for military use on the grounds that it was a two-piece bullet, one of which may get lost. It was, however, the first self-expanding bullet.

The final solution came in 1847 when Captain Minié of the School of Vincennes invented an elongated bullet, which went easily down the barrel when loading, but was made with a flexible wall that expanded into the rifle grooves under the action of gas pressure, Fig. 23.31 (right) A British "Minié" rifle was introduced in 1851, and the "Enfield rifle" pattern in 1853, improved the barrel and the ammunition but retained the "Minié" principle. In 1855 the Enfield rifle replaced the smooth bore Brunswick and with the introduction of a smaller bore of 14.7 mm the range increased.

In 1857 Sir Joseph (then Mr) Whitworth pioneered a 0.45 inch

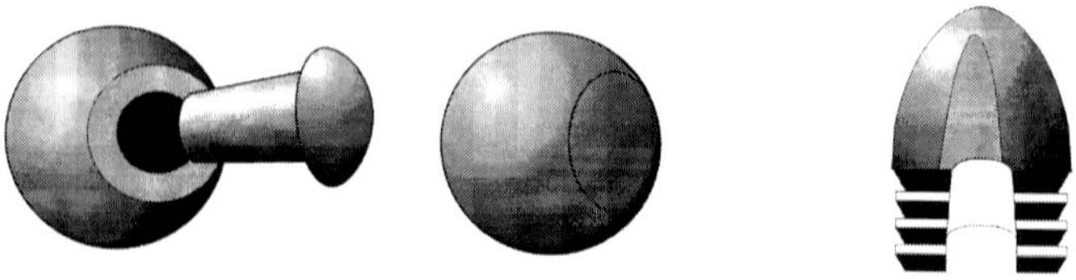

Fig. 23.31 Left and centre: Greener's expanding bullet (1826). Right: Minié's bullet (1847).

(11.4 mm) rapid twist muzzleloader having a hexagonal bore and firing hardened hexagonal bullets, Fig. 23.32. It was made of cast steel, compressed while still in a fluid state which greatly improved the tensile strength. The barrel was strengthened by band *a*, the central and rear part of which is cut away to form a groove in which the breechblock, *c*, slides. The faces of this opening have a series of parallel grooves, *d*, adapted to receive a series of teeth *e* on the breechblock by which it is held for firing. These grooves are slightly inclined to the axis of the bore (like a screw thread) so that as the breechblock is moved into position, by movement of the lever *f*, it pushes against the breech face and the seal *n*. The cartridge chamber *k* is of larger diameter than the bore and tapers to facilitate entry and removal of the cartridge. It was also made as a 2.72 inch (69 mm) calibre field gun from a solid steel ingot without separate reinforcing bands. This was about 2 m long, weighed about 0.4 tonnes and fired about 1 kg of charge.

Whitworth's gun was superior to the Enfield. It penetrated 35 planks whereas the Enfield penetrated only 12 planks. At 800 yards the penetration was 17 planks to the Enfield's 4 planks. It never fouled so as to render loading difficult. General Hayes, speaking at the Institution of Civil Engineers, said he had fired it with common sporting powder, 100 rounds one day, 60 the next, then 40 and so on, and had left the gun without cleaning for ten days, when it fired as well as it did on the first day. For other guns fouling remained a problem that was removed only by the introduction of smokeless powders in the 1880's. Whitworth's gun, however, cost more than the Enfield to produce and the army, despite strong representations, did not adopt it.

There had always been an interest in breech loading as opposed to muzzle loading and, as discussed above, many of the earliest cannon were breechloaders.

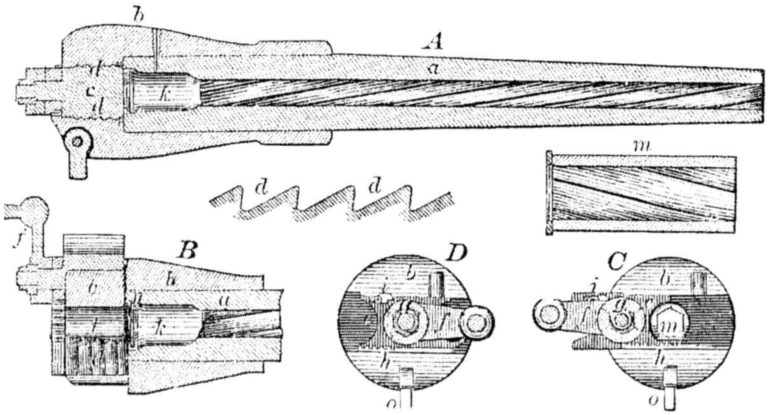

Fig. 23.32 Whitworth's muzzle loader which fired hexagonal hardened bullets. A is a vertical section, *B* a horizontal section, *C* and *D* are end elevations. Knight (1884, 2774).

In the Tower of London there is a sixteenth century breech-loading gun, formerly belonging to Henry VIII, and a breech loading wheel-lock pistol was developed in south Germany in 1540, but these designs were never popular. Probably the most common breech-loading mechanism, and one that flourished from1640 to about 1850 was the so-called "turn off" pistol. In this design the barrel of a pistol could be unscrewed and the gun loaded without having to ram the ball down a barrel. After loading the barrel was screwed back into place. Its advantage was that a rifled barrel could be used and the ball made slightly larger than the bore so that it would seal effectively and engage the rifling when fired. It attained, therefore, a higher muzzle velocity and was more accurate than conventional muzzle-loading pistols. Breech loading, however, did not become common until self-contained cartridges became available and until the problem of gas leakage past the joint faces had been solved. The blowback of gases past the breech face could be very dangerous. As mentioned above, as early as the seventeenth century the powder was measured out beforehand and put into a tubular wooden container or at a later date into a paper wrapping. From about 1800 the paper cartridge contained both the gunpowder and the ball These arrangements speeded up the loading process, yet the shot, powder, and primer had to be loaded separately. Pauly, a Swiss gun maker working in Paris developed one of the first self-contained cartridges in

1812. Ball, gunpowder, and fulminate compound were enclosed in a cardboard cartridge case with a brass base. Production costs, however, were high and Pauly's cartridge did not see much practical application. The task of developing an integrated self-primed self-sealing cartridge occupied the minds of many inventors in the nineteenth century. In the mid 1830's Lefraucheux contrived a pin-fire cartridge. Its paper case had a brass base in which a cap was set and from which a short metal rod protruded laterally. Ignition was caused when the falling hammer struck the rod. This proved very successful but the problem of gas leakage remained and was not solved until the paper case was replaced by an all-metal case. Gas pressure forced the thin metal case outwards against the inside of the barrel creating a near perfect seal. From the 1850's Lefraucheux's pin-fire cartridge became the first, self-sealing, self-primed cartridge to be produced commercially. Their self-sealing properties made such cartridges ideal for breech loading guns and opened the way for numerous other developments such as revolvers, magazine guns, and machine guns. However, metal cartridges introduced a new problem, that of its removal after firing which led to the invention and use of ejector mechanisms.

Cannon in the seventeenth century were very heavy and required large numbers of men and horses for their transport. Thomas Smith in 1628 allowed 90 men, or 16 horses, or 17 yoke of oxen to draw each cannon (see Table 23.4). It was barely possible for such guns to keep up with the infantry let alone the cavalry. Lighter, more manoeuvrable guns were used but, of course, were less effective. Gustav II of Sweden introduced many new tactics during the Thirty Years war. In his linear system he thinned his lines of pikemen and musketeers to five or six to give him a wide front that could not easily be outflanked and the men released by this thinning created an adequate reserve. His opponents typically adopted the more usual large (Swiss) rectangles of pikemen protecting light infantry. These had the capacity for manoeuvre and all round defence but failed against Gustav's relatively thin lines of infantry armed with muskets and harquebuses. Gustav also sought lighter, more manoeuvrable artillery that could be drawn by the infantry. In particular he introduced a light gun having what appeared to be a leather barrel and is usually called a "leather gun". Melchior Wurmbrandt was the designer. Lewendon (1985, 22) has described the gun in detail and

Crossland (1986, 154) considers it to be one of the most innovative ideas of the period. A surviving example is a small one pounder of 5.5 cm bore weighing only 54 kg. It is constructed of thin copper tube tightly wrapped with tinned iron wire, stiffened with four longitudinal wooden lathes, and closely bound with hemp. The lathes were supported at intervals by metal collars, each layer of hemp winding was set in plaster of Paris and the composite barrel was then sheathed in leather. Wire-wound barrels, used to generate a compressive stress at the bore of the gun to counteract the tensile stress generated when firing, have been much used in modern guns. Wurmbrandt's "leather" guns were more mobile than earlier guns and when combined with cavalry they allowed Gustav to develop simple fire and move manoeuvres. After his death his ideas on artillery were ignored and gun development remained stagnant until the war of the Spanish Succession (1702–13), when the Duke of Marlborough's handling of artillery won him many battles.

Six horses per gun were considered reasonable, which meant that a 4.5 inch (11.4 cm) demi-culverin, weighing about 1,360 kg, was the largest practical field gun and this remained the limiting size until horses were replaced by internal combustion engines. Howitzers, essentially lighter than a gun but firing at a reduced muzzle velocity, developed in the eighteenth century and made a useful compromise between the indirect fire of mortar, and the direct fire of cannon. It was a lighter gun that did not sacrifice too much performance. A typical howitzer had a barrel less than half the weight and two-thirds the length of a normal gun barrel and the overall weight of gun and carriage were about half the weight of a normal gun and carriage. The light barrel was made possible not only by its shorter length but also because it used less than half the usual quantity of gunpowder. The result was a weapon with a range of only 1,000 m compared with 1,500–2,000 m of a cannon, but it was quite adequate as a field gun where low weight was important. Even with these improvements field artillery was essentially static and defensive and could not readily be used by attacking troops. The attacking troops, however, had to contend with fire from the defender's field artillery. Field guns were sometimes interspersed at intervals amongst the infantry, if the attackers had a deep formation, and sometimes positioned on the flanks to give a diagonal line of fire along a thinner line of advancing troops. Fire was usually at point blank range and it was thought desirable for the ball to bounce several times before coming to rest. Such bounces reduced the velocity somewhat but the ball remained a fearsome threat and did further damage.

Horse artillery, as opposed to artillery pulled by horses, developed in the eighteenth century as lighter, more powerful guns became available. Frederick the Great was their first exponent. During the Seven Years' War (1756–63) he arranged for his light, yet powerful guns to be mounted on two-wheeled limbers (gun carriages) and for his gunners to be mounted. Ammunition was carried in separate carts. In Germany the Galloper gun developed. This had the trail in the form of shafts for the draught horses, which were unhitched before firing. Various arrangements of horse artillery emerged (the men rode horses, rode on the off-horses, or in special carriages, or in the ammunition wagon). The size of horse team was usually six and this could gallop with a 1,500 kg gun, which became the upper limit in gun design. The British Army formed its first horse artillery in 1793 using the detachment system (all the men rode horses). The remaining artillery was formed into two slower groups: the Foot artillery and the siege train. The latter was often stripped down for transport and the heaviest parts drawn by up to 20 horses.

Artillery ammunition remained spherical although Sir George Cayley, more famous for his experiments with flight, tested a range of more streamlined projectiles in 1804–5, with the idea of improving the range of guns, Fig. 23.33. "My object at the time was to ascertain whether, by obviating as much as possible of the resistance of air, the range of the shot might be so considerably increased as to enable our line-of-battle ships to destroy the French flotilla then collected in the semi-circular basin at Boulogne, and yet keep out of range of the French batteries", quoted by Gibbs-Smith (1962, ch7). For this purpose he was provided with powder and a six-pounder gun. The projectiles, of course, were fired in a smooth bore gun and were provided with fins to stabilise their flight. The pusher *AA* propelled projectile *A* up the bore but it failed because, although made of tough beech wood, it was pushed through the fins by the pressure. Type *B* was fitted with a collar, rather like a modern sabot, and strengthened with an iron ring *BB* but this too failed to improve the range. Type *C* was hollow at the rear, to move the centre of gravity forward, and was fitted with larger fins that were inclined to provide rotation in flight. It was fired with wooden wadding and

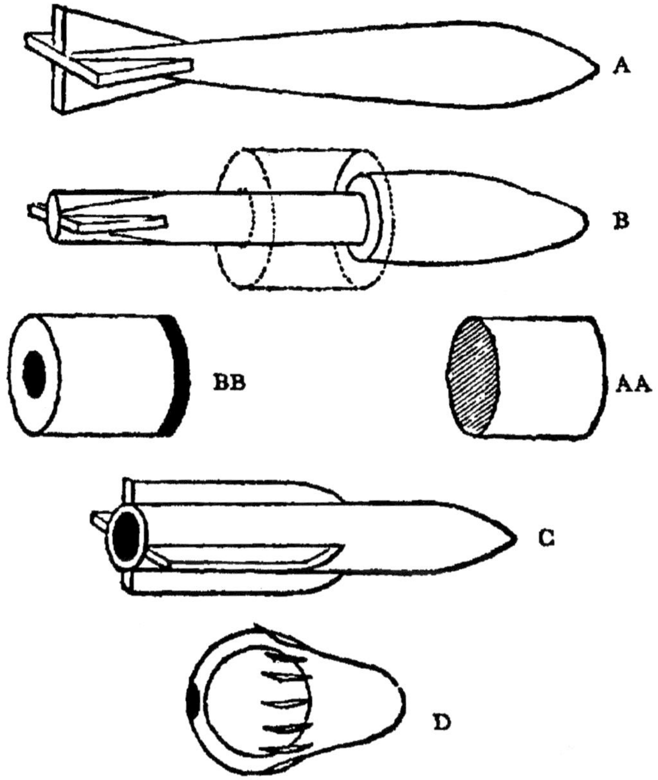

Fig. 23.33 Finned artillery missiles tested by Cayley, 1804–05. After Gibbs-Smith (1962).

proved to be quite successful. It rotated properly in flight and did not vibrate as the long projectiles had. Those that did not vibrate exceeded the range of the round shot by a quarter of a mile (402 m). Type *D* was an ordinary cannon ball, hollowed out, fitted with a solid, pear-shaped nose, and then cut with oblique grooves to provide rotation. Cayley found this to be very steady in flight and the most reliable, but it too increased the range by only a quarter of a mile (402 m). Cayley's experiments were reported to the authorities but no further action was taken. Projectiles of this type were not made until 1846 when the necessary rotation, to ensure stability in flight, was provided by a rifled barrel.

Field artillery did not find an attacking role until early in the nineteenth century. At the siege of San Sebastian in 1813, during the

Peninsular War, British artillery fired over the heads of their own advancing infantry and only ceased firing when they closed for the final assault. This was the first application of what was to become the barrage and the first use of a fire plan in relation to the movement of infantry. Previously artillery simply conducted duels with guns of the enemy or other targets that presented themselves on the battlefield, or fired in defence of their own infantry. Mountainous regions were always difficult for gunners and, as Ramelli has illustrated in Fig. 1.7 of chapter 1, extraordinary measures were often necessary to haul guns over mountain passes. Mountain artillery first came into use during the Peninsular War. Light guns were stripped to their component parts and carried on the backs of mules and this enabled artillery to be placed in otherwise inaccessible places and they played a considerable part in the mountains of Spain.

Breech-loading rifles provided the infantry with weapons having a greater range and accuracy than the smoothbore artillery of the period and by the 1850's it was imperative to develop breech-loading artillery. There were great developments in small arms in the early nineteenth century but artillery, despite Cayley's experiments, remained stagnant. Lieutenant Croly of the British Infantry proposed a rifled breechloader using a shell coated with lead to engage with the rifling, but there was no interest. Several other schemes were proposed throughout Europe but the main objection was the great expense of re-equipping armies already fitted out with serviceable smoothbore guns left over from the Napoleonic wars. The Crimean war changed attitudes and one of the most successful guns of the period was the Armstrong gun (1854), which was taken into service in the British Army and Navy in 1859. Armstrong decided on a wrought iron barrel made with numerous thin grooves of spiral rifling. To get the required strength he made hoops of wrought iron machined precisely, heated, and then shrunk onto the barrel to create a compressive stress at the bore. He repeated this with more hoops near the chamber where pressure was highest. The rear of the barrel had a vertical slot behind which was a hole with a screw thread. A steel block, or vent piece, was dropped into the slot and was clamped tightly to the barrel's rear face by a hollow screw tightened by a handle. The face of the vent piece carried a copper ring, which pressed tightly against the rear face of the barrel to seal the explosion. A vent in the vent-piece allowed ignition of the charge. The shell was coated with a layer of lead and the gunpowder was supplied in a bag and loaded into place through a hole in the clamping screw. On firing the lead coat bit into the rifling grooves and rotated the shell. Armstrong also designed an impact fuse that ignited an explosive filling in the shell when it struck the target.

CHAPTER TWENTY-FOUR

WEAPONS TECHNOLOGY

Introduction

The most important characteristic of weapons is how effective they are in incapacitating an enemy and removing him as a threat. There must be many ways of doing this but the traditional ways, and virtually the only ones that are considered "fair", are by starvation, as in a siege when the enemy will not fight, or if he will fight, by penetrating his vital organs to kill or disable him (usually a man). Thus the most important characteristic of a weapon is its ability to penetrate materials such as human tissue, wood, armour plate, and stonewalls, or how effective they are in preventing such penetration. This chapter, therefore, describes the mechanics of penetration of various materials, for once these are recognized the characteristics of weapons to achieve such penetration are more easily understood. Penetration mechanics is often called terminal ballistics. This is followed by an analysis of flight dynamics, that is, of the path followed by a projectile between launch and hitting its target. This is often called external ballistics and it is important in determining the range at which battles can be fought and the penetration that projectiles can attain when reaching their target. The propulsion systems (bows, ballista, guns, rockets etc) that are used to launch the projectile are then considered. This is sometimes called launch dynamics or internal ballistics. The energy available from a particularly propulsion unit is usually fixed, or has a maximum value, which means that a projectile can have a large mass and a small velocity or a smaller mass and a larger velocity. Most weapon systems have an optimum mass and velocity at which they attain either a maximum range or a maximum penetration at the target so the final section examines this relation between the weapons themselves and their external ballistics and terminal ballistics.

However, before considering these matters in greater detail it is useful to examine the mathematics of a simple battle in which two opposing sides face each other without any geometrical or geographical advantages. This analysis shows that the important characteristics are the number of weapons that can partake, their rate of fire, the probability of hitting the target, and the probability of incapacitation given a hit. Not all these factors are of equal importance. Two simple geometrical arrangements of soldiers, that give a local numerical advantage to one side, are also considered.

Analysis

Analysis of a Simple Battle

The main characteristics of tactics and strategy can be gleaned, in a general way, by mathematical analysis of a simple battle. Mathematical analysis of war and war gaming is particularly useful in training soldiers and very complex computer modelling has been developed, but much can be learned from less complex analysis. In a simple battle between two opposing sides it is often assumed that the side with the larger number wins, other things being equal. Other things, however, are not always equal. One side may be more exposed, may have better armour, may have weapons that are more penetrating or fire at a greater frequency, that is, the probability of incapacitation may be higher for one side than the other.

Rather than discuss the number of men fighting on either side it is more useful to consider the number of weapons being used. Defeat of an enemy depends on the number of weapons in use rather than on the number of men. For example, several men may be required to fire a cannon or a ballista and the loss of a man is unlikely to prevent the weapon from firing. The loss of the weapon, either by direct hit or because a sufficient number of operators are incapacitated, is more significant. Similarly the surrender of a man who drops his weapon and raises his arms reduces the number of weapons in use by one but no one is killed. Nevertheless, the most common way of putting a weapon out of action is to incapacitate its operator.

The probability of incapacitation, in this analysis, is the probability that an enemy weapon is put out of action; it is the product of the probability of hitting the target and the probability of incapacitation given a hit. The rate at which a weapon incapacitates the enemy weapons, k, is the product of its frequency of fire, f, the probability of incapacitation given a hit, $p(I / H)$, and the probability of a hit $p(H)$, thus

$$k = f\, p(I / H)\; p(H) \qquad 24.1$$

The probability of a hit depends on the accuracy of fire and the size and range of the target, whereas the probability of incapacitation given a hit depends largely on the kinetic energy with which a projectile strikes the target and the degree of protection (armour) of the target. The frequency of fire for projectile weapons such as the bow or the gun is fairly obvious, but for shock weapons such as the sword or spear it is not so obvious and must be interpreted as the number of thrusts or slashes per unit time. The rate at which enemy-weapons are incapacitated is the product of the rate at which a single weapon incapacitates the enemy and the number of such weapons in use, thus

$$\frac{dx_1}{dt} = -k_2 x_2 \qquad \frac{dx_2}{dt} = -k_1 x_1 \qquad 24.2$$

The suffixes, 1 and 2, refer to the two sides in the battle, and x is the number of weapons in use. The incapacitation rates, k_1 and k_2, are assumed to be constant although, of course, they may change with increase or decrease in range and with other variables. Differentiating the first equation and substituting from the second eliminates one of the variables to give

$$\frac{d^2 x_1}{dt^2} - m^2 x_1 = 0 \qquad m = \sqrt{k_1 k_2} \qquad 24.3$$

The solution of Equ. 24.3, with the initial conditions $x_1 = X_1$ and $dx_1/dt = -k_2X_2$ *at* $t = 0$ is

$$\frac{x_1}{X_1} = \frac{1-\lambda}{2}\exp(mt) + \frac{1+\lambda}{2}\exp(-mt)$$

$$m = \sqrt{k_1 k_2} \qquad 24.4$$

$$\lambda = \frac{X_2}{X_1}\sqrt{\frac{k_2}{k_1}} = \frac{X_2}{X_1}\sqrt{\frac{f_2\; p_2(I / H)\; p_2(H)}{f_1\; p_1(I / H)\; p_1(H)}}$$

X_1 is the number of weapons in use at $t = 0$ and X_2 is the number of enemy weapons in use at $t = 0$. The ratio λ may be called the lethality ratio; it is the ratio of weapons in use at the start of the battle multiplied by the square root of the incapacitation ratio of the weapons. If the lethality ratio is λ for one side, it is $1 / \lambda$ for their opponents. The main result of this analysis is that the side with $\lambda > 1$ is the victor. The battle must end when the number of weapons in use by one side falls to zero, and at this point, as Fig. 24.1 confirms, the rate of loss of weapons on the winning side is zero. In practice the battle may end when the number of weapons in use by one side falls to, say, a half of the initial number, at which point the remaining troops are likely to surrender or to flee. Alternatively, the commander may put in reserves before this occurs. This may be treated as a step change in x_1 at some time t_1, but it is an option we will not pursue.

The main result of this analysis is that for victory the lethality ratio must be greater than unity. A commander probably has a good knowledge of the size of his own force and that of his enemy but the incapacitation rates are more difficult to assess. If the enemy is twice his size but his enemy's weapons are capable of firing at only half the speed of his own, then he might accept battle on the assumption that the contest is fairly even. He would be wrong. The lethality ratio of his enemy would be 1.414 and his own would be 0.707 so he would lose. The reason for this is that the loss of weapons by both sides is equal initially but the smaller army is less able to stand the loss. Thus if he has 5,000 weapons firing twice the rate of the enemy's 10,000 weapons the rate of weapon loss on both sides is equal and after a certain length of time they may have declined by 1,000 to 4,000 weapons on his own side against 9,000 of his enemy. The ratio of weapons is now 9,000/4,000 = 2.25. This ratio will continue to increase as the battle proceeds until the enemy overwhelms him. In order to give battle and win with weapons firing twice as fast as the enemy's a commander needs to have at least 71% of the enemy weapons at the start.

The lethality ratio is also useful in assessing different weapon systems. Thus two men firing a machine gun firing at 1,000 rounds/min is equivalent to fourteen men firing weapons at 20 rounds/min. There is a saving in men but the consumption of ammunition is increased. This, of course, assumes equal accuracy of fire (equal probability of a hit) and equal probability of incapacitation given a hit,

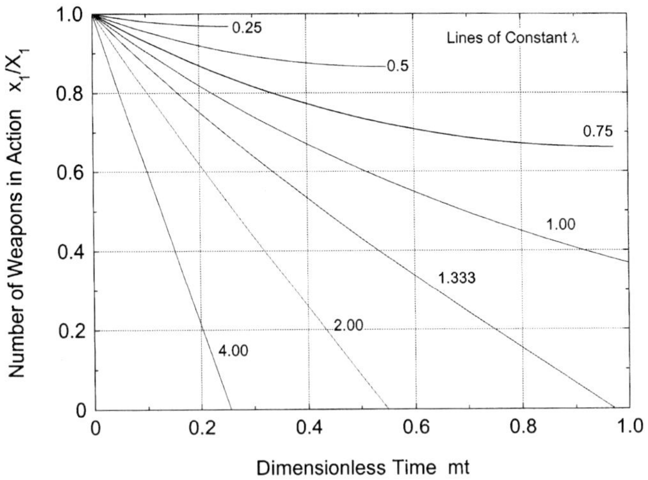

Fig. 24.1 Decline in number of weapons in use during a battle.

but in practice the machine gun is likely to be less accurate. Another useful concept is lethality, L, which may be defined, for each side in the battle, as

$$\begin{aligned} L_1 &= X_1\sqrt{f_1 p_1(I/H)p_1(H)} \\ L_2 &= X_2\sqrt{f_2 p_2(I/H)p_2(H)} \end{aligned} \qquad 24.5$$

Thus the side having the greatest lethality wins. Lethality is not a unique property of any one side but depends also on the characteristics of the enemy because the two probability terms depend not only on the range, accuracy of fire, and projectile energy of one side but also on the enemy's armour protection and target size.

The effect of feeding reserves into the battle at a constant rate proportional to the initial size (a_1X_1 and a_2X_2) is to add these terms to right hand side of the corresponding equations in Eq.24.2. The method of solution is the same as before but the main result is that lethality ratio is then

$$\begin{aligned} \lambda &= \frac{X_2}{X_1}\sqrt{\frac{k_2}{k_1}\left(\frac{1 + a_2/m}{1 + a_1/m}\right)} \\ \lambda &= \frac{X_2}{X_1}\sqrt{\frac{f_2\ p_2(I/H)\ p_2(H)}{f_1 p_1(I/H)p_1(H)}\left(\frac{1 + a_2/m}{1 + a_2/m}\right)} \\ m &= \sqrt{k_1 k_2} \end{aligned} \qquad 24.6$$

Thus the lethality of each side is also proportional to the rate at which reserves can by fed into the battle.

Clearly, in accepting battle, rather than making a tactical retreat, a commander must be satisfied that his own lethality exceeds that of his foe, that is, he has a larger number or troops, weapons and reserves, that he has a higher probability of hit (his foe is more exposed), that he has a higher probability of incapacitation given a hit (his foe has insufficient or ineffective armour in relation to the weapons he intends to use), and that his rate of fire and his reserves of ammunition and troops are sufficient. Other factors may be important, especially if they affect any of these variables. The commander must also try to ensure that a weak opponent is unable to refuse battle by retreating. If a commander is not satisfied that his lethality exceeds that of his enemy then he is likely to refuse battle and to retreat, if this is possible, to find more advantageous conditions.

Flanking Geometry

The distance a weapon can fire (its range) is clearly an important factor in battles but it does not occur explicitly in Equ. 24.4 or Equ. 24.5. This is because it is implied in the variables X_1 and X_2, which are the numbers of weapons within firing range of the enemy at the start of a battle. Clearly, a greater range increases the number of weapons that can be used. In deriving Equ. 24.4 it was assumed that the two sides were drawn up in two lines opposite each other so that neither had a geometrical advantage. In practice, commanders seek to gain a local numerical advantage by a suitable arrangement of their troops. The three common arrangements are illustrated in Fig. 24.2. In Fig. 24.2 (top) the troops are parallel to each other and neither side has an advantage, assuming the troops per unit length of line are the same on both sides. The simplest flanking movement is the sideways extension of one line of troops relative to the other, Fig. 24.2 (middle). This gives a local numerical advantage of up to 2:1, other things being equal, and the short line suffers local defeat at its end and starts to withdraw from this point. The flanking troops are then able to "roll-up" the enemy line from one end. An even more effective arrangement is shown in Fig. 24.2 (lower). Here the opposing troops form a tee shape and the attacking troops from the right have an overwhelming numerical advantage, depending on how close they choose to approach the enemy line. In each case the range of the weapons, R, is the same for each side.

It is clear that in the first case the number of weapons within fighting range is the same on both sides, assuming equal numbers of weapons per unit length of line. This is also true for the second case away from the ends of the line. However, at a distance b from the end of the line, the side with the overlap has a numerical advantage. To determine the numerical advantage at distance b we take the ratio of the number of weapons within range, R, of the enemy. This is shown as X_l and X_2 in Fig. 24.2 (middle). A man at position b can be fired at by all the troops along X_l, and his opposite number can be fired at by all the troops along line X_2. The distance apart of the two lines is a. It is easily shown that the numerical advantage at that point is

$$\frac{X_1}{X_2} = \frac{2}{1 + \dfrac{b + R}{\sqrt{a / R(2 - a / R)}}} \qquad 24.7$$

$$b / R < \sqrt{a / R(2 - a / R)}$$

Fig. 24.3 illustrates how the local numerical advantage changes at the end of the line if there is a simple overlap. At the very end of the line the numerical advantage is always 2.0 if the lines are within range ($a/R < 1$). If the two lines are only just within fighting range ($a/R = 1.0$), then the numerical advantage extends from the end of the line to a distance R along the line. As the distance between the lines decreases ($a/R = 0.1$ say), the numerical advantage extends only about $0.42R$ along the line, and at very short range ($a/R = 0.01$) the numerical advantage extends only about $0.14R$ along the line. It is advantageous to the overlapping line, to keep the distance between the lines at or near the maximum range of the weapons.

In the case of a full flank attack, Fig. 24.2 (bottom), all the troops along the lines of length X_l and X_2 are within range, R, at the start of the battle and thus the numerical advantage is

$$\frac{X_1}{X_2} = \frac{2\sqrt{1 - (a / R)^2}}{1 - a / R} \qquad 24.8$$

$$a / R < 1$$

This ratio is plotted in Fig. 24.4. In this case the numerical advantage is always greater than 2.0. When the two lines are at the extreme

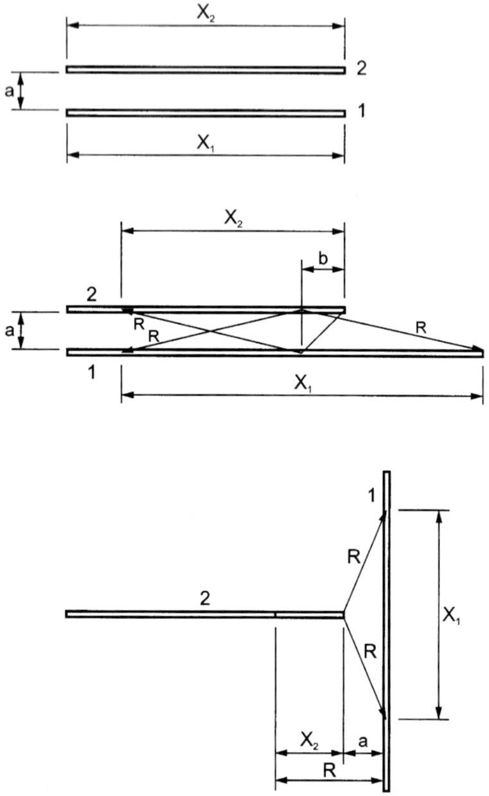

Fig. 24.2 Top: parallel battle lines. Middle: flanking by simple overlap. Bottom: tee flanking.

range of their weapons ($a/R = 1$) there is an infinite advantage but this rapidly reduces such that at 80% range the numerical advantage is 6, at 60% range it is 4, and at zero range it is 2. Such flank attacks give only a local numerical advantage and the troops outside the firing range of the weapons may be considered to be reserves that may be fed into the battle to make good any losses. Although the attacking side has a local numerical advantage, it must also have sufficient reserves if it is to carry local victory through to completion.

Wound Ballistics

An understanding of wound ballistics is essential to the proper appraisal of a weapon's effectiveness and yet, in discussing past weapon sys-

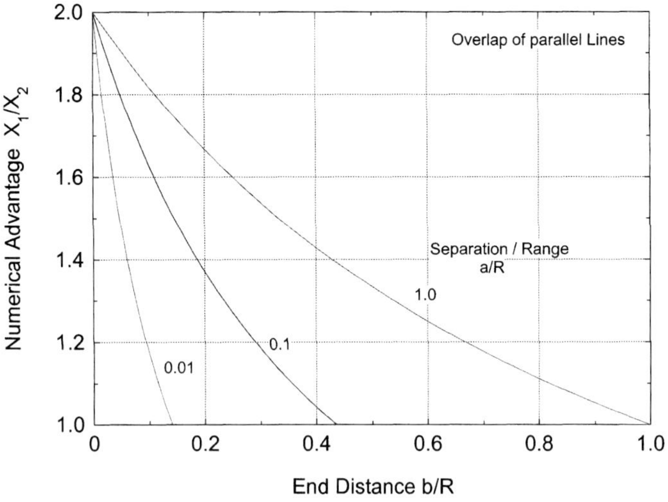

Fig. 24.3. Local numerical advantage with a simple overlap.

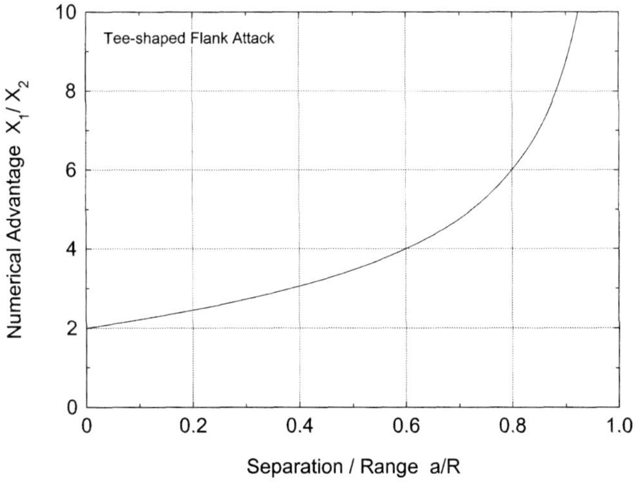

Fig. 24.4 Local numerical advantage during a tee-shaped flank attack.

tems, the findings of wound ballistics are rarely considered. This failure is understandable, perhaps, because it is only in the past decade or so that modern military weapons have been assessed in this way.

Death resulting from high velocity impact by projectiles is usually caused by the collapse of the blood circulation, the destruction of vital regions of the brain, or shock. The immediate collapse of blood circulation produces unconsciousness after about ten seconds because of a lack of oxygen supply to the brain. Penetration of the heart causes immediate collapse in man, perhaps attributable to shock, whereas in animals such as deer, hunting experience shows that the beast may continue running for up to ten seconds after penetration of the heart and may cover up to a hundred metres. A slower loss of blood supply also leads to unconsciousness, the time taken being dependent on the loss rate.

Penetration of the brain cavity, surprisingly, does not always lead to death. Damage to the brain stem, which contains those sections responsible for biological functions such as breathing, heart function, and body temperature control, produces death although blood circulation may continue. On the other hand, damage to the front part of the brain, the cerebrum, can be quite large without necessarily causing death; it would, however, cause severe personality changes. It is found that traverse shots through the cerebrum, using low energy projectiles of less than 500 J, may not cause death whereas the same projectile fired along the front to back axis causes death owing to damage of the cortex.

Shock death is produced by the action of nerves paralysing the heart. It can occur in beasts (e.g. rabbits) when death occurs despite only superficial tissue injury and no damage to vital organs; but it does not normally occur in man. However, wound ballistics is concerned not only with the probability of a projectile causing death but also with the more general probability of incapacitation. By incapacitation, we mean that the target, be it man or beast, is prevented from carrying out its intended function. Thus a beast that is crippled by an arrow in the leg may be prevented from escaping, or attacking, and may be sufficiently incapacitated. When a projectile strikes a target it may fail to penetrate, penetrate but fail to incapacitate, or it may penetrate and incapacitate. The probability of each of these results, at different ranges, is a measure of the effectiveness of a weapon.

In this section we will consider first the motion of the projectile as it passes through the body tissues (and hence the energy absorbed by the tissues), then the ability of a projectile to penetrate naked skin or skin protected by normal clothing, and finally the probability of incapacitation given a hit in a certain area. Ideally we would like such data for every sort of animal against which a projectile is used but such data are not available; the only good data relate to naked or protected human bodies, and much of this is confidential, largely because of the expense involved in collecting the data.

Projectile Motion through Tissue

Fig. 24.5 illustrates the main features of a projectile passing through tissue. A shock wave is generated by the high velocity impact and travels from the impact point, at about 1200 m/s, in the form of a spherical compression wave of decreasing magnitude. Although this could be reflected from bones as a tensile stress, there is no evidence that it is particularly damaging. Sonic pressure waves of very high amplitude are used in the treatment of kidney stones, to break them up, and the waves do not damage the soft tissue through which they must pass. Behind the projectile a permanent cavity is formed by the passage of the projectile, and a temporary cavity, which is usually much larger than the permanent cavity; 6 mm spheres at 1000 m/s produce a temporary cavity of about 66 mm diameter. Bullets that tumble nose-to-tail as they pass through tissue produce much larger temporary cavities and larger permanent cavities, Fig. 24.5 (bottom).

Given the velocity, v_0, of a projectile at entry into tissue then we may assume that its motion is retarded by a drag force, similar to that experienced when the projectile passes through a fluid, and a force required to fracture the tissue, also retards it, see Peters (1990). Thus, we may write the equation of motion as

$$m\frac{dv}{dt} = mv\frac{dv}{dx} = -C_d\rho A\frac{v^2}{2} - \sigma A \qquad 24.9$$

m is the projectile mass, v its velocity, x the distance penetrated, C_d the drag coefficient, ρ the density of the tissue, A the cross-sectional area of the projectile, and σ the fracture stress. It is more convenient to express the fracture stress in terms of a constant velocity u, which is called the tissue penetration velocity, that is

$$\sigma A = C_d\rho A\frac{u^2}{2}$$
$$u = \sqrt{\frac{2\sigma}{C_d\rho}} \qquad 24.10$$

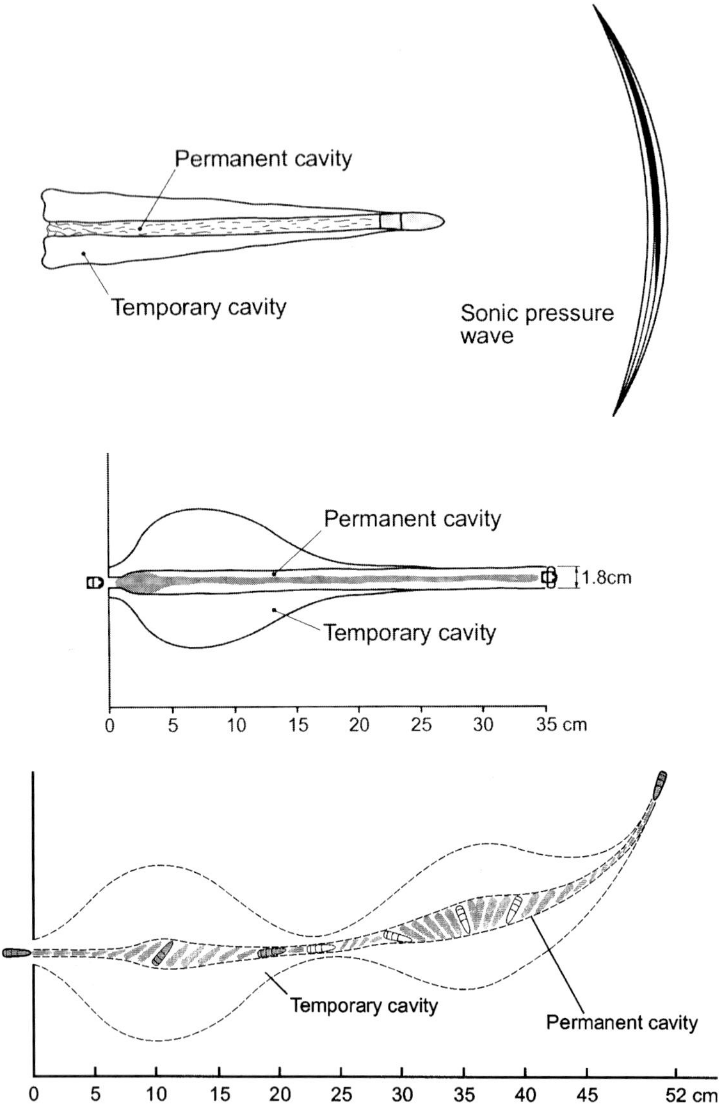

Fig. 24.5 Top: passage of a projectile through tissue. Centre: typical non-tumbling wound profile. Bottom: wound profile with tumbling motion. After Fackler (1989).

Substituting into Equ. 24.9 gives

$$mv = \frac{dv}{dx} = -\frac{C_d \rho A}{2}(v^2 + u^2) \qquad 24.11$$

Drag coefficients have been determined for projectiles fired into the muscle tissue of dogs and pigs and into tissue simulants such as water, gelatin, or soap. The drag coefficients of bullets are usually found in the range $0.3 < C_d < 0.45$ for sub-sonic projectiles and about 1.0 for supersonic projectiles. The drag coefficient of long projectiles, such as arrows, is greater and is in the range $0.6 < C_d < 1.0$. The tissue penetration velocity is found to be in the range $80 < u < 120$ m/s for the tissue of dogs, pigs and the standard tissue simulants such as gelatin or soap, although care must be exercised in using gelatin in the correct proportion and at the appropriate temperature. The NATO standard is to use 20% gelatin in water at 10°C although Fackler (1988) favours 10% gelatin at 4°C.

Integration of Equ. 24.11 has the solution

$$x = \frac{m}{C_d \rho A} \ln \left(\frac{v_0^2 + u^2}{v^2 + u^2} \right) \qquad 24.12$$

v_0 is the projectile velocity at entry to the tissue ($x = 0$). To determine the distance penetrated by the projectile before coming to rest we write $x = L$ when $v = 0$ and so

$$L = \frac{m}{C_d \rho A} \ln[1 + (v_0 / u)^2] \qquad 24.13$$

The penetration distance is plotted in Fig. 24.6 and is found to correlate very well with experimental or clinical data. The significance of the tissue penetration velocity, u, is made clear in Fig. 24.6 which demonstrates that the penetration curve is concave when $v / u < 1$ and convex when $v / u > 1$. This analysis is restricted in application to "chunky" projectiles such as spheres, cubes, short cylinders, or irregular, roughly spherical, fragments. It does not apply to projectiles that are liable to yaw in flight, that is, to tumble over nose to tail. Such tumbling motion increases the drag force quite significantly and stops the projectile much faster than in non-tumbling motion. It also does much more damage to the tissue.

If the initial velocity is small compared to the tissue velocity, that is $v_0 / u << 1$, then the penetration distance, Equ. 24.13, simplifies to the form

$$L = \frac{m v_0^2}{C_d \rho A u^2} = \frac{m v_0^2}{2 \sigma A} \qquad 24.14$$

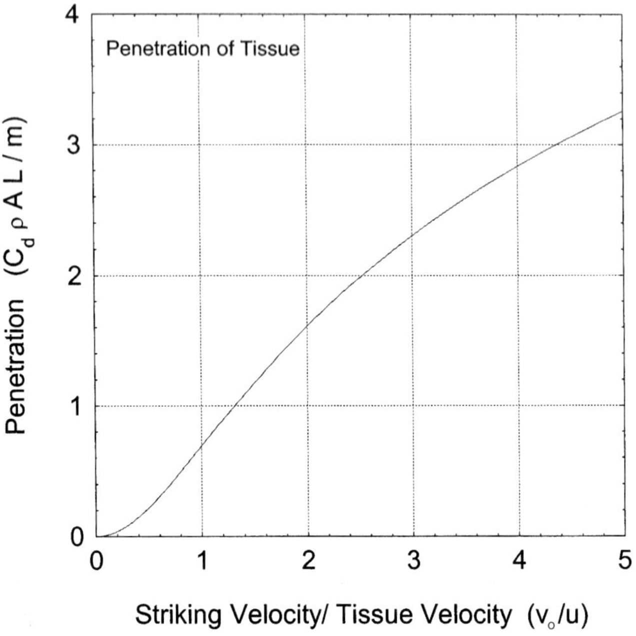

Fig. 24.6 Penetration of tissue by non-tumbling projectiles.

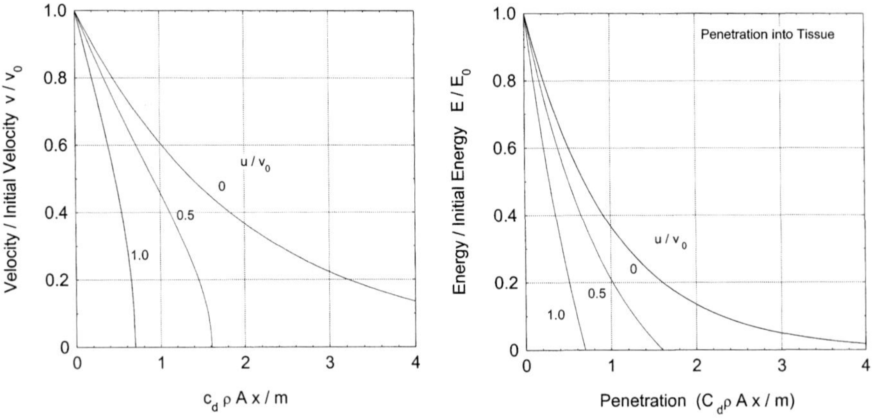

Fig. 24.7 Left: variation of projectile velocity with distance through tissue. Right: loss of kinetic energy by the projectile.

Thus the penetration distance is proportional to the kinetic energy per unit area of projectile, the other quantities being constant for the tissue. This relation is illustrated in Fig. 24.7. For animal tissue the density is 1060 kg/m^3, the drag coefficient is about 0.4, and the tissue penetration velocity is usually close to 100 m/s for 6 mm diameter fragments. Peters (1990) suggests that the tissue penetration velocity be determined from the diameter of the penetrating fragment by

$$u = u_6 \left(\frac{d_6}{d} \right)^{1/3} \qquad 24.15$$

u_6 is the tissue penetration velocity of a 6 mm diameter spherical fragment (about 100 m/s), d_6 is 6 mm and d is the diameter of the penetrating fragment.

Initial Penetration

I shoot the Hippopotamus/With bullets made of platinum,/Because if I use leaden ones/His hide is sure to flatten 'em. Hillaire Belloc's little verse about the hippopotamus reminds us that before a projectile can pass through muscle tissue it must first penetrate the hide. Platinum bullets, however, were chosen more for their rhyming properties than their physical properties. We will return to this problem when we consider the protection offered by armour plate, but even naked tissue offers some protection and this is our concern here.

The initial velocity of the projectile entering the tissue is v_0 and it is tempting to identify it with the velocity with which the projectile strikes the skin surface. This, however, is not the case. A projectile having a velocity less than a certain value cannot penetrate skin, bone, or even tissue simulants. It simply ricochets from the skin causing, perhaps, an effusion of blood beneath the surface. The threshold velocity required to penetrate the surface, v_{th}, does not have a unique value and is not a biological property of that surface; it depends not only on the type of skin attacked but also on the size and mass of the projectile.

For bullets and small fragments of mass 0.05–12 grams and of spherical, round nose, spire point, hollow point, or diabolo geometry, it appears that the projectile energy per unit cross-sectional area, known as the areal energy, E_a, is the critical quantity. It may, in fact, be a biological property of the skin. We have already seen in Equ. 24.14 that this quantity defines the distance a bullet penetrates into the tissue. Measurements, Fig. 24.8, show that the initial penetration of the skin, skin and muscle, and combat clothing, occurs if the areal energy, is greater than about 8, 16 or 47 J/cm^2 respectively, see, for example, Sellier and Kneubuehl (1994), Maynard et al.

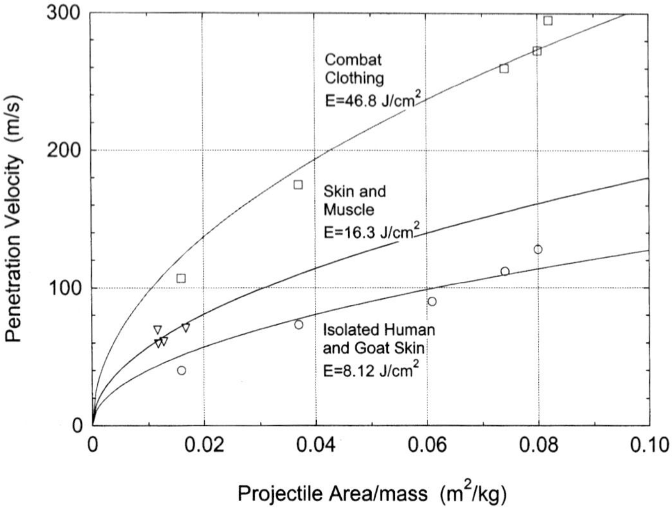

Fig. 24.8 Penetration velocity for skin, skin and muscle, and combat clothing.

(1997). More specifically, a 5 mm diameter projectile requires about 3 J of kinetic energy to penetrate skin and muscle. This is not very much energy but for a light projectile of, say, 5 grams, it represents a velocity of about 35 m/s. Although this is not particularly significant for a bullet travelling at ten times this speed, it is significant for an arrow. From the areal energy the threshold velocity, v_{th}, for penetration is determined as

$$v_{th} = \sqrt{\frac{2\,E_a\,A}{m}} \qquad 24.16$$

Typical values of threshold velocity are in the range 30–200 m/s. Heavy projectiles having a small cross-sectional area have a low threshold velocity and penetrate the skin relatively easily. Arrows have a lower threshold velocity than bullets because they are heavier.

If a projectile strikes the skin surface with a velocity, v_s, and the threshold velocity is v_{th}, then there is a loss of energy at the surface and the velocity of the projectile is reduced to

$$v_0 = \sqrt{v_s^2 - v_{th}^2} \qquad 24.17$$

Thus, when the striking velocity is equal to the threshold velocity the projectile velocity is zero and it fails to penetrate the surface.

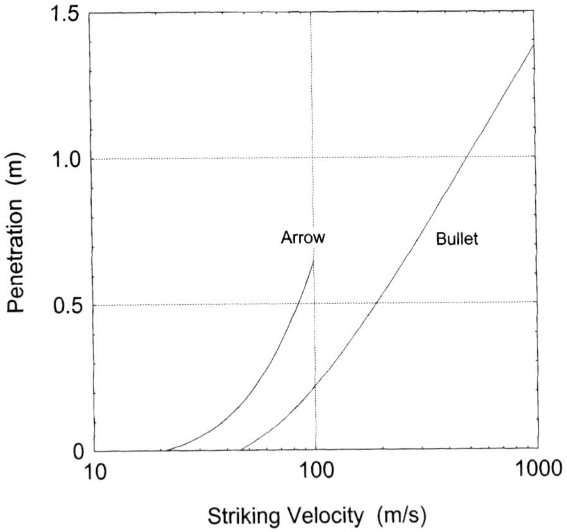

Fig. 24.9 Penetration distance for an arrow and a low velocity bullet.

The penetration distance given by Equ. 24.13, may now be written more usefully as

$$L = \frac{m}{C_d \rho A} \ln\left(1 + \frac{v_s^2 - v_{th}^2}{u^2} \right) \qquad 24.18$$

Table 24.1 Data used in calculating the penetration of arrows and bullets.

	Arrow	*Bullet*
Diameter mm	8.6	11.4
Area mm^2	58.1	102.1
Threshold Velocity m/s	20.9	45.3
Tissue Velocity m/s	88.7	81.0
Projectile Mass grams	40	14.9
Drag Coefficient	0.8	0.5
Tissue Density kg/m^3	1060	1060
Striking Velocity m/s	71	265
Calculated Penetration mm	374	666
Measured Penetration mm	> 300	655

The penetration distances for typical arrows and bullets are shown in Fig. 24.9 and the data used in preparing the curves are displayed in Table 24.1. The arrow, it should be noted, is more effective in penetrating tissue than the bullet. This is mainly because the arrow is heavier in relation to its cross-sectional area and thus has a lower threshold velocity. The arrow is fired from a modern longbow that gives it a velocity of 71 m/s and achieves a penetration depth more than 0.3 m in gelatin. The bullet is a 0.45 inch automatic, full-metal-jacketed construction having a rather low velocity of 265 m/s with which it attains a penetration of 0.655 m into gelatin. The observed penetration in gelatin, it will be noted, is in good agreement with the computed curves for both the arrow and the bullet.

Example: Penetration of an arrow from a longbow
Determine the penetration of an arrow into tissue using the data specified in Table 24.1 concerning the performance of a Delta V longbow, Missliwetz and Wieser (1985). The arrow mass is 0.04 kg, the diameter is 8.6 mm, the striking velocity is 71 m/s at zero range and 48 m/s at maximum range, the drag coefficient is 0.8, and the tissue density is 1060 kg/m^3.

From Equ. 24.16, the threshold velocity is $v_{th} = 20.9$ m/s.
From Equ. 24.17, the entry velocity is $v_0 = 67.9$ m/s.
From Equ. 24.15, the tissue velocity is $u = 88.7$ m/s.
From Equ. 24.13, the penetration distance is $L = 0.374$ m at zero range, and at max range ($v_s = 48$ m/s, $v_0 = 43.2$ m/s) the penetration is $L = 0.173$ m.

For this arrow the measured penetration into gelatin at zero range is greater than 0.3 m.

Probability of Incapacitation Given a Hit, P(I/H)
The energy of a projectile necessary to incapacitate an unprotected man is usually in the range from 40 to 80 J for small projectiles such as bullets or fragments. Although it is still somewhat controversial, the probability of incapacitation is usually expressed as a function of the "expected kinetic energy", EKE, absorbed by the tissue from the projectile.

For example, Janzon et al., (1997) have measured the mass of damaged tissue that the surgeon must excise from a bullet wound and have correlated this with the kinetic energy transferred by steel spheres and bullets, Fig. 24.10. There is a lot of scatter in these data but it does appear that the amount of damaged muscle, per unit of energy lost by the projectile, is about 0.3 to 0.4 g/J. This approach

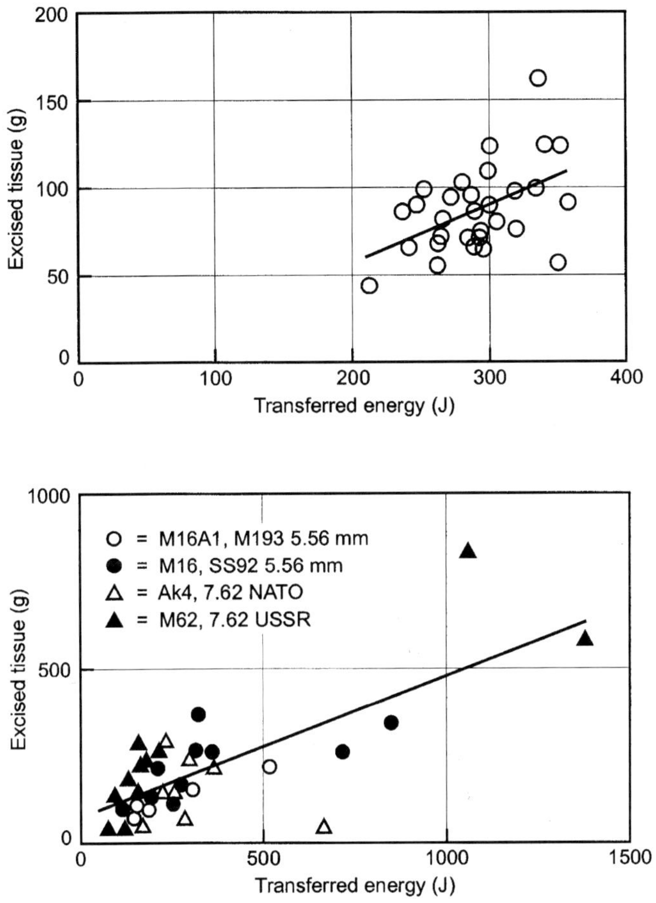

Fig. 24.10 Correlation between excised pig-muscle and energy transferred. Top: 6 mm steel spheres at about 1000 m/s. Bottom: 5.56 mm and 7.62 mm assault rifles. After Janzon et al. (1997).

has been severely criticised by Fackler (1989) who points to particular high-energy bullets, which in practice do less damage to the tissue than less energetic bullets. This fact is confirmed by the scattered data in Fig. 24.10, but nevertheless, the "expected kinetic energy" method is considered the most useful; its value is calculated from the work done by the projectile.

$$EKE = \int_0^L F(x)dx = \int p(x)F(x)dx \qquad 24.19$$

L is the penetration distance or the thickness of tissue, depending on which is the smaller. Of course, the thickness of tissue varies from person to person and varies with the direction of the shot at entry. To allow for this the function $p(x)$ is introduced. In the simple case $p(x) = 1$ if $0 < x < L$ and $p(x) = 0$ otherwise. But to allow for person to person thickness variation and directional variation, the value of $p(x)$ is allowed to fall from 1 to 0 over a range of penetration distances, as illustrated in Fig. 24.11.

The probability of incapacitation is the product of the probability of a hit and the probability of incapacitation given a hit; the latter is determined by the expression

$$P(I / H) = \frac{\lambda}{1 + \exp\left(\alpha - \beta \ln\left(EKE - \gamma\right)\right)} \qquad 24.20$$

λ, α, β, and γ are numerical constants that have been determined for each part of the body and for each direction by teams of experienced medical practitioners. Typical of the sort of analysis that emerges from such a study are the curves plotted in Fig. 24.12. Here it is clear that the probability of incapacitation increases with the

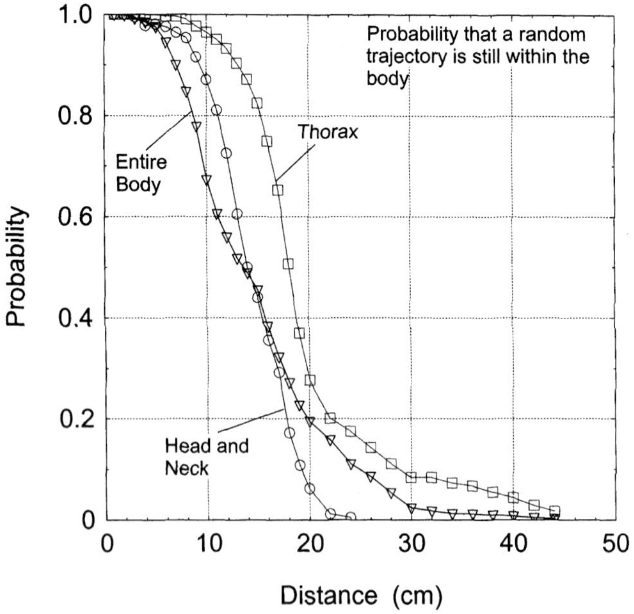

Fig. 24.11 Probability function $p(x)$ for various parts of the human body. Sturdivan (1981).

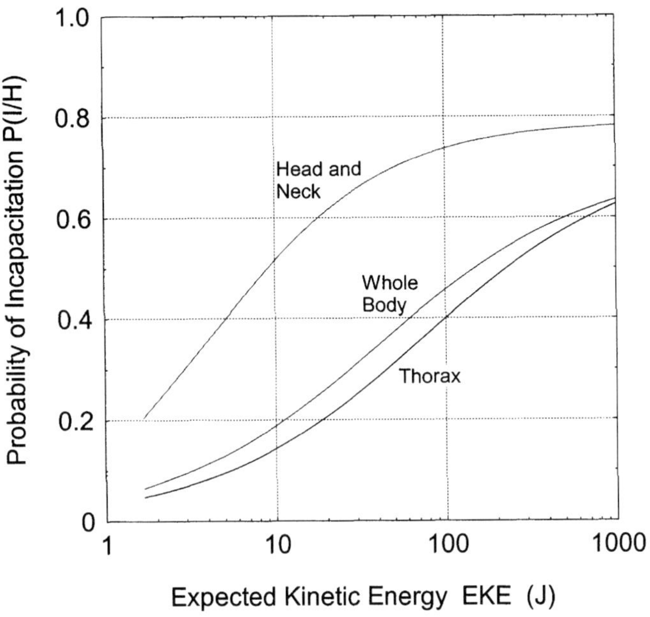

Fig. 24.12 Probability of incapacitation for strikes in the head and neck, the thorax, and the whole body. Sturdivan (1981).

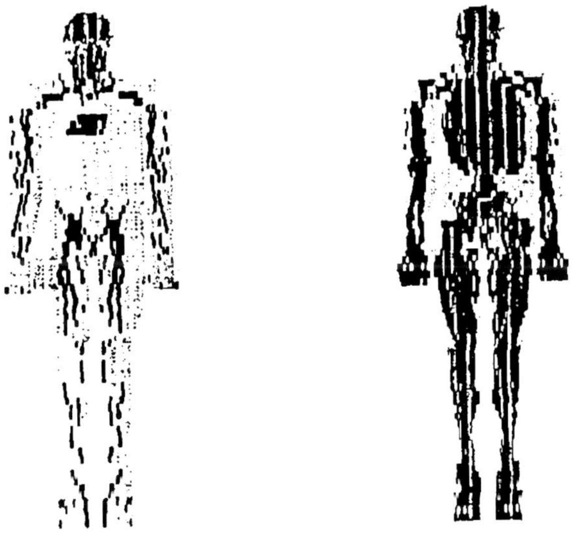

Fig. 24.13 Area of incapacitation, shaded dark, in a naked male. Left: 0.14 g fragment at 300 m/s (6.3 J) and Right: 1.0 g fragment at 900 m/s (405 J). Rickter (1989).

expected kinetic energy deposited by the projectile. The 50% incapacitation level may be taken as a measure of the required energy. Thus for a hit to the head and neck region only about 9 J are required to incapacitate, for 50% of such hits, whereas a hit to the thorax requires about 200 J to incapacitate at the same level. Even small projectile energies, as low as 2J, can cause incapacitation in 20% of cases. This may occur, for example, by a direct hit in the eye. The probability of incapacitation is low because the chance of a hit to the eye is low. If the expected kinetic energy deposit is very high, above 1000 J, then $P(I/H) = \lambda$, which has a value somewhat less than 1.0. Thus even a very large kinetic energy will not guarantee incapacitation; the projectile must strike in a vulnerable area.

The vulnerable areas of a naked human when struck by low-energy and high-energy projectiles are well illustrated in Fig. 24.13. The low-energy projectile is a 0.14 g metal fragment travelling at 300 m/s and having 6.3 J whereas the high-energy projectile is a 1 g metal fragment travelling at 900 m/s and possessing 405 J. The increased area of incapacitation of the high-energy projectile is clear but even in this case there remain significant portions of the body where a projectile strike would not result in incapacitation. Not surprisingly, the increased kinetic energy of the projectile increases the size of the target and makes incapacitation more likely. The advantage of this type of analysis is that it is numerate and it enables an accurate assessment to be made of protective armour against specific threats.

Penetration Mechanics

The theory developed above for the penetration of projectiles into tissue may be adapted to determine the penetration of projectiles through a wide variety of materials. Poncelet in 1829 first proposed the theory and wrote the equation of motion in the form

$$mv\frac{dv}{dx} = -i\,(a + bv^2) = -Ai\,(\sigma + C_d\rho v^2 / 2) \qquad 24.21$$

m is the projectile mass, v the velocity, x the distance travelled, and I, a, and b are constants. To give the equation physical significance it may be compared with Equ. 24.9 and the constant a is then the product of fracture stress and cross-sectional area, and the constant b is half the product of the drag coefficient, density, and cross-sectional area. The quantity i is a non-dimensional constant that depends on the shape of the projectile. For a spherical projectile $i = 1$ and for ordinary shaped shells it has a value of $i = 0.667$. The solution of this equation, when integrated, may be written in the form

$$\frac{AiL}{m} = \frac{1}{C_d\rho}\ln[1 + (v_0 / u)^2]$$
$$u = \sqrt{\frac{2\sigma}{C_d\rho}} = \sqrt{\frac{a}{b}} \qquad 24.22$$

A is the cross-sectional area, L is the penetration distance and ρ is the density of the material. In 1839–40 Didion, Morin, and Piobert determined the values of the constants a and b when firing projectiles into various materials and their values, converted to SI units and adapted to the physical constants, are given in Table 24.2.

Poncelet's equation is valid when the projectile strikes the surface at right angles and it is assumed that the projectile does not tumble in flight and that the surrounding material does not distort it. All these assumptions are valid if the striking velocity is small, less than a few hundred metres/second, but long, rather than spherical projectiles are liable to tumble at higher velocities, and the impact forces associated with higher velocities are quite capable of distorting a projectile. Distortion usually appears as a shortening in the length of a projectile and an increase in its diameter. This increases the drag force and reduces the penetration. Medieval bullets were often made of lead and even at the low velocities then prevailing such bullets could simply squash against armour rather than penetrate. As with human tissue, it requires a certain amount of energy to penetrate the surface of most materials and if the projectile does not have this energy when it strikes then it simply re-bounds without penetration. However, the threshold energy required for penetration is usually ignored.

Experimental results, when obtained under closely controlled conditions can be quite accurate and confirm the form of Poncelet's equation, see for example the results for firing projectiles into sand and gravel Fig. 24.14 (left). Other data, such as that for loam and sandy loam, also shown in Fig. 24.14 (left), give more scatter because, in this case, the data were obtained from the penetration of bombs dropped into various soils that could be described as loam or sandy loam, but were not identical. Fig. 24.14 (right) illustrates the penetration of rigid projectiles into various soils and banking materials. Poncelet's equation for the penetration of harder materials such as wood, brickwork, masonry, and limestone, is illustrated in Fig. 24.15.

When a projectile strikes a surface a compressive stress is set up and travels through the material at its sonic speed. In air the sonic velocity is about 340 m/s, in wood about 1440 m/s, in stone about 2000–6000 m/s depending on the density and elasticity of the stone and in iron or steel about 5130 m/s. The magnitude of the stress wave depends on the density and sonic velocity of the impacting materials and on the striking velocity. This stress wave is reflected from the surface opposite the impact point as a tensile stress and, as the tensile strength of many materials is much less than the compressive strength, the material fails at this point. Thus small pieces

Table 24.2 Penetration coefficients for various materials. After Didion, Morin, and Piobert (1839–40), see Cranz and Becker (1921).

	Material	$10^6 a$	$10^6 b/a$	$C_d \rho$ (kg/m^3)	u (m/s)	σ (MPa)
Walling	*Limestone*	12	15	3532	258	118
	Strong Masonry	5.52	15	1625	258	54.2
	Ordinary Masonry	4.4	15	1295	258	43.2
	Brickwork	3.16	15	930	258	31
Banking	*Sand and Gravel*	0.435	200	1707	70.7	4.27
	Clay	1.045	35	718	169	10.3
	Loose Earth (grassed)	0.7	60	814	129	6.87
	Loose Earth (clay and sand)	0.461	60	543	129	4.52
	Damp Clay	0.266	80	418	112	2.61
Wood	*Oak, Beech, Ash*	2.085	20	818	224	20.5
	Elm	1.6	20	628	224	15.7
	Pine, Birch	1.16	20	455	224	11.4
	Poplar	1.09	20	428	224	10.7

of material break off the surface at a point opposite the impact point. In concrete this starts to occur when the penetration of the projectile is about 70% through the wall thickness.

The limit of validity of Poncelet's equation is illustrated in Fig. 24.16 for the case of high velocity bullets penetrating sand and beech. The measurements were made by Cranz and reported in Cranz and Becker (1921). The different bullet velocities were obtained by reducing the mass of gunpowder in the charge. The bullets were recovered from the sand or wood for inspection and confirm that so long as the bullet was substantially undamaged (scratched but not deformed) Poncelet's equation remained valid and penetration increased with velocity. However, when the bullet was fired at 579 m/s into sand it began to deform (length shortened and diameter increased) and the depth of penetration reduced. At velocities above 815 m/s the bullet fragmented. Similar observations were made for bullets fired into beech except that deformation occurs at velocities above 700 m/s.

Example. Iron cannon ball striking oak or masonry
Find the penetration of an iron cannon ball of 13.3 cm diameter and 8.6 kg mass if it strikes (i) an oak wall and (ii) an ordinary masonry wall, with a velocity of 175 m/s.

For oak $C_d\rho$ = 818 kg/m^3 and $u = 224$ m/s (Table 24.2). From Equ. 24.22 $LAi/m = 5.825 \times 10^{-4}$, the cross-section area is $A = 0.01389$ m^2, $i = 1$ and thus the penetration is $L = 0.361$ m.

For ordinary masonry $C_d\rho = 1295$ kg/m^3 and $u = 258$ m/s (Table 24.2). From Equ. 24.13 $LAi/m = 2.923 \times 10^{-4}$, the cross-section area is $A = 0.01389$ m^2, $i = 1$, and thus the penetration is $L = 0.181$ m (18.1 cm).

The cannon ball would penetrate the sides of all but the largest oak ships but would not penetrate far into a masonry wall protecting a city or castle. Many such strikes, or a heavier projectile, would be required.

Fig. 24.17 illustrates the damage done by penetration of a single cannon ball into ordinary masonry. Joshua Sprigge (*England's Recovery*, 1647) reports that in June 1645 the army of Sir Thomas Fairfax, on route from Lechlade to Wanborough, drew up before the Royalist garrison at Highworth, Wiltshire. This consisted only of the parish

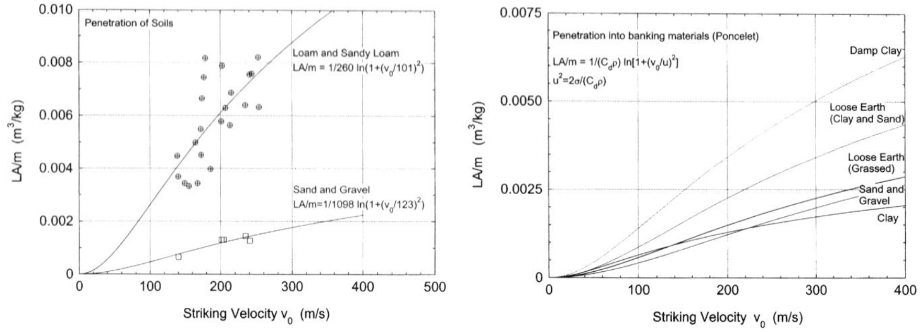

Fig. 24.14 Left: Penetration of rigid projectiles in soils. (RMCS R/79/1329). Right: Poncelet's penetration equation for different soils.

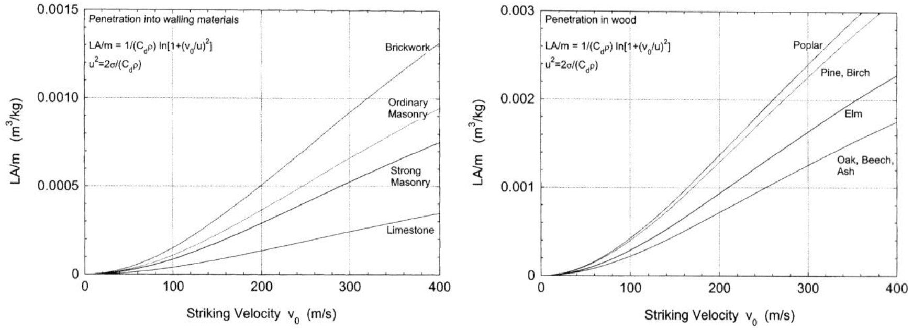

Fig. 24.15 Left: Poncelet's equation for penetration into rock and masonry. Right: Poncelet's equation for penetration into wood.

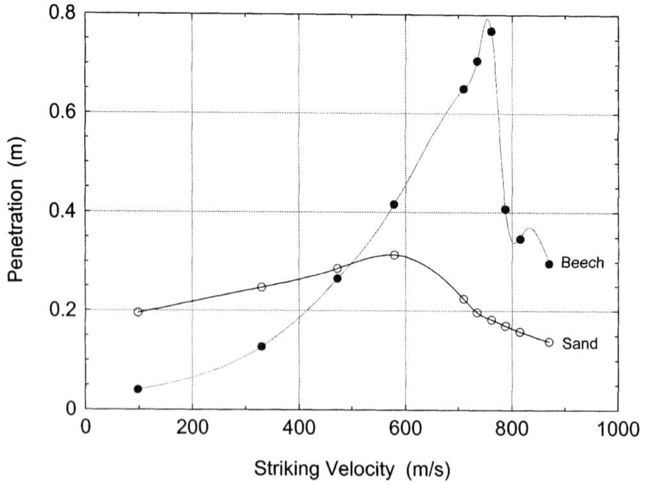

Fig. 24.16 Penetration of sand and beech by high velocity bullets.

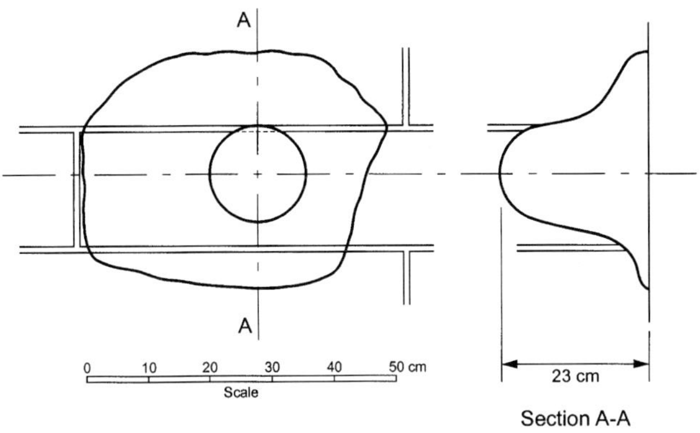

Fig. 24.17 Penetration of a masonry wall by a cannon ball in the English Civil War. St Michael's Church, Highworth, Wilts.

church fortified by a line of bulwarks. Major Henne, the Royalist governor refused to yield, so Fairfax planted his ordnance and prepared to storm the fortified church. Apparently at least one shot was fired and struck the church a little to the left of the west door producing the damage illustrated in Fig. 24.17. Major Henne yielded and 70 prisoners and 80 arms were captured. What is said to be the cannon ball that struck the church is still displayed in one of its chapels. The numerical calculation (above) confirms that the hole could have been made by a culverin of the period and is based on the performance specified by Thomas Smith (The Complete Souldier, 1628) of a 5.5 inch calibre culverin firing an iron cannon ball of 5.25 inch (13.3 cm) diameter with a muzzle velocity of 175 m/s (1920 m maximum range). The calculation suggests that such a culverin might be expected to penetrate 18.1 cm into ordinary masonry. The observed penetration, at the present day, is 23 cm into a wall about 1.5 m thick. The difference in penetration is not too great but it suggests that either Fairfax's culverin was more efficient than that specified by Thomas Smith, or the walls of Highworth church are somewhat weaker than "ordinary masonry".

For penetration of wrought iron or mild steel armour the fracture strength, and hence the value of u, is very large in comparison to the striking velocity and assuming Equ. 24.14 is valid we obtain

$$\left(\frac{L}{D}\right) = \frac{2}{\pi\sigma}\left(\frac{mv_0^2}{D^3}\right) \qquad 24.23$$

L is the penetration, D is the projectile diameter, σ is the fracture strength, m is the projectile mass, and v_0 is the striking velocity. The drag coefficient, density, and projectile geometric factor have cancelled from the equation and are no longer relevant. This equation may be derived more simply by equating the kinetic energy of the projectile, $mv_0^2/2$, to the work done in making the hole, $\sigma\pi D^2L/4$, which assumes that failure is by compression. If failure is by shear then the kinetic energy of the projectile, $mv_0^2/2$, is equated to the work done by shear, $\sigma\pi DL^2$, to give

$$\left(\frac{L}{D}\right)^2 = \frac{1}{2\pi\tau}\left(\frac{mv_0^2}{D^3}\right) \qquad 24.24$$

As failure is due partly to compression and partly to shear the penetration equation is likely to lie part way between Equ. 24.23 and Equ. 24.24 and this, in fact, turns out to be the case so long as the striking velocity is not too high. Cranz and Becker (1921) report that in 1880 the German firm of Krupp used a somewhat similar empirical equation to determine the thickness of armour plate that can be penetrated assuming a direct impact a right angles to the surface and that Krupp's wrought iron plate is used without any support in the rear. This relation is

$$\left(\frac{L}{D}\right)^{4/3} = \frac{1}{1541x10^6}\left(\frac{mv_0^2}{D^3}\right) \qquad 24.25$$

Equ. 24.23 and 24.25 are identical except for the index of 4/3 in Equ. 24.25. The implied strength (σ) in Equ. 24.25 is 981 MPa (63 Ton/in^2) and is much higher than the normal compressive strength of wrought iron or mild steel and indicates, perhaps, the effect of a high strain rate. The deformation during impact occurs in a few milliseconds whereas in the usual tensile test machine it occurs over perhaps a minute. If impact takes place at an angle α with the surface then Equ. 24.25 becomes

$$\left(\frac{L}{D}\right)^{4/3} = \frac{1}{1541x10^6}\left(\frac{mv_0^2}{D^3}\right)\cos^2\alpha \qquad 24.26$$

Penetration relations of this form are usually referred to as the Milne de Marr equation. It is reasonably accurate if the impact velocity is less than about 300 m/s and the impact angle is close to the normal.

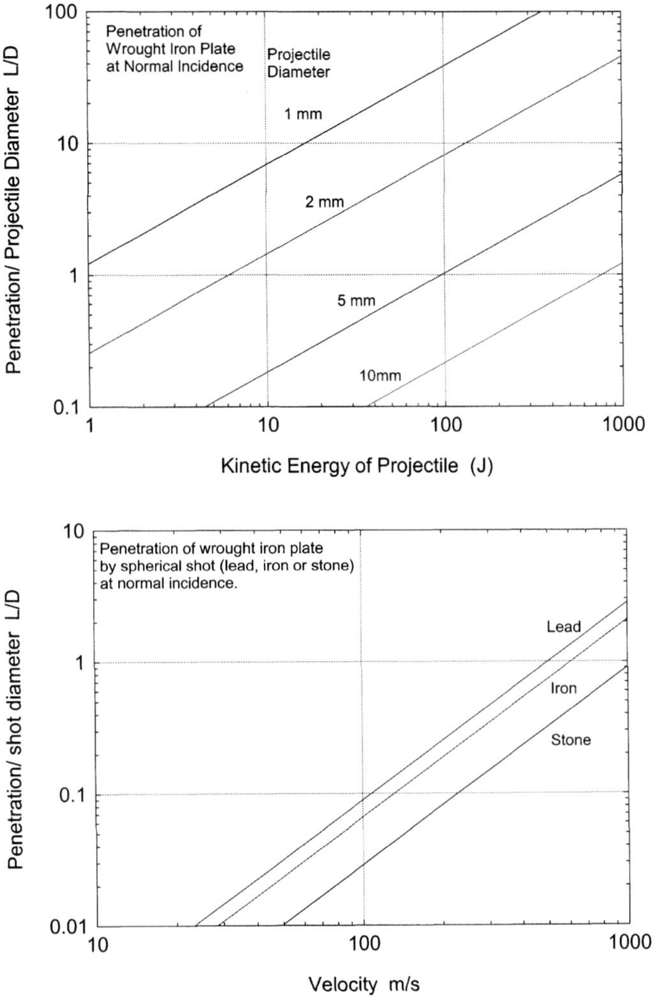

Fig. 24.18 Left: penetration of wrought iron armour, showing influence of projectile energy and diameter. Right: penetration of wrought iron by stone, iron or lead spheres.

It becomes increasingly inaccurate for angles greater than about 20 degrees and at some angle, perhaps about 30–40 degrees, the projectile will ricochet from the surface without significant penetration. Equ. 24.26, for normal incidence ($\alpha = 0$) is plotted in Fig. 24.18 and illustrates the influence of projectile energy and diameter on penetration (top) and the penetration achieved by stone, iron or lead spherical shot (bottom).

Example. Weight and protection of armour
Determine the minimum striking velocity required to penetrate 2 mm of wrought iron armour using (i) a spherical lead projectile of 8 mm diameter and (ii) an 8 mm diameter arrow of 40 grams mass. The density of lead is 11,680 kg/m^3 and the impact is normal to the plate.

The mass of the 8mm lead sphere is 3.132 grams. From Equ. 24.25 the required velocity is 199.2 m/s. From Equ. 24.25 the required velocity for the 8 mm diameter arrow is 55.7 m/s.

Thus an arrow fired from a 2 m long bow would penetrate 2 mm of wrought iron armour but the same diameter of lead bullet requires a muzzle velocity much greater than most medieval guns could attain. To penetrate 2 mm of armour, larger diameter bullets were used.

The weight of 2 mm armour to cover the whole body, assuming a surface area of 1.8 m^2, is 28.3 kg (62.4 lb). It is comparable to the weight a modern soldier might be expected to carry into battle.

In this example an arrow is able to penetrate armour with a lower velocity than the bullet because although it has the same diameter it has a much greater mass (it is longer than the spherical bullet). Equ. 24.25 makes it clear that, for a given diameter and velocity the penetration into armour increases with the mass of the projectile. If the projectile is spherical then the only way to increase its mass is by using a denser material, hence the preference for lead, but if the projectile is made longer than its diameter then the mass increases and the bullet attains a better penetration. Long bullets, however, are liable to be unstable in flight and to topple nose over tail. If the projectile hits the target sideways on then penetration is diminished rather than enhanced. For arrows this problem was overcome from the earliest times by means of flight feathers, usually goose feathers, attached to the tail. Such a solution is not appropriate for a bullet although many modern projectiles for large guns are, in fact, fin-stabilised. The solution for non-spherical bullets is to stabilise them by spin-rotation in flight, for example by the use of rifled barrels.

Stability

The stability of a projectile during flight is an important matter because unstable projectiles yaw and tumble, which increases the drag and reduces the range and velocity. On the other hand, a slight angle of incidence creates lift and increases the distance that spears or javelins may be thrown. On most early projectiles, such as spears and arrows, a heavy head and a light shaft ensured a degree of stability, see Fig. 24.19. The heavy (stone or metal) head mounted on a light wooden shaft meant that that the centre of gravity, G, was well forward. If the projectile yawed through an angle α then a lift force N, normal to the shaft is generated and this acts through the centre of pressure. This produces a pitching moment about the centre of gravity that rotates the projectile. Light flight feathers at the rear of the shaft increased the normal force and also moved the centre of pressure towards the rear consequently the centre of gravity is ahead of the centre of pressure and the pitching moment tends to reduce the yaw angle and the yaw is eliminated. If the centre of pressure is ahead of the centre of gravity, as in Fig. 19 (lower) then the pitching moment rotates the projectile away from the flight direction and the projectile is unstable.

Some experimental drag, lift and pitching moment curves for modern javelins are shown in Fig. 24.20. The experimental lift and drag for all three javelins tested are close together. There is little to choose between them in terms of pitch and drag but there are significant differences in the measured pitching moments. Hubbard and Rust (1984) point out that the theoretical aerodynamics behind the measured pitching moment curves are not well understood. There are three regions of equilibrium, one unstable and two stable, that is, the javelin is unstable when the pitching moment is positive. The three regions of equilibrium show that the position of the centre of pressure changes with the angle of incidence, α. At some angles, for H-70, (0°–12°, 30°–50°) it is behind the centre of gravity and at others (12°–30°) it is ahead of the centre of gravity and therefore unstable. Computer calculations and athletes performance give a maximum range of about 110 m if the launch velocity is 30.5 m/s (100 ft/s). The theoretical range, ignoring lift and drag forces is $v^2/g = 94.7$ m. Thus the actual range attained is greater than the range that could be achieved in a vacuum. This is because at the second equilibrium point the javelin is able to generate large amounts of lift, like an aircraft wing, by flying at large angles of incidence, from 30° to 35°, without generating a restoring moment.

In spherical projectiles such as cannon balls and musket shot, the centre of pressure and centre of gravity coincide and the sphere may rotate, but such rotation is irrelevant and does not affect either its flight or its impact, so long as the rotation is slow. High rotational speed generates a Magnus lift force that would, of course, affect the flight path. Long, thin projectiles, as shown above, achieve better penetration because they have a large kinetic energy per unit cross-sectional area, but such projectiles are usually unstable unless they spin. In Fig. 24.19 (bottom) the centre of pressure of the unstable projectile is ahead of its centre of gravity and thus there is a pitching moment about the centre of gravity tending to increase the yaw angle, consequently the projectile tumbles in flight. To counteract this the projectile may be spun about its longitudinal axis. If the projectile is spinning then a moment generated by the yaw angle causes the projectile to precess such that its nose moves at right angles to the applied moment. Thus the nose moves in a series of circles around the direction of flight rather than tumbling nose over tail. It can be shown that the minimum rotational speed necessary to achieve stable flight is

$$\omega_0 = \frac{2}{I_x}\sqrt{I_y \frac{dM}{d\alpha}} \qquad 24.27$$

ω_0 is the spin-speed (radians/s), I_x is the inertia of the projectile about the direction of motion, I_y is the inertia of the projectile about a line

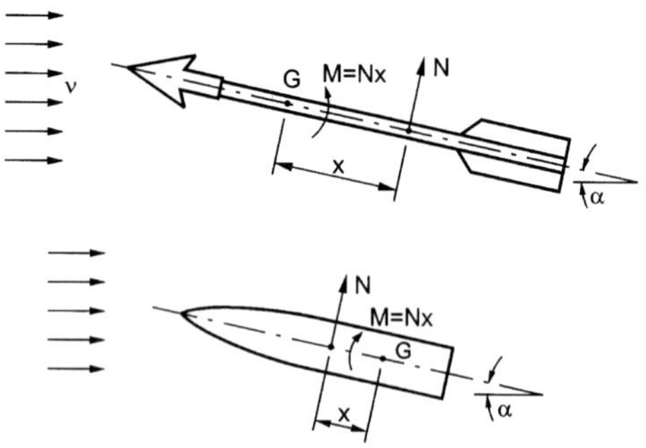

Fig. 24.19 Stability of projectiles in flight. Top: a stable arrow. Bottom: an unstable shell.

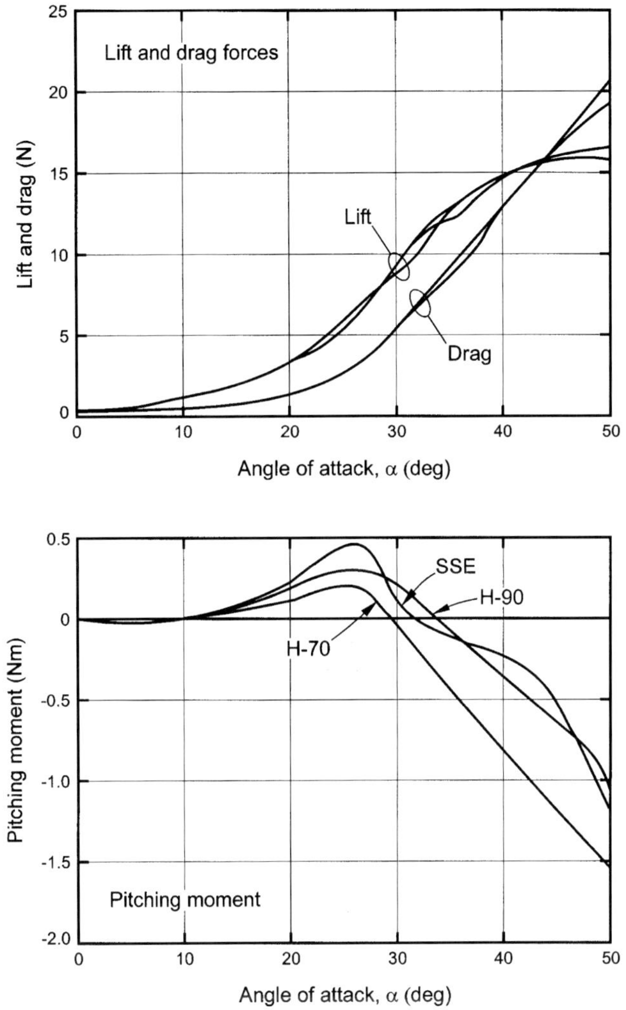

Fig. 24.20 Experimental drag, lift and pitching moments of three modern javelins measured at 30.48 m/s [100 ft/s]. After Hubbard and Rust (1984).

normal to the direction of motion passing through its centre of gravity and $dM/d\alpha$ is the rate of change of the applied moment with yaw angle. The actual spin speed of modern projectiles is usually chosen to be about 20% greater than the minimum speed. If the spin speed is much greater than the minimum then it may be over stabilised with the result that the axis of rotation remains constant and does not follow the flight path. When spin-stabilised projectiles pass through tissue the density of the tissue is about 800 times greater than the density of air and the spin velocity rapidly slows, consequently the projectiles becomes unstable and topples nose over tail. This unstable behaviour is illustrated in Fig. 24.5 above.

Temperature of Armour Plate in Sunlight

The main disadvantages of using armour plate are that it is heavy and becomes very hot when exposed to bright sunlight. This is a particular problem in hot lands and it is probably the reason why crusader knights from Europe used a reflective surcoat over their armour when fighting in the Middle East. Even in England on a hot summer day the temperature of metal exposed to sunlight for any length of time can become hot enough to produce pain when touched and may produce skin burns if contact is prolonged.

The average distance between the earth and the sun is 150 million km and the solar radiation normal to the sun's rays at this radius is 1370 W/m^2. This is called the solar constant although it undergoes a small periodic fluctuation of about $\pm 3.3\%$, largely because of variations in the earth's radius. However, some of this radiation is absorbed, reflected or scattered as it passes through the atmosphere and there are large seasonal variations caused by the tilt of the earth's axis ($\pm 23.5°$). Fig. 24.21 illustrates the variation in the intensity of sunlight for latitude 40° North at mid-winter and mid-summer. It is assumed that there is no cloud cover and the surface is normal to the direction of the sunlight. The intensity of the mid-day sun, it will be noticed, is the same in winter as in summer and the main difference is due to the longer or shorter length of the day.

Sunlight is emitted at the sun's temperature (about 6000 K) and is mostly of relatively short wavelengths in the visible part of the spectrum (about 0.6 microns). On striking an opaque surface a portion of this is reflected and the remainder is absorbed causing the surface temperature to rise. The heated surface also radiates heat depending on its temperature and surface emittance. Because the surface temperature is about 300–400 K it emits radiation of long wavelength (infrared radiation of about 10 microns). Under normal circumstances, when the wavelengths are similar, Kirchhoff's laws of radiation apply and the absorptance and emittance of opaque surfaces are equal, but because the surface is absorbing short wavelength radiation and emitting long wavelength radiation these two

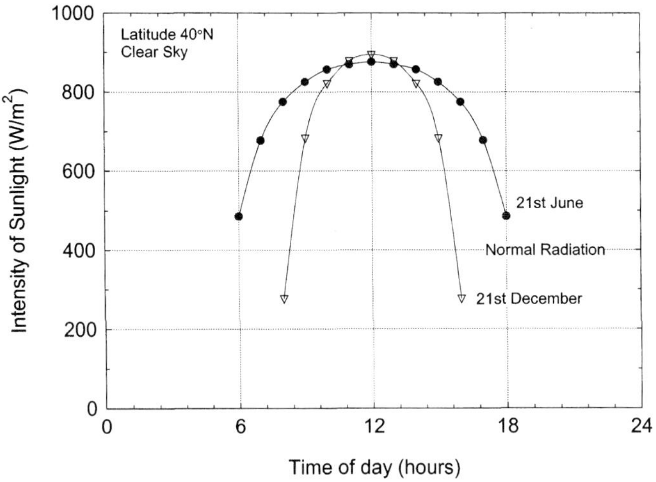

Fig. 24.21 Intensity of normal sunlight in mid-winter and mid-summer at latitude 40° North.

quantities are not necessarily equal. Thus if solar absorptance is high and infrared emittance is low the surface will become very hot. On the other hand, if solar absorptance is low (the surface is a good reflector) and infrared emittance is high then the surface will be cool. Typical values of the solar absorptance and infrared emittance for various surfaces are shown in Table 24.3.

Table 24.3 Solar absorptance and infrared emittance of various surfaces.

Surface	*Solar Absorptance* α_s	*Infrared Emittance* ε	α_s/ε
Black	0.90	0.90	1.00
Red	0.74	0.90	0.82
Green	0.50	0.90	0.55
White	0.25	0.90	0.28
Polished Copper	0.18	0.04	4.5
Oxidised Copper	0.64	0.6–0.9	1.03–0.71
Steel	0.8	0.2	4.0

In addition to radiation, the surface is subjected to natural or forced convection to the surrounding atmosphere, and the surface of armour plate also has to dissipate the heat generated by body metabolism.

Thus, for a surface normal to the direction of sunlight the thermal energy balance is

$$\alpha_s q_s + q_m = \varepsilon\sigma(T^4 - T_a^4) + h(T - T_a) \qquad 24.28$$

The heat absorbed from the sunlight plus that absorbed from contact with a human body is balanced by that lost from the surface by radiation and convection. α_s is the solar absorptance of the irradiated surface, q_s is the intensity of sunlight reaching the surface (W/m^2), q_m is the metabolic rate of the occupant (W/m^2), ε is the infrared emittance of the surface, σ is Stefan-Boltzmann's constant (5.67×10^{-8} W/m^2K^4), T is the surface temperature, T_a is the atmospheric temperature, and h is the convective heat transfer coefficient (W/m^2K).

If the temperature difference $\Delta T = T\text{-}T_a$ is small compared to the ambient temperature (293K) than Equ. 24.28 may be linearized to give

$$\Delta T = T - T_a = \frac{\alpha_s q_s + q_m}{4\varepsilon\sigma T_a^3 + h} \qquad 24.29$$

Equ. 24.29 is illustrated in Fig. 24.22. The pain threshold for contact with good thermal conductors such as most metals has been taken to be 46°C, which corresponds to a temperature rise of 21°C above ambient temperature assumed to be 25°C. Higher temperatures than this cause skin burns. The metabolic rate of the man inside the armour is assumed to be 55 W/m^2 and the heat transfer coefficient between the armour plate and the surrounding air is assumed to be 10 W/m^2K. The latter value can vary in the range 5–30 W/m^2K depending on wind conditions. Clearly white painted surfaces remain below the pain threshold even on the sunniest days when radiation exceeds 1000 W/m^2, but black painted surfaces reach the pain threshold when radiation is only 300 W/m^2 and oxidised steel surfaces when radiation is only 210 W/m^2. The validity of these results may be confirmed by anyone who touches the metal parts of a car that has been left in the midday sun for any considerable length of time. Other results could be obtained from different assumptions but the main conclusion remains and it is hardly surprising that crusader knights wore a white surcoat over their armour.

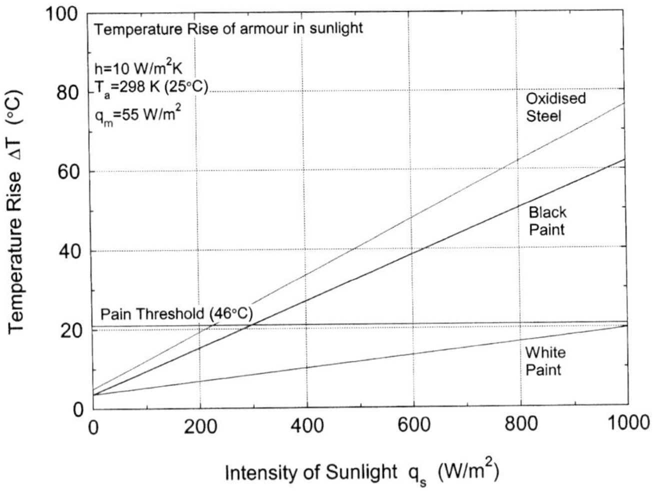

Fig. 24.22 Steady state temperature rise of armour plate subjected to sunlight and cooled by natural convection to air at 25°C.

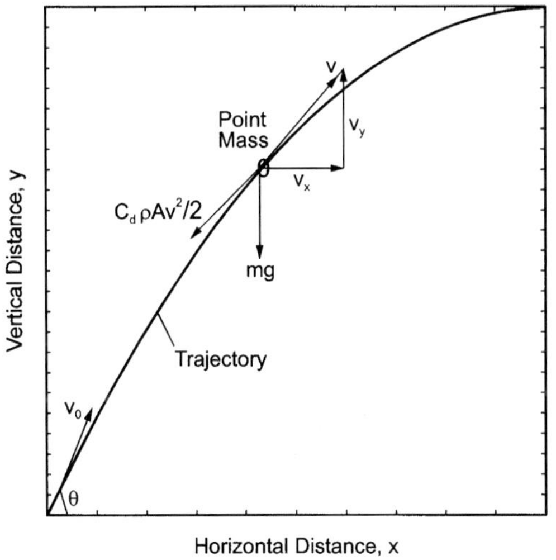

Fig. 24.23 Forces acting on a point mass travelling in a gravitational field.

External Ballistics

The equations of motion for the centre of gravity of a point-mass travelling in the x, y plane through a stationary fluid, illustrated in Fig. 24.23, may be written as

$$m\frac{dv_y}{dt} = mv_y\frac{dv_y}{dy} = -\frac{C_d\rho Avv_y}{2} - m\,g$$
$$m\frac{dv_x}{dt} = mv_x\frac{dv_x}{dx} = -\frac{C_d\rho Avv_x}{2} \qquad 24.30$$
$$v^2 = v_x^2 + v_y^2$$

m is the projectile mass, v_x and v_y are the velocity components in the x and y directions, C_d is the drag coefficient, ρ is the fluid density, A is the cross-sectional area of the projectile, v is the resultant velocity and g is local gravity (9.81 ms^{-2}). The lift force, generated by a javelin, for example, has been ignored. If the initial velocity of the projectile is v_0 and the initial angle of fire is θ from the horizontal, then the initial velocities in the x and y directions at time $t = 0$ are

$$v_x = v_0 \cos\theta$$
$$v_y = v_0 \sin\theta \qquad 24.31$$

We will examine three solutions to these equations corresponding to an elementary solution, in which the drag forces are ignored, a solution for short range, high velocity projectiles in which motion in the y-direction may be ignored, and finally the full numerical solution for the maximum projectile range. In all these solutions the drag coefficient, the fluid density, the projectile area, the projectile mass and local gravity, are assumed to be constant. Moreover it is assumed that the rotation of the Earth during the time of flight, winds, changes in air density with altitude, forces generated by projectile spin, and lifting force that causes a projectile to rotate about its centre of gravity, can be ignored.

Elementary Solution

For motion through a vacuum or for low velocity motion of a heavy projectile through a low-density fluid, the drag force may be ignored and the solutions to Equ. 24.30 and 24.31 are

$$v_x = v_0 \cos\theta$$
$$v_y = v_0 \sin\theta - gt \qquad 24.32$$

The velocity in the x-direction is constant and in the y-direction it varies linearly with time. Integration of Equ. 24.32 gives the x, y ordinates of the projectile's trajectory, namely

$$x = v_0 t \cos\theta$$
$$y = v_0 t \sin\theta - \frac{gt^2}{2} \qquad 24.33$$

The time taken for the projectile to return to $y = 0$ is the flight time, t_f; this may be determined by putting $t = t_f$ and $y = 0$ in Equ. 24.33, thus

$$t_f = \frac{2v_0 \sin\theta}{g} \qquad 24.34$$

The maximum distance travelled by the projectile in the x-direction, its range R, is determined by the product of flight-time and the velocity in the x-direction, thus

$$R = \frac{v_0^2}{g} 2\sin\theta\cos\theta = \frac{v_0^2}{g}\sin(2\theta) \qquad 24.35$$

The maximum range occurs when $\sin(2\theta) = 1$, that is, when the angle of fire is 45 degrees, and so the maximum range is given by the simple equation

$$R_{max} = \frac{v_0^2}{g} \qquad 24.36$$

For targets at distances less than the maximum there are two solutions to Equ. 24.33. If the angle of fire is less than 45° then it is referred to a *direct fire* whereas if the angle of fire is greater than 45° then it is called *indirect fire.* Direct and indirect fire is illustrated in Fig. 24.24. Note that with an angle of fire of 15 degrees the trajectory is nearly horizontal but despite this the projectile attains 50% of its maximum range.

Short Range, Direct Fire

If the target is quite close, less than half the maximum range, then there is insufficient time for significant motion in the y-direction and the projectile may be assumed to travel in a horizontal straight line. The equation of motion in the x-direction is then

$$m\frac{dv_x}{dt} = mv_x\frac{dv_x}{dx} = -\frac{C_d\rho A v_x^2}{2} \qquad 24.37$$

The solution is

$$v = v_0 \exp[-(C_d\rho Ax) / (2m)] \qquad 24.38$$

v_0 is the initial velocity of the projectile. The effect of the drag force is to retard the projectile such that the velocity falls exponentially from its initial value.

General Solution for Maximum Range

The general solution to Equ. 24.30 with the initial conditions of Equ. 24.31 is not conveniently solved in an explicit, analytical form, but the equations may be made non-dimensional and solved numerically using a suitable computer program. In non-dimensional form the equations may be written as

$$v_y^*\frac{dv_y^*}{dy^*} + v^*v_y^* + N = 0 \qquad 24.39$$

$$\frac{dv_x^*}{dx^*} + v^* = 0$$

$$v^{*2} = v_x^{*2} + v_y^{*2}$$

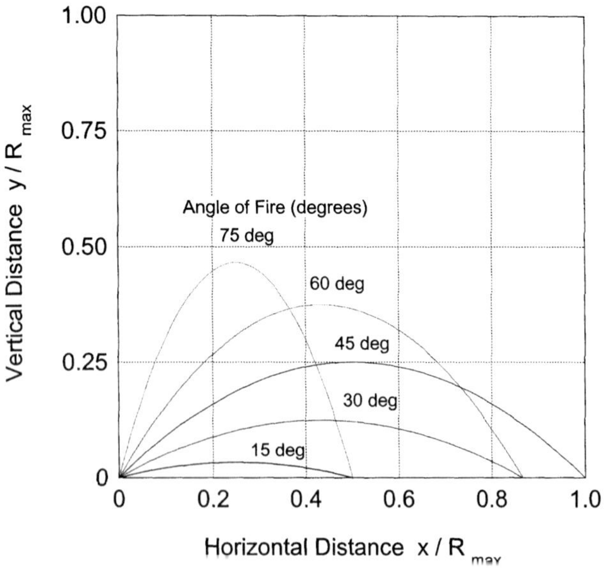

Fig. 24.24 Direct and indirect flight trajectories ignoring aerodynamic drag.

The * indicates a non-dimensional quantity, that is

$$v_x^* = \frac{v_x}{v_0} \quad v_y^* = \frac{v_y}{v_0} \quad v^* = \frac{v}{v_0}$$

$$x^* = \frac{x}{L} \quad y^* = \frac{y}{L} \qquad 24.40$$

$$L = \frac{2m}{C_d \rho A} \quad N = \frac{2mg}{C_d \rho A v_0^2}$$

L is a characteristic length and N is the ratio of projectile weight to drag force at $t = 0$. The initial conditions ($t = 0$) of Equ. 24.39 are then

$$v_x^* = \cos\theta \quad v_y^* = \sin\theta \quad v_0^* = 1 \qquad 24.41$$

The drag coefficients for various spheres and cylinders are shown in Fig. 24.25. Clearly the drag coefficient is not independent of velocity

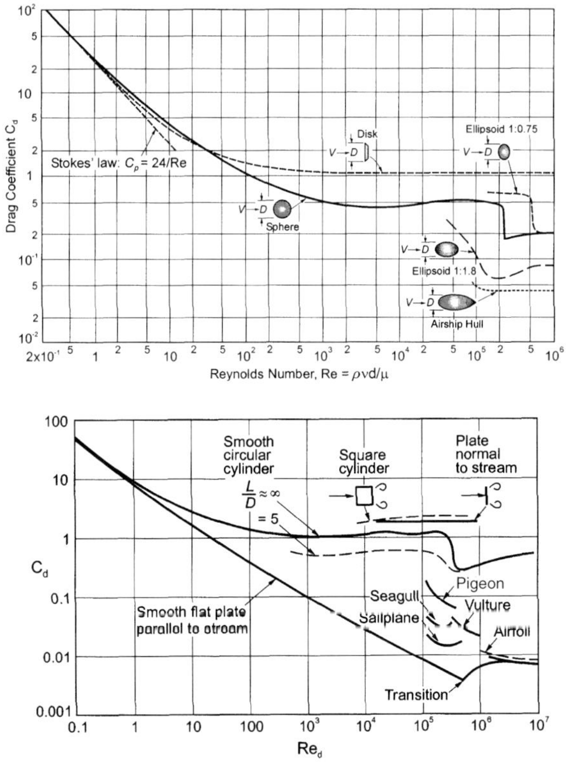

Fig. 24.25 Drag coefficients for spheres, cylinders. After Evett and Liu (1989).

(Mach number, or Reynolds number) but over small ranges of velocity it may be approximated as a constant. Equ. 24.39 and Equ. 24.41 may be solved numerically using a simple computer program and some solutions, giving the maximum projectile range and firing angle, are shown in Fig. 24.26. As we have seen, Equ. 24.36, the maximum range, when drag force is zero and the angle of fire is 45°, is $Rg / v_0^2 = 1$. Fig. 24.26 (left) shows that this solution remains reasonable so long as the weight/drag ratio of the projectile is greater than about 10. But for lighter projectiles the range decreases, as a proportion of the maximum range, and the angle of fire needs to be reduced. For very light projectiles with large drag forces the angle of fire for maximum range may be 30° or less, and the maximum range may be less than 20% of the range that could be attained when shooting into a vacuum. Extreme examples of objects having a low weight/drag ratio or high weight/drag ratio are feathers and cannon balls. The computed data plotted in Fig. 24.26 are also shown in Table 24.4. The fitted equation shown in Fig. 24.26 (left) is accurate to better than 1% over most of the range and may be applied to most pre-industrial weapons, including guns, so long as the air density and drag coefficient remain constant along the flight path. As a good approximation the maximum range, R, may be determined from the equation

$$\frac{Rg}{v_0^2} = \frac{1}{1 + \exp(-0.9(\ln(N) + 0.47))} \qquad N = \frac{2mg}{C_d \rho A v_0^2} \qquad 24.42$$

Table 24.4 Effect of weight/drag ratio on Range.

Weight/Drag Ratio N	*Initial Angle θ degrees*	*Range Rg/v_0^2*
0.1	33	0.168
0.2	35.5	0.266
0.5	39	0.442
1	41.2	0.595
2	43	0.736
5	44.3	0.870
10	44.8	0.932
20	45	0.965
50	45	0.987
100	45	0.994

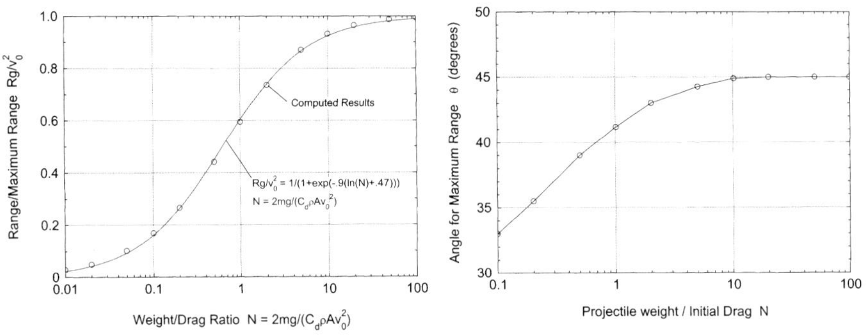

Fig. 24.26 Left: influence of drag force on maximum range of projectiles. Right: angle for maximum range.

At first sight it may appear that the external ballistics of a projectile is independent of the propulsion system (bow, ballista, gun etc) so long as it provides the requisite initial velocity. However, most propulsion systems have a more or less fixed energy available that must be shared between the inevitable system losses and the projectile mass. So not only does the range of a projectile depend on its own weight and initial velocity but also depends on parasitic losses of the weapon and its efficiency. A heavy projectile cannot be given a large velocity and consequently does not achieve a great range. On the other hand, a very light projectile can be given a large velocity but because weight to drag ratio is low it too cannot attain a great range. The maximum range or maximum striking velocity is attained with some intermediate projectile weight, and this will be determined after the characteristics of propulsion systems have been described.

Bow and Arrow

The force, F, required to draw a bow through a given distance, x is illustrated in Fig. 24.27. In general the relation between these quantities is non-linear and the shape shown in Fig. 24.27 is typical. For small draw-distances the force increases linearly with distance, and the slope of the force-distance curve is the stiffness of the bow. As draw distance increases the bow stiffness decreases but as the string approaches its maximum draw the stiffness increases quickly and the bow becomes very difficult to bend. When the arrow is released the force reduces to zero by a different path, as indicated in Fig. 24.27, because of hysteresis losses in the bow stave. The work done in drawing the bow is the area under the upper curve, but the work after release is the area beneath the lower curve. This work can be expressed as some fraction, η_i, of the maximum draw force, F_0, and the draw distance, x_0, and is equal to the kinetic energy of the arrow when it is released plus a small amount of parasitic energy loss. This loss arises from a variety of sources such as the viscous drag as the bow and string travels through the air, and the vibrational energy retained in the bow and string after the arrow has departed. When a bow is drawn and released without an arrow then the work done in drawing the bow is entirely expended against these losses. It is usual to lump all these losses together as a parasitic mass m_p and to write

$$W = \int_0^{x_0} F \, dx = \eta_i F_0 x_0 = \frac{(m + m_p)v^2}{2} \qquad 24.43$$

m is the mass of the arrow, m_p is the parasitic mass, v is initial velocity, F_0 the maximum force exerted by the bowstring, x_0 the draw distance, and η_i is the intrinsic efficiency of the bow, that is, it is the proportion of the maximum work that is realised when the arrow is released. Clearly, for maximum arrow-velocity and maximum range, the area under the force-draw curve should be a maximum.

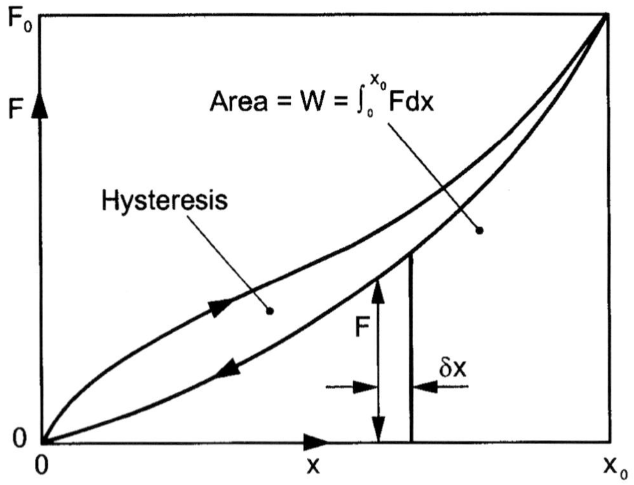

Fig. 24.27 Draw force and work of a typical bow.

Restrictions on the performance of a bow are related to the strength and geometry of the human body. Generally, the length of the extended arm dictates the maximum distance between the stave and the string. If the string is drawn to touch the nose, then this distance is about 0.6 m, whereas if it is drawn past the ear, the distance is about 0.8 m. The greatest force that can be exerted by the arms is usually less than about 800 N. Thus, the maximum energy that may be imparted to an arrow, assuming an efficiency of about 50%, is $0.5 \times 0.8 \times 800 = 320$ J. This is sufficient to cause incapacitation in man and in many animals. A striking energy of about 80 J is often sufficient to incapacitate man but, of course, a lot depends on where the projectile strikes and his degree of protection, as described in the sections on wound ballistics and penetration of armour. If the arrow mass is 0.1 kg and the parasitic mass is 0.05 kg then the launch velocity is about 65 m/s and the maximum range, neglecting drag forces on the arrow, is about 435 m. Drag forces on the arrow will reduce this to, perhaps, 300 m, and the kinetic energy of the arrow at this range is reduced to about 140 J, which is still sufficient to incapacitate. These results represent the best that can be attained from a simple hand operated bow. But not all bows can attain this level of performance and to understand the bow's historical development it is necessary to understand the conditions under which this performance may be attained. The real efficiency of a bow may be defined as the ratio of the kinetic energy of the arrow to the maximum work input

$$\eta = \frac{mv^2}{2F_0x_0} \qquad 24.44$$

From Equ. 24.43 and 24.44 the relation between the actual efficiency and the intrinsic efficiency is

$$\eta = \frac{\eta_i}{1 + m_p / m} = \frac{m\,\eta_i}{m + m_p} \qquad 24.45$$

The parasitic mass of a bow is difficult to determine theoretically but its value is roughly equal to or less than the mass of the arrow and so the actual bow efficiency is about half its intrinsic efficiency. The parasitic mass and intrinsic efficiency is usually determined experimentally by firing a number of arrows of different weights. Bergman et al. (1988) have tested replicas of several types of ancient bow and have measured the launch velocities using the sky screen technique, frequently used to measure the velocity of bullets and artillery shells. They tested the traditional English yew longbow, which is a simple self bow, and two composite bows of ancient design, namely the Egyptian composite bow and the Crimean composite bow. Composite bows are made from several materials but the Crimean bow, in addition to being composite is also recursive, that is, when unstrung it bends the opposite way. Although other bows were also tested these three are the only ones tested using arrows of different masses; and this is a prerequisite if the intrinsic efficiency and parasitic mass is to be determined. To Bergman's observations we may add similar observations made by Pratt (1992) using a 70 lb longbow. The test results are shown in Table 24.5.

Table 24.5 Technical data for bows.

Bow	*Length m*	*Draw Force N*	*Draw Length m*	*Arrow Mass kg*	*Arrow Velocity m/s*
**Sioux*	1.117	244	0.533	0.03	30
**African*	1.727	235	0.508	0.04	35
				0.09	43
				0.07	47
**Yew Longbow*	1.93	355	0.813	0.067	49
				0.065	51
				0.05	53
**Apache*	1.193	169	0.609	0.028	43
				0.09	32
**Egyptian*	1.530	282	1.016	0.05	40
Composite				0.04	47
				0.25	52
				0.05	51
**Crimean*	1.500	267	0.813	0.03	58
Composite				0.025	60
**Crossbow*	–	400	–	0.013	62
**Spear Thrower*	–	–	–	0.195	23
				.0162	64.1
				.024	60.1
***70lb Longbow*	–	310	0.67	.0479	46.5
				.0574	43.6
				.0736	38.7

* Bergman et al., (1988), ** Pratt (1992).

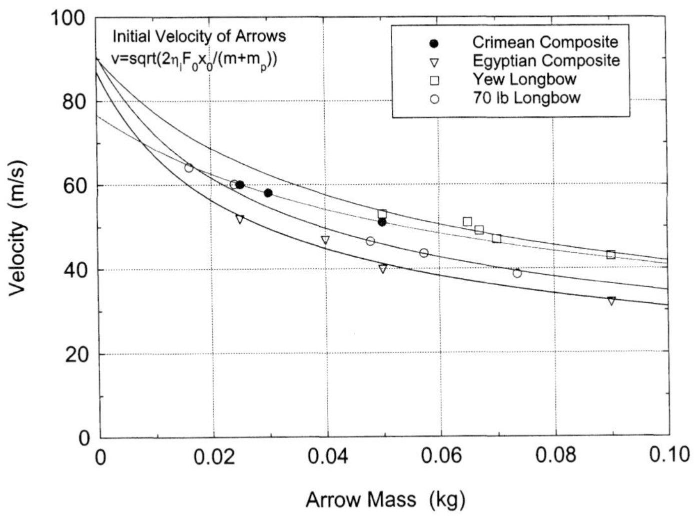

Egyptian bow:	$\eta_i = 19.0\%$.	$m_p = 14.4$ grams	$F_0x_0 = 286.5$ J
70lb Longbow:	$\eta_i = 33.9\%$.	$m_p = 17.1$ grams	$F_0x_0 = 207.7$ J
Yew Longbow:	$\eta_i = 38.7\%$.	$m_p = 27.3$ grams	$F_0x_0 = 288.6$ J
Crimean Composite bow:	$\eta_i = 53.8\%$.	$m_p = 53.5$ grams	$F_0x_0 = 217.1$ J

Fig. 24.28 Variation of initial velocity with arrow mass for various bows.

By rearranging Equ. 24.44 the initial velocity of the arrow may be expressed as

$$v = \sqrt{\frac{2\eta_i F_0 x_0}{m + m_p}} \qquad 24.46$$

Thus plotting the measured velocity against the arrow mass, and knowing the draw force and draw distance, it is possible to fit Equ. 24.46 to the observed data in Table 24.5 and determine the values of the intrinsic efficiency and the parasitic mass for each bow. This is shown in Fig. 24.28. Clearly, Equ. 24.46 fits the observed data quite well and the initial velocity of the arrow declines as its weight increases. Velocities between 30 m/s and about 65 m/s were achieved. The Egyptian composite bow was the least effective and the Yew (1.93 m) longbow was the most effective although it was only marginally better than the Crimean composite bow. The intrinsic efficiency varied from 19% for the Egyptian bow to about 54% for the Crimean composite bow and the parasitic mass varied from 14.4 grams for the Egyptian bow to about 54 grams for the Crimean composite bow.

It seems, from these results, that the parasitic mass increases linearly with the intrinsic efficiency, but this may not be true in general.

In computing the range attainable by projectiles, as in Fig. 24.26 or Table 24.4, it is assumed that the drag force and the mass are independent variables. Such an assumption is reasonable but in prac-

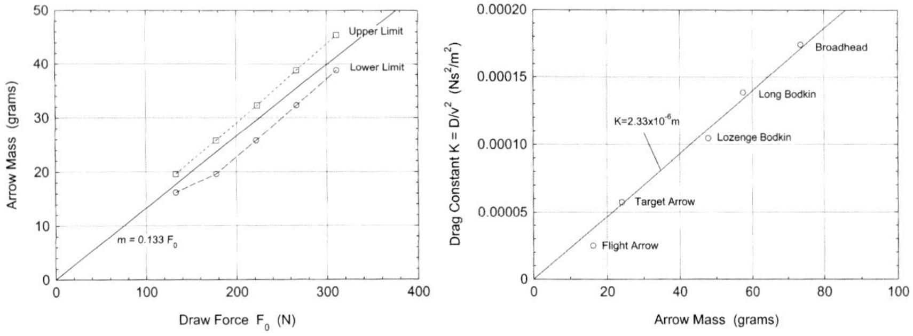

Fig. 24.29 Left: Mass of modern arrows. Right: Measured drag of arrows. The drag constant is $K = C_d\rho\ A\ /\ 2$. After Pratt (1992).

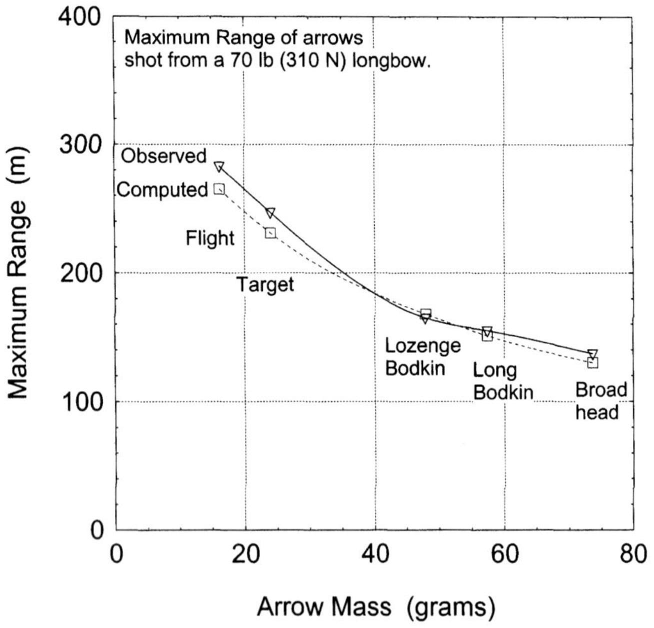

Fig. 24.30 Comparison of Computed and measured maximum range of modern arrows. Experimental data measured by Pratt (1992).

tice changing one variable, the arrow mass for example, will usually change the cross-sectional area, and hence the drag. In modern arrows, and no doubt in older arrows also, the weight is related to the maximum draw force of the bow, see Fig. 24.29 (left), and heavier arrows usually create more drag because they require larger feathers to retain stability in flight. Pratt (1992) has measured the relation between drag force and velocity for several arrows, see Fig. 24.29 (right), and it appears that the drag constant (drag force divided by velocity squared) increases linearly with arrow mass.

Taking the efficiency and parametric mass of the 70 lb longbow and assuming that the drag constant is proportional to arrow mass, then it is possible to compute the initial velocity from Equ. 24.46 and the maximum range from Table 24.4 for the arrows tested by Pratt. The results are plotted in Fig. 24.30 and compared with the observed maximum range. The result is quite accurate for heavy arrows but is about 6% in error for the target arrow and the flight arrow.

Example. Initial velocity and range of a Yew Longbow
Determine the initial velocity and maximum range of a 73.6 gram arrow shot from a yew longbow having an intrinsic efficiency of 38.7% and a parasitic mass of 27.3 grams if the draw force is 355 N and the draw distance is 0.813 m.

From Equ. 24.46 the initial velocity is 47.05 m/s.
From Fig. 24.29 (right) the drag constant is 0.0001715 Ns^2/m^2
The weight/drag ratio from Equ. 24.42 is N = 1.902
Thus the range from Table 24.4 or from the equation in Equ. 24.42 is 165 m.

The performance of a modern longbow (Deltas V) has been measured by Missliwetz and Wieser (1985) and is summarised in Table 24.6. The kinetic energy of the arrow at launch was 97.7 J, the efficiency was 0.372 and the drag coefficient of the arrow passing through air was about 0.8, based on a diameter of 8.6 mm and an air density of 1.205 kg/m^3. The theoretical maximum range, assuming optimum arrow mass, is 398 m, which compares well with the measured range of 337 m so the arrow mass is close to its optimum. The ideal bow efficiency and the parasitic mass cannot be determined from tests on a single arrow, nevertheless it will be noticed that the efficiency of this modern longbow is greater than that of the yew longbow. The penetration of this weapon into wood is 4.6 cm and it passed through a 30 cm thickness of gelatine (20% gelatine, NATO standard). The gelatine is supposed to offer the same resistance to projectiles as human flesh so it is supposed that the arrow will penetrate in excess of 30 cm through flesh at short range.

Table 24.6 Performance of a Delta V longbow, Missliwetz and Wieser (1985).

Mass of Bow (kg)	3.58	*Arrow mass (kg)*	0.0393
Draw Force (N)	356	*Arrow length (m)*	0.768
Draw Length (m)	0.737	*Arrow Dia. (mm)*	8.6
Launch Velocity (m/s)	70.5	*Drag coefficient*	0.8
Maximum Range (m)	337	*Energy at maximum range (J)*	46.6
Penetration (m)	4.6 cm wood >30 cm gelatine	*Velocity at maximum range (m/s)*	48.7
Launch Energy (J)	97.7	*Flight time (s)*	8.6

Efficiency = 37.2%

The Constant Torque Ballista

In a constant torque catapult, Fig. 24.31, the arm is pulled back through θ_0 degrees from the vertical and then released. At some angle θ degrees to the vertical, the arm is stopped and the projectile, of mass *m*, is released. Ignoring drag resistance the range attained by

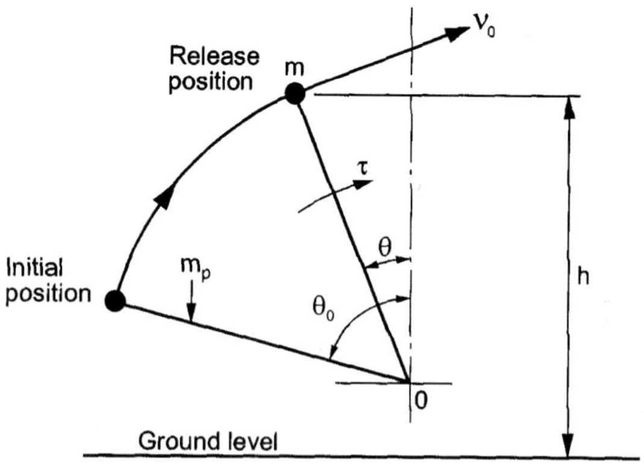

Fig. 24.31 Essential geometry of a constant torque Ballista

a heavy missile depends only on its launch velocity, v_0, its launch angle, θ, and the release height H above the impact point. The range is

$$R = \frac{v^2}{g}[\sin(2\theta) + \frac{h}{R}\cos^2\theta] \quad 24.47$$

The maximum range, if $h = 0$, occurs when $\theta = 45$ degrees. The analysis of the ballista is similar to that for a bow except that the work input is written in terms of torque rather than force. A proportion η_i of the work done is transferred to the kinetic energy of the arm and the projectile and the remainder is expended in friction (rubbing of the torsion springs) thus

$$\eta_i \tau(\theta_0 - \theta) = (m + m_p)\frac{v_0^2}{2} \quad 24.48$$

τ is the torque, m_p is the parasitic mass, due mainly to the inertia of the throwing arm. Eliminating the velocity from the above two equations, and assuming that H/R is small, gives an expression for the range, which may be written as

$$R = \frac{2\eta_i \tau}{(m + m_p)g}(\theta_0 - \theta)\sin 2\theta \quad 24.49$$

Fig. 24.32 illustrates the variation in dimensionless range with the initial angle, θ_0 and the release angle θ. Clearly the initial angle should be as large as possible and 70 to 90 degrees is readily attained by most ballistae. For a 90 degree initial angle, that is, the arm is initially horizontal; the greatest range is attained when the release angle is about 30 degrees from the vertical. The optimum release angle, it should be noted, depends only on the initial angle and is independent of the applied torque, the projectile mass, and the parasitic mass.

The effectiveness of a ballista against a target, such as a vertical wall, depends on the striking velocity, the angle of impact, and the mass of the projectile. If air resistance is ignored and the target is at the same height as the ballista then the striking angle will be the same as the release angle, θ, and the striking velocity will be the same as the launch velocity. Such assumptions are reasonable for heavy ballistae where aerodynamic drag is insignificant. If the release angle is 30 degrees from the horizontal then the striking angle will be 30 degrees from the horizontal and so the range may be the maximum but the effectiveness of the projectile impact will be reduced to some degree compared to a normal (90 degree) impact. It may be better to sacrifice range to improve impact performance, but this may not always be possible.

The non-dimensional performance characteristics of a model ballista are shown in Fig. 24.33. The release angle was varied from about 13° to about 38°. The arm radius was about 11 cm and the torque was provided by twisted elastic bands that were twisted sufficiently to produce a substantially constant torque of about 32.4×10^{-3} Nm. Ball bearings from 6 mm to 12 mm diameter were thrown and the height of the ball at the release point was about 12 cm. A range of between 50 and 120 cm was attained and the balls landed in soft sand so the exact range was easily measured to the first impact point. The non-dimensional axes used in Fig. 24.33 are based on Equ. 24.49 and the input energy, used in computing the vertical axis, is $E = \tau\ (\theta_0 - \theta)$. The non-dimensional range increase linearly with the mass ratio and the constant of proportionality is $sin(2\theta)$, where θ is the release angle. The experimental results lie quite close to the theoretical lines if the parasitic mass of the model is taken to be $m_p = 2.5$ grams, and the intrinsic efficiency is $\eta_i = 0.65$.

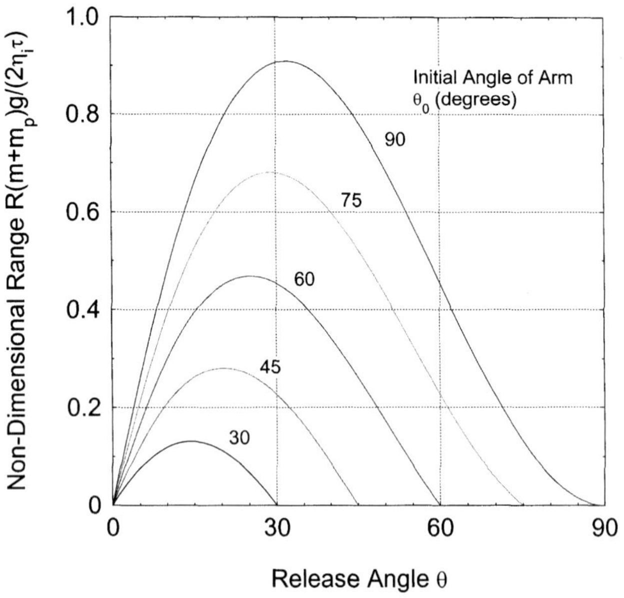

Fig. 24.32 Range of a constant torque ballista, illustrating the optimum release angle.

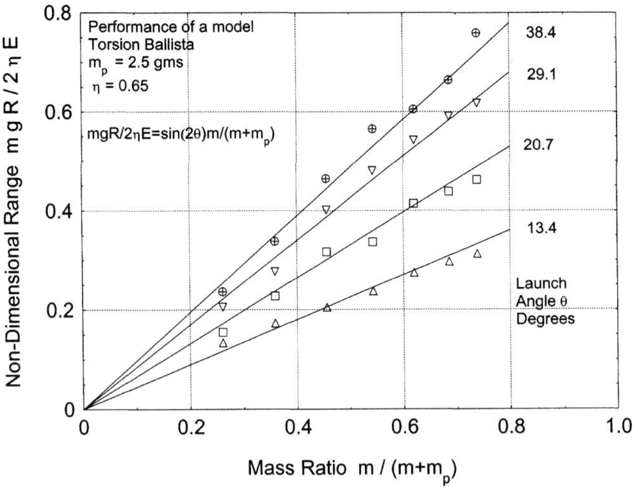

Fig. 24.33 Comparison of the actual and the theoretical performance of a model Ballista.

These results and equations may be used for other, larger, ballistae if the parasitic mass and efficiency is known.

Example. A Roman Catapult
Josephus claimed that a Roman catapult could throw a 25 kg stone a distance of 366 m. Determine the torque to do this assuming an intrinsic efficiency of 65%, parasitic mass of 50 kg, release angle $\theta = 30°$ and initial angle $\theta_0 = 90°$. Take $g = 9.81\ m/s^2$.

From Equ. 24.49 the torque is $\tau = 228{,}383$ Nm (23.28 tonne m) This is a load of 5.82 tonnes at a radius of 4 m, or 2.9 tonne at 8 m etc.

Performance of Guns

Fig. 24.34 shows the geometry of a typical gun barrel and the pressure distribution along the barrel as the projectile moves from its initial position at the breech to the muzzle. The real pressure-distance curve may be approximated with reasonable accuracy by constant volume, constant pressure and polytropic curves as illustrated. The gunpowder is assumed to burn at constant volume (the projectile remains stationary) until the maximum pressure is reached, after which it burns at constant pressure until combustion is complete. During this

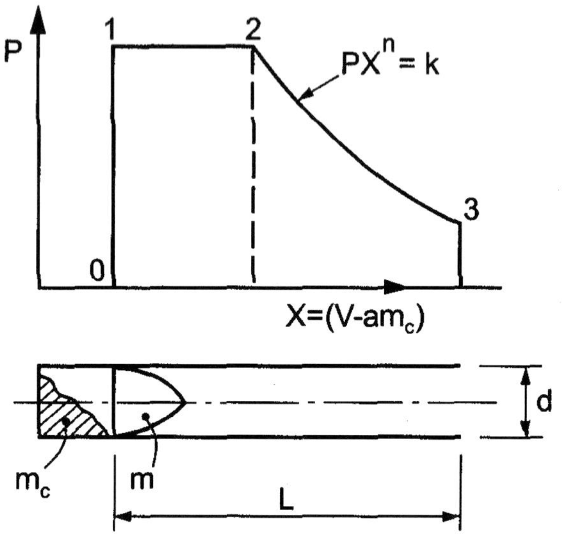

Fig. 24.34 Geometry of a gun barrel and pressure variation along barrel.

constant pressure period the projectile necessarily moves down the barrel and after the gunpowder has burned to completion the projectile continues to accelerate down the barrel causing the pressure to fall. The fall in pressure may be assumed to be polytropic, that is, the relation between the pressure and the volume of the gas is $p(V\text{-}am_c)^n = constant$, where n is a constant close to the ratio of specific heats of the gas; for most gun propellants the value of the index of expansion is $n = 1.3$. The constant a is the co-volume and a typical value is 0.001 m^3/kg, p is the pressure, and V is the volume of gas between the projectile and the breech. The shot travel distance is L, the bore diameter is d, the charge mass is m_c and the projectile mass is m.

The area under the pressure-volume curve is the work done by the expanding gas, thus

$$W = p_1(X_2 - X_1) + \frac{p_2X_2 - p_3X_3}{n-1} \qquad 24.50$$

W is the work done, p is gas pressure, X is gas volume minus the product of charge mass and co-volume, and the suffices are defined in Fig. 24.34. For a polytropic process

$$p_2X_2^n = p_3X_3^n \qquad 24.51$$

Thus Equ. 24.50 may be written as

$$\frac{W}{p_1 X_3} = \frac{n\lambda - \lambda^n}{n-1} - \frac{X_1}{X_3} \qquad \lambda = \frac{X_2}{X_3} \qquad 24.52$$

λ is the position of "all-burnt", that is, it defines the position along the barrel at which all the solid propellant charge, usually gunpowder, has burnt to form a gas. Its value may vary between $0 < \lambda < 1$. In gunpowder and other solid propellants there is insufficient oxygen to give complete combustion of the carbon and hydrogen so combustion of un-burnt hydrogen and carbon monoxide restarts when these gases emerge from the muzzle, giving rise to the characteristic muzzle flash. The value of X_1/X_3 is usually quite close to zero, partly because X_1 is very small ($a = V_1/m_c$, approximately) and partly because X_3 is large. Thus the ratio X_1/X_3 is ignored and the value of X_3 is the bore area times the shot travel distance. Thus

$$\frac{W}{p_1 A} = \frac{p_m}{p_{\max}} = \frac{n\lambda - \lambda^n}{n-1} \qquad A = \frac{\pi}{4} d^2 L \qquad 24.53$$

p_m is the mean pressure driving the projectile down the barrel, $p_{max} = p_1$ is the maximum pressure, and A is the cross-sectional area of the barrel. The assumed pressure-volume relations, Fig. 24.34, always give an overestimate of the work done by the expanding gas but experience with modern weapons shows that the ratio of the real work to the ideal work, called the piezometric efficiency, η_p, is about 90–95% and so the relation between mean pressure and maximum pressure is more accurately written as

$$\frac{p_m}{p_1} = \frac{\eta_p (n\lambda - \lambda^n)}{n-1} \qquad 24.54$$

The mean pressure may be found also by equating the work done by the expanding gas to the kinetic energy of the projectile and the kinetic energy of the gas. The velocity of the gas is assumed to be linear with distance travelled down the barrel and its kinetic energy thus varies as the square of distance, consequently we may write

$$p_m \frac{\pi}{4} d^2 L = \left(m + \frac{m_c}{3} \right) \frac{v_0^2}{2} \qquad 24.55$$

m is the projectile mass, m_c is the charge mass and v_0 is the muzzle velocity. Re-arranging we obtain the muzzle velocity as

$$v_0 = \sqrt{\frac{\pi d^2 L\, p_m}{2(m + m_c / 3)}} \qquad 24.56$$

The mean pressure, of course, is determined from Equ. 24.54. The relation between mean pressure, maximum pressure and position of "all burnt" is shown in Fig. 24.35. For an all-burnt ratio of 0.3 the mean pressure is about 60% of the maximum pressure.

In the constant-volume, constant pressure, polytropic expansion model of gun internal ballistics it may be assumed that some proportion α of the charge mass is burnt at constant volume and the remainder $(1-\alpha)$ is burnt at constant pressure. The gas equation at point 1 of Fig. 24.34, including an allowance for co-volume, is

$$p_1(V_1 - a\alpha m_c) = \alpha m_c R T_1 \qquad 24.57$$

If the loading density is defined as the ratio of charge mass to volume at point 1, $\rho_1 = m_c / V_1$, and we write the force constant of the propellant as $F = RT_1$ then Equ. 24.57 is

$$\alpha = \frac{p_1}{\rho_1 (F + a p_1)} \qquad 24.58$$

The charge mass may be determined from the energy equation for constant pressure combustion between points 1 and 2, thus

$$(1 - \alpha) m_c E = \alpha m_c C_p T_1 - m_c C_p T_2 \qquad 24.59$$

Recognising that T_1 is the propellant's flame temperature, the energy content is $E = RT_1 / (\gamma - 1)$, and the specific heat is $C_p = \gamma R T_1 / (\gamma - 1)$, where γ is the ratio of specific heats, gives

$$\frac{T_2}{T_1} = \alpha + (1 - \alpha) / \gamma \qquad 24.60$$

The gas equation at point 2 is

$$p_2 X_2 = m_c R T_2 \qquad 24.61$$

Substituting from Equ. 24.60, putting $p_2 = p_1$, $X_2 = \lambda X_3$, $X_3 = \pi d^2 / 4$ and $F = RT_1$, gives a useful equation for the charge mass

$$m_c = \frac{\lambda p_1 \pi d^2 L}{4F[\alpha + (1 - \alpha) / \gamma]} \qquad 24.62$$

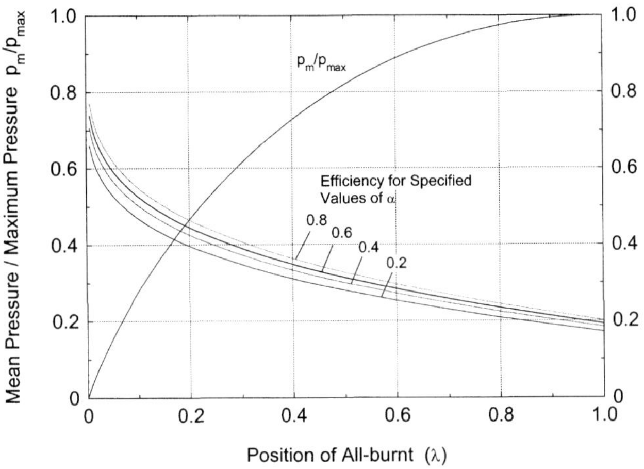

Fig. 24.35 Relation between mean pressure and position of "all burnt".

Defining the thermal efficiency, η, of a gun as the ratio of work W to the energy in the charge, $m_c E$, gives

$$\eta = \frac{\gamma - 1}{n - 1} \frac{n\lambda - \lambda^n}{\lambda} \left(\alpha + \frac{1 - \alpha}{\gamma} \right) \qquad 24.63$$

Although the charge mass is an important parameter in gun design the thermal efficiency is relatively unimportant and is rarely determined but it is, nevertheless, a useful quantity that deserves to be better known and used. Equ. 24.63 is plotted in Fig. 24.34 where it is apparent that for an all-burnt ratio of 30% the thermal efficiency is about 40%, depending on the proportion of propellant that burns at constant volume and defined by Equ. 24.58.

Gas leakage past the projectile has been ignored in this analysis. This is reasonable for modern projectiles that are a close fit in the barrel and are provided with driving bands to provide a seal, but for pre-industrial guns the clearance between a cannon ball and the bore was typically about 2 cm and a significant amount of gas flowed past the cannon ball, lowering the gas pressure and the muzzle velocity. The flow through the minimum area between the cannon ball and the barrel will be choked, because the pressure ratio is always greater than the critical pressure ratio, and so if the ratio of specific heats is 1.25 the average flow rate m_L/t is given by

$$\frac{m_L\sqrt{RT_m}}{tA_L p_m} = 0.655 \qquad 24.64$$

R is the gas constant, T_m is the mean temperature, A_L is the leakage area, and p_m is the mean pressure. The time taken for the projectile to reach the muzzle, t, may be determined approximately by equating the impulse to the change in momentum

$$t = \frac{mv_0}{Ap_m} \qquad 24.65$$

A is the cross-sectional area of the cannon ball, m is its mass, and v_0 is the muzzle velocity. Thus the gas that leaks past the cannon ball between ignition and shot exit is approximately

$$m_L = 0.655m\frac{A_L}{A}\sqrt{\frac{v_0^2}{RT_m}} \qquad 24.66$$

Taking R = 380 J/kgK, T_m = 1000K, v_0 = 150 m/s, A_L/A = 0.17 then m_L/m = 0.027 and the gas leakage is less than 3% of the projectile mass and is small enough to be ignored. For higher muzzle velocity, about 750 m/s, the proportion rises to about 13% of the projectile mass.

Lamé and Clapeyron in 1833 derived the equation relating the internal pressure of an elastic cylinder to the maximum tensile stress at the bore surface; see for example Timoshenko (1955). Lamé's equation may be written as

$$\frac{\sigma_t}{p} = \frac{\varsigma^2 + 1}{\varsigma^2 - 1} \qquad \varsigma = \frac{r_2}{r_1} \qquad 24.67$$

σ_t is the tangential (tensile) stress at the bore surface, p is the internal pressure and ς is the ratio of the outer radius to the inner radius. This equation may be re-arranged to the form

$$\varsigma = \sqrt{\frac{\sigma_t / p + 1}{\sigma_t / p - 1}} \qquad 24.68$$

Example: Gun Internal Ballistics

In 1742 Robins, see Crossland (1986), determined that the maximum pressure in a 0.75 inch (0.019 m) bore gun having a barrel length of 45 inches (1.15 m) is about 101 MPa. The gun fired a spherical lead bullet at a velocity of 1668 ft/s (757 m/s) and the charge mass was half the bullet mass. Determine the mean pressure and the maximum pressure assuming that "all burnt" is at 20% of the shot travel, the piezometric efficiency is 0.9, and the index of expansion is 1.3.

If the bullet diameter is 0.019 m and the density of lead is 11,680 kg/m^3, then the bullet mass is 0.042 kg and the charge mass is 0.021 kg.

From Equ. 24.55 the mean pressure required to give a muzzle velocity of 757 m/s is 43.0 MPa.

From Equ. 24.54 the ratio of mean pressure to maximum pressure is 0.41 if $\lambda = 0.2$, $n = 1.3$, and $\eta_p = 0.9$. Thus the maximum pressure is 105 MPa. This compares with Robins' value of 101 MPa.

Example: Barrel Stress and Strength
Determine the minimum outside diameter of the barrel of the above gun (19 mm bore and 105 MPa maximum pressure) if the yield strength of wrought iron is 210 MPa. What is the minimum outside diameter if the factor of safety is 1.5? If the tensile strength of wrought iron is 345 MPa determine the bursting pressure.

If σ_t is 210 MPa and p is 105 MPa then from Equ. 24.68 $\varsigma = 1.732$. Thus the minimum outside diameter is 0.0329 m (33 mm).

If σ_t is 210 MPa and p is $1.5 \times 105 = 157.5$ MPa then from Equ. 24.68 $\varsigma = 2.646$. Thus the minimum outside diameter is 0.0502 m (50 mm).

If $\varsigma = 2.646$ and σ_t is 345 MPa then from Equ. 24.67 the bursting pressure is 259 MPa.

Where medieval guns are concerned some rather strange ideas are put forward from time to time by otherwise excellent historians. Guilmartin (1974) for example says "as the cannonball accelerates down the barrel, a point is reached where the projectile is expanding the volume behind it faster than the decomposition of black powder can keep it pressurized. Extending the length of the barrel beyond this point will result in an actual reduction in the muzzle velocity and range of cannon." This is wrong; it is the pressure that reduces not the projectile velocity. For a cannonball to lose velocity the pressure ahead of it must be greater than the pressure behind it, and this never occurs. He also writes, "cannon were cast with long barrels not for ballistic reasons, but for structural ones—to give them added strength and safety." This too is incorrect for culverins had a longer barrel and higher muzzle velocity than cannon and Equ. 24.67 confirms that the strength of a barrel depends on the ratio of internal to external diameter but does not depend on its length. He was also of the opinion that sixteenth century cannon founders had found a way to overcome stress concentrations caused by sharp corners and that the outer 50% of the thickness of a cannon barrel made little or no contribution to the cannon's strength, neither of which are true.

Fixed Energy Propulsion and Optimum Performance

The energy given to a projectile is often limited by the propulsion system (bow, ballista, gun etc.). Thus a projectile of fixed energy may be of low mass and high velocity, or it may have a high mass and a low velocity. By a proper choice of projectile mass and velocity the performance of the system may be optimised. We will determine the optimum performance of a propulsion system firing firstly in short range, horizontal mode, and secondly when firing to give maximum range.

Direct Fire

A propulsion system is assumed capable of supplying to the projectile a quantity of energy, ηE_0, where E_0 is the energy supplied to the propulsion system and η is the efficiency of the system. This can be equated with the initial kinetic energy of the projectile to give

$$\eta E_0 = \frac{mv_0^2}{2} \qquad 24.69$$

Usually, the efficiency is not constant but varies with the projectile mass. This behaviour may be modelled by an intrinsic efficiency η_i, and a parasitic mass m_p, that is

$$\begin{aligned} \eta_i E_0 &= \frac{(m + m_p)v_0^2}{2} \\ \eta &= \frac{\eta_i}{1 + m_p / m} \end{aligned} \qquad 24.70$$

Thus the real efficiency approaches the intrinsic efficiency as the projectile mass increases and the parasitic mass decreases. This models

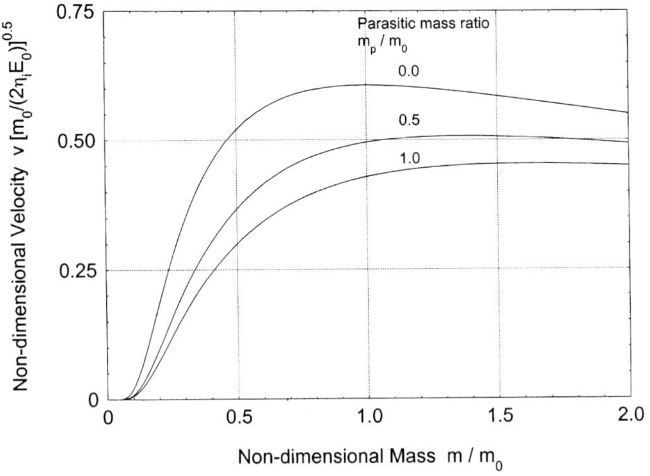

Fig. 24.36 Influence of projectile mass and parasitic mass on striking velocity at range x. Direct fire.

the behaviour of bows and other throwing engines such as ballistas and trebuchets, although the parasitic loss must be determined experimentally, and it also models the behaviour of guns if the parasitic mass is taken to be a third of the charge mass. If the system energy, the parasitic mass, and ideal efficiency are fixed then we may choose a large projectile mass with a small initial velocity or a smaller mass with a larger initial velocity. To optimise the performance in high velocity, short-range fire, we eliminate the initial velocity from Equ. 24.38 using Equ. 24.70 to give

$$v = \sqrt{\frac{2\eta_i E_0}{m(1 + m_p / m)}} \exp\left(-\frac{C_d \rho A x}{2m}\right) \qquad 24.71$$

In non-dimensional form

$$v\sqrt{\frac{m_0}{2\eta_i E_0}} = \sqrt{\frac{m_0}{m + m_p}} \exp\left(-\frac{m_0}{2m}\right) \qquad 24.72$$

$$m_0 = C_d \rho A x$$

Equ. 24.72 is plotted in Fig. 24.36 where it is evident that the impact velocity, at any specified range, has a maximum value. Differentiating Equ. 24.72 with respect to the projectile mass and equating to zero to give the mass for optimum striking velocity

$$\frac{m_{opt}}{m_0} = \frac{1}{2} + \sqrt{\frac{1}{4} + \frac{m_p}{m_0}} \qquad 24.73$$

$$m_0 = C_d \rho A x$$

m_{opt} is the projectile masses that maximises the striking velocity at range x. There is a different optimum mass for each target range and so, in general, a suitable average range must be chosen. If the mass is less than the optimum then the initial velocity will be high but the drag forces will retard the projectile so that the striking velocity at range x is less than the optimum. Alternatively, if the mass is greater than the optimum then the initial velocity is too low and the striking velocity at range x is again less than the optimum. This result is particularly useful if the flight-time is to be short or if the striking velocity is to be a maximum. If the parasitic mass is small in comparison to the mass m_0 then the optimum mass is approximately

$$m_{opt} = m_0 + \frac{2}{3} m_p = C_d \rho A x + \frac{2}{3} m_p \qquad 24.74$$

Fig. 24.36 also shows the effect of parasitic mass. The best results, of course, are obtained when the parasitic mass is zero and the optimum projectile mass is then equal to the mass m_0. Increasing the parasitic mass reduces the striking velocity of the projectile and also increases the magnitude of the optimum mass.

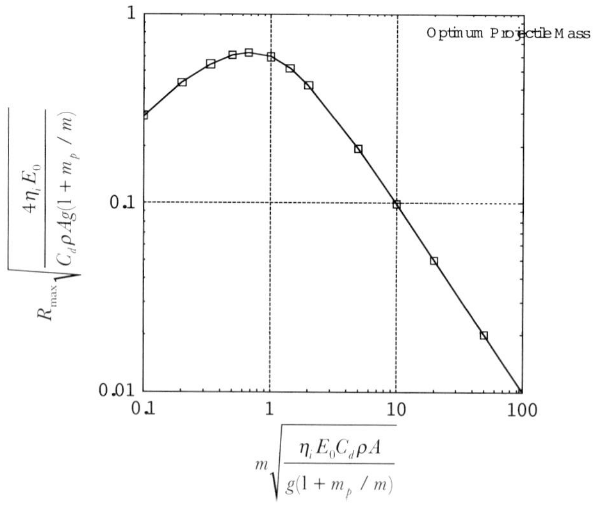

Fig. 24.37 The influence of projectile mass on maximum range.

Maximum Range

In a similar way, if the propulsion energy is fixed, it is possible to choose a suitable projectile mass and velocity to maximise a projectile's range. This is illustrated in Fig. 24.37, which is calculated by the numerical solution of Equ. 24.39 and Equ. 24.40 subject to the restriction of Equ. 24.41. The calculated results are shown in Table 24.7. In Fig. 24.37 there is a well-defined maximum range when the non-dimensional mass is 0.667. If the mass is smaller than this then the initial velocity is high but the drag forces are also high so the maximum range is reduced. On the other hand, if projectile mass is larger than the optimum then initial velocity is low and the maximum range is also low. From Fig. 24.37, and Table 24.7, the maximum range occurs when the projectile mass is chosen to be

$$m_{opt} = \frac{2}{3}\sqrt{\frac{\eta_i E_0 C_d \rho A}{g(1 + m_p / m)}} \qquad 24.75$$

The maximum range is then

$$R_{opt} = \frac{5}{4}\sqrt{\frac{\eta_i E_0}{C_d \rho A g(1 + m_p / m)}} \qquad 24.76$$

From Table 24.7 the firing angle for this condition is $\theta = 42°$.

Example. Maximum range of a yew longbow.

Determine the maximum range of the yew longbow specified in Table 24.5. The arrow mass is 50 grams, the ideal efficiency is 0.368, the bow energy (E_0 = FX) is 288 J, the drag coefficient is 0.8. The air density is 1.2 kg/m^3, the arrow diameter is 9 mm, local gravity is 9.81 m^2/s, and the parasitic mass of the bow is 23 grams.

Table 24.7 Effect of projectile mass on maximum range and firing angle.

*Non-Dimensional Projectile Mass**	*Non-Dimensional Range** and Firing Angle (deg)*	*Non-Dimensional Projectile Mass**	*Non-Dimensional Range** and Firing Angle (deg)*
0.100	0.290 (44.0°)	1.5	0.501 (38.4°)
0.200	0.434 (43.7°)	2.000	0.421 (36.0°)
0.333	0.543 (43.2°)	5.000	0.194 (30.1°)
0.500	0.607 (42.8°)	10.00	0.099 (26.3°)
0.667	0.625 (42.1°)	20.00	0.050 (——)
1.000	0.594 (40.1°)	50.00	0.020 (——)

* $m\sqrt{\frac{\eta_i E_0 C_d \rho A}{g(1 + m_p / m)}}$ ** $R_{max}\sqrt{\frac{4\eta_i E_0}{C_d \rho A g(1 + m_p / m)}}$

The non-dimensional mass is calculated to be 1.377. This is about twice as heavy as the optimum mass (0.667), but the loss of range is not too significant. From Fig. 24.37 (or Table 24.7) the non-dimensional range for this mass is about 0.52 and so the maximum range, when the bow is fired at about 39 degrees to the horizontal, is 261 m. The initial velocity is 55 m/s so the theoretical maximum range, ignoring aerodynamic drag, is 286 m.

Rockets

Fig. 24.38 shows the essential elements of a simple, solid propellant rocket. The warhead may be explosive or a simple incendiary composition. The propellant used for the earliest rockets was usually gunpowder, compressed so that gas could not penetrate and thus combustion occurred only at the surface. The propellant is shown here as a solid so it burns from one end, rather like a cigarette, but it could be made tubular so it burns from the inside of the tube, or in other shapes. The gas released by the very rapid combustion causes a pressure rise in the rocket and the gas accelerates to the throat (the minimum area). If the pressure rise is greater than the critical value, about 2 bar, then the nozzle is choked, the Mach number at the throat is unity, and the mass flow rate is then proportional to the upstream stagnation pressure. If a diverging nozzle is fitted then the gas continues to accelerate to exit producing a supersonic flow.

The thrust produced by the expanding gas is the sum of the momentum thrust and the pressure thrust at exit

$$F = wc_e + (p_e - p_a)A_e \qquad 24.77$$

w is the burning rate of propellant, kg/s, c_e is the exit velocity of the propellant gas, m/s, p_e is the exit pressure, N/m^2, p_a is the pressure of the atmosphere, and A_e is the nozzle area, m^2, at exit. Applying the steady flow energy equation between the stagnation conditions upstream of the nozzle and the nozzle exit, and assuming no heat loss, no work transfer, and no change in potential energy, gives

$$C_pT_0 = C_pT_e + \frac{c_e^2}{2} \qquad 24.78$$

C_p is the specific heat of the gas, T_0 is the burning temperature of the propellant and T_e is the exit temperature of the gas. It is con-

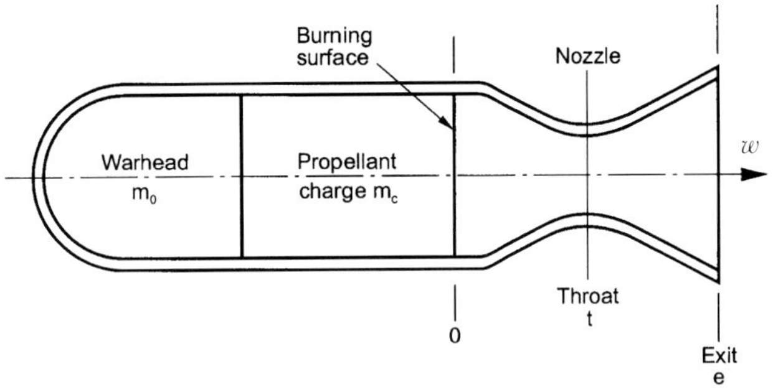

Fig. 24.38. Cross-section of a simple rocket.

ventional to replace the specific heat by its equivalent, $\gamma R/(\gamma-1)$, where R is the gas constant and γ is the ratio of specific heats, and for isentropic flow $T_e = T_0(p_e/p_0)^{\gamma/(\gamma-1)}$, thus the steady flow energy equation may be written as

$$c_e = \sqrt{\frac{2\gamma}{\gamma-1} RT_0\left\{1-(p_e/p_0)^{(\gamma-1)/\gamma}\right\}} \qquad 24.79$$

The thrust of the rocket is then

$$F = w\sqrt{\frac{2\gamma}{\gamma-1} RT_0\left\{1-(p_e/p_0)^{(\gamma-1)/\gamma}\right\}} + (p_e - p_a)A_e \qquad 24.80$$

In a well-designed rocket nozzle the gas is expanded from its stagnation pressure, p_0, to atmospheric pressure p_a by having a suitably sized exit area, thus the pressure thrust is zero. Also, in most rockets the mass of propellant greatly exceeds the fixed mass so the important characteristic is not the thrust but the thrust developed per unit of propellant flow rate, F/w. This is called the specific impulse, I; it has the same units as velocity.

$$\frac{F}{w} = I = \sqrt{\frac{2\gamma}{\gamma-1} RT_0\left\{1-(p_a/p_0)^{(\gamma-1)/\gamma}\right\}} \qquad 24.81$$

It is, of course, identical to the exit velocity, Equ. 24.79, when the pressure thrust is zero. The specific impulse is easily measured in static trials but it is not a unique property of a propellant. The gas constant, flame temperature T_0 and the ratio of specific heats depend only on the propellant but the stagnation pressure at which the propellant burns depends on the ratio of the burning area to the nozzle throat area. However, this area ratio has only a slight effect and the specific impulse is treated as if it were a unique property of the propellant.

Ignoring gravity and assuming a constant burning rate w (kg/s) the equation of motion for a solid propellant rocket may be written as

$$(m_0 + m_c - wt)\frac{dv}{dt} = Iw = C_d \rho A \frac{v^2}{2} \qquad 24.82$$

m_0 is the fixed mass, m_c is the initial mass of propellant v is the velocity of the rocket, t is time, I is the specific impulse of the propellant charge, C_d is the drag coefficient, ρ is the atmospheric density, A is the cross-sectional area. This may be written in dimensionless form using the specific impulse I as the characteristic velocity and the burn time, t_b as the characteristic time.

$$\frac{dv*}{dt*} = \frac{\beta^2 - v*^2}{\beta^2(1 + \alpha - t*)}$$
$$v* = v / I \quad t* = t / t_b \quad t_b = m_c / w \qquad 24.83$$
$$\beta^2 = \frac{2w}{C_d \rho A I} \qquad \alpha = \frac{m_0}{m_c}$$

β is a measure of the thrust to drag ratio and may be called the thrust to drag coefficient, and α is the ratio of the rocket's fixed mass to its initial charge mass. In integral form Equ. 24.83 is

$$\int_0^1 \frac{dt*}{1 + \alpha - t*} = \int_0^{v_b^*} \frac{\beta^2 dv*}{\beta^2 - v*^2} \qquad 24.84$$

The limits of integration correspond to the start and stop of propellant burning and $v_b* = v_b/I$ and is the rocket's velocity at burnout. The solution of Equ. 24.84 is

$$\ln\left(\frac{1 + \alpha}{\alpha}\right) = \frac{\beta}{2}\ln\left(\frac{\beta + v_b*}{\beta - v_b*}\right) \qquad 24.85$$

This may be rearranged to give the burnout velocity directly

$$v_b* = \frac{v_b}{I} = \beta\frac{K - 1}{K + 1}$$
$$K = \left(\frac{1 + \alpha}{\alpha}\right)^{2/\beta} \qquad 24.86$$

$$\beta = \sqrt{\frac{2w}{C_d \rho A I}} \qquad \alpha = \frac{m_0}{m_c}$$

Equ. 24.86 is plotted in Fig. 24.39 and illustrates the influence of the fixed mass to charge mass ratio and the thrust to drag coefficient on the velocity at burnout. Three approximations of Equ. 24.80 are useful. If the fixed mass (delivery mass) is small compared to the mass of propellant charge then $v_b^* = \beta$. On the other hand, if the fixed mass is large compared to the charge mass then $v_b^* = 1/\alpha$, regardless of the thrust to drag coefficient. Finally, if the thrust to drag coefficient is large, as it is for a rocket travelling through a vacuum, then $v_b^* = ln(1 + 1/\alpha)$.

The velocity at burn-out plotted in Fig. 24.39 is not the maximum possible velocity of the rocket but is simply the velocity attained when the propellant is used up. Greater velocity can be attained by carrying and burning more fuel but the limiting velocity occurs when the rocket ceases to accelerate, $dv^*/dt^* = 0$. From Equ. 24.83 this occurs when $v^* = \beta$, and thus the limiting velocity, assuming sufficient propellant is

$$v_{\max} = \sqrt{\frac{2wI}{C_d \rho A}} \qquad 24.87$$

The low velocities attained by the earliest rockets, usually less than 200 m/s, occurred not because the specific impulse, I, of the propellant, usually gunpowder, was low, or because the drag forces were high, but because its burning rate, w, was low. Burning rate increases with the pressure inside the rocket chamber and it increases with the surface area of the propellant that is burning. For a high velocity the chamber pressure and the burning area of the propellant must be large and on the earliest rockets this was not the case.

Example 1. Rocket Flight

A thin-skinned rocket of 100 mm diameter burns a propellant having a specific impulse of 800 m/s. If the linear burning rate is 5 mm/s and the burning area is equal to cross-section area determine the maximum possible velocity, assuming there is sufficient propellant. The drag coefficient is 0.3 the air density is 1.2 kg/m^3 and the propellant density is 1000 kg/m^3.

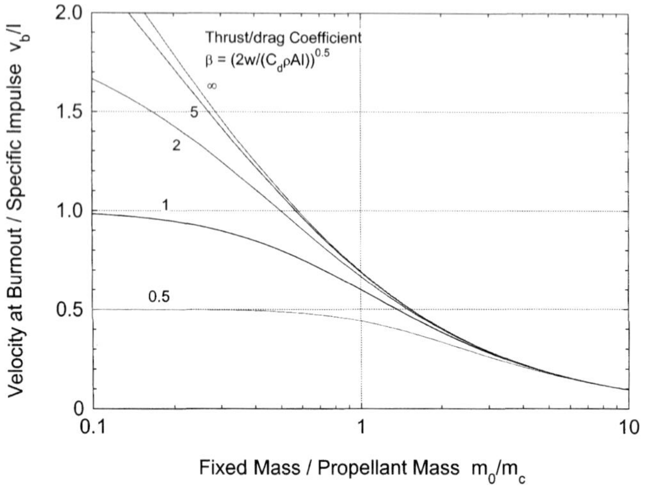

Fig. 24.39 Velocity of a rocket at burnout.

The area is $A = 3.142 \times .01^2/4 = 0.00786$ m^2
The burning rate is $1000 \times .00786 \times 0.005 = 0.0393$ kg/s
From Equ. 24.87 the maximum velocity is 149 m/s.

Ignoring drag the maximum range is $149^2/9.81 = 2.2$ km.

Example 2. Rocket Flight
If the above rocket has a propellant mass of 0.2 kg and a fixed mass of 1 kg estimate its velocity at burn-out.

From Equ. 24.86
$\alpha = 1 / 0.2 = 5$; $\beta = 0.186$; $K = 7.08$; Burn-out velocity is $v_b = 112$ m/s

BIBLIOGRAPHY

Agricola, G., "De Re Metallica", Trans. Hoover, H.C., and Hoover, L.H., Dover Publications, Inc., New York, 1950, originally published 1556.

Ambers, J., "Radiocarbon, Calibration and Early Mining", in Crew, P. and A. (Editors), "Early Mining in the British Isles", Proc. of a Workshop on Early Mining, National Park Study Centre, Plas Tan-y-Bwich, 1990.

Anon., "Wilton Windmill", Wiltshire County Council and Wilton Windmill Society, Trowbridge, Wilts, 1979.

Anon., "Pocket World in Figures", The Economist in association with Profile Books Ltd., 1997.

Alexander, M., "The Earliest English Poems", Penguin Books, 1980.

——, "Beowulf", Penguin Books, 1973.

Archard, J.F., "Contact and Rubbing of Flat Surfaces", J. Appl. Phys., 24, 981,1953.

——, and Hirst, W., "The Wear of Metals Under Unlubricated Conditions", Proc. Roy. Soc., London, Ser. A, 236, 397, 1956.

Ardrey, R., "African Genesis", Collins, London, 1965.

Armstrong, J., "Transport and Trade", in Pope, R., (Editor), "Atlas of British Social and Economic History Since 1700", Routledge, London, 1989.

Astill, G.G., "A Medieval Industrial Complex and its Landscape: the Metalworking Watermills and Workshops of Bordesley Abbey", Council for British Archaeology Research Report 92, 1993.

Atkinson, F., "The Horse as a Source of Rotary Power", Trans. Newcomen Soc., 33, 31–55, 1960–61.

Atkinson, J.C., "Stonehenge", Hamish Hamilton, London, 1956.

Aubrey, J., "Aubrey's Natural History of Wiltshire", David and Charles Reprints, Newton Abbot, Devon, re-printed 1969.

Austin, J., "Liverpool and the Atlantic Ferry", Proc. I. Mech. E, 127, 83–117, 1934.

——, "The Machinery of the Queen Mary", Proc. I. Mech. E, 135, 3–32, 1937.

Avellone, E.A., and Burmeister, T., (Editors) "Marks' Standard Handbook for Mechanical Engineers", McGraw-Hill, New York, 1996.

Backinsell, W.G.C., "The Medieval Clock in Amesbury Abbey", Historical Monograph 3, South Wiltshire Industrial Archaeology Society, Salisbury, 1979.

——, "Medieval Windlasses at Salisbury, Peterborough, and Tewkesbury", Historical Monograph 7, South Wiltshire Industrial Archaeology Society, Salisbury, 1980.

Bagwell, P.S., "The Transport Revolution", Routledge, London, 1988.

Baines, E., "History of the Cotton Manufacturers", Fisher, Fisher and Jackson, 1835.

Barraclough, K.C., "Sheffield Steel", Sheffield City Museums, 1989.

Baron, R.C., (Editor), "The Garden and Farm Books of Thomas Jefferson", Fulcrum, Golden, Colorado, 1987.

Bass, G.F. (Editor), 'A History of Seafaring Book Club Associates, London, 1974.

Bayliss-Smith, T., "Energy Use, Food Production, and Welfare: Perspectives on the Efficiency of Agricultural Systems", in Harrison, G.A. (Ed), "Energy and Effort", Symposia of the Society for the Study of Human Biology, XXII, Taylor and Francis Ltd., London, 283–303, 1992.

Baynes, K., and Pugh, F., "The Art of the Engineer", Lutterworth Press, Guildford, Surry, 1981.

Beffeyte, R., "Les machines de Guerre au Moyen Age", Edilarge SA, Rennes, 2000.

Bekker, M.G., "Off-the-Road Vehicles", University of Michigan Press, 1956.

Bennett, R., and Elton, J., "History of Corn Milling", Volumes 1, 2, and 3, Simpkin, Marshall and Co. Ltd., London, 1898.

Beresford, M., "The Lost Villages of England", Lutterworth Press, London, 1965.
Bergman, C.A., McEwen, E., and Miller, R. "Experimental Archery: Projectile Velocities and Comparison of Bow Performances", Antiquity, 62, 658–670, 1988.
Bienkowski, P., "The Sotiel Coronada Achimedes Screw in Liverpool Re-examined", Philipp von Zabern, Mainz, 135–40, 1987.
Bird, R.B., Stewart, W.E., and Lightfoot, W.E., "Transport Phenomena", John Wiley and Sons, Inc., New York, 1960.
Biringuccio, V., "The Pirotechnia of Vannoccio Biringuccio", Translated and Edited by Smith, C.S., and Gnudi, M.T., Dover Publications, New York, 1990, originally published 1540.
Blanchard, I.S.W., "Lead Mining in Medieval England and Wales", in Crossley, D.W. (Ed.), "Medieval Industry", Research Council Report 40, Council for British Archaeology, London, 1981.
Boulton, W.H., "The Pageant of Transport Through the Ages", Sampson Low, Marston and Co., London, 1931.
Bowen, R.L., "Egypt's Earliest Sailing Ships", Antiquity, XXXIV, 117–31, 1960.
Bowman, W.M., "Five Thousand Acres of Old Ashton", M.T.D. Rigg Publications, Leeds, 1990.
Brandon, S.G.F., Editor, "Ancient Empires", Newsweek Books, New York, 1973.
Breasted, J.H., "A History of Egypt", Charles Scribner's Sons, New York, 1912.
Brett, G., "Byzantine Water Mill", Antiquity, XIII, 51, 354–356, 1939.
Bridbury, A.R., 'Medieval English Clothmaking': An Economic Survey', London, 1982.
Brigg, T.H., "Haulage by Horses", Trans ASME, 14, 1014–1065, 1893.
Brody, S., "Bioenergetics and Growth", Hafner Publishing Co. Inc., New York, 1968.
Brown, D.K., "Before the Ironclad", Conway Maritime Press Ltd., London, 1990.
Brown, J., "Farm Machinery 1750–1945", B.T. Batsford Ltd., London, 1989.
Buckland, J.S.P., "Technical Notes on 16th and 17th Century London Windmills", Trans Newcomen Society, 60, 127–136, 1989.
Buller, A.J., "The Contractile Behaviour of Mammalian Skeletal Muscle", Oxford Biology Readers, Oxford University Press, 1975.
Bull, S., Edited by North, A., "An Historical Guide to Arms and Armour", Cassell, Studio Editions Ltd., 1991.
Butterworth, J., "History and Description of the Town and Parish of Ashton-u-Lyne", E.J. Morten, Didsbury, Manchester, 1972. Originally published 1823.
Button, K.J., "Transport Economics", Edward Elgar Publishing Ltd., Cheltenham, 1993.
Burford, A., "Heavy Transport in Classical Antiquity", Economic History Review, XIII, 1, 1–18, 1960.
Campbell, G.F., "China Tea Clippers", Adlard Coles Ltd., London, 1974.
Carpenter, T.H., "Art and Myth in Ancient Greece", Thames and Hudson, London, 1996.
Carslaw, H.S., and Jaeger, J.C., "Conduction of Heat in Solids", Oxford University Press, 1990.
Carus-Wilson, E.M., "An Industrial Revolution of the Thirteenth Century", Economic History Review, 12, 39–60, 1941.
—— and Coleman, O., 'England's Export Trade, 1275–1542', Oxford, 1963.
Casson, L., "Ships and Seamanship in the Ancient World", Princeton University Press, 1971.
——, "Ships and Seafaring in Ancient Times", British Museum Press, London, 1994.
Cerretelli, P., "Exercise and Endurance", in Larson, L.A., (Editor), "Fitness, Health, and Work Capacity: International Standards of Assessment", Macmillan Publishing Co., New York, 1974.
Chadwick, H, "The Early Church", Penguin Books, London, 1990.
Chagnon, N., "Yanomamö: The Fierce People", Holt, Rinehart and Winston, New York, 1983.

Chambers, W. and Chambers, R., "Information for the People", Vol. II, Edinburgh, 1842.
——, "Information for the People", Vol. 1, Edinburgh, 1874.
Chandler, J. (Editor), "John Leland's Itinerary: Travels in Tudor England", Sutton Publishing, Stroud, 1993.
—— (Editor), "Travels Through Stuart Britain: The Adventures of John Taylor, The Water Poet", Sutton Publishing, Stroud, 1999.
Chandler, D., "Oxford Illustrated History of the British Army", Oxford University Press, 1994.
Champion, T., Gamble, C., Shennans, S., and Whittle, A., "Prehistoric Europe", Academic Press, London, 1984.
Chew, K., and Wilson, A., "Victorian Science and Engineering Portrayed in the Illustrated London News", Science Museum, Alan Sutton, London, 1993.
Childe, G., "What Happened in History", Penguin Books, London, 1942, reprinted 1957.
Clark, J.G.D., "Neolithic Bows from Somerset, England, and the Prehistory of Archery in North-western Europe", The Prehistoric Society, 29, 50–98, 1963.
Clarke, R., "The Science of War and Peace", Jonathon Cape, London, 1971.
Clayton, P.A., "Chronicle of the Pharoahs", Thames and Hudson, London, 1994.
Collins, D., Editor, "The Origins of Europe", George Allen and Unwin Ltd., London, 1975.
Collon, D., "Ancient Near Eastern Art", British Museum Press, London, 1995.
Constantinou, D., "Ancient Copper Mining in Cyprus", in "Cyprus, Copper and the Sea", The Government of Cyprus, 1992.
Cooper, G.J., Dudley, H.A.F., Gann, D.S., Little, R.A., and Maynard, R.L., (Editors), "Scientific Foundations of Trauma", Butterworth-Heinemann, Oxford, 1997.
Cotterell, B., and Kamminga, J., "Mechanics of Pre-industrial Technology", Cambridge University Press, Cambridge, 1992.
Crew, P. and A. (Editors), "Early Mining in the British Isles", Proc. of a Workshop on Early Mining, National Park Study Centre, Plas Tan-y-Bwich, 1990.
Crossland, B., "From Cannon to Diamonds: A History of High Pressure Engineering", Trans. Newcomen Society, 57, 149–168, 1986.
Crossley, D.W., "Cannon Manufacture at Pippingford", Post-Medieval Archaeology, 9. 1–37, 1975.
——, "Medieval Iron Smelting", in Crossley, D.W. (Ed.), "Medieval Industry", Research Council Report 40, Council for British Archaeology, London, 1981.
Cruden, S.H., "Click Mill, Orkney", Ancient Monuments and Historic Buildings, Ministry of Works, HMSO, Edinburgh, 1949.
Davidson, A.E., "Adventure", President's Address, Proc. I. Mech. E, 130, 337–369, 1935.
Davidson, C.St.C., "Transporting Sixty-Ton Statues in Early Assyria and Egypt", Technology and Culture, 2, 11–16, 1961.
Davies, O., "Roman Mines in Europe", Clarendon Press, Oxford, 1935.
Davies, V., and Friedman, R., "Egypt", British Museum Press, London, 1998.
Defoe, D., "A Tour Through the Whole Island of Great Britain", Penguin Books, 1971.
Denny, Sir Maurice, "BSRA Resistance Measurements on the *Lucy Ashton*", part 1, "Full Scale measurements", Trans. Instn. of Naval Architects, 1951.
Dennt, N., and Flimer-Sankey, J., "The Travels of Sir John Mandeville", Collins, London, 1973.
Derry, T.K., and Williams, T.I., "A Short History of Technology", Oxford University Press, Oxford, 1960.
Dickinson, H.W., and Straker, E., "The Shetland Watermill", Trans. Newcomen Society, XIII, 89–94, 1932–3.
Dickinson, H.W., "Origin and Manufacture of Word Screws", Trans. Newcomer Soc., XXII, 79–85, 1941–2.
Diesner, H.J., "The Great Migrations", Orbis Publishing, London, 1982.

Diderot, D., "A Diderot Pictorial Encyclopaedia of Trades and Industry", Edited by Charles Gillespie, Vols. 1 and 2, Dover Publications Inc., New York, 1959. Originally published 1751.

—— and d'Alembert, J.L., "Encyclopèdie", Paris, 1772.

Dolínek, V., and Durdík, J., "The Enclyclopedia of European Historical Weapons", Hamlym, London, 1993.

Douglas, D.C., and Myers, A.R., Editors, "English Historical Documents, 1327–1485", Eyre and Spottiswoode, London, 1969.

Doxiadis, C.A., and Papaioannu, J.G., "Ecumenopolis: The inevitable City of the Future". Atene, 1974.

Drachmann, A.G., "The Mechanical Technology of Greek and Roman Antiquity", Munksgaard, Copenhagen, 1963.

Drews, R., "The End of the Bronze Age", Princeton University Press, New Jersey, 1993.

Dugas, R., "A History of Mechanics", Dover Publications, New York, 1988.

Earl, B., "Cornish Mining", Cornish Hillside Publications, St Austell, 1994.

Ekelund, S., "The Analysis of Factors Influencing Rolling Pressure and Power Consumption in the Hot Rolling of Steel", Steel, 93, 8–14, 1933.

Ellicott, J., "An Account of Some Experiments, made by Benjamin Robins Esq; FRS, Mr Samuel Da Costa, and Several other Gentlemen, in order to discover the Height to which Rockets may be made to ascend, and to what Distance their Light may be seen." Proc. of the Royal Society, London, 578–585. 13th December 1750.

Ember, C.R., and Ember, M., "Violence in the Ethnographic Record: Results of Cross-Cultural Research on War and Aggression", in Martin, D.L. and Frayer, D.W., (Ed), "Troubled Times: Violence and Warfare in the Past", Gordon and Breach Publishers, Amsterdam, 1997.

Espitallier, G.E., "La Technique du Ballon", Octave Doin, Paris, 1907.

Evelyn, John, 'Diaries', Routledge, London, 1987.

Evett, J.B., and Lui, C., "Fluid Mechanics and Hydraulics", McGraw Hill, Inc., New York, 1989.

Fackler, M.L. et al., "Ordnance Gelatin for Ballistic Studies", Am. Forens. Med. Pathol., 9, 1988, 218–219.

—— "Effects of Small Arms on the Human Body", Proc. 11th International Symposium on Ballistics, Brussels, American Defence Preparedness Association, Washington DC, 3, 1–8, 1989.

Farrington, B., "Greek Science", Penguin Books, 1961.

Finley, M.I., (Editor), "Classical Slavery", London, 1987.

——, "Was Greek Civilization Based on Slave Labour", Historia, 8, 145–164, 1959.

—— (Editor), "Slavery in Classical Antiquity", Heffer and sons Ltd., Cambridge, 1960.

——, "Technical Innovation and Economic Progress in the Ancient World", Economic History Review, 18, 29–45, 1965.

Foley, V., Soedel, W., Turner, J., and Wilhoite, B., "The Origin of Gearing", History of Technology, 7, 101–29, 1982.

Foote, P.G., and Wilson, D.M., "The Viking Achievement", Book Club Associates, London, 1979.

Forbes[1], R.J., "Studies in Ancient Technology", Vol. I, E.J. Brill, Leiden, 1955.

Forbes[2], R.J., "Studies in Ancient Technology", Vol. II, E.J. Brill, Leiden, 1955.

Forbes[3], R.J., "Studies in Ancient Technology", Vol. III, E.J. Brill, Leiden, 1955.

Fossier, R., (Editor), "The Middle Ages I 350–950", Cambridge University Press, Cambridge, 1989.

——, (Editor), "The Middle Ages II 950–1250", Cambridge University Press, Cambridge, 1998.

Fox, R., "Aggression Then and Now", in Robinson, M.H., and Tiger, L., (Editors) "Man and Beast Revisited", Smithsonian Institution Press, Washington, 1991.

Frayer, D.W., "Ofnet: Evidence for a Mesolithic Massacre", in Martin, D.L. and

Frayer, D.W., (Ed.), "Troubled Times: Violence and Warfare in the Past", Gordon and Breach Publishers, Amsterdam, 181–213, 1997.
Freris, L.L. (Editor), Wind Energy Conversion Systems, Prentice Hall, New York, 1990.
Froude, W., "On Experiments with H.M.S. *Greyhound*", Trans. Instn. of Naval Architects, 1874.
Fussell, G.E., "Farming Systems and Society: A Historical Survey", Trans. Newcomen Society, 39, 47–64, 1967.
Garlan, Y., "War in the Ancient World", Translated by Lloyd, J., Chatto and Windus, London, 1976.
Gerald of Wales "The Description of Wales", Translated by L. Thorpe, Penguin Books, London, 1978.
Gertler, M., "A Re-Analysis of the Original Test Data for the Taylor Standard Series", Navy Department, Washington DC, 1954.
Gibbon, E., Abridged by Lowe, D.M., "The Decline and Fall of the Roman Empire", Book Club Associates, 1972.
Gibbons, B., "On the Ventilation of Mines", Proc. I. Mech. E, 12, 8–33, 1850–51.
Gibbs-Smith, C.H., "Sir George Caley's Aeronautics, 1796–1855", HMSO, 1962.
Gillespie, C. (Ed.), "A Diderot Pictorial Encyclopedia of Trades and Industry", Denis Diderot, Vols. 1 and 2, Dover Publications Inc., New York, 1959. Originally published in 1751.
Gimpell. J., "The Medieval Machine", Pimlico, London, 1993.
Glotz, G., "Un Transpot de Marbre pour le Portique d'Eleusis", Revue des études Grecques, XXXVI, 26–45, 1923.
——, "Ancient Greece at Work", Routledge and Kegan Paul Ltd., London, 1965.
Graudi, M.T. in Ramelli, A., 'The Various and Ingenious Machines of Agustine Ramelli's Dover Publications, 1976.
Graves, R., "The Greek Myths", Combined Edition, Penguin Books, 1992.
Green, B., "Guide to Grime's Graves", HMSO, London, 1977.
Green, W.G., "Theory of Machines", Blackie and Son Ltd., London, 1960.
Greener, W.W., "The Gun and its Development", Arms and Armour Press, London, 1910.
Grübler, A., and Nakicenovic, N., "Evolution of Transport Systems: Past and Future", International Institute for Applied Systems Analysis, Laxenburg, Austria, RR-91-8, June, 1991.
Guilmartin, J.F., "Gunpowder and Galleys", Cambridge University Press, 1974.
Gurney, O.R., "The Hittites", Penguin Books, London, 1990.
Haddock, M.H., "Mine Ventilation and Ventilators", Charles Griffin and Co. Ltd., London, 1924.
Haden, W., "Letter to Board of Education", Wiltshire Record Office, 1325, 1928.
Hadfield, C., "The Canal Age", The David and Charles Series, Pan Books, London, 1968.
Hale, J.R., "The Lost Technology of Ancient Greek Rowing" Scientific American, 66–71, May 1966.
Hall, A.R., "Architectura Navalis", Trans. Newcomen Society, 51, 157–174, 1980.
Hall, A.S., Holowenko, A.R., and Laughlin, H.G., "Theory and Problems of Machine Design", Schaum's Outline Series, McGraw-Hill Book Company, New York, 1980.
Halliday (1990) in Freris, L.L. (Editor), Wind Energy Conversion Systems, Prentice Hall, New York, 1990.
Harland, J., and Myers, M., "Seamanship in the Age of Sail", Conway Maritime Press, London, 1984.
Harris, D.G., "F.H. Chapman: The First Naval Architect and his Work", Conway Maritime Press, 1989.
Harris, T.R., "Williams' Perran Illustrated Catalogue", The Trevithick Society reprint, Redruth, 1994.

Harrison, G.A. (Ed.), "Energy and Effort", Symposia of the Society for the Study of Human Biology, XXII, Taylor and Francis Ltd, London, 283–303, 1982.
Hartenberg, R.S., and Schmidt, J., "The Egyptian Drill and the Origin of the Crank", Technology and Culture, 10, 155–65, 1969.
Hart, I.B., "The World of Leonardo da Vinci", Macdonald, London, 1961.
Harvey, Sir Paul, "The Oxford Companion to Classical Literature", Oxford, Clarendon Press, 1962.
Hassan, U. and Sykes, D.M., "Wind Structure and Statistics", in Freris, L.L. (Editor), Wind Energy Conversion Systems, Prentice Hall, New York, 1990.
Hassall, W.O. (Compiler), "They Saw it Happen", Vol. 1, 55 B.C.–1485, Basil Blackwell, Oxford, 1973.
Herodotus "The Histories", Translated de Sélincourt, A., Penguin Books, 1996.
Hesiod, Translated Evelyn-White, H.G., "The Homeric Hymns and Homerica", William Heinemann, London, 1919.
Hill, A.V., "Muscles and Nerves: the Maintenance of Posture, the Development of Power, and the Transmission of Messages in the Body", Proc. I. Mech. E, 131, 353–81, 1935.
——, "The Heat of Shortening and the Dynamic Constants of a Muscle", Proc. Roy. Soc., Series B, London, 126, 136–195, 1938.
Hills, R.L., "Power from the Wind", Cambridge University Press, 1994.
Hinsley, F.B., "The Development of Coal Mine Ventilation in Great Britain up to the End of the Nineteenth Century", Trans. Newcomen Society, London, XLII, 25–39, 1969–70.
Hockey, S.F., "Quarr Abbey and its Lands", Leicester University Press, 1970.
Hodgen, M.T., "Domesday Water Mills", Antiquity, XIII, 51, 261–279, 1939.
Hoerner, S.F., "Fluid-Dynamic Drag", Hoerner Fluid Dynamics, Albuquerque, NM, 1965.
Hogg, I.V., "The Guinness Encyclopedia of Weaponry", Guinness Publishing, London, 1992.
Hogg, O.F.G., "English Artillery, 1326–1716", Royal Artillery Institution. London, 1963.
Holman, B.W., "Heat Treatment as an Agent in Rock Breaking", Transactions Institute Mining and metals, 36, 239–245, 1927.
Holmes, G., "The Later Middle Ages", Cardinal, London, 1974.
Homer, "The Illiad", Translated Rieu, E.V., Penguin Books, London, 1950.
——, "The Odyssey", Translated Rieu, E.V., Penguin Books, Harmondsworth, 1946.
Hood, S., "The Home of Heroes: The Aegean Before the Greeks", Thames and Hudson Ltd., London, 1974.
Hopkins, K., 'Conquerers and Slaves', Sociological Studies in Roman History, Vol. 1, Cambridge, 1978.
Hopper, R.J., "The Attic Silver Mines on the Fourth Century B.C.", Annual of the British School at Athens, 48, 200–254, 1953.
Hornell, J., "Water Transport, Origins and Early Evolution", Cambridge University Press, 1946.
Howarth, D., "Sovereign of the Seas", William Collins and Sons, Glasgow, 1974.
Hubbard, M., and Rust, H.J., "Javelin Dynamics with Measured Lift, Drag, and Pitching Moment", Trans. ASME, 51, 406–8, June 1984.
Humphrey, J.W., Oleson, J.P., and Sherwood, A.N., "Greek and Roman Technology: A Sourcebook", Routledge, London, 1999.
Ingstad, H., "Westward to Vinland", Jonathan Cape, London, 1969.
Ishiko, T., "The Organism and Musclular Work", in Larson, L.A., (Editor), "Fitness, Health, and Work Capacity: International Standards of Assessment", Macmillan Publishing Co., New York, 1974.
Jackman, W.T., "The Development of Transportation in Modern England", Frank Cass and Co. Ltd., 1962.

James, D., "Prehistoric Copper Mining on the Great Orme's Head", in Crew, P. and A. (Editors), "Early Mining in the British Isles", Proc. of a Workshop on Early Mining, National Park Study Centre, Plas Tan-y-Bwich, 1990.
Janeway, "Ride and Vibration Data Manual", SAEJ6a, Society of Automotive Engineers, 1965.
Janzon, B., Hull, J.B., and Ryan, J.M., "Projectile-Material Interactions: Soft Tissue and Bone", in Cooper, G.J., et al., (Editors) "Scientific Foundations of Trauma", Butterworth-Heinemann, Oxford, 37–52, 1997.
Jenkins, R, "The Collected Papers of Rhys Jenkins, M.I. Mech E.", Newcomen Society at the University of Cambridge Press, 1936.
——, "The Slitting Mill", The Engineer, 445–46 and 487–89, 24th May and 7th June 1918. Reprinted in "The Collected papers of Rhys Jenkins, M.I. Mech. E.", Newcomen Society at the University Press, Cambridge, 1936.
Jennings, H., "Pandaemonium", Papermac, London, 1995.
Jocelin of Brakelond, "Chronicle of the Abbey of Bury St Edmunds", Translated by Greenway, D., and Sayers, J., Oxford University Press, Oxford, 1989.
Jones, A., "The Art of War in the Western World", Harrap, London, 1988.
Jones, A.M.H., "Slavery in the Ancient World", The Economic History Review, 2nd Ser., 9, 185–199, 1956.
Jones, P., in Hardy, R., 'Langbow: a Social and Military History', Patrick Stephens Ltd., Yeavil, 232–36, 1992.
Josephus, "The Jewish War", Translated Williams, G.A., Revised Smallwood, E.M., Penguin Books, 1981.
Juvinal, R.C., "Fundamentals of Machine Component Design", John Wiley and Sons, New York, 1993.
Karvinen, E., and Komi, P.V., "Neuromuscular Performance", in Larson, L.A., (Editor), "Fitness, Health, and Work Capacity: International Standards of Assessment", Macmillan Publishing Co., New York, 1974.
Keegan, J., "A History of Warfare", Hutchinson, London, 1993.
Keeley, L.H., "Frontier Warfare in the Early Neolithic", in Martin, D.L. and Frayer, D.W., (Ed.), "Troubled Times: Violence and Warfare in the Past", Gordon and Breach Publishers, Amsterdam, 303–19, 1997.
Keil, I., "Building a Post Mill in 1342", Trans Newcomen Soc., XXXIV, 151–4, 1962.
Keller, A.G., "A Theatre of Machines", Chapman and Hall, London, 1964.
Kemp, P., "The History of Ships", Black Cat, Macdonald and Co. (Publishers) Ltd., London, 1988.
Khisty, C.J., "Transportation Engineering", Prentice-Hall, New Jersey, 1990
King, P.W., "Sir Clement Clerke and the Adoption of Coal in Metallurgy", Trans. Newcomen Society, Advance copy, 14th Nov. 2001.
Kirkby, H.R., (Editor), "Personal Travel Budgets", Transportation Research, 15A, January 1981.
Kirkham, N., "Early Lead Smelting in Derbyshire", Trans. Newcomen Soc., XLI, 119–138, 1968–9
Klemm, F., "A History of Western Technology", George Allen and Unwin Ltd., London, 1959.
Knight, E.H., "The Practical Dictionary of Mechanics", Volumes 1–5, Cassell & Co. Ltd., London, 1884.
Knuttgen, H.G., "The Development of Work Capacity and Physical Fitness", in Larson, L.A., (Editor), "Fitness, Health, and Work Capacity: International Standards of Assessment", Macmillan Publishing Co., New York, 1974.
Kühn, H., Translated by Brodrick, A.H., "The Rock Pictures of Europe", Sidgwick and Jackson, London, 1956.
Lambert, P.M., "Patterns of Violence in Prehistoric Hunter-Gatherer Societies of Coastal Southern California", in Martin, D.L. and Frayer, D.W., (Ed.), "Troubled

Times: Violence and Warfare in the Past", Gordon and Breach Publishers, Amsterdam, 77–103, 1997.
Lancaster Burne, E., Russel, J., and Wailes, R., "Windmill Sails", Trans. Newcomen Soc., XXIII, 147–161, 1943–5.
Landström, B., "Sailing Ships", George Allen and Unwin, London, 1969.
Lane, F.C., "The Economic Meaning of the Invention of the Compass", American Historical Review, 68, 605–617, 1963.
Lansdown, Marquis of, (Editor), "The Petty Papers", 1927, re-printed Augustus M. Kelly, New York, 1967.
Larson, L.A., (Editor), "Fitness, Health, and Work Capacity: International Standards of Assessment", Macmillan Publishing Co., New York, 1974.
Lawson Dick, O., "Aubrey's Brief Lives", Mandarin, London, 1949.
Leach, G., "Energy and Food Production", IPC Science and Technology Press, Guildford, 1976.
Lefebvre des Noëttes, R.J.E.C., "Le Chevalde Selle à Travers les Ages: Contribution à l'Histoire de l'Esclavage", Paris, 1931.
Leupold, J., "Theatrum Machinarum", 10 vols., Leipzig, 1724–39.
Lewendon, R.J., "The Leather Gun of Gustavus Adolphus of Sweden in the Rotunda Collection at Woolwich", Journal of the Royal Artillery, 112, 22–25, 1985.
Lewis, A., "Underground Exploration of the Great Orme Copper Mines", Crew, P. and A. (Editors), "Early Mining in the British Isles", Proc. of a Workshop on Early Ming, National Park Study Centre, Plas Tan-y-Bwich, 1990.
Links, J.G., "Travellers in Europe", The Bodley Head, London, 1980.
Littauer, M.A. and Crouwel, J.H., "Wheeled Vehicles and Ridden Animals in the Ancient Near East", E.J. Brill, Leiden, 1979.
Liversidge, J., "Britain in the Roman Empire", Book Club Associates, London, 1973.
Lofts, N., "Domestic Life in England", Weidenfield and Nicolson, London, 1976.
Lopez, R.S., "The Evolution of Land Transport in the Middle Ages", Past and Present, 9, 17–29, 1956.
Lorenz, K., "On Aggression", Trans. Latke, M., Routledge, London, 1996.
Low, D.A., "The Regency Underworld", Alan Sutton Publishing, Trowbridge, reprinted 2000.
Lucas, F.L., (Editor), "Greek Poetry for Everyman", J.M. Dent and Sons Ltd., London, 1951.
Lu, G.-D., Salmon, R.A. and Needham, J., "The Wheelwright's Art in Ancient China. 1. The Invention of Dishing", Physis (Florence), 1, 103–26, 1959.
Madden, J., "Slavery in the Roman Empire: Numbers and Origins", Classics Ireland, 3, 1–9, 1996.
Major, J.K., "The Horse Engine in the 19th Century", Newcomen Society Trans. 60, 31–48, 1988–89.
Marchetti, C., "Anthropological Invariants in Travel Behaviour", Technological Forecasting and Social Change, 47, 75–88, 1994.
Margaria, R., "Biomechanics and Energetics of Muscular Exercise", Clarendon Press, Oxford, 1976.
Marieb, E.N., "Human Anatomy and Physiology", The Benjamin/Cummings Publishing Company, Inc., Redwood City, Cal., 1992.
Marsden, E.W., "Greek and Roman Artillery: Historical Development", Oxford at the Clarendon Press, 1969.
——, "Greek and Roman Artillery: Ancient Technical Treatise", Oxford at the Clarendon Press, 1971.
Martin, D.L. and Frayer, D.W., (Ed.), "Troubled Times: Violence and Warfare in the Past", Gordon and Breach Publishers, Amsterdam, 1997.
Mathias, P., "The First Industrial Nation", Routledge, London, 1995.

May, T., "Britain's Earliest Iron Furnaces and Moulding Floors", The Iron and Coal Trades Review, 427–432, Aug. 11th, 1905.
McCutcheon, W.A., "Water Power in the North of Ireland", Trans. Newcomen Society, 32, 67–94, 1966–7.
Mertens, J., "Römische Skulpturen von Buzenol, Provinz Luxemburg", Germania, XXXVI, 386–92, 1958.
Meyer, C., "Bir Umm Fawakhir: Insights into Ancient Egyptian Mining", Journal of Metals, 49, 3, 64–68, 1997.
Missliwetz, J., and Wieser, I., "Medizinische und Technische Aspekte der Waffenwirkung, I. Bogen und Armbrust, Beitr. gerichtl. Med. 43, 437–444, 1985.
Mitcham, C., "Three Ways of Being-With Technology", in "From Artefact to Habitat", Ed. Ormiston, G.L., Associated University Presses, Inc., Cranbury, NJ, 1990.
Mitchell, B.R., and Deane, P., "Abstract of British Historical Statistics", Cambridge University Press, 1962.
Mitchell, B.R., "European Historical Statistics 1750–1970", The Macmillan Press, London, 1978.
Modelski, G., and Morgan, P., "Understanding Global War", Journal of Conflict Resolution, 29, 473–502. 1985.
Modelski, G., and Thompson, W.R., "Long Cycles and Global War", in "Handbook of War Studies", Midlarski, M.I., (Ed.), Unwin Hyman, Boston, 1989.
Moriarty, C., "The Voice of the Middle Ages", Lennard Publishing, 1989
Morrison, J., and Coates, J., "The Athenian Trireme", Cambridge, 1986.
Morrissey, S.J., and Liou, Y.H., "Metabolic Cost of Load Carriage with Different Container Sizes", Ergonomics, 27, 8, 847–853, 1984.
Morsley, C., "News from the English Countryside, 1750–1850", Harrap, London, 1979.
Mossé, C., "The Ancient World at Work", Translated Lloyd, J., Chatto and Windus, London, 1969.
Muckle, W., and Taylor, D.A., "Muckle's Naval Architecture", Butterworths, London, 1987.
Mumford, L., "The Myth of the Machine: Technics and Human Development", Secker and Warburg, London, 1967.
Nash-Williams, V.G., 'The Roman Gold Mines at Dolaucothi', Bull. of the Board of Celtic Studies, xiv, 79–84, 1950.
Neale, M.J., Editor, "Tribology Handbook", Newnes-Butterworths, London, 1973.
Needham, J., Ping-Yu, H., Gwei-Djen, L., and Ling, W., "Science and Civilisation in China", 5, pt 7, "Military Technology; The Gunpowder Epic", Cambridge University Press, Cambridge, 1986.
Newby, P.H., "Warrior Pharaohs", Book Club Associates, London, 1980.
Oakeshott, R.E., "The Archaeology of Weapons", Lutterworth Press, London, 1960.
Ohlgren, T.H., (Editor), "Medieval Outlaws", Sutton Publishing, Stroud, 1998.
Overton, M., "The Diffusion of Agricultural Innovations in Early Modern England: Turnips and Clover in Norfolk and Suffolk 1580–1740", Trans. Institute of British Geographers, new series, 10, 205–21, 1985.
——, "Agricultural Revolution in England", Cambridge University Press, 1996.
Paine, A.B., "Mark Twain's Notebook", Harper and Brothers Publishers, New York, 1935.
Palmer, R.E., "Notes on Some Ancient Mine Equipments", Trans. Institute of Mining and Metals, 36, 299–336, 1927.
Pandolf, K.B., Givoni, B., and Goldman, R.F., "Predicting Energy Expenditure with Loads while Standing or Walking very Slowly", Journal of Applied Physiology: Respiratory, Environmental, and Exercise Physiology, 43, 577–581, 1977.
Parsons, A.W., "A Roman Water-Mill in the Athenian Agora", Hispera, V, 70–90, 1936.
Parsons, W.B., "Engineers and Engineering in the Renaissance", MIT Press, Cambridge, Mass., (1968).

Payne-Gallwey, R., "The book of the Crossbow", Longmans Green Co, London, 1903. Re-printed by Dover Publications, New York, 1995.
Peacock, D.P.S., "Pottery in the Roman World: an Ethnoarchaeological Approach", Longman, London, 1982.
Peele, R. (Ed.), "Mining Engineers' Handbook", Vols. 1 and 2, John Wiley and Sons, New York, 1927.
Pelham, R.A., "The Distribution of Early Fulling Mills in England and Wales", Geography, 13, 52–56, 1944.
Pendred, L.St.L., "A Survey of Ships and Engines", 7th Thomas Lowe Lecture, Proc I Mech E, 128, 79–118, 1934.
Pennant, T., "A Tour in Scotland, and Voyage to the Hebrides", 1772, Quoted in Jennings, H., "Pandaemonium", Papermac, London, 1995.
Peters, C.E., "A Mathematical-Physical Model of Wound Ballistics", J. Trauma (China), 6 Suppl., 303–318, 1990.
Philipson, J., "The Art and Craft of Coachbuilding", George Bell and Sons, London, 1897.
Piggott, S., "The Earliest Wheeled Transport", Thames and Hudson, 1983.
Pimentel, D. and Pimentel, M., "Food, Energy and Society", Arnold, London, 1979.
Pleiner, R., "The Celtic Sword", Clarendon Press, Oxford, 1993.
Pliny the Elder, "Natural History", Translated Healy, J.F., Penguin Classics, London, 1991.
Plutarch, Translated Scott-Kilvert, I., "The Rise and Fall of Athens: Nine Greek Lives", Penguin Books, London, 1960.
Pole, W., (Editor), "The Life of Sir William Fairbairn, Bart", David and Charles Reprints, Newton Abbot, Devon, 1970.
Pope, R., (Editor), "Atlas of British Social and Economic History Since 1700", Routledge, London, 1989.
Postgate, J.N., "Early Mesopotamia", Routledge, London, 1992.
Potts, R., "Untying the Knot", in Robinson, M.H., and Tiger, L., (Editors), "Man and Beast Revisited", Smithsonian Institution Press, Washington, 1991.
Pratt, P.L., "The Arrow", in Hardy, R., "Longbow", Patrick Stephens Ltd., Yeovil, Somerset, 1992.
Pryce, W., "Mineralogia Cornubiensis: A Treatise on Minerals, Mines, and Mining", London, 1778.
Pugh, A.B. and Crittall, E., (Editors), "A History of Wiltshire", The Victoria History of the Counties of England, University of London, Institute of Historical Research, Oxford University Press, 1955.
Pugh, R.B., "Court Rolls of the Wiltshire Manors of Adam de Stratton", Wiltshire Records Society, XXIV, Devizes, 1970.
Rabinowicz, E., "Friction and Wear of Materials", John Wiley and Sons, New York, 1965.
——, "Wear coefficients", in Booser, F.R., (Editor), "Handbook of Lubrication", Vol. 2, CRC Press Inc., Boca Raton, Florida, 201–08, 1984.
Rahtz, P.A., "Medieval Milling", in Crossley, D.W., (Editor), "Medieval Industry", Research Report No. 40, Council for British Archaeology, 1–15, 1981.
Ramelli, A., "The Various and Ingenious Machines of Agostino Ramelli", Translated Gnudi, M.T., Annotated Ferguson, E.S., Dover Publications, New York, 1994. Originally published 1588.
Randall, H.A., "Some Mid-Gloucestershire Engineers and Inventors", Newcomen Society Trans., 38, 89–96, 1966.
Rankine, W.J.M., "Useful Rules and Tables Relating to Mensuration, Engineering, Structures, and Machines", Revised by Millar, W.J., Charles Griffin and Co., London, 1889.
Reece, A.R., "Principles of Soil-Vehicle Mechanics", Proc. Instn. Mech. Engrs, 180, Part 2A(2), 45–67, 1964.
Renfrew, C., "The Autonomy of the South-East European Copper Age", Proc. Prehistoric Society, N.S. 35, 12–47, 1969.

——, and Bahn, P., "Archaeology: Theories and Methods", Thames and Hudson, London, 2000.
Renström, R., *"Fyrpilen", "The Fire Arrow". Logotype of the Ballistics-92*, 13th International Symposium on Ballistics, Stockholm, Sweden, American Defence Preparedness Association, Washington, 1992.
Reynolds, J., "Windmills and Watermills", Praeger Publishers, New York, 1970.
Reynolds, O., "On the Theory of Lubrication and its Applications to Mr Beauchamp Tower's Experiments", Phil. Trans. Roy. Soc., London, 177, 157–234, 1886.
Richardson, L.F., "Statistics of Deadly Quarrels", Wright, Q. and Lienau, C.C., (Ed.), The Boxwood Press, Pittsburgh, Quadrangle Books, Chicago, 1960.
Rieu, E.V. (Trans.), "Homer: The Odyssey", The Penguin Classics, London, 1946.
Richter, L.R., "Selections from the Notebooks of Leonardo da Vinci", Oxford University Press, London, 1966.
Rickter, B., "Use of Computer Man Model in the Design of Personnel Armour", Proc 11th International Symposium of Ballistics, Brussels, American Defence Preparedness Association, Washington DC, Vol. 3, 53–62, 1989.
Ridley, A., "An Illustrated History of Transport", Heinemann, London, 1969.
Robins, B., "New Principles of Gunnery", Introduction by Curtis, W.S., Richmond Publishing, Richmond, England, First published 1741, republished 1972.
Robinson, J.H., "Readings in European History", Vol. 1, Ginn and Co, Boston, 1904.
Robinson, M.H., and Tiger, L., (Editors), "Man and Beast Revisited", Smithsonian Institution Press, Washington, 1991.
Rostovtzeff, M., Translated by Duff, J.D., "A History of the Ancient World", Vols. 1 and 2, HMSO and Oxford University Press, 1945.
Rowe, G.E., "Principles of Industrial Metalworking Processes", Edward Arnold, London, 1977.
Russell, B., "Sceptical Essays", W.W. Norton and Co., Inc., New York, 1928.
Salzman, L.F., "English Industries of the Middle Ages", H. Pordes, London, 1964.
Saunders, P.R., "Causewayed Enclodure at Crofton (Great Bedwyn)", The Wiltshire Archaeological Magazine, Devizes, 70–71, 124–26, 1975–76.
Schafer, A., and Victor, D., "The Past and Future of Global Mobility", Scientific American, 37–39, October 1997.
Schlaifer, R., 'Greek Theories of Slavery from Homer to Aristotle', Harvard Studies in Classical Philology, 165–209, 1936.
Schofield, R.B., "The Construction of the Huddersfield Narrow Canal 1794–1811", Trans. Newcomen Soc., 53, 17–38, 1981–82.
Schubert, H.R., "The First Cast-Iron Cannon Made in England", Journal of the Iron and Steel Institute, CXLVI, 137, 1942.
——, "History of the British Iron and Steel Industry", Routledge and Kegan Paul, London, 1957.
Sellier, K.G. and Kneubuehl, B.K., "Wound Ballistics and the Scientific Background", Elsevier, Amsterdam, 1994.
Seppings, R., "On a New Principle of Constructing His Majesty's Ships of War", Phil. Trans. (Royal Society), 285–302, 1814.
Sibley, C., and Ahlquist, J.E., "The Phylogeny of the Hominoid Primates, as Indicated by DNA-DNA Hybridization", Journal of Molecular Evolution, 20, 2–15, 1984.
Singer, C., Holmyard, E.J., Hall, A.R., and Williams, T.I., "A History of Technology", Vol. II, *c.* 700B.–*c.* 1500 A.D., Clarendon Press, Oxford, 1956.
Sissom, L.E., and Pitts, D.R. "Elements of Transport Phenomena", McGraw-Hill Kogakusha Ltd, Tokyo, 1972.
Skempton, A.W., "Engineering on the Thames Navigation, 1770–1845", Trans. Newcomen Society, 55, 153–176, 1984.
Skjölsvold, A., "The Stone Statues and Quarries of Rano Raraku" in Heyerdahl, T., and Ferdon, F.N., (Eds.), "Archaeology of Easter Island", Vol. 1, George Allen and Unwin, London, 339–79, 1961.

Smeaton, J., "Experimental Enquiry Concerning the Natural Powers of Wind and Water to Turn Mills and other Machines Depending on a Circular Motion", I. and J. Taylor, London, 2nd Ed., 1796.
Smiles, S., "Industrial Biography: Iron Workers and Tool Makers", John Murray, London, 1882.
——, "Men of Invention and Industry", John Murray, London, 1884.
Smiley, J., (Preface), "The saga of the Icelanders", Viking Penguin, London, 2000.
Smith, Adam, "The Wealth of Nations", Book II, Dent (Everyman's Library), London, 1954, First Published 1776.
Smithells, C.J.S. (Editor), 'Metals Reference Book', 5th Edition, Butterworthes, London, 1976.
Smollett, T., "Travels Through France and Italy", Edited Felsenstein, F., Oxford University Press, 1992.
Spedding, C.R.W., "The Comparative Economics of Agricultural Systems", in Harrison, G.A. (Ed.), "Energy and Effort", Symposia of the Society for the Study of Human Biology, XXII, Taylor and Francis Ltd., London, 303–317, 1992.
Spencer, J.R., "Filarete's Description of a Fifteenth Century Iron Smelter at Ferrièr", Technology and Culture, 4, 2, 201–6, 1963.
Spengler, O., "Man and Technics", Alfred A Knopf, Borzoi Books, New York, 1932.
Spooner, H.J., "Machine Design", Longmans, London, 1960.
Steeds, W., "A History of Machine Tools, 1700–1910", Clarendon Press, Oxford, 1969.
Stern, Van Doren, P., "Prehistoric Europe", Allen and Unwin Ltd., London, 1970.
Stevenson, D.A., "English Lighthouse Tours", London, 1946.
Storr, A., "Human Aggression", Allen Lane, The Penguin Press, London, 1968.
Stowers, A., "Observations on the History of Water Power", Trans. Newcomen Society, XXX, 239–56, 1957.
Sturdivan, L.M., "Handbook of Human Vulnerability Criteria", chapter 2, "Spheres, cubes, and fragments", US Army Research and Development Command, Chemical Systems Laboratory, Aberdeen Proving Ground, Maryland 21010, USA, 1981.
Swanton, M., "The Deeds of Hereward", in Ohlgren, T.H., "Medieval Outlaws", Sutton Publishing, Stroud, 1998.
Tanner, J.C., "Factors Affecting the Amount of Travel", Rd. Res. Tech. Paper No 51, HMSO, London, 1961.
Taylor, D.W., "Speed and Power of Ships", US Shipping Board, revised, 1933.
Taylor, T.J., "On the Progressive Application of Machinery to Mining Purposes", Proc I Mech E., 10, 15–35, 1859.
Templeton, W., "The Millwright and Engineer's Pocket Companion", Simpkin Marshall, London, 1856.
Thompson, E.A., "A Roman Reformer and Inventor", Oxford at the Clarendon Press, 1952.
——, "Slavery in Early Germany", Hermathena, 89, 17–29, 1957.
Thorn, C., and Thorn, F., "Domesday Book: Wiltshire", History from the Sources, General Editor Morris, J., Phillimore, Chichester, 1979.
Thorpe, L., Translator, Gerald of Wales "The Description of Wales", Penguin Books, London, 1978.
Thrupp, G.A., "The History of Coaches", Kerby and Endean, London, 1877.
Timberlake, S., "Review of the Historical Evidence for the Use of Fire setting", in Crew, P. and A. (Editors), "Early Mining in the British Isles", Proc. of a Workshop on Early Mining, National Park study Centre, Plas Tan-y-Bwich, 1990.
Timoshenko, S., "Strength of Materials", Part I, D. Van Norstrand Co., Princeton, NJ, 1955.
Toynbee, A.J., "A Study of History", Vol. 9, Oxford University Press, London, 1954.
Tylecote, R.F., Austin, J.N., and Wraith, A.E., "The Mechanism of the Bloomery Process in Shaft Furnaces", J of The Iron and Steel Institute, 204, 342–363, 1971.

Tylecote, R.F., "The Medieval Smith and his Methods", in Crossley, D.W., "Medieval Industry", Research Report No. 40, Council for British Archaeology, 42–50, 1981.
Usher, A.P., "A History of Mechanical Inventions", McGraw-Hill Book Company, New York, 1929.
Vincent, B., "Haydn's Dictionary of Dates and Universal Information", Ward, Lock, and Co, London, 1881
Vincent, E.T., Hicks, H.H., Oktar, E.I., and Kapur, D.K., "Rigid Wheel Studies by Means of Dimensional Analysis", US Army Land and Locomotion Laboratory, ATAC, Report No 7841, LL-79, Warren, Mich., 1978.
Vinci, Leonardo da, "The Notebooks of Leonardo da Vinci", Translated by MacCurdy, E., Reynal and Hitchcock, New York, 1938.
Wadsworth, A.P., and de Lacy Mann, J., "The Cotton Trade and Industrial Lancashire", Manchester University Press, 1931.
Wailes, R., "Horizontal Windmills", Trans. Newcomen Society, XL, 125–145, 1967–8.
Walton, J.R., "Mechanisation in Agriculture: a Study of the Adoption Process", in Fox, H.S.A., and Butlin, R.A. (Editors), "Change in the Countryside: Essays on Rural England, 1500–1900", Institute of British Geographers Special Publication 10, 1979.
Washburn, S.L., "Biochemical Insights into Our Ancestry", in Robinson, M.H., and Tiger, L., (Editors), "Man and Beast Revisited", Smithsonian Institution Press, Washington, 1991.
Webb, F.J., "The Age of Bede", Penguin Classics, London, 1985.
Weir, J., "New Methods for Calculating the Metabolic Rate with Special Reference to Protein Metabolism", J. Physiol., 109, 1–9, 1949.
Westbrook, J.H., "Temperature Dependence of the Hardness of Pure Metals", Trans. ASME, 145, 221–47, 1953.
Westermann, W.L., "Athenaeus and the Slaves of Athens", Harvard Studies in Classical Philology, Supp. Vol., 451–470, 1941.
——, "Industrial Slavery in Roman Italy", Journal of Economic History, 2, 149–163, 1942.
White, K.D., "Agricultural Implements of the Roman World", Cambridge University Press, Cambridge, 1967.
White, L., "Eilmer of Malmesbury, An 11th Century Aviator", The Society for the History of Technology, December 1960, UCLA Press, 1978.
Whitehouse, R., "Later Prehistory", in Collins D., (Editor), "The Origins of Europe", George Allen and Unwin Ltd., London, 1975.
Whitlock, R., "The English Farm", J.M. Dent and Son Ltd., London and Melbourne, 1983.
Whitton, D., "The Society of Northern Europe in the High Middle Ages", in Holmes, G. (Ed.), "The Oxford Illustrated History of Medieval Europe", Guild Publishing, London, 1988.
Wick, B.H., "Study of Sub-Sonic Forces and Moments on an Inclined Plate of Infinite Span", NACA Technical Note 3221, 1954.
Wiedemann, T.E.J., "Slavery. Greece and Rome", New Surveys in the Classics No. 19, Oxford, 1987.
Wikander, Ö. (Editor), 'Handbook of Ancient Water Technology', Brill, Leiden, 377, 2000.
Williams, C.J., "Great Orme Mines", British Mining Monograph No. 52, Northern Mine Research Society, Keighley, May 1995.
Wilson, D., "The Anglo-Saxons", Thames and Hudson, London, 1960.
——, "The Anglo-Saxons", Penguin Books, London, 1981.
Wilson, P.N., "The Waterwheels of John Smeaton", Trans. Newcomen Society, XXX, 37–38, December 1955.
Wolf, A., "A History of Science, Technology, and Philosophy in the 16th and 17th Centuries", Vol. II, Peter Smith, Gloucester, Mass., 1968.

Wolfegg, C., Graf zu Waldburg, "Venus and Mars: The World of the Medieval Housebook", Prestel, Munich, 1998.
Wong, J.Y., "Terramechanics and Off-Road Vehicles", Elsevier, Amsterdam, 1989.
——, "Theory of Ground Vehicles", John Wiley and Sons, Inc., New York, 1993.
Wood, E.S., "Field Guide to Archaeology", Collins, London, 1967.
Wood, M., "In Search of the Dark Ages", BBC, London, 1981.
Woosnam, M., "Eilmer, 11th Century Monk of Malmesbury, The Flight and The Comet", Friends of Malmesbury Abbey, Malmesbury, Wilts., 1986.
Wright, T., "Thomas Young and Robert Seppings: Science and Ship Construction in the Early Nineteenth Century", Trans. Newcomen Soc., 53, 55–72, 1981.
Wright, Q., "A Study of War", University of Chicago Press, Chicago, 1942.
Young, A., "General View of the Agriculture of Oxfordshire", 1813, reprinted: Augustus M. Kelly, New York, 1969.
Young, T., "Remarks on the Employment of Oblique Riders", Phil. Trans. (Royal Society), 303–336, 1814.
Zahavi, Y., "The UMOT [Unified Mechanism of Travel] Project", Report for the US Dept of Transportation, Research and Special Programs Administration, and the Federal Rep. of Germany, Ministry of Transport. DOT-RSPA-DPB-20-79-3, August 1979.
Zonca, V., "Nova Teatro di Machine et Edificii, Padua (1607).

SUBJECT INDEX

abrasive wear, 106
Ackermann's hinged axle, 451
Actium, battle of, 532
adze, 151, 157, 401
  tempering, 963, 1071, 1108
aeroplane, 642
  Cayley's, 640
  pedal-powered, 212
afterdamp, 862
aggression, 1053, 1054
agricultural implements,
  Romano-British, 728
agricultural revolution, 694
agriculture
  ancient, 699
  early modern, 737
  Hittite, 710
  mediaeval, 729
  twelfth century Welsh, 729
aircraft, 631
  heavier than air, 640
  need for petrol engine, 643
Alexander, tactics of, 1128
Alexandria, Museum in, 665
alloy
  copper and arsenic, 772, 779
  copper and zinc (brass), 920
  lead, ductility at low temperature, 988
  quenching, 963
American War of Independence, 1196
Amontons, laws of friction, 11
Amsterdam dredger, 176
anchor, 530
ancient civilisations, 775
Anglo-Saxon Chronicle, 541, 543
animal power, 167
annealing, need for, 935
Archimedean screw, 717, 795, 796, 799
  driven by a windmill, 805
  efficiency and power, 804
  for raising water, 15
Archimedes
  theory of levers, 5
  worm, 18
Aristotle, contempt for artisans, 659
armour, 1079, 1098, 1112, 1155
  bronze, 1117
  fashion in, 1160
  temperature in sunlight, 1115
  weight of, 1160
army
  Celtic, 1139
  Roman, 1131
arquebus. *See* harquebus
arrow, 1093, 1100
  shaft smoothers and straighteners, 1106
arrowhead, 1102
artemon, 534
artillery
  better than rockets, 1194
  British, 1202
  Gustav II's use of, 1200
  Ramelli's method for hauling, 12
Assyria, 163, 412, 882, 1122
Athena, invention of useful arts, 655
Athens, subsistence in, 717
ATP (adenosine triphosphate), 183
axe
  Abbevillian and Achueulian, 1062
  flint, 765
  flint-knapped, 768
  Hittite, 1121
  metal, 768
  shaft-hole, 772
  stone and bronze contrasted, 772
  stone, fully polished, 763
  striking, 151
  tempering, 963
  tempering colour, 964
  use in classical warfare, 1097
  use of, 157
axle
  of tympanum, 718
  radius of, 25
axle-grease, unusual use of, 72

Bacon, Francis, empiricism of, 687
balance, 7, 33
  Roman, 35
ballista, 671, 1135, 1149
  analysis of, 1257
  constant torque, 1256
ballistic pendulum, 1186, 1187

ballistics, 1180
of projectiles, 1242
balloons, 635
need for independent propulsion, 637
notable flights, 640
typical construction of, 638
typical hydrogen, 636
uses of, 637
band brake, 83
barouche, 447
Barton Aqueduct, 387
barulkos, 62
basinet, 1159
battery mill, 936
battle car, 401
battle cart, 1109
Battle-axe culture, 775
Bayeux tapestry, 544, 1147
bayonet, socket, 1175
beam-press, 9
bearing modulus, 100
bearings, 70, 71
boundary lubrication, 99
Collinge, 444
dry rubbing, 102
frictional losses, 71
lubricated, 104
lubrication of, 73
mail, 444
roller, 72
theory of, 98
wooden, 104
Beaufort scale, 322
Belisarius, defence of Rome, 673
Bell Beaker culture, 775
bellows, 151, 162
cam-operated, 916
effort required to use, 921
foot-operated in New Kingdom, 886
not used in Egyptian Old Kingdom, 885
operated by a man's weight, 162
operation by cams, 68
operation by man-power, 907, 910
powered, to melt iron, 912
to increase furnace temperature, 161
with trip-hammer at Bordesley Abbey, 688
bell-pits, 817
belly bow, 1131–2
belt and rope drives, 55
analysis of, 74
maximum force limitation, 77
strength of, 78
belt claw, 1155
Beowulf (poem), manufactured goods, 673
bicycle, efficiency of, 203
bireme, 527
Black Death, labour shortage, 909
blackdamp, 862
blacksmith, 904, 938
patron saint of, 677
Roman, 157, 161, 757, 896–7
blackwork, 906–8
blast furnace, 909–11, 925
charge for, 912, 921
coke-fired, 921, 948
compared with bloomery, 910
early, 911, 914
length of campaign, 928
number in England and Wales, 945
use described, 943
blister steel, 957
block brake, 81
block trail carriage, 1170
bloomery, 905, 910–12
boats
ancient, mast position of, 513
early, 508
early modern, 553
in Greece and Rome, 526
Medieval and Renaissance, 538
of ancient civilisations, 510
on the Euphrates, 520
skin-covered, 537
bombard, 1165
boring machine, 626, 965–6
early, 962
for bronze cannon, 979
horizontal cylinder, 984
problems of, 983
Wilkinson's, 983
bow, 1103, 1125, 1126
composite, 1105
decline of in Greece, 1125
deemed dishonorable, 1054
efficiency of, 1250–52
recursive, 1252
reflex, 1109
use from a vehicle, 403
bow and arrow, analysis of, 1247
bow shapes (ship), 565
bow stave, hysteresis loss in, 1248
bow wave, 583
bowl furnace, 882–3
bows of a cog, 548
brake, 56, 291
band, 58
block, 59
brass, 689, 901, 933, 952

Briare canal, 384
bricole, 1158
Bridgewater Canal, 386
bronze, 538, 779, 780, 781, 782, 879, 884, 890, 892, 898, 912, 937, 945, 979, 989, 1053, 1112, 1121, 1125, 1167
  casting, 652
  Chimera of Arezzo, 892
  Colossus of Rhodes, 891
  flanged axe, 774
  ineffective against iron, 1082
  porous, 104
  pot, 892
  statue, making, 893
  working, spread of, 656
Bronze Age, 167, 415, 416, 1058
  boats in, 509
  casting technique, 880, 882
  copper mines, 769
  end of, 1123
  flint mines of, 764
  importance of smith and artisan, 650
  metal-workers, hobbling gait of, 658
  mining in, 409
  rapiers in, 1119
  status of smiths, 653
  transition to Iron Age, 523
  use of sulphide ore, 879
bronze axe, 773
Brougham, 446
Brown Bess, 1199
bucket train, 719
bucket winch, 852
bullet
  drag coefficient, 1217
  dynamics of, 1186
  elongated, 1197
  hexagonal, 1198
  high velocity, 1231
  iron compared with stone, 1180
  mediaeval, 1229
  oval, 1197
  penetration by, 1219, 1220, 1222, 1228, 1231, 1234
  round and square, 1185
  stability of, 1237
  threshold velocity, 1221
  to shoot hippopotamus, 1224
  tumbling, damage caused by, 1218
  velocity of, 1184
buss (ship), 546

calico, 1033, 1034, 1043
cam, 22, 66, 69, 70, 93, 95, 97, 234, 245, 247, 301, 305, 977
  dynamics of, 95
  in corn mill, 156
  in Diderot's slitting machine, 939
  kinematics of, 91
  operating bellows, 914
camel, domestication of, 417
campaign (of a blast furnace), 933
canal, 12, 354, 355, 357
  British, 358
  construction of, 383
  impact on trade, 390
  network, 389
  through Nile cataracts, 516
  traffic on, 371
Canal du Midi, 384
canaries, 862
cannel, 818
cannon, 1169
  accuracy of, 555
  axle wheel of, 485
  bore clearance, 979
  boring mill for, 980
  casting, 948
    danger of, 952
  competition with trebuchet, 1158
  dished wheels for, 488
  early history of, 1170
  effective against walled city, 1171
  export of, 945
  heavy, effect on warship design, 551
  iron, 944
    need for, 933
  making, 950
  performance of, 1180
  properties and construction, 1172
  pulling up a hill, 11
  rate of fire, 1171
  transport of, 1200
cannon ball
  damage to masonry, 1231
  leakage past, 1263
capstan, 24, 29, 55, 151, 165, 166, 176, 201
  as drive for dredger, 559
  for wire drawing, 939
caracole, 1173
caravan, 371
caravel, 551
carbon
  dioxide, 184
  monoxide
    effect on canaries, 862
  oxides, poisonous levels of, 792
carding machine, 1038
carpentum, 725

carrack, 549, 551
carriage
  excessive cost of, 422
  excessive tolls for passing locks, 381
  of mail, 443
carriages
  access by ladies, 451
  design of, 446
  gun, 985, 1206
  numbers in use, 431
  stability of, 456
  travel by, 369
carruca, 424, 425
cart
  Chersiphron's, 421
  Colonel Blunt's, 433
  crank-axle, 447
  Greek country, 417
  load for, 420, 426
  Roman, 423
  Viking ceremonial, 426
cartridge, 1176
  Pauly's, 1199
cartwheel, dished, 488
cassiterite (stream tin), 787
cast iron
  axle box, wrought iron preferred to cast iron, 444
  conversion to malleable iron, 934
  conversion to steel, 954
  displaces bloomery, 910
  for agricultural implements, 757
  impurities in, 924
  kinds of, 927
  spokes, reason for curving, 156
  strength of, 90
  uses for, 949
  white and grey, 927
casting, in dry sand, 951
castration, 705
catapult, 1135, 1139, 1141, 1143, 1256
  analysis of, 1148
cattle
  domestication of, 703
  to provide power, 399
cave paintings, 1069
cavesson, 403
cavitation, 810
celt, bronze, 880
cementation furnace, 957, 963
cementation process, 961
centrifugal fan, 841, 869
chafery furnace, 914, 929
chain drives, 55
chain mail, 1152
chanfron, 1165
charcoal, 344, 697
chariot, 404, 409, 418, 463, 1111, 1114, 1117
  Amenhotep III's, 1112
  Assyrian, 1122
  development of, 406
  Egyptian, 407, 467
  Hittite, 405
  of Ramesses II, 1120
  with wheel scythes, 671
chariot wheels, 403
chariot-scythe, 1053
Charles, The (ship), 557
chill casting, 757
Chimera of Arezzo, 892, 894
chiton, 1116
chivalry, decline of, 1086
choke damp, 817
Christianity, early, attitude to slavery, 668
chronometer, Harrison's, 563
cinnabar, 802
cire perdue, 882
cisium, 424
Cistercians
  leading iron producers, 684
  use of technology, 680
clay-mill, 175
clippers, 572
  American and British compared, 573
clocks, 63
clockwork, 301
clod-crusher, 757
cloth, manufacture in Malmesbury, 691
coaches, 428
  Blunt's design, 433
  design of, 446
  Paris, 433
  specification, 449
  speed of, 436
  sprung suspension, 434
  steel springs for, 439
  suspension of, 442
  turning, 451
coal mining, 817, 818
Coalbrookdale Ironworks of Darby, 949
cog (ship), 548, 549
coke
  used to make iron, 938, 949
colossi, stone, 163
Colossus of Rhodes, 896

column drums, transport of, 421
combing, 1015
comfort at sea, 616
compass, 16, 546, 548
composite bow, 1105, 1126
Cooks Kitchen Mine, drainage of, 837
copper, 410, 522, 529, 538, 712, 768, 770, 772, 775, 779, 787, 836, 853, 878, 879, 880, 882, 884, 886, 892, 894, 919, 952, 986, 1053, 1108–9
  bottoming (sheathing), 622
  made cheap chisels, 160
  metallurgy of, 779, 781
  smelting, 924
Copper Age, 879
copyhold, 737
coracle, 508, 521
Corded-Ware culture, 775
corn mill, 156
corrosion, 107, 580, 945
cotton, 1021, 1035
  principal UK export, 1032
  wearing forbidden, 1033
cotton industry, 1005
Coulomb, laws of friction, 11
coulter, 715, 721
cowl ventilator, 859
crane, 24, 387
Crete, 9, 228, 403, 655, 656, 772, 1079
Crimean composite bow, 1252
cross staff, 561
crossbow, 1147, 1153, 1154
  deemed dishonorable, 1154
  performance of, 1155
  re-introduction by Richard I, 1154
crowds, 455, 462
  walking in, 461
crown-and-lantern gear, 247
crucible, 895
crucible steel, 956, 959
crupper, 1160
Crusades, proclaimed to avoid internecine war, 1084
C-spring carriage, 442
Cullen (Cologne) steel, 956
culverin, 1182, 1189
cuneiform, 37, 708
curved cranks, 156
Cutty Sark (ship), 573

Daedalus, myth of, interpreted as spread of metal-working, 656
dagger, 772, 1079
Dark Ages
  decline of slavery, 148, 362, 365, 426, 647, 670, 673
  manufacture of swords, 675
demi-culverin, 1185, 1201
depopulation, 735
deserted villages, 735
Diderot's Encyclopaedia, 179
diligence, 368
dipper wheel, 825, 835, 858
distribution of settlements, 350
dividing engine, Ramsden's, 973
dividing head, 976
diving bell, 556, 625
  Halley's, 627
  uses of, 626
diving suit, Borelli's, 626
division of labour, Greek attitude to, 660
dog as power source, 178
drainage, 695, 788, 817, 824, 837, 863
draught animals
  limits to perfomance, 470
  power of, 850
drawloom, 1025
drawplate, 934
dredger, 559, 560, 671
  horse driven, 559
drill, 160, 847, 965
  Egyptian, 160
drill (seed-sowing), 741, 744
drives, flexible, 55
drop hammer, 937
drop shuttle box, 1036
Durham oven, 907
dynamometer, 452

edging ore, 894
education, 144, 213
efficiency, 48
  of gunpowder engine, 309
  of pulley, 49
  of screw thread, 48
Egypt, 7, 154, 160, 359, 399, 403, 410, 517, 520, 523, 648, 676, 702, 709, 710, 714, 728, 763, 776, 793, 803, 884–8, 916, 920, 924, 1089, 1100–01, 1120–25
  agriculture of, 708
  New Kingdom, 778
  Old Kingdom, 776
  urbanisation of, 1107
Egyptian composite bow, 1252–3
Eilmer, flight by, 632
elasticity of demand, 347, 392

electrum, 884
elliptical spring, 444, 452
emigrants, conditions provided for, 570
enclosure, 737
encyclopaedia, 978
endurance, 196
energy, 3, 113, 120, 131, 151, 181, 183, 191, 216, 243, 270, 283, 305, 309, 344, 352, 390, 455, 457, 480, 499, 697, 870, 1143, 1151, 1207, 1225, 1234, 1245, 1261
  consumption of, 115
energy, agricultural demand for, 119
Enfield rifle, 1197
engineer, 28, 176, 263, 384, 516, 648, 650, 655, 658, 660, 822, 939, 1053, 1061, 1186, 1199
  lowly status of, 658
  Swiss (Maritz), 983
engineers, early, 648
English naval cannon, superiority of, 945
English yew longbow, 1252
epinetron, 1014
equirotal carriage, 451
erosion, 106
essedum, 424
Essenes, attitude to slavery, 668
Etruscan, 894, 1082
Evelyn, John
  dines on galley, 550
  on smoke from coal, 697
  pallbearer at Shish's funeral, 558
  tries diving bell, 556, 625
Exeter canal, 384
Experiment (ship), 560
explosion, 843, 861, 1157
explosive, 183, 184, 283, 308, 630, 792, 838, 1163, 1189, 1273

factories, 661, 697, 757, 1005–6, 1020, 1038
Falkirk, battle of, 1156
fantail, 296
farming in SE Europe, 704
farming, spread of, 701
fatigue, 184
feathering, 299
feudal system, 732
field artillery, 1201
file-cutting, 976
  Card and Studley's machine, 976
finery furnace, 912, 929
fire, 784, 806, 924
fire engine. See steam engine
fire setting, 789
  analysis of, 806
  Roman use of, 784
firearms
  deemed dishonorable, 1059
  handheld, 1170
firedamp, 862
fishhooks made from bone, 507
flanking geometry, 1212
flax, 729
flight, 632, 1188, 1219, 1246, 1255
flint-knapping, 768
flintlock, 1174, 1175
flintlock musket, firing, 1176
flints, 3, 763, 764, 766, 768, 1102, 1125
floating mill, 241
flying shuttle, 1039
flywheel, 153, 966
food, 113, 118, 212, 546, 570, 660, 696, 699, 710, 1057, 1061, 1064, 1066, 1076–7, 1097
  energy to produce, 120
force pump, 795
forge hammer, 70, 915
forging, 885, 895, 903, 912, 916, 937, 959, 963, 987, 990
foundries, introduction of, 954
fragile goods, transport of, 422
France, 301, 367, 368, 375, 434, 637, 705, 903, 905, 909, 914, 1006, 1022, 1025, 1029, 1036, 1088, 1090, 1095–6, 1126, 1146, 1154, 1163, 1165, 1179, 1196
fretting, 107
friction, 11, 39
  angle of, 43
  coefficients of, 42
  laws of, 40
frictional forces, 43
frigate, Blackwall, 572
Froude number, 584
fulling, 68, 115, 143, 223, 243, 245, 254, 679, 681, 1005, 1019, 1021–2
  'walking on cloth', 245
  revolution, 1022
fulling mill, 1023
fulminates to ignite gunpowder, 1194
fur, 1006, 1064
furnace, 68, 680, 697, 845, 879, 880, 882–3, 885, 889, 893–5, 896, 898, 900, 902–3, 906, 911, 914, 918, 919, 921–30, 935, 941, 943, 1140
fustian, 1031

gad, 800
galena, 924
galley, rowed, 550
Galloper gun, 1202
gas, 635, 693, 843, 845, 846, 861, 862, 923–7, 1169, 1187, 1197, 1199, 1258, 1270
gas turbine, 303, 354
gastrophetes, 1131, 1133
gate-road, 844
Gaussian distribution of wind speed, 321
gear, 29, 55, 59, 61, 86, 153, 240, 243, 491, 971, 975–80
gear-cutting machine, 977
gears, 59
  crown, 66
  helical, 91
  lantern, 65
  spur, 65
  theory of, 86
  tooth-shapes, 89
  velocity fluctuations in, 91
gearwheel, 18
genetic distance between man and primates, 1059
Germany, 447, 574, 705, 785, 822, 847, 937, 939, 1071, 1089, 1093, 1096, 1161, 1171, 1196, 1199, 1202
gin, 167
gladius, 1140
glass, 42, 304, 432, 434, 439, 625, 662, 663, 920, 937
Globular-Amphora culture, 775
glue, 21
gold, 7, 136, 148, 263, 404, 519, 541, 776, 784, 790, 804, 857, 864, 886, 887, 916, 918, 934, 936, 1100, 1114, 1153
  extraction of, 857
  in Egypt, 885
Golden Devil (ship), 556
Golden Fleece, myth of, 788
goldsmiths' workshop, Roman, 893
Goliath, bronze armour of, 1079
goods, carriage of, 371
governor, 214, 266, 296
Grace Dieu (ship), 553
grain mill, Roman, 728
granite, Egyptian, transport of, 516
Great Britain (ship), 574
Great Harry (ship), 554
Great Michael (ship), 553
Great Orme, copper mine, 768
Great Pyramid, 511
Great Republic (ship), 573, 591
greaves, 1118
Greece, 130, 137, 350, 358, 359, 362, 391, 417, 428, 526, 658, 662, 687, 691, 703, 714, 734, 785, 1011, 1082, 1090, 1127, 1130, 1140
Greek fire, 1149
Greek manufacture, small scale of, 662
Greek mill, 268, 276
Greenwich, 937
Grimes Graves, 765
Guild system, 1027
gun
  breech-loading, 1168, 1199
  gas leakage in, 1200, 1263
  invention of, 1161
  performance of, 1180, 1259
  pressure inside, 1187
  revolving, 1184
  rifled, 1196
  Whitworth's, 1197
gun barrel
  composite, 1201
  possible reason for name, 1169
gun carriage, 1170, 1201
gun-boring machine, 979, 981–3
  at Royal Arsenal, 982
gunner's quadrant, 1180
gunpowder, 283, 302, 308, 310, 311, 336, 630, 808, 844, 847, 848, 849, 1053, 1153, 1161–3, 1171, 1178, 1187, 1191, 1194, 1199, 1224, 1258
  invention of, 1161
  making, 1163
  used for blasting, 847
Gunpowder Engine, 308, 336, 337, 338
Gustav II, tactics of, 1200

H.L. Hunley (torpedo), 631
Hackney carriage, 431
halberd, invention of, 772
hammer, 70, 679, 680, 689, 768, 800, 847, 909, 912, 913, 929, 934, 937, 938, 955, 976, 1168, 1200
hammer-axe, 878
hammer-welding, 899
hard surfaces, wheels on, 480
Harfleur, capture of, 1166
harness
  ancient, ineffective, 168
  mediaeval improvement of, 170
harquebus, 1170, 1172
harrow, 170, 179, 722, 723, 732
heaving, 849

heddle, 120, 1013
heels on soft surfaces, 476
hemp, 76, 78, 1015
Henri Grace à Dieu (ship), 554
Hephaestus, smith-god, attitudes to, 653
Hereward the Wake, chivalrous character of, 1085
hieroglyphic writing, 1106
Highworth Church, damaged by cannon ball, 1231
Hittite estate, described, 707
Hittite Law Code, 710
Hittites, 128, 129, 405, 522, 780, 781, 782, 887, 888, 1011, 1108, 1112, 1113, 1117, 1118, 1119, 1120, 1124
hodometer (road-measurer), 18
hogging, 515, 553, 594
hominids, earliest, 1061
hoplite, 1127
horizontal treadle loom, 1029
horizontal-axis waterwheel, mine drainage by, 786
horse
  choked by harness, 421, 1112
  dismantled for shoeing, 679
  hoeing by, 743
  performance of, 217
  source of power, 751
  use in agriculture, 747
horse- and ox-ploughs, compared, 170
horse artillery, 1202
horse harness, 169, 174, 415, 426
horse whim, 852
horse-driven machines, 174, 176
horse-hoe, nineteenth century, 746
horsepower, related to the unit, 181
horseshoes, 171, 423, 428, 1145
hot wire, 1170
howitzer, 1197
Huddersfield Canal (Standedge Tunnel), 383
hull, 173, 510, 512, 515, 518, 520, 529, 535, 536, 540, 541, 548, 553, 556, 564, 567, 574, 579, 586, 607, 618, 623, 625, 630
  rectangular cross-section, 618
  triangular section, 619
hulls, clinker- and carvel-built, 538
human characteristics, 1061
Hundred Court, 351
Hungary, 428, 847
hunting, 403, 415, 510, 703, 705, 729, 1006, 1054, 1064, 1071, 1100, 1109, 1125, 1132, 1149, 1154, 1196, 1214
hushing, 790
Iceman, 1008
impact fuse, 1201
incapacitation
  by impact, 1227
  probability of given a hit, 1220, 1224, 1226
inclined plane, 10, 37
  at Fusina, 12
  forces on, 43
India, 399, 550, 569, 572, 1033, 1044, 1095, 1172, 1192
industrial revolution, 119, 693
industrialisation, limited by low food production, 696
inertia chain, 152
inland waterways, 358
instrument, 12, 15, 26, 62, 755, 1180
invariants of transport, 344
Ireland, 228, 508
iron, 76, 91, 107, 141, 159, 160, 171, 243, 254, 256, 262, 296, 306, 343, 371, 406, 421, 423, 444, 451, 473, 486, 489, 541, 553, 555, 565, 571, 574, 590, 623, 629, 667, 672, 677, 680, 684, 694, 718, 719, 720, 755, 757, 782, 786, 789, 793, 799, 803, 804, 819, 828, 829, 836, 847, 852, 880, 881, 887, 888, 894, 895, 896, 903, 904–24, 933–65, 980–89, 1003–5, 1015, 1032, 1053, 1078, 1082–3, 1115–21, 1140,–42, 1166–9, 1180, 1184, 1201, 1230–36, 1265
  as precious metal, 896
  carburising, 956, 1003
  cast, uses for, 943
  Dudley makes using coal, 947
  Hittites skill, 783
  import and export of, 961
  in Domesday Survey, 909
  in Rome, 895
  in Sussex and Kent, 906
  making using coke, 921, 933, 998
  mediaeval manufacture, 680
  not to be used for weapons, 1082
  welding of, 1140
  wrought. *See* wrought iron
Iron Age, 523, 1126
iron weapons very effective against bronze, 1082
ironworks, English, 944
irrigation, 311, 702
Italy, 137, 143, 224, 243, 360, 362, 426, 528, 670, 680, 703, 720, 728, 772, 785, 787, 1006, 1025, 1029, 1032, 1089, 1096, 1129, 1174

Jacquard loom, 1029
javelin, 1126
jenny, spinning, 1040–44
jigging, 855
juice extraction, 18

Kadesh, battle of, 1114
kibble, 901
kiln, 133, 523, 544
King Alfred, navy of, 541
knife, 721, 884, 885, 916, 1062, 1100, 1119
Kondratieff cycles, 1093
Kurgan (Pit Grave) culture, 775

ladder, 796
Lake Nemi, ships on, 533
lakeside village, 507, 1007
lamp, 728, 793, 862, 889
lance, 1113, 1125, 1126, 1136, 1145, 1157, 1173, 1189
land transport, prehistoric and ancient, 399
landau, 447
large-scale statues, Roman, 891
Late Bronze Age, 781
lateen sails
  Roman use of, 535
  use of, 551
lathe, 423, 717, 933, 941, 943, 966–76, 984, 1201
  foot-operated, 966
  mandrel, 966, 970, 972
  pewterer's, 967
  screw-cutting, Maudslay's, 975
LBK, 1067
lead, 343, 421, 529, 537, 559, 622, 626, 629, 782, 790, 796, 882, 886, 892, 894–5, 901, 903, 906–8, 998, 1171, 1180–81, 1188, 1197, 1204, 1229, 1234, 1263
lead shot, making, 925
lead smelting, mediaeval, 906
leather, 9, 43, 76, 78, 308, 405, 432, 439, 442, 452, 468, 496, 790, 796, 800, 828, 829, 836, 850, 886, 938, 1008, 1108, 1112, 1127, 1147, 1156, 1160, 1200
leather gun, 1200
Leonardo, mechanics of, 27
lethal competition, 1055
lethality ratio, 1208
lever, 4, 7, 14, 20, 24, 58, 83, 160, 161, 199, 225, 245, 621, 634, 665, 708, 756, 938, 1198
  bent, 10
liburnium, 532
life expectancy, 118
lift pump, 809
limber, 1202
linen, 68, 405, 793, 1005, 1011, 1012, 1015, 1031, 1033, 1043
Linienbandkeramik Culture, 1071
liquation process, 920
literacy, 118
load-carrying, metabolic cost of, 208
loaming, 735
lock, 251, 379, 380, 381, 384, 1194
lockwool, 1017
locomotives, horse-driven, 178
logistic growth curve, 356, 396
long coach, 371
longbow, 1125, 1155–7, 1163, 1222
  modern, 1254
  performance of, 1155
  proved at Falkirk, 1156
  superiority of, 1157
longitude, determination of, 563
loom, 129, 245, 1007, 1011–16, 1019–21, 1028–30, 1034–36, 1041, 1046–49
  horizontal treadle, 1019
  warp weighted, 1016
lost wax process, 882, 892, 989
lubrication, 42, 55, 73, 99, 102, 106, 108, 1002
  hydrodynamic theory of, 100
luddites, 750, 1043

machine gun, 1200, 1210
machine tool, 301, 965
man power, 151
mangonel, 1149
mangonneau, 1153
manorial system, 732
manual labour, Aquinas excuses monks from, 682
manufacture, 647
  in Greece and Rome, 658
manure, 734, 743
map, 384, 436, 1023
marching, military, 463
market size, influenced by transport cost, 390
marling, 735
Mary Rose (ship), 555
matchlock, 1170
  musket, firing, 1175
mathematics, Sumerian, 38
McAdam's method of road construction, 370
measurement, 563

mechanic, 181, 691, 1038
mechanics, 3
  theoretical, foundations of, 693
Mediterranean shipbuilding practice, 535
Megiddo, Battle of, 1113
merchant ships, excavation of, 529
mercury, 802, 857, 1161, 1194
Mercury (newspaper), 179, 750, 768
Mercury, statue of, 892
Mesolithic, 507, 699, 763, 1006, 1065, 1067, 1070–72, 1100
Mesopotamia, 7, 126, 400, 410, 520, 702, 710, 741, 780, 1108, 1110
  agriculture, 708
  sea-borne trade, 522
metabolic rate, 200, 202, 208, 212, 217, 221, 460
metacentre, 615
metacentric height, to find, 619
metal, 65, 72, 108, 157, 161, 232, 255, 260, 301, 308, 344, 402, 407, 409, 574, 625, 654, 656, 664, 712, 719, 723, 757, 763, 768
  hot and cold working, 985
  spinning, 967
  strength and hardness, 985
  workers crippled, 658
metal ores, sources of, 780
metallurgy
  Anglo-Saxon, 803
  Balkan, 771
  development of, 770
  in Europe, 907
  in Mesopotamia, 779
  Italian, 772
metalworkers, patron saint of, 677
metal-working
  ancient and medieval, 911
  Ancient Greek, 888
  mediaeval, 895
micro-burin, 712
mild steel, 1233, 1234
mill dam, obstacle to navigation, 249
milling machine, 985
millpond, 249
millwright, 836
mina, 782
mine, 60, 126, 130, 143, 147, 166, 223, 314, 358, 386, 409, 519, 559, 662, 680, 685, 689, 763, 1157
Miner's Friend, 312
miner's needle, 849
mines
  "collapse of mountains", 789
  accidents in, 819
  air needed for ventilation, 846
  air temperature distribution in, 870
  British, 784
  coal, 817
  deep vein, 790
  drainage of, 822
  early European, 823
  flint, 764
  flooding, 794
  forced convection in, 868
  gold, 788
  gold, in Egypt, 793
  gold-bearing quartz, 803
  Greek, 784
  gunpowder compared with fire setting, 808
  harmful gases in, 792
  limits of practicability, 864
  natural convection in, 863
  prehistoric, 763
  productivity of, 822
  Roman, 784
  tin, 815
    blamed for environmental damage, 816
  toxicity of gases, 862
  ventilation, 845, 861
  by convection, 793
  ventilation, 838
Minié rifle, 1197
mining, 126, 135, 343, 371, 656, 687, 697, 717, 763, 768, 769, 776, 779, 780, 781, 784
Minoan civilization. *See* Crete
mirror, 562, 689, 698
modern longbow, 1251
monasteries, 300, 676, 677, 680, 690, 907, 935, 950, 1017
monks reviving technology, 676
Mons Meg, 1167
mortar, 175
mould, 755, 880, 883, 907, 943, 951
mould board, 758, 761
mule (animal), 123, 424, 653, 863
mule (machine), 1044, 1045, 1046
muscle, 115, 120, 151, 181, 216, 283, 353, 579, 1217
  animal, 216
  chemistry, 183
  endurance, 209
  fatigue, 198
  human, 200
  mechanical model of, 189
  mechanics of, 186
  physiology, 182

power and efficiency of, 192
training, 205
muscle power, reactions producing, 184
musket, 558, 1173–76, 1194, 1196, 1200, 1236
percussion, 1194–5

Nautilus (submarine), 630
naval cannon, recoil of, 552
navigation, 372, 374, 377, 520, 548, 642
Navy, Roman, 530
Neolithic, 417, 699, 763, 884, 1006, 1062, 1066, 1071–2, 1102, 1104
Neolithic farming, 702
Netherlands, 429, 692, 982, 1093
Newry Canal, 385
Nile
canal to Red Sea, 520
cargo boats described by Herodotus, 519
ease of sailing on, 512
shipbuilding on, 511
noria, 67, 718, 795, 796
Norman invasion of England, fleet for, 544
Normans, arming of, 1085, 1146
Norse mill, 225, 232, 268, 278
Nubia, 778

oar, 518, 528, 546, 600
obelisks, 519
Ockham's razor', 686
octant, 562
oil, 40, 42, 50, 72, 80, 99, 426, 444, 452, 522, 838, 895, 963, 1012, 1015, 1141
olive oil, 15, 664, 729
omnibus, 448, 449
onager, 400
onos, 1014
ore
cleaning and concenrating, 800
dressing, 853
transportation of, 774
ornithopter, Swedenborg's, 635
Ötzi, 1008
overshot wheels, power output, 271
oxen proposed for marine propulsion, 173

paddlewheel, 166, 228, 558
paddling, 164
painting, 304, 655, 699, 706, 888, 981, 1006, 1064, 1072, 1101, 1120, 1126
Palaeolithic, 399, 507, 699, 763, 1062, 1064, 1067, 1101–03
Panama Canal, 577
paper, 292, 635, 680, 848, 1022, 1029, 1171, 1176, 1188, 1195, 1199
papyrus, 56, 167
parachute
early practical, 637
Leonardo's, 633
patent, 156, 295, 300, 312, 441, 744, 747, 799, 946–8, 955, 957, 1035, 1038, 1040, 1041–3, 1184, 1194
pattern welding, 1140, 1142
paving, 365
Peasants Revolt, 737
peltast, 1127
Pelton wheel, power output, 276
pendulum, 301, 563, 1186
penetration
by projectile, 1217
mechanics of, 1207, 1228
Milne de Marr equation, 1234
Pennines, transport across, 382
pepperbox, 1184
perch, 442, 443, 452
percussion to ignite gunpowder, 1194
Persia, 359, 700, 1082, 1090, 1129
petrariae, 1149
pewter, 967–9, 920
peytral, 1160
Phaeton, 447
Phoenicians
exploration by, 525
trading activities of, 523
pick, 765, 786, 790, 847, 901, 906
pierrière, 1153
pig iron, 912, 929, 946, 949, 954, 961
pikemen, 1147, 1161, 1173, 1200
pile, 43
pin-fire cartridge, 1200
Pippingford Park, making iron at, 952
pistol, 1162, 1173, 1176, 1199
pitch, 143, 799
pitch (aircraft), 1235, 1238
pithead baths, 786
pitting, 788
placer, 787
plague, 735
plane, 43
plane, inclined, 10, 13, 37, 48, 874, 1110
planing machine, 984
plate armour, 1079, 1157, 1160, 1162
Plato, snobbery of, 659
plough, 170, 179, 349, 411, 707, 711, 715, 720, 729, 752, 762
double furrow, 757
mediaeval, 731

mouldboard, 715
Roman, 720
Rotherham, 754
seventeenth century, 738
Small's, 755
turning, 715
turnwrest, 756
types of, 752
ploughing, 716
analysis of, 758
Finlayson's rules for, 755
ploughman, mediaeval, 730
Poll Tax, 737
Poncelet's equation, 1225
population, 116, 131, 141, 146, 243, 346, 395, 538, 682, 695, 702, 729, 734, 771, 775, 887, 1025, 1057, 1066–7, 1070, 1076, 1079, 1083, 1088, 1093, 1099
population growth, 395
Portugal, 1090, 1095
post mill, 171, 289
postal service, 437
posting, 371
potter's wheel, 123
pottery, 9, 133, 147, 395, 511, 521, 526, 694, 898, 1016
pound lock, 384
power, 3, 17, 28, 76, 113, 151, 181, 192
generated by man, 202
generation, 111
transmission, 32, 55
power loom, Cartwright's, 1046
power source, 114
gravity, 300
hot air, gunpowder and steam, 302
wind, 283
press, 9, 15
pressure cooker, 311
pressure-sinkage relation, 456, 474
Preussen (ship), 574
prime movers using hot air or steam, 693
Prince Rupert, method for softening iron, 955
projectile
maximum range of, 1243, 1246
motion through tissue, 1215
optimisation of, 1266
stability of, 1235
Prometheus, attitudes to, 651
puddling furnace, 962
pulley, 11, 23, 74, 641, 938, 985, 1020
compound, 25
efficiency of, 49
mechanical advantage of, 51
medical uses, 25
simple, 152
pulley ropes, stiffness of, 51
pump, 828
punic cart, 725
Punt, 517
Pyramid, 10, 160

quadireme, 528
quadrant, 561, 1180, 1184
quarrel, 1155
quarrying, 343
Queen Boudicca, use of chariots, 1122
Queen Hatshepsut, trading activities of, 517
quenching of steel, 963
quern, 159, 232, 243, 728
conflict with water mills, 234
suppression of, 232
quffa, 521
quinquireme, 528

rag and chain water engine, 836
rail, 345, 353
railway, 353, 358, 389, 400, 439, 448, 851
reaper, Bell's, 747
reaping machine, 723
reclamation, 735
rectae, 1015
redundant work, 991
reed, 1013, 1020
Renaissance, workshops in, 689
reservoir, 249, 719
reverberatory furnace, 919, 947, 949, 954
conversion of iron to steel, 957
revolver, Colt's, 1185
Reynolds number, 586
rheda, 424
ridging, 722
riding
early, 415
in Persia, Greece and Rome, 417
medieval, renaissance, and early modern, 426
rifle, 1196
rifling, 1197
right wing, advantage of, 1127
rivers, problems of reconciling uses, 378
road, 18, 232, 344, 358, 360, 419, 434, 456, 461, 462, 485, 495, 776, 850, 905, 1031, 1092
alpine, 364

damage to, 435
dangers of, 366
early, 358
Highway Act of 1555, 365
improved construction, 370
in Anglo-Saxon England, 362
infrastructure, development of, 358
limitations on vehicles, 435
mediaeval, 363
opposition to improved, 369
poor condition of, 366
repair and construction, 434
Roman, 360
surface, representation of, 499
rocker-crank, 154
rocket, 1268
analysis of, 1268
Congreve's, 1193
early, 1189
military attraction of, 1193
motor, 1191
multi-stage, 1191
rocket spear, 1189
roller, 291, 414
roller bearing, 11, 72
roller crusher, 855
roller spinning, 1038
rolling, 995
rolling mill, 936–9, 949, 962
rolls, crushing, geometry of, 872
Roman, 17, 35, 129, 138, 141, 158, 172, 176, 224, 231, 268, 279, 343, 360, 365, 395, 422, 529, 601, 647, 661, 691, 718, 784, 926, 947
Roman Empire, proposals for reform, 671
Roman ship proposed to be powered by oxen, 536
Romantic Movement, 648
Rome, ports of, 534
rotary grinding wheel, 728
rotary quern, 159
rowing, 164, 353, 515, 526, 540, 579, 601, 637
energy needed for, 206
theory of, 599
Royal Mail, 440
Royal Society, 691
Royal William (ship), first Atlantic crossing under steam, 576
rubber, 76, 451
rudder, 530, 549
position of, 546
Ruin, The, poem speculating on Roman remains, 672
running, 184, 201, 207, 211
energy requirement, 459
speed of, 455
rutting, 1056

sabot, 1191
saddle, 405, 1124, 1137, 1145, 1147, 1158
saddle (lathe), 970
saddle quern, 804
saddle stone, 154, 710, 714, 727
safety valve, 309
sail, 292
sailing, 152, 283, 372, 527, 530, 540, 548, 569, 602, 604, 1147
sailing ship, 293, 317, 344, 510, 534, 584
sailing speed related to Beaufort scale, 614
sailing, theory of, 605
sainfoin, 741
Salamis, battle of, 527
saller, 841
saltpetre, 849, 1161, 1163, 1177, 1179, 1194
making, 1177
saluting, origin of, 1159
Sankey Brook navigation, 385
Savannah, paddle steamer, 576
saw, 247, 965–8
sawing machine, water-driven, 247
sawmill, 968
science, 658, 665, 687, 691, 1161, 1185
Scotland, 245, 371, 374, 382, 436, 553, 750, 755, 757, 785, 924
scouring, 1022
screw, 15, 630
for raising water, 16
of Vitruvius, 17
screw (prison warder), 163
screw cutter, internal, 20
screw jack, 21
screw manufacture, 717
screw propeller, 537, 637
screw thread, 11
efficiency of, 48
screw-cutting, 22
Leonardo's method, 970
screw-press, 20
scythe, 723, 747, 919
sea transport, 374
sea-coal, 373, 818
sea-going ships, 515
seed drill, 744, 752
Cooke's, 744

Tull's, 741
reluctance to use, 743
seeder, 179
self-bow, 1104, 1125
Seneca, disdain for technology, 662
serfdom, 1025
seric iron, 956
serpentine, 1171
Seville statement, 1060
sextant, 563
shaduf, 4, 5, 702, 704, 717
shear steel, 958
shedding, 1013, 1020
sheep, 400, 418, 544, 700, 736, 1009, 1011, 1015
shekel, 782
shell, 529, 535, 776, 1110
shell (missile), 1204
shingling, 962
ship, 15, 56, 172, 292, 347, 353, 372, 427, 511, 530, 729, 796, 816, 835, 946, 1129, 1184, 1185
shipbuilding, 25, 511, 524, 535, 557, 947
ships
ancient, speed of, 530
fouling and corrosion, 622
hulls, strength of, 589
hydrodynamics of, 564
improved design of, 572
longitudinal rigidity of, 567
metacentric height of, 566
power to propel, 580
resistance
Beaufoy's experiments, 567
Froude's and Gertler's equations, 581, 600
sailing
decline of, 577
metacentric height of, 617
stability
at large angles of heel, 620
stability, 580, 614
stiffness of, 592
technology of, 579
transatlantic speeds, 570
wooden hulled, limits of, 553
shoe brake, 81
shot, casting, 954
sickle sword, 1118
siege engines, performance of, 891, 1091, 1109, 1113, 1122, 1134, 1149
siege techniques, Assyrian, 1122
silicosis, 863
silk, 690, 1005, 1029, 1032
rubberised, 636
silver, 7, 132, 135, 404, 425, 541, 641, 662, 677, 772, 776, 780, 794, 802, 879, 883, 892, 901, 920, 933, 987, 1114, 1129, 1140
simple battle, analysis of, 1208
Single-Grave cultures, 775
slavery, 4, 113, 123, 124, 647, 660, 666
among barbarians, 668
ancient, 125
decline of, 647
Greek, 129
Hittite, 128
Roman, 140
sledge, 12, 168, 404, 413, 463, 793, 823
slide rest, 971
slitting and rolling mill, 933, 937, 939
slitting mill, 937
slow match, 1170
smallware loom, 1036
smelting, 774, 815, 877, 878, 896
copper, 878–9
iron, 343, 783
lead, 924
shortage of wood for, 697
silver, 134, 662, 802
use of bellows, 161
Smollett
on Channel crossing, 375
on coach travel, 367
snaphance lock, 1174
snubbing, 29, 32, 160
Society for the Improvement of Naval Architecture, 566, 567
Society of the Mineral and Battery Works, 935, 937
soft surfaces, wheels on, 476
soil mechanics, 474
solder, 884, 934
Sovereign of the Seas (ship), 556
Spain, 148, 537, 549, 727, 772, 785, 787, 798, 895, 909, 924, 1006, 1065, 1072, 1089, 1095–6, 1102
spangenhelm, 1157
spar torpedo, 630
Spartan army, 1080, 1127
spear, 415, 1065, 1067, 1077–9, 1085, 1097, 1100–2, 1108, 1110, 1126, 1145, 1209
spear thrower, 1102, 1136
specialisation, Greek attitude to, 660
speed governors, 265
spinning, 55, 80, 544, 966–9, 1005, 1007, 1009, 1012, 1014, 1016, 1019, 1027
frame, 1039
jenny, 1040–46
machine, James Taylor's, 1040

mule, 1044
wheel, 966, 1019, 1027
spinning (motion), 1196, 1238
spring, 439, 456, 502, 504
springing of vehicles, 502
springs, elliptical, or laminated, 502
St Brendan, voyages of, 543
stability, 32, 403, 456, 486, 555, 556, 566, 635, 1191, 1196, 1203, 1235, 1238, 1253
stamp mill, 305, 853
Standard of Ur, 1109
Stanhope Gig, 446
statue, 8, 154, 161, 412, 422, 517, 648, 776, 888, 891–3, 1115–6, 1166
Statute of Labourers, 736
Statute of Winchester, 362
steam, 124, 178, 253, 283, 296, 344, 375, 559, 571, 576, 630, 693, 747, 824, 836, 845, 912, 921, 951, 983, 1005, 1046–7
steam digester, Papin's, 311
steam engine
early, 309, 311
Newcomen's, 315
Savery's, 312
steam gun, Leonardo's, 306
steam hammer, 912
steam power, 3
steam turbines, early, 305
steel, 42, 76, 228, 434, 439, 451, 553, 571, 590, 596, 895, 914–6, 923, 933, 954–65, 1103, 1140–42, 1160–61, 1174, 1195, 1230, 1233–34
crucible, 956
German, 952
hardening and tempering, 959
making, 922, 1140
tempering colours, 963
steel mill, mine lighting by, 846
steelyard, 8
steering, 456, 491, 510, 558, 642, 1109
Stele of Vultures, 1108
Step Pyramid, 648
stern wave, 583
stirrup, 934, 1145
stithe, 817
stone, 8, 14, 32, 157, 159, 163, 225, 254, 283, 358, 369, 391, 482, 486, 489, 517, 672, 768, 772, 786, 918, 1100
Stone Age, 699
stone axe, 772, 774, 1100
utility of, 764
stone missile, 1135–38, 1143, 1149, 1153, 1166, 1169
stone tools, 1061, 1063
stone-crushers, 213
stonedamp, 862
stones, pulling heavy, 164, 412
strategy, 1097
raiding or persisting, 1099
striking tools, 157
stripper, Gallic, 723
submarine, 558, 625
Bushnell's, 629
too unfair to use, 630
Suez Canal, 577
Sumerian agriculture, 710
surcoat, need for, 1115
suspension, 433, 442, 456, 494, 496
Sutton Hoo, boat found at, 541
Sweden, 555, 564, 592, 935, 1088, 1089, 1090, 1103, 1190, 1200
swimming, energy needed for, 206
Swiss square, 1161
Switzerland, 1090
sword, 674, 887–8, 1118–1126, 1140–48, 1157, 1209
Celtic, manufacture of, 1140
thrusting and slashing, 1118
sword (watermill), 228
sword (weaving), 1021
Syria, 402, 523

tamping, 848
tappet, 1049
taxation, 1027
technology
ambivalent attitude to, 698
ancient objections to, 666
attitudes to, 648
damage caused by, 697
fostered by monks, 684
Greek myths about, 650
low status of, 647
monastic enthusiasm for, 677
Renaissance attitude to, 686
Roman attitude to, 663
value of, 676
tentering, 296
teredo, 622
terminal ballistics, 1207
textiles, 1005
effect of Black Death and Peasant's Revolt, 1025
in Egypt and Mesopotamia, 1007–8
manufacture in Europe, 1018
mediaeval, 1016
prehistoric, 1006
vertical warp-weighted loom, 1016

Thames
  navigation on, 380
  Roman ships found in, 535
Theodosian Code, 426
Theophilus encourages manufacture, 679
Third Crusade, fleet for, 546
three-field system, 735
threshing, 120, 708, 710, 725, 749, 752
threshing machines, 750
throstle, 1043
tide mill, 251, 286
tilbury, 446
tilt hammer, 912–14, 937, 962
tin, 537, 538, 763, 774, 779, 781, 816, 836, 847, 853, 879–80, 892, 924, 933, 968, 987
tin smelting, 920
tool, 5, 22, 301, 757, 763, 896, 902–3, 906, 950, 964–6, 970–3, 983–5, 1020, 1100, 1161–64
  Anglo-Saxon, 731
tooth pitch, 90
Tordesillas, Treaty of, 1095
torsion spring, 1134
tower mill, 276, 290
town, 141, 317, 362, 428, 431, 445, 511, 546, 661, 673, 691, 707, 780, 790, 803, 900, 1025, 1134, 1153, 1166
traction, 12, 167, 411, 456, 470
traffic, goods, statistics, 371
transport, 8, 12, 115, 124, 164, 167, 202, 251, 343, 420, 523, 758, 788, 817, 849, 941, 1023, 1031, 1064, 1099, 1200, 1202
  by water, 507
  competition between modes, 353
  efficiency and energy, 352
  importance of, 343
  of ore, 849
  parameters of, 345
transport systems, evolution of, 354
travel time to work, 348
travel, problems of, 367
treadmill, 32, 53, 73, 145, 153, 156, 163, 167, 178, 182, 209, 212, 215, 719, 786, 827, 979, 1151
  horizontal, 153
  uses of, 213
trebuchet, 1151
  preferred to cannon, 1153
trip-hammer, 972
triplane, 643
trireme, 527, 603
tunnel, 383, 806
tunnelling, 846
turbine, 235, 262, 265, 303, 305, 307
turnpike, 357, 367, 389, 434, 444, 495
  development of, 369
turnspit, Leonardo's, 58, 303
two-field system, 734
tympanum, 717
tyre, 402, 409, 423, 435

undershot wheels, power output, 269
United States of America, 149

vallus, 725
Vasa (ship), 555
vehicles
  Ackermann steering gear, 493
  early, 401
  four-wheeled, 468
  lateral stability of, 485
  position of centre of gravity, 466
  power and maximum velocity, 483
  springing of, 502
  steering, 491
  suspension and ride, 494
  two-wheeled, 463
ventilation, 571, 791, 838, 861, 872
vertical-axis waterheels, power output, 276
Victory (ship), 557, 613
Viking longboat, 541
Vikings, N. Atlantic voyages of, 544
Vulcan, attitudes to, 654

wagon, 29, 59, 371, 401, 410, 425, 428, 435, 483
wales, 534
Wales, 369, 382, 508, 675, 784, 846, 903, 924, 945–7, 951, 961
walking, 178, 201, 207, 221, 350, 367, 455, 719, 982, 1038, 1061, 1063
  energy requirement, 457
  speed of, 455
war
  for possession of the Lelantine plain, 1081
  Lydian-Milesian, 1081
  magnitude of, 1086
  prehistoric stategy and tactics, 1100
  settled by single combat, 1079–85
  statistical distribution of, 1086
warfare, 1053

biological, Leonardo's proposal, 1153
early, 1106
Greek and Roman, 1125
in primitive societies, 1074
mediaeval, 1145
prehistoric, 1065
ritual, 1076–79
survives as mediaeval chivalry, 1083
siege, 1099, 1106, 1109, 1110
warrior, Homeric, arms of, 1126
warship, 523, 524, 525, 550, 605, 613, 626, 630, 671, 1094–6, 1129
water frame, 1042, 1046
water mill, 224, 226, 229, 236, 241, 244, 251, 268, 270, 288, 728
water mills
distribution in England, 232
floating, 241
horizontal axis (Roman), 235
in Domesday Book, 231
in England, 243
major source of power in eleventh century, 679
Mithridates', 224
of Vitruvius, 238
overshot and undershot, 240
pressure to use, 160
sixteenth century uses of, 254
vertical axis (Norse or Greek), 223
vertical axis, example of, 225
water plane, inclined, separation on, 874
water supply, 56, 238, 673, 838
water transport, 507
development of, 372
very efficient, 353
water turbine
Barker's, 262
Fourneyron's, 262
Pelton's, 263
waterpower, 223, 243, 909, 910, 918, 934, 1021, 1023, 1031
waterwheel
apparent perpetual motion, 256
at London Bridge, 251
Branca's, 235
Hadley's, 253
power output, 267
preferred to steam, 257
Smeaton's analysis, 256
undershot and overshot, 241
ventilated buckets, 258
vertical axis (Fourneyron), 262
wealth, 113, 120, 123, 131, 139, 144, 345, 534, 734, 1053
wealth related to energy consumption, 115
weapon, 526, 629, 684, 1052–1274
weapon systems, 1098, 1137
weapons
bronze and iron, 1112, 1115
early, 1100, 1102
mediaeval, 1145
metal, 1108
stabbing, 1119
unfair, disapproval of, 1085
wear, 23, 55, 66, 72, 83, 99, 102, 106, 720, 761, 853, 879, 1105
adhesive and abrasive, 106
coefficients of, 108
corrosive, 107
corrosive and surface fatigue, 107
wedge, 14, 19, 46, 56, 99, 248, 755, 847, 1102
force on, 46
Weibull distribution of wind speed, 321
weir, 12, 254, 381
well, 126, 290, 291, 719, 778, 793, 795, 835
well sweep, 4, 5
wheat, 120, 121, 126, 154, 156, 679, 700, 703, 728, 735, 743, 804
wheel, 11, 57, 165, 406, 410, 417, 436, 456, 465, 474, 667, 718, 719, 738, 973
dished, 488
early, 401
manufacture of, 490
on hard surfaces, 480
wheel lock, 1172
wheel shuttle, 1035
wheelbarrow, 165, 850
whim, 167
whitedamp, 862
Whitworth, rifle of, 1197
Wilts and Berks Canal, 382
winch, 24, 29, 152, 160, 201, 381, 559, 823, 1133
horse driven, 72
wind, 115, 283, 284, 315
Beaufort scale, 614
difficulty of sailing into, 512
distribution of, 315
structures to withstand, 324
wind power, 223, 283, 286, 579
wind rose, 548
wind tunnel, 607, 1189

windflow, Prandtl's and von Karman's theories, 315
windlass, 29, 32, 143, 800, 934, 980
  equilibrium of, 52
windlass crossbow, 1155
windmill, 172, 283, 799, 805
  American, 298
  at Highworth, 289
  Bacon's development of, 291
  horizontal axis, 286
  location of, 315
  objection to, 288
  power of, 325
  range of wind speed, 323
  siting, 320
  Smeaton's development, 293
  speed control (Cubitt), 299
  speed governor, 296
  stability of, 334
  stopping, 58
  transport of, 172
  turret, 291
  twist of sails, 332
  vertical axis, 284
  wind speed to start, 320
Windmill Hill culture, 705
window, 304, 889
wind-tunnel, 327
winnowing, 712, 733, 751
wiper, 1049
wire, 933
  drawing, 917, 933, 998
  gold, 933
  iron, 933
wireworks, English, 935
wood screws, early, 968
woodland
  depletion of, 781, 946
wool
  English, taxes on, 1027
  production in England, 1016
woollen industry, 682, 1033, 1036
wootz, 956
worms, marine boring, 622
wound ballistics, 1214
wrist guard, 1106
wrought iron, 444, 680, 719, 757, 881, 915–16, 943, 954–7, 1167–9, 1184, 1204, 1233–5, 1265
  conversion to steel, 915–16, 954–7
  making, 943

NAME INDEX

Abbé Soumille, 179
Achilles, 654, 888
Ackermann, 451, 456, 492, 494
Adams, Bridges, 451
Adcock, 89, 90
Addaeus of Macedonia, 717
Aelfric, Abbot of Eynsham, 730
Aglaia, 655
Agricola, 58, 60, 69, 72, 114, 174, 687, 822, 824, 825, 826, 827, 828, 829, 830, 831, 832, 833, 834, 839, 840, 841, 850, 852, 853, 854, 856, 857, 858, 935, 941, 1144
Alaric, 670
Alcibiades, 1129
Alexander, 139, 359, 526, 625, 673, 674, 1128
Amal, 548
Amenemhet II, 777
Amenhotep II, 7
Amenhotep III, 650, 887, 1112
Amontons, 11
Anketil, 677
Antipator of Thessalonica, 673
Antisthenes, 140
Apellis, 26
Aphrodite, 653, 654, 658
Aquinas, 682, 686
Arbuthnot, 752, 753
Arcadius, 241
Archard, 108
Archimedes, 5, 15, 18, 26, 62, 66, 306, 660, 664, 667, 1053
Archytas, 632, 659
Ardrey, 1061, 1100
Ares, 655
Arete, Queen of the Phaeacians, 1012
Aristarchus, 136, 1012, 1013
Aristotle, 4, 5, 138, 139, 529, 579, 625, 659, 660, 667, 691, 877
Arkwright, 55, 1038, 1042, 1043, 1044, 1046
Artemon, 658
Ashurbanipal, 407, 408, 416
Ashurbanipal II, 415
Ashurnasirpal II, 406, 415
Athena, 655, 656, 715
Atkinson, 114, 173, 176, 179, 412
Attwood, 566, 567, 621
Aubrey, 383, 384, 634, 687, 692
Augustus, 141, 144, 146, 532, 897, 1083
Aulus Gellius, 631

Bacon, Francis, 291, 292, 293, 625, 687
Bacon, Roger, 625, 1161, 1163, 1165
Bagley, 952, 954
Bagwell, 366, 371, 373, 436
Bahn, 1008, 1009
Baker, 1196, 1197
Balen, 981, 982
Balsan, 640
Banks, 1192
Baptista Porta, 693
Barber, 1046
Bass, 529, 555
Bauer, Georg. See Agricola
Bawde, 940, 941
Bayliss-Smith, 121, 122
Beauchamp Tower, 100
Beaufoy, Colonel, 567
Becker, 1230, 1231, 1234
Beckman, 223
Beighnton, 253
Bekker, 414, 477, 478, 480
Belisarius, 241, 673, 674
Bell, 363, 556, 747, 748, 817
Benedict of Nursia, 676
Bennett, 156, 158, 159, 173, 212, 223, 231, 235, 292, 362
Beowulf, 673, 674
Bergman, 1251, 1252
Bernoulli, 280, 325, 564, 860
Berry, 386
Berthelot, 157
Besson, 69, 284, 285, 309, 740
Betz, 327
Bezant, 441
Bird, 317, 922
Biringuccio, 68, 162, 485, 488, 489, 687, 689, 697, 822, 823, 824, 838, 846, 847, 850, 933, 934, 936, 937,

941, 942, 943, 944, 955, 964, 979, 980, 1162, 1163, 1177, 1179, 1180, 1184, 1189, 1190
Biton, 1131
Black Prince, 915
Bladud, 632
Blanchard, 637, 906, 907, 908
Blith, 738, 739, 740
Blunt, Colonel, 433, 444
Bond, William, 554
Boonen, 429
Boswell, 113, 1061
Bouchon, 1029
Bouguer, 564, 565, 566, 592, 611
Boulton, 113
Bourne, 625
Bowen, 511, 512, 513, 514, 523
Brachal, 976
Bradshaw, 818
Bramah, 975, 976, 984
Branca, 235, 237, 305, 306, 307, 693
Breasted, 7, 510, 517, 524, 776, 778, 886, 1008, 1113, 1121
Brett, 240, 241, 242
Breughel, 981, 982, 983
Brigg, 452, 453, 471
Brigges, 383, 384
Brindley, 386
Brittannicus, 901
Brody, 208, 217, 218, 220
Brooke, 957
Brougham, 446, 451
Brouncker, 384
Brown, 566, 568, 617, 621, 622, 751, 757
Brown, John, 955
Bruce, Sir George, 835
Buller, 192
Bulmer, Bevis, 935
Buret, 1006, 1007
Burford, 403, 419, 420
Burghley, 379
Burne, 293
Bushnell, 629

Caliph Umar I, 665
Callistus of Rome, 669
Campbell, 563, 573, 574, 575
Canova, 655, 657
Card, 977, 978
Cardan, 305, 309
Carslaw, 806
Cartwright, 1046, 1047, 1048, 1049
Carus-Wilson, 243, 247, 681, 1022, 1023, 1024, 1026
Casson, 516, 523, 528, 530
Castle, 385, 944, 1155, 1157, 1166, 1167
Cato, 141, 142, 159, 426
Cayley, 640, 641, 642, 643, 1202, 1203, 1204
Cerretelli, 207, 211, 459, 460, 461
Cervantes, 1081
Chadwick, 668, 669
Chagnon, 1076, 1077
Chamberlayne, 431, 432
Chambers, 748, 754, 755, 921, 925, 962
Chandler, 177, 344, 363, 372, 380, 816, 835, 1030
Chapman, 564, 565, 566, 592
Chares of Lindos, 891
Charistion, 8, 35, 36
Charlemagne, 732, 1083, 1146
Charles, Jacques Alexander, 384, 635, 636, 692
Charles I, 949
Chersiphron, 421
Chillingham, 750
Cicero, 424
Circe, 1013
Clark, 1072, 1101–1105, 1125
Clarke, 1054, 1057, 1078
Claudius, 144, 901, 904
Cleinias, 1129
Clement, 22, 929, 945, 984, 985
Clyburn, 452, 453
Coates, 605
Colbert, 566
Coleridge, 157
Collier, 828, 1185
Collins, 507, 701, 771, 775, 808, 1006, 1007, 1102, 1107
Colt, 1185
Columbus, 541, 551
Columella, 721
Congreve, 1047, 1193, 1194
Constantine, 242, 669, 671
Constantinou, 781, 947
Conviva, Aulus Vettius, 893
Cooke, 744, 745, 746
Copernicus, 686, 687
Cort, 961, 962
Cosmo de Medici, 312
Cotterell, 7, 8, 51, 412, 413, 460, 603, 604
Coulomb, 11, 42, 45, 98, 99, 100
Coxwell, 637
Cranz, 1230, 1231, 1234
Crassus, 143, 144
Croce-Spinelli, 637

Croly, 1204
Crompton, 1044, 1045, 1046
Cromwell, 384, 948, 949, 1175
Crosskill, 757, 758
Crossland, 1168, 1169, 1170, 1181, 1201, 1264
Crossley, 902, 903, 909, 910, 911, 952, 953, 980, 981
Crouwel, 400, 401, 403
Ctesibius of Alexandria, 1131
Culhare, Willliam, 817
Cyrus, 359, 418, 420

d'Alembert, 70, 293, 430, 490, 825, 859, 940, 973, 974, 982
da Costa, 1192
Daedalus, 632, 655, 656, 657
Damian, 632
Danti, 632
Darby, 933, 944, 951
Darius I, 359, 417, 520
Darwin, 455, 692, 1061
Dauphin, 1167
Davidson, 226, 412, 413, 414
Davies, 769, 777, 784, 785, 790, 795, 796, 800, 887
Davis, John, 561, 801
Davy, Sir Humphrey, 622
Dayraven, 674
de Abbetoft, Sir Walter, 905
de Caus, 305, 309, 693
de Ferrara, 916
de Gennes, 1046
de la Vaulx, 640
de Millemete, 1165
de Son, 559, 628
de Vaucanson, 1030
Defoe, 1031
della Porta, 305
Deloney, 689, 690
Delvigne, 1197
Demosthenes, 132, 133, 661
Derry, 265, 486, 635, 702, 743, 748, 802, 956
Desaguliers, 264
Descartes, 693
Dickinson, 225, 227, 968
Diderot, 22, 70, 114, 179, 430, 490, 824, 825, 857, 859, 939, 940, 941, 973, 974, 982, 983
Diodorus, 15, 148, 527, 793, 796, 803, 896, 1083, 1129
Diogenes, 140, 666
Dionysius the Elder, 1129
Dionysius, Tyrant of Syracuse, 527
Dioscorides, 129, 889
Dircks, 312
Djawhari, 632
Dolinek, 1174
Domitian, 140
Dorking, 751
Douglas, 1086
Doxiadis, 350
Drachmann, 14, 15, 20, 23, 24, 25, 26, 59, 62
Drews, 1115, 1117, 1118, 1121
Duckett, 757
Dudley, 933, 947, 948, 949
Dugas, 4, 8, 11
Duke of Marlborough, 1201
Durdik, 1174
Duverger, 976

Earl of Mar, 836
Earnshaw, 1038
Echemus, 1082
Edward I, 362, 378, 1026, 1156
Edward III, 1026, 1163
Eilmer, 632, 633
Ekelund, 997
Eldred, 1180
Elliot, Obadiah, 443, 444
Ellis, William, 744, 752
Ellyott, 957
Elton, 156, 158, 159, 173, 212, 223, 231, 235, 292, 362
Ember, 1074, 1075
Epictetus, 140
Erghum, 63
Eric the Red, 544
Ericsson, 976
Erik XIV, 1190
Ernstinger, 12
Errand, 69
Escher, 261, 262
Ethelbert, 231
Euclid, 665
Eudoxus, 659
Euler, 74, 75, 293, 566, 569, 592
Eustace, 428
Evelyn, 550, 556, 557, 558, 560, 625, 697, 1179

Fabius Ambustus, 1083
Fackler, 1218, 1219, 1225
Fairbairn, 90, 258, 261, 262
Falcon, 1025, 1029
Fardouet, 976
Farmer, 710, 744, 751, 753
Farrington, 665

Favorinus, 631, 632
Fergusson, 59, 558
Festus, 1015
Filarete, 680, 697, 911, 912, 1169
Finley, 129, 131, 135, 224, 660, 661, 784
Firnas, 632
Fitzherbert, 254, 255, 256, 740, 743
FitzOsbert, William, 905
Fitzralph, Sir William, 1158, 1159
Foljambe, 754
Foote, 1016
Forbes, 5, 8, 168, 284, 358, 417, 714, 743
Forsyth, 1190, 1191
Fouke, 428
Fourneyron, 262, 263, 265
Fox, 984, 1060
Franzén, 556
Frayer, 1067, 1070, 1071
Frederick the Great, 1202
Freeling, 751
Friedman, 777, 887
Froissart, 1155, 1156, 1165
Froude, 567, 581, 582, 584, 585, 586, 587, 588, 600
Fry, Elizabeth, 212
Fulton, 629, 630, 631

Galileo, 11, 687, 692, 693
Gallwey, 1132, 1133, 1134, 1151, 1152, 1153, 1154, 1155, 1156
Gamain, 976
Gambetta, 637
Garlan, 1082, 1083
Garrett, 749, 751
Gelius, 632
Geoffrey de Vinsauf, 1149
Gerald of Wales, 729, 1155
Gertler, 581, 582, 600
Gibbon, 670, 671, 676
Gibbons, 843, 844, 845, 846
Gibbs-Smith, 640, 1202, 1203
Gilbert, 386, 471
Gimpel, 251, 679, 1018, 1025
Glaisher, 637
Glaukos of Chios, 1140
Glotz, 129, 131, 132, 139, 419, 422, 530, 661, 717, 1012
Gnudi, 55, 59, 153, 166, 485, 558, 944, 952
Godfrey, Thomas, 562
Graunt, 692
Graves, 358, 650, 651, 658
Green, 100, 494, 764, 765, 766, 767
Greener, 1085, 1169, 1196, 1197, 1198
Gregory of Nyssa, 676, 685
Grenbergerus, 634
Grendel, 674
Grosse-Lordemenn, 203
Grübler, 354, 355, 356, 357
Guilmartin, 945, 1172, 1265
Gurney, 128, 405, 522, 523, 706, 710, 780, 781, 782, 783, 1011, 1108, 1113, 1120
Gustav I, 1190
Gustav II, 1200
Guyton de Morveau, 637

Haden, 213, 214
Hadley, 251, 253, 562
Hale, 164, 382, 1194
Hall, 100, 101, 386, 447, 553, 565
Halley, 627, 628
Halliday, 325
Hannibal, 794
Hargreaves, 1040, 1041, 1042, 1043, 1044, 1046
Harland, 611, 614
Harris, 129, 260, 314, 564, 565
Harrison, George, 952
Harrison, John, 562, 563
Hart, 11, 72, 301, 303, 304, 305, 966
Harvey, 650
Hasker, 443, 444
Hassall, 170, 546, 730, 1149
Hassan, 321, 324
Hattusilis III, 783, 1119
Hayes, 1198
Henne, Major, 1233
Henry II, 681, 906
Henry IV, 1081, 1166
Henry of Lewes, 906
Henry V, 551, 1165, 1166
Henry VIII, 555, 933, 940, 1199
Hephaestus, 138, 653, 654, 655, 656, 658, 698, 888
Heraclitus, 666
Hereward the Wake, 428, 1085
Hero, 66, 67, 235, 262, 283, 305, 664
Herodotus, 136, 137, 404, 519, 520, 521, 525, 527, 1080, 1081, 1082, 1129
Heron, 5, 18, 19, 20, 23, 29, 31, 62, 63, 970, 1131, 1134
Hesiod, 138, 654, 666, 715, 716, 763, 890
Hill, 181, 187, 188, 189, 191, 193, 200, 213, 433

Hills, 286, 287, 291, 294, 295, 296, 298, 300, 322, 533, 534, 1179
Hipparchus, 665
Hiram of Tyre, 525
Hockey, 243, 252, 681, 1017
Hoerner, 605, 608
Hogg, 428, 952, 1165, 1166, 1167, 1180, 1192, 1193
Hogge, 944, 945
Holm, 108
Holman, 786
Homer, 654, 655, 658, 888, 963, 965, 1011, 1126, 1130
Honorius, 241
Hooke, 91, 92, 181, 312, 691, 693
Hopkins, 141
Hopper, 135
Hornell, 508, 509, 537, 539
Horrocks, 1049
Hubbard, 1237, 1239
Hughes, 382
Hulsbos, 799, 805
Humfrey, 935
Humphrey, 168, 622, 632, 718, 719, 793, 879, 889, 896, 1013, 1015
Hunley, 630, 631, 632
Huntsman, 959
Hussey, Obed, 747
Huygens, 308, 309, 564, 566, 567, 611, 693
Hybertszoon, 555
Hyllus, 1081, 1082

Icarus, 655, 656, 657, 658
Imhotep, 648, 649, 650

Jackman, 360, 378, 379, 381, 383, 429, 431, 434, 435, 436, 437
Jacobson, 555
Jacquard, 1029, 1030
Jaeger, 806
James, 1041
James I, 1086
James, Duncan, 769
Janzon, 1224, 1225
Jefferey, 1048, 1049
Jefferson, 149
Jeffries, 637
Jenkins, 156, 251, 253, 312, 314, 315, 939, 941
Jennings, 159, 181, 245, 728
Jocelin of Brakelond, 287
John of Salisbury, 1084
Johnson, 1001
Johnson, Benjamin, 744, 745
Johnson, Robert, 985
Johnson, Samuel, 113, 1057
Jones, 129, 131, 132, 133, 135, 143, 146, 147, 148, 1097, 1160, 1173, 1176
Jones, Daniel, 751
Jordanus, 10
Josephus, 668, 1135, 1136, 1137, 1138, 1150, 1259
Julius Caesar, 537, 896

Kallikles, 658, 659
Kamel el-Mallakh, 511
Kamminga, 7, 8, 51, 412, 413, 460, 603, 604
Kay, 1034, 1035, 1036, 1038, 1042, 1046
Keeley, 1071, 1072
Keller, 12, 251, 252, 937
Kemp, 551, 625, 629, 631, 632
Khisty, 461
King, 948
King Agesilaus, 1129
King Alfred, 541
King Arthur, 674
King Cocalus, 656
King Demetrius, 891
King Djoser, 648
King Edgar, 677
King Menes, 1106
King Minos, 655
King of Aleppo, 522
King of Assyria, 887
King of Der, 522
King of Diniktum, 522
King of Mittani, 887
King Saul, 1079
King Solomon, 525
King, Gregory, 123, 740
Kinloch, 1049
Kirkham, 912, 913, 914
Klemm, 169, 176, 244, 308, 626, 664, 665, 677, 679, 682, 684, 800, 965, 1024
Knight, 175, 235, 237, 253, 263, 265, 267, 295, 300, 704, 846, 886, 890, 938, 974, 976, 977, 978, 1186, 1199
Komi, 204, 206
Kühn, 411, 538, 706, 1074

Lambert, 1067, 1068, 1069, 1070
Landgraf of Hesse, 1196
Landström, 518, 527, 540, 545, 548, 556, 576
Lane, 546, 548

Larson, 199, 203, 204, 206, 207, 209, 211, 459, 460, 461
Lassie, 637
Lawson-Dick, 383, 634, 687, 692
Leach, 121
Lee, 294, 295, 296, 379
Lefebrvre des Noëttes, 168, 414, 420, 421
Lefraucheux, 1200
Leibniz, 693
Leif Ericsson, 544
Leighton, 946
Leland, 343, 344, 363, 364, 365, 373, 690, 816, 818, 1023, 1030
Lenard, 950
Leonardo, 3, 11, 22, 27, 28, 55, 56, 57, 58, 66, 72, 156, 301, 302, 303, 304, 305, 306, 307, 309, 625, 626, 632, 633, 634, 687, 693, 936, 937, 966, 967, 968, 970, 971, 972, 976, 1027, 1151, 1152, 1172
Leupold, 114
Lewendon, 1200
Lilienthal, 632
Links, 432
Littauer, 400, 401, 403
Liversidge, 243, 423, 728, 784, 1015, 1017
Livy, 1083
Löelholtz, 1172
Longfellow, 877
Lopez, 360, 361, 364, 422, 426
Lorenz, 1055, 1057, 1058, 1059, 1060
Louis XIII, 1196
Lucas, 130, 140, 717, 955
Lucillius Domitius, 665
Lysias, 132, 890
Lysippus, 891

MacAdam, 262
Madden, 129, 146
Manlius, 1083
Mann, 1031, 1036, 1038
Mansfield, 124
Marchetti, 350
Marcus Aurelius, 141
Margaria, 186, 197
Maritz, 983
Mark Anthony, 532
Marquis of Worcester, 311, 313, 314, 315, 693
Marsden, 1131, 1132, 1134, 1135, 1137
Marshall, 745, 751, 754, 756
Mathew, 366, 384, 980
Maudslay, 22, 975, 976
Maximilian of Bavaria, 1196
McAdam, 370
McCormick, 747, 748
McCutcheon, 245, 246
McKee, 102
McLintock, 631
Meikle, 749, 750
Menzies, 750
Metagenes, 421, 422
Metcalfe, 370
Meyer, 803, 804
Meysey, 957
Miller, 1049
Milne de Marr, 1234
Minié, 1197, 1198
Missliwetz, 1224, 1255, 1256
Mitcham, 666, 685
Mitchell, 354, 375, 376, 577, 784, 946, 961, 1032, 1033
Mithridates, 224
Modelski, 1093, 1094, 1095, 1096
Monad, 199, 200, 209
Montgolfier, 635
Moore, Sir Jonas, 445, 446, 634
Moray, Sir Robert, 842
Morgan, 1093, 1158
Moriarty, 1166
Morice, 251
Morrison, 605
Morsley, 179, 441, 744, 750, 751
Moxon, 956
Müller, 203
Mumford, 1053, 1066, 1067
Muntluc, 1085
Murphy, 1035, 1046, 1049
Mursilis II, 1120
Myers, 611, 614

Nakicenovic, 354, 355, 356, 357
Napoleon, 630, 961, 1096, 1193, 1196
Narmer, 1106, 1107
Necho, 520, 525
Need, 662, 1043
Neilson, 925
Neville, 946
Newcomen, 315
Newton, 28, 562, 564, 691, 693
Nicholas of Oresme, 686
Nomarck Kheti, 517
Norden, 945, 946
Numa Pompilious, 1083

Oakeshott, 1084, 1085
Octavius, 532

Odysseus, 126, 129, 130, 963, 1011, 1012, 1013, 1032
Ohlgren, 428
Oldknow, 1046
Oreibasios, 25, 26
Orosius, 141
Oswy, King of Kent, 900
Overton, 695, 736, 737, 740, 746

Packet, 1190
Paconius, 422
Palladius, 240, 723
Pandolf, 208, 460
Papaioannu, 350
Papin, 309, 310, 311, 312, 313, 693
Pappus, 11, 14, 15, 664, 665
Parent, 292, 332
Parsiphaë, 656
Parsons, 238, 239, 240
Pascal, 448
Paul, 231, 384, 719, 1037, 1038, 1043
Pauly, 1149, 1200
Pausanias, 360, 794, 1130, 1140
Payne, 156, 157, 1132, 1133, 1134, 1151, 1152, 1153, 1154, 1155, 1156
Peel, 1040
Peele, 819, 865
Pegolotti, 1017
Pelham, 248, 1022, 1023
Pelton, 228, 263, 265, 266, 268, 276, 277, 278
Pembridge, Sir Richard, 916
Penelope, 130, 1132
Pennant, 159, 245, 366, 728
Pepys, 251, 432, 433, 434
Perehu, 517
Petro, 99, 101, 102
Pett, Phineas, 557
Petty, 553, 559, 592, 691, 692
Philipson, 442, 443, 449, 452, 453, 470, 491, 504
Philo, 668
Philon, 1131, 1137
Piggott, 399, 401, 410
Pimentel, 121
Pitaggatallis, 1120
Pitt, 630
Plato, 131, 138, 140, 658, 659, 664, 665, 666, 686, 691
Platt, Sir Hugh, 738
Plattes, 738, 740, 743
Pleiner, 1083, 1140, 1141, 1142
Pliny, 9, 67, 72, 159, 224, 537, 663, 668, 683, 686, 720, 721, 722, 723, 786, 787, 788, 789, 790, 793, 794, 802, 891, 892, 894, 895, 924, 963, 1082
Plot, 935
Plutarch, 138, 142, 143, 144, 658, 659, 660, 667
Poisson, 997, 1090, 1091, 1092, 1093
Pole, 258, 983
Pollux, 1014
Polyaenus, 1142
Polybius, 144, 1135
Pompey, 143, 537
Poncelet, 262, 263, 265, 1228, 1229, 1230, 1231, 1232
Pope Adrian IV, 677
Porsenna, 1082, 1083
Posidonius, 137, 1083
Postegate, 400
Postgate, 38, 126, 128, 400, 521, 522, 708, 711, 712, 779, 780, 783, 1009, 1010, 1108, 1109, 1110
Potts, 1061, 1063, 1064
Prandtl, 315, 316
Pratt, 1252, 1254, 1255
Prince Rupert, 847, 955
Prometheus, 651, 652
Prudentius, 241
Pryce, 824, 836, 837, 838, 841, 842, 843
Ptolemy, 665
Puckle, 1184, 1185, 1186
Pugh, 289
Pythagoras, 457, 666

Queen Boudicca, 1122
Queen Elizabeth, 379, 429
Queen Hatshepsut, 517
Queen Hetepheres, 776

Rabinowicz, 108
Rahtz, 228, 229, 232, 241
Raleigh, 946
Ramelli, 11, 12, 16, 17, 18, 21, 29, 31, 55, 56, 58, 59, 61, 66, 67, 69, 72, 73, 88, 89, 114, 152, 153, 155, 156, 166, 170, 171, 174, 175, 234, 235, 236, 247, 250, 256, 257, 289, 290, 291, 301, 302, 303, 428, 558, 687, 1204
Ramesses, 7, 405, 519, 520, 523, 524, 632, 779, 887, 1120, 1122
Ramesses III, 7, 519, 523, 524
Ramsden, 973, 974
Rankine, 122, 123, 164, 172, 202, 235, 268, 274, 333, 334, 419, 482

Raul, 976
Rea., 951
Rekhmire, 6, 7
Renfrew, 771, 1008, 1009
Renström, 1189, 1191
Restitutus, Aulus Vettius, 893
Reynolds, 100, 101, 102, 581, 585, 586, 684, 870, 1248
Richard I, 911, 906, 1150, 1154
Richard of Devizes, 546, 1149
Richardson, 1086, 1087, 1088, 1089, 1090, 1091
Richter, 27
Rieu, 126
Riquet, 384
Rivault, 309, 693
Robert de Busli, 911
Robins, 642, 1185, 1187, 1188, 1189, 1192, 1264, 1265
Robinson, 446, 670, 976
Rostovtzeff, 715, 893, 1123
Rovenzon, 943
Rowe, 991, 994, 997, 1001
Rudolf of Nuremberg, 918
Rust, 1237, 1239

Salvian, 670
Salzman, 817, 818, 900, 903, 905
Sargon II, 406
Savery, 311, 312, 314, 315, 693
Schafer, 345, 346, 348, 349
Scherrer, 199, 200, 209
Schlaifer, 129
Schubert, 787, 881, 883, 899, 900, 903, 906, 909, 910, 914, 915, 916, 926, 928, 929, 935, 937, 944, 945, 946, 951, 954, 958
Schwarz, Berthold, 1161
Selby, Sir George, 824
Seneca, 147, 662, 663
Senenmut, 519
Sennacherib, 5, 8, 161, 163, 407, 415
Senusret I, 520, 776, 777
Senusret II, 777
Senusret III, 517, 520
Seppings, 567, 568, 569, 573, 592, 599
Servius, 1015
Seti I, 778
Shaen, 384
Shakespeare, 1085
Shalmaneser III, 406, 415, 1124
Shaw, Captain, 471
Shephard, 637
Shillibeer, 449, 450
Shun, 632
Simineowicz, 1192
Simon Magnus, 632
Singer, 799, 805
Sival, 637
Skjölsvold, 412
Small, James, 755
Smalley, 1042
Smallwood, 1150
Smeaton, 253, 256, 259, 268, 293, 294, 332, 642, 983
Smiles, 22, 681, 877, 946, 947, 959, 984
Smith, Adam, 63, 372, 428, 429, 485
Smith, R., 1168
Smith, Thomas, 1180, 1182, 1184, 1185, 1200, 1233
Smithells, 1003
Smollett, 367, 368, 369, 377, 436, 437, 551
Socrates, 132, 138, 140, 1013
Solinus, 538
Solomon, 405
Solon, 130, 138, 658
Sommereld, 98, 100, 101
Sorocould, 836
Spartacus, 143
Spooner, 65, 91, 104, 105
St Brendan, 543, 544
St Dunstan, 677
St Egwin, 900
St Eloy, 677, 678, 679
Stanhope, 446, 447, 567
Steeds, 982, 984, 985
Steers, 385, 386
Stell, 1036
Stern, 597, 1006
Stevenson, 176
Stevin, 692
Stirling of Dumblane, 750
Storr, 1055
Stow, 251, 939
Stowers, 244, 251, 254, 257
Strabo, 147, 148, 224, 788, 791, 802, 1081, 1083
Straker, 225, 227
Strutt, 1043
Studley, 977, 978
Stumpe, 690, 691
Sturtevant, 947
Swaine, Monamy, 613
Swanton, 1085
Swift, 366
Sykes, 321, 324

Taccola, 909
Talos, 655, 658
Tantalus, 658
Targett, 702
Tartaglia, 1180
Taylor, 581, 585, 789, 823, 824, 827, 835, 836, 837, 845, 846
Taylor, James, 1040
Taylor, John, 372, 380, 383, 428, 431, 835, 854
Taylor, Joseph, 819
Taylor, Samuel, 157
Tebi, 516
Telford, 370
Templeton, 293, 294, 332
Thâbit ibn Kurrah, 8, 35
Theodosius I, 665
Theophilus, 679, 915, 916
Theophratus, 879
Thetis, 888, 889
Thiout, 976
Thompson, 166, 173, 265, 537, 671, 1093, 1094, 1095, 1096
Thrupp, 366, 368, 424, 429, 433, 441, 447, 449
Thucydides, 134, 418, 662
Tilgath-Pileser III, 406
Timberlake, 806, 808
Tissandier, 637
Toynbee, 693, 1093, 1094, 1095
Trew, 384
Tull, Jethro, 740, 741, 742, 743, 745
Tutankhamun, 405, 1118, 1119
Tuthmosis I, 517
Tuthmosis III, 6, 7, 886, 1113, 1214
Tylecote, 675, 921, 922, 923
Tyler, Wat, 234

Unferth, 674
Uni, 516, 517
Urban II, 1084

Valentinian, 671
van Balen, 981
van Collen, 944
Varro, 159, 721, 725
Vegetius, 1138, 1154
Venus, 654
Verantius, 251, 252, 286
Verbruggen, 983
Vespasian, 896, 1136, 1137, 1150
Victor, 345, 346, 348, 349, 683
Vidler, 441
Vigor, 751
Villard de Honnecourt, 247, 249, 300, 302
Vincent of Beauvais, 1084
Virgil, 138, 654, 655, 720, 723
Vitruvius, 15, 16, 17, 18, 23, 24, 25, 62, 66, 235, 238, 240, 421, 422, 673, 717, 718, 719, 796, 798, 799, 804, 1131
von Hohenklingen, 1159
von Karman, 315, 316
Vulcan, 138, 159, 654, 655, 697

Wade, 954
Wadsworth, 1031, 1036, 1038
Washburn, 1058
Watson, 631
Watt, 113, 115, 983, 1046
Webb, 543, 572
Weidemann, 129
Weidl, 847
Wenamon, 524
Westermann, 129, 141
Wheeler, 1185
White, 632, 633, 714, 720, 721, 723, 725, 727
Whitehouse, 701, 771, 775, 1006, 1107
Whitworth, 23, 1197, 1198, 1199
Wick, 327, 329
Wieser, 1224, 1255, 1256
Wikander, 228, 230
Wilkinson, 983, 984
William II, 1147, 1154
William of Malmesbury, 632
William of Ockham, 686, 687
William of Poitiers, 903
William of Worcester, 1086
William the Conqueror, 903
Williams, 265, 486, 635, 702, 743, 748, 802, 956, 1150
Wilson, 256, 259, 362, 729, 1016
Winchcombe, 690
Wise, 640
Woosnam, 632, 633
Wright, 593, 596, 597, 744, 954, 1086, 1088, 1090, 1091, 1092
Wright Brothers, 643
Wurmbrandt, 1200, 1201
Wyatt, 1038

Xenophon, 132, 135, 136, 138, 418, 426, 434, 660, 661, 662, 1012, 1128
Xerxes, 527, 1082

Young, 598
Young, Arthur, 367, 744, 745, 751, 753
Young, John, 836
Young, Thomas, 555, 569, 592, 593, 594, 596, 597, 598, 599
Zeno, 241
Zeus, 138, 653, 654, 656, 716, 1013
Ziegler, 983
Zonca, 114, 174, 937, 1288

TECHNOLOGY AND CHANGE IN HISTORY

This new series of scholarly surveys is intended to offer an updating of the discussion of questions regarding the nature of technology and technological change first broached in the nine-volume survey by R. Forbes: *Studies in Ancient Technology.* The series will however take in not only the original scope of Forbes' work namely the Ancient Near East and the Greco-Roman world, but will extend beyond this to cover the medieval and early modern periods.

The volumes in the series will be in English, of 300-800 pp., divided into 10-15 topical chapters and aim to present to scholars, graduate students and to non-specialist scholars the current state of knowledge in the various fields in the history of technology.

1. Astill, G. and Langdon, J. (eds.). *Medieval Farming and Technology.* The Impact of Agricultural Change in Northwest Europe. 1997. ISBN 90 04 10582 4
2. Wikander, Ö. (ed.). *Handbook of Ancient Water Technology.* 2000. ISBN 90 04 11123 9
3. Squatriti, P. (ed.) *Working with Water in Medieval Europe.* Technology and Resource-Use. 2000. ISBN 90 04 10680 4
4. G.R.H. Wright: *Ancient Building Technology. Volume 1: Historical Background.* 2000. ISBN 90 04 10680 4
5. Robert I. Curtis: *Ancient Food Technology*. 2001. ISBN 90 04 09681 7
6. Bryan Lawton: *Various and Ingenious Machines. Volume 1: Power Generation and Transport & Volume 2: Manufacturing and Weapons Technology*. 2004. ISBN 90 04 13609 6
7. G.R.H. Wright: *Ancient Building Technology. Volume 2: Materials.* 2004. ISBN 90 04 14007 7

ISSN 1385-920X

Portland Community College Libraries